THE GREAT DEVONIAN CONTROVERSY

A modern aerial view of the coast of north Devon,
showing the contrast between the clear exposures of
strata in the cliffs and on the shore, and the gentle
landscape inland with exposures too small and
scattered to be visible from a distance. The strata in
this view (near Hartland Point) were classed in the
early nineteenth century as "Greywacke" or "Culm";
their strong zigzag folding is visible in the cliffs.

Martin J.S. Rudwick

THE
GREAT DEVONIAN
CONTROVERSY

**The Shaping of Scientific Knowledge
among Gentlemanly Specialists**

The University of Chicago Press

Chicago and London

The University of Chicago Press, Chicago 60637
The University of Chicago Press, Ltd., London

©1985 by The University of Chicago
All rights reserved. Published 1985
Printed in the United States of America
94 93 92 91 90 89 88 87 86 85   54321

Library of Congress Cataloging in Publication Data

Rudwick, M. J. S.
   The great Devonian controversy.

   (Science and its conceptual foundations)
   Bibliography: p. 463
   Includes index.
      1. Geology, Stratigraphic—Devonian.   2. Science—
History.   3. Geology—History.   4. Geology—England—
Devon   I. Title.   II. Series.
QE665.R83   1985   551.7′4   84-16199
ISBN 0-226-73101-4

MARTIN J. S. RUDWICK has taught the history of
science at universities in several countries and has
written extensively on the conceptual history of
evolution and the earth sciences and on
palaeontology. He is the author of *The Meaning of
Fossils*, also available from the University of Chicago
Press.

For Tricia

That much good ensues, and that the
science is greatly advanced, by the
collision of various theories, cannot
be doubted. Each party is anxious to
support opinions by facts. Thus, new
countries are explored, and old
districts re-examined; facts come to
light that do not suit either party;
new theories spring up; and in the
end, a greater insight into the real
structure of the earth's surface is
obtained.

Henry Thomas De la Beche,
*Sections and Views illustrative
of geological Phaenomena*
(1830)

Contents

List of Illustrations

NOTE: Diagrams, sections, and maps described in the footnotes as *traced* from original sources have been redrawn to eliminate irrelevant detail for the sake of clarity, to highlight particular features, or to render colored originals suitable for black-and-white reproduction. No features or words have been introduced that are not strictly contemporary with the original: for example, the geological maps depict early nineteenth-century, *not* modern, knowledge of the distribution of the rocks.

Permission to reproduce illustrations from original documents is gratefully acknowledged. Credits are given above. All other photographs, and all line drawings, are by the author.

The following abbreviations are used in the diagrams:

Cam	Cambrian (system)
C(arb)	Carboniferous (system)
Car	Caradoc (part of Lower Silurian)
CM	Coal Measures
D(ev)	Devonian
Gw	Greywacke
LC	Lower Cambrian
LG	Lower Greywacke (or Cambrian)
Lla	Llandeilo (part of Lower Silurian)
LS	Lower Silurian
lst.	limestone(s)
Lud	Ludlow (part of Upper Silurian)
ML	Mountain Limestone
OR	Old Red Sandstone
S(il)	Silurian (system)
sh.	shale(s)
sst.	sandstone(s)
UC	Upper Cambrian
UG	Upper Greywacke (or Silurian)
unconf.	unconformity
Wen	Wenlock (part of Upper Silurian)

Preface

The Devonian controversy was "great" to the scientists who participated in it. While it was still in progress, Charles Lyell—who was no mean judge of such matters—said he thought it was one of the most important theoretical issues ever to be discussed at the Geological Society of London, which at the time was the leading forum for geological debate. Yet today the controversy is virtually unknown to the geologists who are Lyell's intellectual descendants, or even to those historians who study the past development of the earth sciences. The paradox has a simple explanation. The controversy has slipped out of sight for the good and adequate reason that the problems it raised were eventually resolved in a way that satisfied almost all participants. The Devonian controversy is important today because it was a *characteristic* piece of scientific debate: in the judgment of the relevant scientists, it resulted in a significant new piece of reliable knowledge about the natural world. By contrast, the controversies that remain in the consciousness of scientists, and provide much of the staple diet of historians of science, are often those that are *not* characteristic, because they have not been resolved and are not consensually regarded as having added to the stock of natural knowledge. If we wish to gain a better understanding of how scientific knowledge is shaped, in the present as well as in the past, our attention must focus on those episodes that the practitioners of a science quickly forget, not on those they remember. The Devonian controversy has been forgotten because geologists now take for granted, and use in a routine manner, the intellectual construct that embodies its consensual outcome: the Devonian system and period.

That the Devonian controversy was an important and characteristic piece of scientific argument would not by itself justify devoting a rather long book to a single episode that spanned less than a decade, a century and a half ago. What does justify the time and effort of both author and readers is, I believe, the fact that this particular episode can be traced and analyzed with a richness of revealing detail that is probably unrivaled in *any* period of history, and that could certainly not be matched in any study of modern scientific research. The grounds for that apparently implausible claim are reviewed in the first chapter, but it can only be substantiated fully by the narrative that comprises the bulk of the book. In view of the exceptional nature of the documentation on which the account can be based, I have written this book in a way that will, I hope, make the Devonian controversy accessible to a wide range of readers. At the risk of explaining what some may consider too obvious and elementary to be worth mentioning at all, I have tried to write an account that can be followed by those with little or no previous knowledge of geology or of the early nineteenth century. This book has in fact three overlapping aims and is addressed to three corresponding groups of readers.

First, it is addressed to practitioners of *Wissenschaftsforschung* or Science Studies, whether they label themselves sociologists or historians or philosophers of science. For them the book is designed to portray in detail a piece of scientific knowledge *in the making,* as a contribution to the empirical study of the practice of scientific research. It displays the processes of scientific knowledge making as ineluctably and intrinsically *social* in character, not (or not primarily) in the sense of the pressures of the wider social world, but in the sense of intense social interaction among a small group of participants. But it also shows that the knowledge produced through this interaction is not "merely" a social construction, and that the concrete natural world does have an identifiable input, constraining though not determining the eventual outcome of the research. Whether this view of the research process is valid for other branches of science in other periods of history remains to be seen. But only an array or repertoire of equally detailed case studies will ever allow analysts of science to avoid the idealized abstractions favored (at least until recently) by philosophers of science, and to reach higher-level conclusions that are soundly based on empirical evidence of what scientists really do in their research.

Second, the book is addressed to historians of science. It is an account of one important episode within the development of geology in the nineteenth century, in the course of which that science assumed much of its modern conceptual shape, if not its detailed modern content. More broadly, for historians of ideas, it concerns one episode within the longer process by which intellectual members of western societies turned the traditional "closed world" of nature into the modern "infinite universe," in a temporal as much as in a spatial sense. For social historians it is an account of science during a period in which the polymathic scientific virtuoso or savant was fading from the scene, but also in which the professional connotations of the modern term "scientist" were as yet equally out of place. It is an attempt to convey what it was like to *do* science of a certain kind in the early nineteenth century. This was a time when much of the best research, especially but not only in England, was in the hands of a gentlemanly social group, intensely concerned with building careers and enhancing their social status through their practice of specialist science, but not yet primarily dependent on that science for their livelihood.

Third but not least, this book is addressed to geologists, palaeontologists, and other natural scientists. They will recognize in the word "Devonian" a valuable conceptual category for understanding and describing the history of the earth and of life. This is the story of how that concept was constructed by exploration and argument among a group of geologists that included some of the sharpest minds the science has ever attracted. However remote from modern experience the social setting may seem to be, I hope the story will not only interest and entertain its scientific readers, but also cause them to reflect on the nature of their own research activity and on the dynamics of the controversies in which they are currently engaged.

The research for this book has now covered a longer period than the scientific work that it describes and analyzes. I was first alerted to the importance of the Devonian controversy many years ago, when I was a professional palaeontologist, and a historian of science only in my spare time. I had been asked to advise on a private collection of the scientific manuscripts of

George Bellas Greenough, the first president of the Geological Society of London. When I first saw them they were in chaos, with letters from different correspondents and different decades mixed up without a trace of order. It was therefore all the more striking to come across one bundle of letters carefully tied with red tape, and labeled in Greenough's distinctive handwriting, "Great Devonian Controversy." I was already aware that the origin of the modern concept of the Devonian system in the geological record had involved some argument in the early nineteenth century. But it was Greenough's label that first suggested that the Devonian controversy, although less well known than other conflicts in the history of geology, might have particularly deserved the epithet "great," and not only in the eyes of those who took part in it.

Most of the letters in Greenough's bundle were from Henry Thomas De la Beche, the first director of the Geological Survey in Britain. But their contents soon showed that the Devonian controversy had involved all the most active British geologists of an exceptionally creative generation. They included not only Greenough and De la Beche, but also Adam Sedgwick, the professor of geology at Cambridge; Roderick Impey Murchison, his early collaborator and later his implacable antagonist; Charles Lyell, the self-appointed chief theoretician of geology at that period; William Buckland, Sedgwick's counterpart at Oxford and geology's most entertaining publicist; and several others. Not only did the letters reveal this star-studded cast; they also made it clear that the Devonian controversy had been one in which intellectual and social components had been so inextricably intertwined that, in the jargon of the time, neither "internalist" nor "externalist" history alone would do it justice.

Over the subsequent years, in the interstices of other research projects, I accumulated a pile of transcripts of manuscript letters about the Devonian controversy. By allowing the correspondence itself to indicate who the participants had been, the network grew slowly outward from where I had happened to encounter it, as I searched through the widely scattered manuscript collections of other participants, until it became clear that the Devonian controversy might be one of the best-documented episodes of its kind in any part of the history of science. But it is only more recently that I have had the opportunity to piece together the different sides of the correspondence, to amplify it with the evidence of field notebooks, and to set all this manuscript material within the framework of the published articles and books that formed, as it were, the previously visible tip of the controversy.

Since I began to draft a narrative of the controversy and to analyze its shape, I have learned much from discussions that have followed presentations of various parts of the material to groups of historians and sociologists of science and others at several universities—Bath, Brandeis, Cambridge, Edinburgh, Harvard, Minnesota, Pennsylvania, Princeton, and Syracuse—and at the Institute for Advanced Study at Princeton. Some other occasions deserve particular mention. Giving the opening lecture to the International Symposium on the Devonian System in Bristol in 1978 helped me to sort out the shape of the controversy from the perspective of modern geologists and palaeontologists. A Special University Lecture, given at University College London in 1981, first enabled me to attempt an overall interpretation of the episode. And invitations to contribute to the ERISS symposium ("Epistemologically Relevant Internalist Sociology of Science"), organized by the Maxwell School of Syracuse University in 1981, and to the first Israel symposium on the history, philosophy, and sociology of science, in 1982, gave me invaluable opportunities to develop the wider implications of my interpretation. I am very grateful to those who contributed to the discussions on all these occasions, and for the invitations that made those discussions possible. On a more individual level, I am indebted in very diverse ways to David Bloor, John Bowker, Gerald L. Geison, Charles C. Gillispie, Howard E. Gruber, Mary Hesse, Frederic L. Holmes, Thomas S. Kuhn, Patrick de Maré, Paul McCartney, J. B. Morrell, Dorinda Outram, Roy Porter, Sylvan S. Schweber, James A. Secord, Steven Shapin, and John Thackray. I am particularly grateful to Jack Morrell for urging me—in the blunt but friendly manner of his fellow Yorkshireman Adam Sedgwick—to get on with the Devonian and for scolding me whenever I seemed to be slackening my pace.

Friends and colleagues such as these have sustained this project through some difficult times, both during and since my years at the sadly misnamed Free University in Amsterdam. In particular, after my abrupt departure from Holland in a forlorn protest at a disgraceful case of political intolerance, this book might have sunk without a trace had it not been for the moral support and stimulating company of colleagues at the Science Studies Unit at Edinburgh, at the Wellcome Institute in London, and in the Program in History of Science at Princeton University. Since that time, the completion of the project has been made possible by the exceptional generosity of Trinity College, Cambridge, and of the Institute for Advanced Study at Princeton. I am very greatly indebted to the Master and Fellows of Trinity, and to the faculty of the School of Historical Studies at the Institute, for giving me such congenial environments for writing that I had no further excuse for not finishing the book.

I am grateful to the librarians and archivists at all the institutions listed in my note on manuscript sources for access to their collections, and to the following for permission to quote from documents in their care: the Curator of the Sedgwick Museum, University of Cambridge; the Library, St Andrews University; the Master and Fellows of Trinity College, Cambridge; the Somerset Archaeological and Natural History Society; the Library, University College London; the Director, British Geological Survey; the National Museum of Wales; the American Philosophical Society; the Syndics of Cambridge University Library; University Museum, Oxford; and the Rt Hon. the Lady Lyell and the Lord Lyell. A travel grant from the Royal Society of London enabled me to add several important collections to my list of sources. For access to parts of the most important private archive relevant to the Devonian controversy, I am particularly grateful to Lady Lyell. The Stiftung Volkswagenwerk gave me generous financial support as a Volkswagen Fellow during my first period at the Institute in Princeton. Janice Rowe taught me to use the CPT word processor at Trinity, and with great accuracy transferred my heavily corrected drafts onto discs. Professor Ron Oxburgh kindly allowed me to use the facilities of the Department of Earth Sciences at Cambridge for drawing the maps, sections, and other diagrams. Last in time but not least in importance, David Hull relieved me of one of my worst worries by reacting to a very thick pile of typescript with both enthusiasm and understanding.

Finally, more than the customary note of thanks to my wife. She has driven me around the lanes of Devonshire, while I scanned the countryside for clues to nineteenth-century perceptions of its geology. She spent much of 1982, while we were in Jerusalem and Princeton, typing a first draft from a manuscript so messy that only she could decipher it. In our final weeks at Cambridge in 1983 her work on the Trinity word processor made it possible to meet an inflexible deadline. But above all, when circumstances made it hard to keep the project going at all, she gave me the kind of support without which this book would certainly never have been completed.

LIST OF
Footnote Abbreviations

AM	*Annales des Mines*
ANH	*Annals of Natural History*
ASB	Académie des Sciences de Belgique
ASP	Académie des Sciences, Paris
Ath.	*Athenaeum* (newspaper)
BAAS	British Association for the Advancement of Science
BSGF	*Bulletin de la Société géologique de France*
CPS	Cambridge Philosophical Society
CRAS	*Comptes-rendus hebdomédaires de l'Académie des Sciences* (Paris)
ENPJ	*Edinburgh new philosophical Journal*
ER	*Edinburgh Review*
GDNA	Gesellschaft Deutscher Naturforscher und Ärzte
GSL	Geological Society of London
LG	*Literary Gazette*
MGS	*Memoirs of the Geological Survey of Great Britain*
n.d.	no date (on letter)
PGS	*Proceedings of the Geological Society of London*
PMJS	*Philosophical Magazine and Journal of Science*
pmk	postmark date (only given where manuscript does not provide day or month).
QR	*Quarterly Review*
RBAAS	*Reports of the British Association for the Advancement of Science*
SGF	Société géologique de France
TGS	*Transactions of the Geological Society of London*
wmk	watermark year (only given where all other direct evidence of date is lacking).

NOTE: All quotations are reproduced as accurately as is consistent with clarity: the punctuation, however unconventional or informal, is original, apart from a very few modifications that are essential to preserve the sense. *Italicized* words are those underlined in the original; those printed in *ITALICIZED CAPITALS* are those underlined two or three times in the original. All editorial additions and explanations are enclosed in square brackets. The sources of quotations (and other references) are given in the footnotes, and refer to the manuscript and printed sources listed in the bibliography.

Dramatis Personae

This is a list of all the persons mentioned in the narrative (excluding those named only in footnotes, and a few marginal men such as Prime Ministers), arranged in alphabetical order for convenience of reference. The numeral in parentheses after each name denotes the person's age in the year 1834, when the Devonian controversy erupted (or the year of death, if before 1834). The section of the narrative in which the person is first mentioned, in some cases with further biographical information, is given in parentheses at the end of each entry. Descriptions of occupation refer *only* to the period of the controversy, i.e., about 1834 to 1842 (except of course in the case of those who died earlier). "Major actors," or participants with "major" involvement in the controversy (see §15.5), are named in boldface, "minor actors," or participants with "middling" involvement, are named in italics, "walk-on parts," or participants with "minor" to marginal involvement, are named in regular type.

Agassiz, Louis (27): professor of natural history at the Academy of Neuchâtel (§5.2).

Ansted, David Thomas (20): student at University of Cambridge; later, Fellow of Jesus College (§9.8).

d'Archiac, Étienne Jules Adolphe Dexmier de St-Simon, Vicomte (32): gentlemanly palaeontologist in Paris (§14.1).

Asmus, Hermann: professor of zoology at the University of Dorpat (§14.2).

Austen, Robert Alfred Cloyne (26): country gentleman, living near Newton Abbot (Devonshire), later near Guildford (Surrey) (§7.1). He is commonly indexed under the name Godwin-Austen, the name he adopted in later life.

Baring, Sir Francis Thornhill (38): Member of Parliament for Portsmouth; Secretary to the Treasury (later, Chancellor of the Exchequer) in Lord Melbourne's second administration (§5.8).

Beyrich, Ernst (19): student in Berlin; later, geologist there (§10.7).

Bilton, Revd William (*c.* 36): clergyman living in London and near Bideford (Devonshire) (§8.2).

Boase, Dr Henry (35): physician in Penzance (Cornwall) (§4.6).

Brochant de Villiers, André Jean François

Marie (62; d. 1840): professor of geology and mineralogy at the École des Mines in Paris (§4.6).

Brongniart, Adolphe (33): professor of botany at the Muséum nationale d'Histoire naturelle in Paris; son of Alexandre Brongniart (§5.3).

Brongniart, Alexandre (64): professor of mineralogy at the Muséum nationale d'Histoire naturelle in Paris (§3.2).

Blumenbach, Johann Friedrich (82; d. 1840): professor of medicine at the University of Göttingen (§4.1).

von Buch, Christian Leopold (60): gentlemanly geologist in Berlin (§6.3).

Buckland, Revd William (50): reader in mineralogy and geology at the University of Oxford; canon of Christ Church (§4.2).

Cauchy, François Philippe (39): professor of mineralogy and metallurgy at the Athenaeum in Namur (§12.4)

Colby, Colonel Thomas (50): director of the Ordnance Trigonometrical Survey of Great Britain (§4.6).

Cole, William Willoughby, Viscount (27): gentlemanly fossil collector; later, third Earl of Enniskillen (§6.3).

Conrad, Timothy Abbott (31): geologist in Pennsylvania; later, palaeontologist on the New York State Geological Survey (§14.2)

Conybeare, Revd William Daniel (47): gentlemanly geologist and theologian; rector of Sully (Glamorgan) and later vicar of Axminster (Devonshire) (§4.2).

Cuvier, Georges Léopold Chrétien Frédéric Dagobert (d. 1832): professor of animal anatomy at the Muséum nationale d'Histoire naturelle in Paris (§3.2).

Darwin, Charles Robert (25): gentlemanly geologist and naturalist attached to H.M.S. *Beagle;* later, in London (§4.2).

Davy, Sir Humphrey, Bart (d. 1829): professor of chemistry at the Royal Institution in London (§4.1).

von Dechen, Heinrich (34): mining geologist and mines administrator in Berlin, later in Bonn (§12.1).

De la Beche, Henry Thomas (38): gentlemanly geologist, attached to the Ordnance Survey of Great Britain; later, director of the Geological Survey and of the Museum of Economic Geology in London (§4.2).

Deshayes, Gerard Paul (37): palaeontologist in Paris (§14.1).

Dufrénoy, Ours Pierre Armand (42): professor of mineralogy at the École des Mines in Paris (§4.6).

Dumont, André-Hubert (25): mining geologist in Liège; later, professor of mineralogy and geology at the University of Liège (§6.3).

Eaton, Amos (58): geologist and professor at the Rensselaer Institute in Troy, New York (§4.4).

Egerton, Sir Philip de Malpas Grey, Bart (28): Member of Parliament for south Cheshire; gentlemanly fossil collector (§6.4).

Ehrenberg, Dr Christian Gottfried (39): naturalist and microscopist in Berlin; later, professor of medicine at the University of Berlin (§13.4).

Élie de Beaumont, Jean-Baptiste Armand Louis Léonce (36): mining geologist at the École des Mines in Paris; professor at the Collège de France (§4.6).

Featherstonhaugh, George William (54): British geologist resident in the United States (§11.4).

Fitton, Dr William Henry (54): gentlemanly geologist in London (§5.3).

Forbes, James David (25): gentlemanly physicist; professor of natural philosophy at the University of Edinburgh (§4.3).

Gardner, James: mapmaker and mapseller in London (§4.6).

Goldfuss, Georg August (52): professor of zoology

and mineralogy at the University of Bonn (§10.1).

Greenough, George Bellas (56): gentlemanly geologist and geographer in London (§4.1).

Griffith, Richard (50): mining engineer and consultant geologist in Dublin (§4.4).

Hall, James (23): geologist on the New York State Geological Survey (§13.6).

Harding, Major William (42): country gentleman, living at Ilfracombe (Devonshire) and later at Tiverton (Devonshire) (§7.1).

Harcourt, Revd William Vernon (42): gentlemanly naturalist and canon of York.

Hausmann, Johann Friedrich (52): professor of mineralogy and technology at the University of Göttingen (§12.2).

Hennah, Revd Richard (78): chaplain to the military garrison at Plymouth (Devonshire) (§4.6).

Henslow, Revd John Stevens (38): professor of botany at the University

of Cambridge, and Fellow of St John's College (§11.1).

Herschel, John Frederick William (42): gentlemanly mathematician and astronomer, working in Cape Colony, later living at Slough (Buckinghamshire) (§10.1).

von Humboldt, Friedrich Heinrich Alexander, Freiherr (65): traveler, universal man of science, and Prussian courtier in Berlin (§13.2).

Jameson, Robert (60): professor of natural history at the University of Edinburgh (§4.5).

von Keyserling, Count Alexander (19): law student in St Petersburg; later, gentlemanly naturalist there (§13.4).

König, Charles Dietrich Eberhard (60): curator of mineralogy at the British Museum in London (§8.7).

de Koninck, Laurent Guillaume (25): chemist at the University of Ghent; later, professor of chemistry at the University of Liège (§14.2).

Lansdowne, Henry Petty-Fitzmaurice, Marquis of (54): Lord President of the Council in Lord Grey's and Lord Melbourne's administrations (§5.8).

Lewis, Revd Thomas Taylor (33): curate of Aymestrey (Shropshire) (§4.3).

Lindley, John (35): professor of botany at the London University (later, University College); assistant secretary at the Horticultural Society (§5.1).

Lonsdale, William (40): curator and librarian at the Geological Society of London (§5.1).

Lyell, Charles (37): gentlemanly geologist and author in London (§4.3).

MacCulloch, John (61; d. 1835): geologist in the pay of the Treasury (formerly, attached to the Ordnance Trigonometrical Survey of Great Britain) (§7.5).

von Meyendorf, Baron Alexander (36): Russian government official.

von Meyer, Christian Erich Hermann (33): gentlemanly palaeontologist in Frankfurt-am-Main; later, administrator in the Bundestag (§12.1).

Miller, Hugh (32): bank official in Cromarty; later, journalist and popular writer in Edinburgh (§11.1).

Moore, Dr Edward: physician in Plymouth (Devonshire) (§7.1).

zu Münster, Georg, Graf
(58): gentlemanly
geologist in Bayreuth
(§12.1).

Murchison, Charlotte
(45): wife of Roderick
Murchison (§4.1).

Murchison, Roderick
Impey (42): gentlemanly
geologist in London
(§4.1).

Nöggerath, Jacob (46):
professor of mineralogy
and mining at the
University of Bonn
(§12.1).

Northampton, Spencer
Joshua Alwyne
Compton, Marquis of
(44): gentleman of
science in London (§7.3).

von Oeynhausen, Karl
(39): mining geologist at
Bonn (§12.2).

d'Omalius d'Halloy,
Jean-Baptiste (51):
gentlemanly geologist
living near Namur (§6.3).

Pander, Christian
Heinrich (40):
gentlemanly naturalist
living near Riga (§14.2).

Pattison, Samuel Rowles
(25): solicitor in
Launceston (Cornwall)
(§7.4).

Peach, Charles William
(34): customs official in
London (§14.5).

Philipps, Trenham:
curator of the Museum of
Economic Geology in
London (§8.5).

Phillips, John (34):
curator at the Yorkshire
Philosophical Society in
York; professor of
geology at King's
College, London;
nephew of William
Smith (§4.5).

Phillips, William (d.
1828): publisher in
London (§4.2).

Portlock, Captain Joseph
(40): military engineer in
charge of the Ordnance
Trigonometrical Survey
in Ireland (§10.1).

Prévost, Constant (47):
professor of geology at
the Sorbonne in Paris
(§6.3).

Römer, Carl Ferdinand
(16): student in Berlin;
later, geologist in Bonn
(§14.5).

Rogers, Henry Darwin
(26): state geologist in
Pennsylvania (§5.2).

Rozet, Claude Antoine
(36): member of the
corps royal of military
surveyors in Paris
(§10.2).

Sandberger, Johann
Philipp: teacher at the
Gymnasium in Weilburg
(Nassau) (§12.1).

Scrope, George Julius
Poulett (37): Member of
Parliament for Stroud
(Gloucestershire);
political economist and
gentlemanly geologist in
London and Wiltshire
(§10.6).

Sedgwick, Revd Adam
(49): professor of geology
at the University of
Cambridge, and Fellow
of Trinity College;
prebendary of Norwich
(§4.1).

Silliman, Benjamin (55):
professor of chemistry
and natural history at
Yale University, New
Haven; consultant
chemist and geologist
(§14.2).

Smith, William (65; d.
1839): consultant land
and mineral surveyor,
living at Scarborough
(Yorkshire) (§3.2).

Sowerby, James de Carle
(47): natural history artist
and engraver in London
(§9.3).

Spring-Rice, Thomas
(later, Lord Mounteagle)
(44): Member of
Parliament for
Cambridge, and
Chancellor of the
Exchequer in Lord
Melbourne's second
administration (§5.8).

Steininger, Johann (40): teacher of mathematics and physics at the Gymnasium in Trier (§12.1).

Stokes, Charles (51): stockbroker in London (§10.4).

Turner, Edward (36; d. 1837): professor of chemistry at the London University (later, University College) (§5.2).

de Verneuil, Philippe Édouard Poulletier (29): lawyer and gentlemanly geologist in Paris (§10.2).

Warburton, Henry (*c.* 50): Member of Parliament for Bridport; political economist in London (§8.5).

Weaver, Thomas (61): consultant mining geologist in Dublin, later living at Wrington (Somerset) (§4.4).

Whewell, Revd William (42): Fellow and Tutor (later, Master) of Trinity College, Cambridge; later, professor of moral philosophy at the University of Cambridge (§7.4).

Williams, Revd David (42): rector of Bleadon (Somerset) (§6.2).

Werner, Abraham Gottlob (d. 1817): mineralogist and mining geologist at the Bergakademie in Freiberg (§3.3).

Wollaston, William Hyde (d. 1828): chemist and mineralogist in London (§4.1).

Young, Revd George (57): Presbyterian minister at Whitby (Yorkshire) (§10.1).

Part One

SETTING THE SCENE

Chapter One

SCIENTIFIC RESEARCH UNDER A HISTORICAL MICROSCOPE

1.1   Introducing the Devonian Controversy

In the early nineteenth century, geology was a new, exciting, and fashionable science.[1] It was experiencing its first and greatest boom in conceptual innovation, empirical expansion, and public approval and interest. It attracted some of the most talented in the scientific world, particularly those with a taste for travel and the outdoor life rather than for mathematics. But those with lesser talents or more limited opportunities could also hope to gain recognition and respect, for it was felt that any worthwhile grand conclusions in geology had to rest on the foundations of local details that required much time and patience to acquire. The empiricism esteemed by geologists bound leaders and locals into a symbiotic partnership that was seen as a model of the ideal community of science.

The most spectacular achievement of this fashionable new science was its disclosure of the vast history of the earth and of life on its surface. Human

1. There is no adequate general account of the earth sciences in the early nineteenth century. Porter (1977) gives an important analysis of the preceding century and a half, with due attention to empirical, intellectual, and social aspects, but his account is confined to Britain. Gillispie (1951) gives a classic interpretation of the perceived implications of the new science of geology in early nineteenth-century Britain, but he does not set out to analyze the technical content of geological research in any detail. Rudwick (1972) describes the palaeontological branch of the science over a wider period, and Greene (1982) the structural or tectonic; both attempt to transcend the British bias of much historical writing in English, but neither deals with the central *stratigraphical* tradition within which the Devonian controversy developed.

beings were discovered to be the merest newcomers at the end of a long saga of prehuman life. Bizarre extinct monsters were pieced together from scattered bones. But while that required all the skills of a trained anatomist, even the country lady or the schoolboy could hope to find traces of lesser denizens of "former worlds." Ammonites and trilobites might be less spectacular than mammoths and mosasaurs, but they were equally strange, and anyway more appropriate in a drawing-room cabinet. This new and exciting vista into the history of the earth was no fantasy devised by skeptical philosophes to confound the faithful; it was vouched for by leading geologists of impeccable piety as a disclosure that ought to evoke an enlarged sense of awe at the scale and diversity of the created world.

Geology was also popular and important from a quite different point of view. The growth of new forms of industrial production, though hardly yet perceived as the "industrial revolution" it was to become in retrospect, was transforming parts of Britain and spreading to parts of the Continent too. Heavy industry was increasingly dependent on expanding supplies of coal and metal ores; mines were meeting the demand by driving ever deeper below the surface and further from the shallow workings that had been sufficient in previous centuries. In doing so, the owners and managers of mines found their traditional empirical rules of operation increasingly inadequate. The extension of old workings and the discovery of new ones demanded new methods of prediction and exploration; geology was looked to as a potential source of methods that would be soundly based on scientific principles. Geology was therefore esteemed as much for its prospective contribution to the economy as for its enlarged view of the natural world.

The popular interest in geology, in both its theoretical and practical forms, was rooted securely in foundations laid by its leading practitioners. They claimed that the history of the earth and its inhabitants could be deciphered by learning to read "the record of the rocks," and that the succession of layers or strata in the earth's crust were Nature's own historical documents. Imperfect the record might be, but deceptive it was not. Correctly interpreted, much of the history seemed plain and straightforward. Each "formation" of strata was like a chapter in the volume of Nature's history; the sequence of chapters was clear, and each was characterized more or less by distinctive types of rock and by distinctive fossils. By hammering along a line of coastal cliffs or by going from quarry to quarry across country, the geologist could plot local sequences of strata and collect their fossils. By combining and coordinating such local sequences, the broad outlines of a global history could be discerned with ever increasing clarity, and predictions could be made about the likely extension of economically valuable deposits such as coal.

At the lower end of the whole pile of strata, however, the rocks that recorded the earlier chapters in the earth's history proved less easy to decipher. The geological record became ever more puzzling, the formations apparently chaotic, and their fossils rare and obscure. Yet it was here if anywhere that scientific knowledge might hope to reach back to some record of the creation of life itself; it was also here that many deposits of coal, and most of the valuable metal ores, were known to be situated.

By the 1830's this problem was attracting some of the best geologists of that generation. They began to unravel the sequence of these ancient strata, to discern what they believed to be the earliest history of life, and to infer the principles on which mineral exploration might be based. But in the midst of this work a controversy arose that threatened the stability and success of the whole enterprise. The Devonian controversy began as a dispute about the identification and correct sequence of strata in the county of Devon in southwest England. From the start, however, its implications were seen to be international, and indeed global. As it developed, the controversy drew in most of the leading geologists in Britain and increasingly those in other countries too. It was only resolved—to the satisfaction of most if not all of the participants—when it was exported from Britain to France, Belgium, and the German states, and finally to Russia, North America, and the rest of the world. The vast expansion of research in the century and a half since the controversy subsided has been regarded by geologists as confirming the validity of that solution in ways that would have delighted, but not surprised, those who first worked it out.

The Devonian controversy exposed the procedural roots of geological practice and subjected them to more probing scrutiny than ever before. The successful resolution of the controversy endowed geological practice with a new confidence in the reliability of its conclusions—a confidence it retains to the present day. This is the episode in scientific research practice that in this book will be put under a historical microscope. The remainder of this chapter explains programmatically what is meant by that metaphor. Readers who are more interested in the story itself than in the principles that the storytelling entails can turn to chapter two without further ado.

1.2   Watching the Natives at Work

Nearly twenty years ago Sir Peter Medawar urged those interested in understanding science to study in detail "what scientists *do*" in their research.[2] As a distinguished practicing scientist, he himself was well aware that what scientists *say* about their activity can never be taken at face value: their accounts are invaluable as source materials, but not necessarily reliable as interpretative conclusions. He might have added that what philosophers say should be handled even more gingerly, since at least until recently they seemed to be interested only in prescribing what scientists ought to be doing, and they showed indifference if not hostility to any truly empirical study of scientific activity. Medawar claimed in fact that "only unstudied evidence will do—and that means listening at a keyhole." The metaphor conveyed vividly the sense of intimate knowledge that he realized was needed, but it also implied a degree of observational and interpretative distance—not to mention improper intrusion—which was not really intrinsic to his suggestion.

Those analysts of science who have followed Medawar's suggestion, whether knowingly or not, have rightly chosen to open the laboratory door and to establish themselves as observers on the inside. They have done so in two distinct styles, each with its own limitations. Some have studied modern scientific research by being accepted as participant-observers within a labo-

2. Medawar (1967, p. 151).

ratory. They have used the perspective and even the techniques of the anthropologist, treating the scientists as exotic natives with strange and puzzling customs. These ethnographers or microsociologists of science have given some detailed and illuminating accounts of routine procedures in scientific research. But with few exceptions they have described the research in a static manner, failing to show how the procedures are used in a *temporal process* to develop some new scientific conclusion. Furthermore, they often show an extreme skepticism—or at least an extreme agnosticism—about the status of the knowledge the scientists claim to be producing. In minimizing if not discounting its reference to any "real" external world of nature, their accounts of science open up a gulf in self-understanding between themselves and the scientists they observe—a gulf which surely no modern anthropologist would find tolerable in the interpretation of exotic cultures.[3]

Historians of science, on the other hand, have not in recent decades been neglectful of the "natives' point of view." Those who have studied past scientific knowledge in the making have not needed to listen at a keyhole or to peep through it: rather they have looked over the shoulder of their chosen subject and retraced the course of the research from laboratory notebooks, letters, and other texts.[4] But most historical studies of what scientists *did*, in some specific setting in the past, are focused on one person. The biographical genre rightly remains popular with historians of science, although it has become unfashionable among other historians. But even the best biographies are bound to distort to some extent the processes of scientific activity. By focusing on the work of one individual, they inevitably give less than adequate attention to the complex web of social and cognitive interactions that bind even the most distinguished or reclusive scientist into his or her immediate network of colleagues, in collaboration or rivalry or both.

What are needed, for a fuller understanding of the processes by which scientific knowledge is shaped, are empirical studies of science in the making—whether in the past or the present is of lesser consequence—which focus not on one individual scientist but on a specific scientific *problem* that brought together some *group* of individuals in an interacting network of exchange. Such studies need to pay attention to the role of all the participants, however minor their contributions may seem to be; to follow the dynamics of interaction with equal attention to exchanges both public and private, formal and informal, ritualized and spontaneous; and above all, by something akin to Gilbert Ryle's "thick description," to try to discern the *meaning*, for the participants themselves, of the social drama that is scientific research. For as Clifford Geertz commented of anthropological understanding, "the trick is not to get yourself into some inner correspondence of spirit with your informants . . . [but] to

3. Latour and Woolgar (1979) and Knorr (1981) are representative examples of the "ethnographic" approach: both are studies of research in large biological laboratories in California. For important collections of essays summarizing this kind of work more generally, see Knorr, Krohn, and Whitley (1981) and Knorr-Cetina and Mulkay (1983). The importance of studying temporal processes in (even) conventional ethnography has been stressed particularly by Victor Turner (see, e.g., Turner 1974).

4. Of many fine examples, the following recent studies are representative: Gruber's (1981a) and Kohn's (1980) reconstructions of Charles Darwin's early theoretical development; Holmes's (1974) analysis of some of Claude Bernard's laboratory research; and Westfall's (1980) biographical study of Isaac Newton.

figure out what the devil they think they are up to."[5] This book is a contribution to that empirical study of scientific research, seen as the work of an interacting social group. It is an attempt in the first place to look over the shoulders of nineteenth-century geologists as they wrestled with the problems raised by the Devonian controversy, and to figure out what the devil they thought they were up to.

1.3   Applying a Historical Microscope

The historian of science who studies any period before the twentieth century is deprived of one of the sociologist's classic forms of evidence, the interview. But historians of science who have had to deal with its nearest equivalent, namely the written or taped reminiscences of scientists who are no longer alive, generally feel that such deprivation is less than disastrous. For the recollections of scientists—whether spoken or written; prompted or spontaneous—are notoriously and systematically unreliable, even when they are made only weeks or months after the events being recalled, let alone years or decades later. This is certainly not a matter of dishonesty or even primarily of imperfect memory. It is an inevitable consequence of the ever-changing *contexts* of meaning and use within which the events are retrospectively set, even by those with the most reliable memories. Charles Darwin's well-known statement that as a young man he had collected facts and worked "without any theory," referring to a period when he had actually compiled some of the most creatively theoretical notebooks known in the history of science, is but one outstanding example of a pervasive problem.[6] The historian of science is therefore on safer ground if able, by the nature of the record, to focus attention on documentary material that is strictly contemporary with the events being reconstructed and analyzed.

The character of that material determines what may be termed the "graininess" or degree of temporal "resolution" that can be reached in any given historical study. As the study of human anatomy can benefit from every degree of optical resolution, from naked-eye dissection to electronmicrography, so the empirical study of scientific activity needs research at every degree of temporal graininess. At one extreme is the historian's study of the *longue durée* of scientific practice over the centuries—a study, it should be said, that has hardly yet begun. At the other extreme are such disparate studies as the experimental psychologist's analysis of scientific thinking in a task measured in minutes, and the ethnomethodologist's analysis of scientists' conversations at the laboratory bench. Between those extremes lies what is perhaps the most

---

5. Geertz (1976, p. 224). On "thick description," see Geertz (1973, chap. 1), and Elkana (1981). The rejection by "discourse analysts" of any such interpretative goal (see, e.g., Mulkay and Gilbert 1982) seems to be based on the assumption that all other analysts of scientific work do, or must necessarily, take the "accounts" of participants at face value, and make an arbitrary selection from those accounts if they are inconsistent. No historian could possibly accept the validity of that assumption.

6. Barlow (1958, p. 119). For analyses of the theoretical thinking in the notebooks, see, e.g., Gruber (1981a) and Kohn (1980). For a balanced appreciation of the oral testimony of living scientists, see Holmes (1984).

promising level of description and analysis to which the historian can hope to contribute: namely, the level of reconstruction plotted neither in centuries and decades, nor in hours and minutes, but in years, months, and weeks, and —with a bit of luck—sometimes even in days.[7] This is, as it were, the high-quality "light microscope" of the analysis of scientific research practice. It shows somewhat less detail than the "electron microscope" of still more short-term or fine-grained studies; but what it does show is less confusing and much easier to relate to the larger-scale or long-term features that can be seen with the "dissecting microscope" and "naked-eye" studies of conventional historical analysis. At this degree of relatively fine-grained resolution, the historian can exploit to the full the rich and varied evidence that is sometimes available.

The Devonian controversy is an exceptionally favorable case for the application of this kind of historical microscope. For various contingent historical reasons, the period in which it took place is unusually well endowed with source materials. "Is it not true," Susan Cannon asked rhetorically some twenty years ago, "that for the middle of the nineteenth century we possess the most complete documentation, for selected individuals, not only that ever has existed *but that ever will exist?*"[8] The records of the Devonian controversy show that this perhaps surprising claim may be as valid for selected scientific *problems* as for some outstanding individuals. The contemporary evidence, as for much other scientific research in the past two centuries, ranges from published books and articles, through participants' accounts of meetings and public discussions, to the early drafts of scientific papers, exchanges of private letters, and finally the private notebooks of individuals. That spectrum runs broadly from public to private, but no category of material has any special privileged status above any other. All are needed for any adequate reconstruction of the process of scientific research.

The records of the Devonian controversy are not perfect; there are some important and tantalizing gaps, for example, in what are generally quite complete series of letters exchanged between certain pairs of participants. There is also a systematic bias in the documentation, in that the fullest records are those of the major participants, whereas the notebooks and correspondence of less important figures are generally underrepresented. Nevertheless, the historical traces of the Devonian controversy are so rich that the historian can be fairly confident of having reached the point of diminishing returns, where even the discovery of another bundle of letters or notebooks in someone's attic would be unlikely to cause any major alteration in the reconstruction of the whole controversy.[9]

7. For the concept of "fine-grained" analyses, see Holmes (1981, 1984). The biographical studies mentioned earlier are good examples of this level of resolution, applied to individual scientists. The importance of working at different levels of resolution in the study of individuals is well argued by Gruber (1981b).

8. Cannon (1964, p. 30). Cannon may have had second thoughts about the validity of the claim, for it was omitted from the revised version of this seminal article: Cannon (1978, chap. 8). Although the prediction about the future was rash, the claim was in my opinion justified with respect to the present, provided the documentation is judged not just on bulk but also on the *quality* of the insights that the documents allow. The archives of distinguished twentieth-century scientists may be more voluminous, but also less revealing, than those of some of their nineteenth-century predecessors.

9. See introduction to list of manuscript sources (p. 463).

The dramatis personae in any scientific episode are not, of course, known in advance to the historian. Nevertheless, the records themselves can be made to disclose who, in the opinion of contemporaries, were the most significant actors. The sociologist, investigating some analogous problem in modern science, can detect and map the network of those most intensely involved by using a technique of "snowball sampling," asking each informant to name the others most relevant to the debate, whether as supporters or as opponents.[10] In much the same way the historian can let the documents themselves, and especially the correspondence, lead to a snowballing reconstitution of a cast list that is likewise founded on the actors' own perceptions. That is what has been possible in the case of the Devonian controversy. Furthermore, the historian, unlike the overprudent sociologist, has no need to conceal historical personages behind the obstructive and often ineffectual veil of pseudonymity.[11] The historical natives can be themselves, in all their marvelous particularity, warts and all.

1.4   Research as Skilled Craftsmanship

It has been more than twenty years since Thomas Kuhn's *The Structure of Scientific Revolutions* first made widely known and readably explicit much that reflective practicing scientists had long known intuitively in their bones.[12] Leaving aside Kuhn's striking but more problematic claims about "revolutionary science" and the incommensurability of paradigms, much in his account of "normal science" seemed as descriptively correct and perceptive to practicing scientists as it seemed alarming and threatening to prescriptively inclined or moralizing philosophers. As Kuhn emphasized, the ordinary business of scientific research is carried on within a shared or collective framework of methodological assumptions, heuristic maxims, routine procedures, observational and experimental standards, criteria of interpretative judgment, and much else besides. Even before Kuhn's book was published, Michael Polanyi, drawing on his long, firsthand experience as a distinguished scientist, had likewise emphasized the communal framework of the pervasive *tacit* knowledge that underlies the widest range of human skills. He pointed out that in such skills, including those of scientific research, much cannot be fully specified, at least not by those who practice them. They are the skills of connoisseurship and other forms of personal judgment; like the skills of the craftsman,

---

10. See, e.g., Collins (1975, 1981).

11. The scientists in Latour and Woolgar (1979), as in many other sociological studies, are labeled with letters of the alphabet, although one of the participants in the research that is analyzed was a Nobel prize winner! One major actor ("O") in Collins's (1975) study of gravitational radiation was revealed by name in a follow-up study (1981). Likewise it is said that the pseudonymous characters in Goodfield (1981) are easily identifiable by specialists in the relevant scientific field.

12. Kuhn (1970, first published in 1962). The comment is not intended to detract from the originality of Kuhn's work, but on the contrary to point out that certain components of it were immediately recognized by many practicing scientists as an authentic characterization of their work.

they are learned not from textbooks but by working alongside a more experienced practitioner within a living communal tradition.[13]

This picture of scientific work as skilled craftsmanship, practiced within a shared tradition that is maintained by a social collectivity, jarred not only against the fiercely held convictions of many philosophers but also against the routine assumptions of many historians of science. Even now, its validity would be more widely appreciated if those who analyze scientific work were not generally such narrowly bookish people, and if they had firsthand experience not only of scientific research itself but also of skilled *manual* crafts outside the intellectual or academic sphere altogether. In any case, it is clear that historians have not yet adjusted their modes of writing to take full account of the social dimension of scientific practice. It has become common, and in some circles fashionable, to explore the social impact of scientific work or, conversely, the penetration of wider social interests into scientific ideas and concepts.[14] But historical studies of the *detailed content* of scientific research are still generally individualistic in their approach. The work of individual scientists is often analyzed in admirable detail, and full weight may also be given to the ways in which that work was influenced by others. But few historians of science have set out deliberately to recover what such a network of individuals had in common, particularly what they held *tacitly* in common.[15]

This book is not primarily designed to remedy that defect, even by a single example. Nonetheless, any detailed account of a scientific episode such as the Devonian controversy must be preceded by at least a brief sketch of the social and intellectual framework within which the new scientific knowledge was shaped; making explicit, as far as possible, what for the actors remained largely tacit and taken for granted. To extend the dramaturgical metaphor, the scene must be set and the background of the plot summarized before the play itself can begin. Dramatists have used preludes, choruses, soliloquies, ignorant newcomers, and other devices for this purpose. I have chosen the form of a prelude, which is part one of this book. Here the social scenery of the Devonian controversy is sketched, and the tacit conventions of the actors' scientific practice are briefly made explicit. This leads into the first chapter of part two, in which the background history is summarized and the main actors introduced. The way is then clear for the rest of the narrative to unfold without repeated interruption for the explanations that would otherwise be necessary. But the description of the craft knowledge underlying the Devonian controversy cannot be confined to the introductory part of the book. It is within the main narrative that the collective tacit framework of early nineteenth-century geology must be shown in operation, if its role in the shaping of a new piece of scientific knowledge is to be adequately portrayed.

13. Polanyi (1958, 1966). The term "craft knowledge" seems to have been first used explicitly in this context by Ravetz (1971, chap. 3), but it is certainly implicit in Polanyi's much earlier work. Fleck's prewar study (1935, trans. 1979), although important in retrospect, had little influence at the time.

14. For examples of the latter, based on thorough historical research rather than purely on political conviction, see the essays in Barnes and Shapin (1979) and Shapin's (1982) important review of this research.

15. An outstanding exception is Frank's (1980) fine study of the physiological research of William Harvey's followers in Oxford in the mid-seventeenth century. See also Geison's (1981) important review of the problems of studying research "schools" in science.

1.5   The Revival of Narrative

Narrative is out of fashion among historians generally. It survives in the traditional genres of political history and biography, but number-crunching cliometricians reject it as unscientific, and those who trace the *longue durée* of historical *structure* dismiss it as mere *histoire événementielle*. When Lawrence Stone discerned a recent movement toward a revival of narrative among the so-called new historians, the examples he was able to cite showed that the trend was defined more by the absence of quantification and analysis than by the positive presence of a strict temporal framework.[16] In the history of science, too, narrative is out of favor. Chronicles of disconnected events— usually a series of publications—still pass for the history of science in some quarters, but this only confirms the opinion of more sophisticated practitioners that narrative has no place in progressive historiography.

It is high time for a genuine revival of narrative to be set in train, but it must be narrative with a purpose, and no mere chronicle. In the fine-grained study of scientific research practice, narrative is not so much a literary convenience as a methodological necessity. If scientific knowledge is to be studied *in the making*, the closest attention must be paid to strict chronology, not only in description but also in analysis. If, to use Harry Collins's apt metaphor, the "ship in a bottle" of any established scientific conclusion is to be understood through its mode of construction, the *sequence* of manipulations by which the model is inserted and made to appear so permanent must be a primary object of attention.[17] A casual or anecdotal jumping back and forth over years or even decades, which is common in many accounts of the fortunes of both scientific concepts and scientific institutions, has no proper place in any fine-grained account of scientific knowledge in the making.

Sociologists of science such as Collins have been so concerned with avoiding retrospective judgments of scientific research—and rightly so—that they have deliberately chosen to study problems or controversies that are as yet unresolved, and where the science is therefore still in the making.[18] This is highly effective in preventing any possible use of hindsight, either by the sociologists themselves or by the scientists they interview. But it ensures methodological purity at the cost of comprehensive understanding, for only incomplete episodes can be analysed—at least until many years are past, when repeat interviews may reveal that some conclusion or consensus has at last been reached. This is where the historian of science can make a distinctive contribution; for despite some risk to purity, historical materials allow the study of *completed* episodes of scientific practice.

The risk is that the description and analysis may be irreparably distorted by the historian's or the readers' knowledge of the outcome of the episode or the "correct" solution of the controversy. That risk is not negligible. Many detailed historical studies—some of them otherwise admirable—analyze the earlier phases of specific scientific developments with repeated forward ref-

16. Stone (1979); see also White (1984).
17. Collins (1974); see also Brannigan (1981, chap. 3).
18. See, e.g., Collins (1975, 1981), Pinch (1981), and Pickering (1981).

erence to problems that had not yet arisen, experiments not yet performed, theories not yet devised, and publications not yet composed. Even historians of science who are zealous in sniffing out the "presentist" or "whiggish" heresies of others are themselves often guilty of what may be termed the "second-order whiggism" of retrospective description. This may not be as blatant as the presentist interpretations of some scientists, with their repeated invocation of what "we now know" as an unproblematic standard for understanding the past history of their field. The forward reference may not be to present knowledge, but rather to the later and mature work of the same individual or to the later development of the same discipline—however unmodern that may still have been. But even this precludes any genuine understanding of the *processes* by which new knowledge is shaped.

Narrative in the service of understanding the shaping of knowledge must rigorously and self-consciously avoid hindsight. To paraphrase and extend an earlier quotation, the historian of science must try to figure out what the devil the scientists thought they were up to, not just in the general sense of empathizing with their worldview or comprehending the contemporary state of their discipline, but also in the far more specific sense of following and making sense of what they did and said and wrote and argued about, week by week and month by month. What one knows, as a historian, that certain scientists said or wrote in September must not be allowed to warp one's judgment of what the devil they thought they were up to back in June. As far as is humanly possible (which means imperfectly), the historian must shelve any knowledge of what was for the participants an unknown future, in order to reconstruct the processes by which they were later to reach it. It is not entirely frivolous to add that this feat is greatly facilitated if the historian has a poor memory.

A deliberate and consistent abstention from hindsight is only one way in which a narrative of the shaping of any piece of scientific knowledge is as contrived a literary form as any other kind of history. A proper ambition to tell the story "how it really was," at least in the sense of handling the extant records as conscientiously as possible, does not imply that the resultant narrative will be a plain, unvarnished chronicle. On the contrary, it must aim to be as carefully constructed for its purpose as any well-crafted, traditional novel. For example, if the narrative deals with an interacting group of scientists rather than a single individual, the separate activities of different members must be woven into the main plot in a way that respects the trajectories of their individual lives while breaching the strict chronology as little as possible. Likewise, there is no need for the undoubted tedium of long stretches of scientific work to be reproduced in stretches of equally tedious narrative, as Frederic Holmes too pessimistically concluded.[19] A narrative that does justice to the intricate twists and turns of research does indeed need to be long and detailed. But it can and should reflect the "subjective time" of the scientists themselves, expanding where the action seemed to them most intense and exciting, and contracting where they were bored, frustrated, or just diverted to other lines of work. The chronology that a narrative should punctiliously observe is primarily that of the sequence of events, rather than the lapse of "real time" as measured by the calendar on the laboratory wall. Like nine-

19. Holmes (1974, pp. xvi, xvii).

teenth-century geology, it is more concerned with a relative than with an absolute timescale.

No single narrative account of an episode in scientific research can claim to be definitive, and not only because new evidence may come to light subsequently. In principle the same story could be told from a number of different viewpoints. If pursued consistently this would generate a series of interlocking but separate narratives. In Alan Ayckbourn's dramatic trilogy *The Norman Conquests*, the same plot unfolds and the same actors perform their parts in each of three parallel plays, set in the dining room, sitting room, and garden of the same house over the same weekend.[20] By analogy, it would be possible in principle to write at least three parallel narratives of the Devonian controversy; for example, as it was experienced within the English-, French-, and German-speaking scientific worlds. I have chosen to tell it primarily from the first of those perspectives because that is the setting in which the dramatic action was most intense, and for which the documentation is most complete. It should be emphasized, however, that the possibility of telling the same story from several different perspectives does not reduce each indifferently to a mere "account," in the somewhat pejorative sense of that word favored by some sociologists. The materials for each narrative may be selected according to different criteria of significance, but all the alternative accounts are, or should be, under the same constraints in their proper use of documentary and other evidence.

Part two of this book comprises an attempt to write a narrative of the Devonian controversy in the manner just outlined. The narrative is as rigorously nonretrospective as I can make it, and it keeps as closely to the original sequence of events as is compatible with the simultaneous tracing of many individual trajectories. Generous quotations are included, not least because the tone of the participants' exchanges and the metaphors they chose to use are essential to an understanding of the ways in which the new scientific knowledge was shaped. Interpretative commentary is confined to the exposition of meanings and inferences that the participants might plausibly have recognized for themselves, at that particular moment in the development of the controversy, and to the recall of relevant earlier moments that the participants themselves might plausibly have remembered. The anticipation of later moments that they could not possibly have known in advance is rigorously excluded. The resultant narrative may at first sight seem unduly long. But any substantial contraction of its scale would have reduced this case study to the schematic level, draining it of any value as a portrayal of the complexities of real scientific research. (Readers who find themselves bogged down in the narrative can skip straight to the analysis in chapter fifteen, though they will miss much—in more senses than one—by doing so.)

A nonretrospective narrative is designed, by careful contrivance, to minimize the bias that a knowledge of the eventual outcome can have on the telling of the story. Herbert Butterfield's plea for a narrative political history applies with equal force to the history of science: "we must have the kind of story in which . . . we can never quite guess, at any given moment, what is going to happen next."[21] In the case of the Devonian controversy, an extra bonus comes fortuitously from the very obscurity into which the episode has

20. Ayckbourn (1975).
21. Butterfield (1957, p. 106).

fallen since it was successfully resolved. With luck, most readers of this book will not know who were the goodies and who the baddies—not that such labels have any rightful place in the history of science—nor will they know in advance how the plot was to end. I shall not tell them.

1.6   Beyond Earshot of the Natives

The goal of this study is to contribute toward making "small facts speak to large issues." Any worthwhile speculation about large issues must be empirically grounded in "long-term, mainly (though not exclusively) qualitative, highly participative, and almost obsessively fine-comb field study."[22] Geertz's prescription for anthropological research is directly applicable, *mutatis mutandis*, to the empirical study of scientific research practice and the shaping of scientific knowledge, past and present. It is an accurate characterization of the historical method that will be used in this account of the Devonian controversy. As Geertz recognized for his own discipline, however, the problem is to make the transition from microscopic description to large-scale interpretation. It is not enough to treat a fine-grained study as adequate in itself without further analysis, either as a microscopic "world in a teacup" or as a natural experiment or "sociological cloudchamber." Small facts can only speak to large issues if they are deliberately made to do so.

A nonretrospective narrative of any episode in the history of science should be couched in terms that the historical actors themselves could have recognized and appreciated with only minor cultural translation. But then the risk is that one is left "awash with immediacies" and trapped inside the conceptual world of the natives being studied. Most sociologists and philosophers of science have in practice tilted toward the opposite extreme, couching their conclusions in terms so distant from the natives' own perceptions and experience that one is left "stranded in abstractions."[23] The ideal for the investigator might be to turn the formal hermeneutical circle into a dynamic spiral of involvement and detachment, immediacy and abstraction. For purposes of presentation, however, it may be better to cut through the historian's knot of indecision about the proper balance between description and analysis, and simply to attempt both, separately and in succession. This is what I have chosen to do in this book: the narrative of the Devonian controversy in part two is followed in part three by an analysis of the case study and its implications.

In contrast with the formalistic hypothetical examples still favored by many philosophers, any real piece of scientific research must be described by the historian (or, for that matter, the sociologist) in a style that does justice to its real complexity, muddle, and messiness. As already noted, this means that any adequate narrative is bound to be long and detailed. Furthermore, if the narrative is to be fully nonretrospective, the historian must abstain not only from forward reference to later phases but also from imposing on the narrative the simplifying generalizations that may be apparent in retrospect. The reader may have to be left feeling awash in immediacies, with only such landmarks for orientation as were also available to the historical actors at the

22. Geertz (1973, p. 23).
23. Geertz (1976).

time. The proper moment for rescue from the rising tide of immediacies comes after the narrative and outside its framework. The terms of the discussion can then be those of late twentieth-century analysts of science—historical, socio-logical, and philosophical—rather than those of the early nineteenth-century scientists themselves. Retrospective reflection in the light of the known out-come of the controversy becomes at this point permissible, indeed indis-pensable. The scientific natives are out of earshot.

The analysis in part three of this book is structured by the conviction that the new scientific knowledge produced in most episodes of scientific research practice—including the Devonian controversy— should not be treated only or primarily as the creative achievement of one or a few outstanding individuals. It should be regarded rather as the outcome of processes of in-teraction within a group or cast list that included, in their diverse roles, not only star performers but also minor actors and walk-on parts. The interactions may include those of bitter rivalry and fierce antagonism as well as those of generous giving and amicable collaboration. But whatever their affective qual-ity or moral status, they are the manifestations of what Bruno Latour and Steve Woolgar have aptly termed the "agonistic field" of intensive social negotiation in science.[24] In the course of research it is in the agonistic field that new interpretative schemes may slowly grow in plausibility and perceived solidity, forming the focus for a consensus, while other schemes gradually dissolve into implausibility and fade into oblivion, adhered to only by marginal in-dividuals. This is what the historian of science should be able to trace and analyze in detail, perhaps even more reliably than the participant-observer in current research.

Some philosophers may continue to portray natural science as "a ship of reason powering its own way through a silent sea of social contingencies."[25] Those who are concerned with what scientists really did in the past—or with what they do at present—have rightly rejected that image as incompatible with any truthful description of scientific activity. But some have now swung to the opposite extreme, presenting science as the making of model ships in bottles with an entirely questionable relation to any real ships there may be in the world outside. But if scientific activity is in any sense a social "learning machine," by which human beings can gain some kind of reliable knowledge of the natural world, then the question of external reference cannot be shelved or evaded. Without abandoning the immense gains that have come from fo-cusing on the human bearers of scientific ideas and beliefs, it is now time to find new and more adequate ways of describing the relation between our social constructions in scientific research and whatever constraints may be imposed on those constructions by the natural world itself. I have chosen to refer to the "shaping" of knowledge, in place of either "discovery" or "con-struction," in order to express on the metaphorical level the centrality of that relation.

I hope that this account of the Devonian controversy, although only a single case study, may contribute to a more adequate understanding of the shaping of scientific knowledge. Only an array or repertoire of similarly de-

---

24. Latour and Woolgar (1979, p. 237 and passim).
25. Barnes (1982, p. 117); the comment was made in strong *criticism* of the philosophers' approach.

tailed examples will ever release the analysis of scientific knowledge from being dominated by the purveyors of programmatic generalizations and prescriptive abstractions. At least in the first instance, the empirical study of scientific research must adopt an inductive strategy. Since it should not be naively inductive, however, such a strategy does raise the issue of the applicability of any one example. Early nineteenth-century geology may appear at first sight to be poles apart from, say, modern elementary particle physics. The observational and scarcely quantitative character of the one contrasts with the experimental and highly mathematized character of the other; the gentlemanly individualism of the one contrasts no less strongly with the professionalized teamwork of the other. It may seem at first sight that any conclusions about scientific activity that are based on the first are unlikely to apply at all to the second. Yet it is axiomatic that any one branch of science can be linked to any other, however different, by an indefinitely graded chain of intermediates; an analogous continuity links one historical period to another.[26] If the conclusions of this study of the Devonian controversy are held to have no relevance beyond the world of nineteenth-century gentlemanly geologists, the burden of proof lies squarely with those who impose that restriction.

26. Holmes (1984) rightly emphasises the remarkable degree of continuity in the procedures of scientific research in the near and "deeper" past, at least on the individual level (e.g., between Krebs and Lavoisier).

Chapter Two

ARENAS OF GENTLEMANLY DEBATE

2.1   Gentlemanly Specialists in Geology

The world of science in the early nineteenth century was dominated by gentlemanly specialists. The days of the universal virtuoso or polymathic savant in science were over. Scientific endeavor was divided into a set of relatively discrete fields, each with practitioners, institutions, and journals to match. Those with ambitions to evaluate the whole range of natural knowledge were so exceptional as to be celebrated as living monuments to human genius (e.g., Humboldt, Whewell). For most men of science, acknowledged leadership in a single chosen field was ample recognition. Although the degree of specialization was only moderate by modern standards, most of the scientific achievements of the early nineteenth century were unmistakably the work of *specialists*. They have often been termed "amateurs," too, because many of them were not economically dependent on their scientific work. But that term is highly misleading because it now carries uneliminable overtones of less than professional standards of work. Amateurs they were indeed, in the original sense of the word, but their work was anything but amateurish.[1]

Many of the leaders of science at this period, particularly but not only in England, were also unmistakably *gentlemen*.[2] This social status, sustained

---

1. On this point, see particularly Cannon (1978, chap. 5). In this and the following chapter, individuals cited by name are those who figured in the Devonian controversy—whether centrally or marginally—and about whom further biographical information is given later in this book.

2. See especially Morrell and Thackray (1981). Extending their apt characterization of the

by wealth that had often been inherited or acquired through marriage, enabled them to pursue their scientific interests with an independence that was not enjoyed by their more professionalized successors later in the century. But this did not place them in some imagined paradise of scientific harmony and cooperation. On the contrary, the very absence of narrowly professional rewards gave a sharp edge to their competitive building of less formal careers. Without a recognized ladder of advancement as the anticipated reward for mere diligence, they were left to struggle more individualistically for their due share of less tangible rewards. They were therefore concerned with issues of recognition and scientific priority with an even greater intensity than later generations. The epoch of the gentlemanly specialists was one in which the elements of competition and controversy, which are arguably perennial in the business of science, were exposed with a clarity that has rarely been matched in later periods.[3]

2.2   The Geological Society

When the Devonian controversy first erupted in public in 1834, it did so in a meeting of the Geological Society of London. Throughout the controversy the Society remained the principal arena in which the drama was played out (fig. 2.1).[4] All the major participants and many of the minor ones were members; for most of them the Society was their main scientific affiliation (most of those who were not British were "foreign members" of the Society). The Society's meetings and its publications provided the main institutional framework for the whole debate.

By 1834 the Geological Society was over a quarter-century old. It was a well-established part of the English scientific scene, and by common consent it was the liveliest of all the scientific bodies in London. It was the oldest society in the world devoted specifically to the earth sciences, and it was regarded internationally as the most active center of geological research anywhere. It was fashionable, prosperous, and successful, but it had not always been so. The Society's early years had been undistinguished, and even its survival had at one time been uncertain. By the 1830s it had evolved—more by tinkering than design—a pattern of organization and activity that had become one of the keys to its success. But when it was founded in 1807 the direction of its development could not have been predicted, and, if it had been, would not have wholly satisfied anyone. For around the time of its foundation there were at least three distinct institutional models that the Geological Society might have adopted. Each was favored by some of its

---

"gentlemen of science" who dominated English science at this period, I first used the term "gentlemanly specialists" to describe the scientific milieu of the young Charles Darwin, which was also that of the Devonian controversy (Rudwick 1982b). On gentlemen in geology, see Porter (1978); on the place of men of science within the "clerisy" of English intellectuals, see Heyck (1982).

3. On the importance of scientific individualism at this period, see Morrell (1971).

4. Fig. 2.1 is reproduced from an anonymous and undated sketch in GSL; it probably depicts a meeting held *before* GSL moved to more spacious rooms at Somerset House, where by 1838 there was a "great horseshoe table" (K. Lyell 1881, 2:37).

Fig. 2.1.   A contemporary sketch of the Geological Society of London in session. Note the parliamentary arrangement of the seating. The leading members—De la Beche is easily recognizable by his spectacles—face each other on the front benches, while less important figures are crowded behind them, and only faintly sketched in. The president, flanked by the two secretaries, occupies the Speaker's position. Specimens for discussion are on the central table (in this sketch, an ichthyosaur's head is just to the right of the president's); visual aids are displayed on the far wall (in this sketch, a large traverse section of a mountain range).

founders and early members, but only one was finally realized in anything like the form its supporters had hoped. The three models may be termed for convenience the "mineral resource center," the "scientific dining club," and the "learned society."

At the time the Society was founded there were informal groups in the capital that favored the creation of a body that would act as a center for collecting information on the nation's mineral resources.[5] An English precedent for this utilitarian model, though with more general aims, would have been the Royal Institution, as it was originally conceived at its foundation in 1799. A Continental precedent more closely comparable in its aims was the École des Mines in Paris (founded in 1788), though this was a state institution in which the function of advanced training had been added to that of surveying mineral resources.[6] When the Geological Society was founded, some of its earliest activities did suggest that this utilitarian model had been adopted and that the new body was aiming to become the nation's unofficial mineral re-

5. Weindling (1979, 1983).
6. Berman (1978); Birembaut (1964).

source center. Its very first publication was a booklet entitled *Geological Inquiries,* which was explicitly designed to elicit useful factual information from a broad social spread of scattered local observers. The Society offered itself, in effect, as a national repository for geological data and specimens. The voluminous early correspondence of the first president (Greenough) shows how successful this appeal proved to be.[7] Likewise, the extant early archives show that many mineowners were prepared to deposit with the Society some records of their mine workings, despite the problems of industrial secrecy. Such materials were used in the construction of what was initially conceived as the Society's collective research project: a geological map that would summarize in visual form the pattern of the nation's actual and potential resources.

However, the model of the Society as a mineral resource center was quickly swamped by the demands of the other two models. Those who had most to offer in the way of expert information, namely, the professional mineral and land surveyors, were virtually excluded from the Society on social grounds. The proposed map became in effect the personal research project of the president, although he continued to receive information and assistance from the members of the Society and from provincial observers. The Society lost its early commitment to utilitarian goals, partly as a result of a shift of power from those who had favored that model to those with other interests. But one enduring legacy of the mineral resource center model was the policy—almost, one might say, the ideology—of favoring the accumulation of empirical "facts" in geology over and above any attempt to make theoretical generalizations (see fig. 2.7). This antitheoretical stance continued to be the Society's official policy long after the intellectual and social realities of the Society's activities had overturned it.

While some of the founders and early members had wanted the Society to become a mineral resource center, others wanted it to give permanent form to a group of friends who enjoyed talking about geology over a convivial meal. They wanted, in effect, to add a gentlemen's geological dining club to other such informal groups of London men of science. The supporters of this second model, like those of the first, scored an early success: the Society was formally founded over a private dinner in a London tavern, and it was agreed that similar dinners would be the format for regular meetings.[8] The high price of the meal (fifteen shillings) would have been enough by itself to deter the professional surveyors from thinking of membership, even if they had not been discouraged by more subtle social means. The Society as a scientific dining club was unmistakably a society for gentlemen.

As a dining club the Geological Society posed no threat to the scientific hegemony of the Royal Society. Many of its earliest members were already Fellows of the older body, and they were soon joined by the Royal Society's autocratic president Sir Joseph Banks. Before long, however, the new society showed unmistakable signs of shifting in the direction of the third and last model, that of a learned society. Among other changes, this brought in due course a great increase in the number of members, and the intimacy of the

---

7. Geological Society (1808). Most of the responses to the *Inquiries* were addressed to Greenough personally and are preserved among his correspondence in ULC.
8. Rudwick (1963).

original small circle was lost. Yet the model of the scientific dining club did not disappear, for even in a formal learned society it fulfilled essential social functions. A dining club was established within the Society as early as 1810, and refounded on a more formal basis in 1824 as the Geological Society Club.[9] With a membership limited to forty, the dinner table of the Club became de facto the place where the Society was governed, behind the scenes of the de jure constitutional structure of the Society as a whole.

The first signs of the emergence of the third and last model for the Geological Society became apparent when some of the early members began to talk of having a formal constitution and officers and, especially, of acquiring their own premises and publishing their scientific proceedings. Such talk at once alarmed Banks and others in the Royal Society. They feared that this would be only the first step in the institutional fragmentation of the natural sciences, and that the Royal Society's established position as the primary forum for *all* the natural sciences in England would be eroded and ultimately lost.[10] Banks resigned from the Geological Society and tried to stop it in its tracks, but his action had little effect; the crisis passed and the new body prospered. It acquired its own premises, initially to house its rapidly growing collections of specimens but also later as a place for its meetings. In 1811 it began publishing its own proceedings: not in a form or with contents to appeal to mineowners and surveyors, as the *Journal des Mines* did in France, but in handsome and expensive *Transactions* modeled after those of the Royal Society. Here too the mineral resource center had been submerged by the learned society.

The triumph of this third and last model for the Geological Society was effectively sealed in 1825 with the award of a royal charter. The members were thereafter styled Fellows, like those of the Royal Society (the simpler and original term "members" will be used for convenience throughout this book). Three years later the Royal Society, with which the Geological Society had been fully reconciled after Banks's death in 1820, recommended to the government that the junior society be allotted some of the space in Somerset House for its meeting room, library, and museum. Meanwhile the Geological Society had indeed set a precedent for the foundation of other specialist societies in London, such as the Astronomical (1820), the Zoological (1826), and the Geographical (1830). But this proliferation had not in fact led to the result that Banks had feared; the Royal Society continued to make up in prestige what it lacked in dynamism. All the leading members of the Geological Society were also Fellows of the Royal Society. But they rarely read their papers at its meetings, and they rarely published their research in the senior society's *Philosophical Transactions*. Their own specialist society remained by far their more significant affiliation.

Under the terms of its charter, the Society was run formally by a Council of about twenty persons, including a president, four vice-presidents, two secretaries, a treasurer, and a "foreign secretary" to foster links with scientific bodies in other countries. Newly elected officers took up their positions at

9. Woodward (1907, pp. 65–67). Little is known of the earlier club: Rudwick (1963, p. 341n.32).
    10. Rudwick (1963). The Linnean Society, founded in 1788, restricted itself to descriptive natural history and was not seen as such a threat.

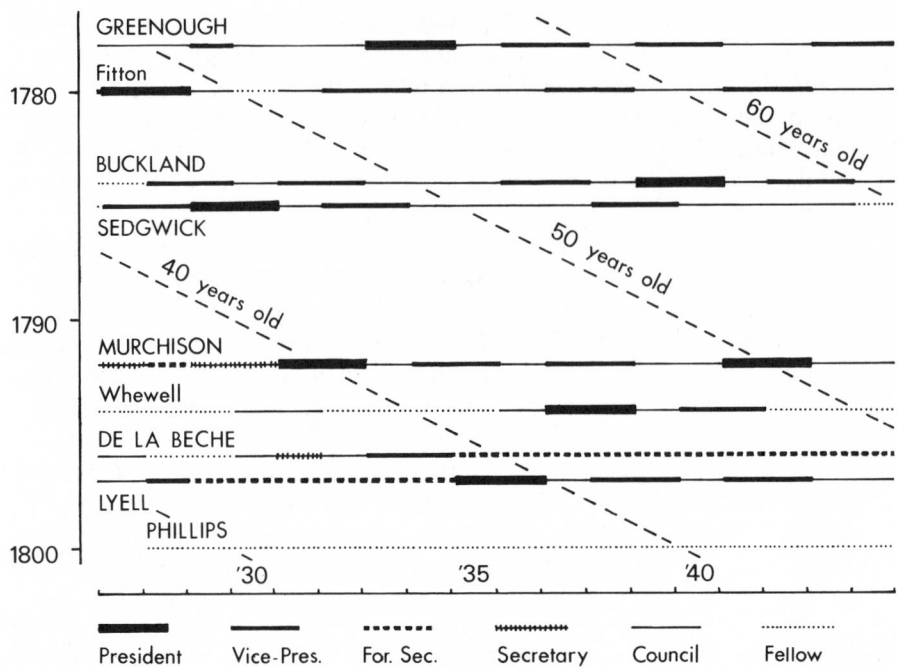

Fig. 2.2.   Officers in the Geological Society of London, 1827–44. This diagram shows
how the geologists who became major figures in the Devonian controversy (names in
capitals) provided the Society with most of its presidents during and before that period,
and how in other years they held other offices or were at least members of the Council.
Individuals are arranged in order of age (year of birth at left); diagonal lines show
ages at time of office holding.

the close of the "anniversary meeting," an annual business meeting always
held on a Friday in February. In informal practice, as already mentioned, the
Society was largely managed by a small coterie of active members who were
able to devote time to its affairs, and often their own money too. These active
members formed the pool from which the officers were generally drawn, and
when not in office they usually remained on the Council (fig. 2.2).[11] In practice,
they decided who should be formally proposed to fill the main offices when
they fell vacant, who should be nominated as members of the Council, and
who should referee whose papers. In the interstices of these informal con-
sultations, however, much of the routine business, including the choice of
papers to be read at the meetings, was in the hands of the president and
secretaries, assisted by a salaried employee of the Society, the "librarian and
curator."

In the two decades before and during the Devonian controversy, the
Society's membership grew steadily, more than doubling between 1825 and
1845. By the time the controversy began, at the end of 1834, it had a total of
745 members.[12] This included three royal members elected to give social
distinction; fifty-seven foreign members to give scientific distinction and to

11. Fig. 2.2 is based on data in annual reports in *PGS*.
12. This and the following figures are given in *PGS* 2:128.

encourage international contacts; and forty-four honorary members, a declining category of those elected in the early years and living outside London, who had not accepted the Society's invitation to become Fellows. The bulk of the membership included two major groups of almost equal size, namely, 313 "contributing" members living in or near London and 328 "non-resident" members living in the provinces or overseas. This last distinction underlines the analogy between the Society and the many London gentlemen's clubs of the period; the Society's apartments at Somerset House provided many of the same facilities as a club, and those living in London were expected to pay more for their greater use of them. Of the 313 resident members, 84 (or 27%) of the most active—or most wealthy—had "compounded" their subscriptions for life by paying a substantial lump sum.[13] Of the impressive total membership of the Society, only a minority was at all active, and attendance at meetings was usually small enough to preserve an informal atmosphere. No great achievements in geology were required for membership: an expressed interest in the science was sufficient, provided the candidate had contacts with existing members who would propose him and vouch for his social respectability. As in a London social club, women were not admitted, even as members' guests.

The meetings took place on Wednesday evenings, usually every two weeks from early November until early June: in total about fifteen meetings in each session. In sharp contrast to the Royal Society, but like the Society of Antiquaries (also at Somerset House), the meeting room was arranged like a small-scale House of Commons, with two sets of seats facing each other (fig. 2.1).[14] If on any issue there were conflicting groups or at least rival individuals, as there were during the Devonian controversy, they could easily choose to sit on opposite sides, confronting each other like parties or politicians in Parliament, and thereby heightening the sense of polarized conflict. By custom the leading figures sat on the front benches, again as in Parliament; the younger or less active members crowded into the rows behind them; the president occupied a position corresponding to the Speaker's in Parliament, thereby functioning implicitly as an impartial arbitrator. The crucial importance of visual communication in geology was recognized in the meetings as well as in the Society's publications; large-scale versions of diagrams destined for publication were either drawn by the authors themselves or commissioned from a professional artist, and a blackboard was also available.[15] Relevant specimens would be displayed on a table set between the front benches for members to inspect before or after a paper had been read.

The division of the year into a seven-month session and a five-month

13. Nonresident members paid a higher admission fee (ten guineas) than residents (six guineas), but were excused the annual contribution (three guineas). The fee for "compounding" contributions was thirty guineas, so that those who could afford the outlay had to wait ten years to break even (GLS, byelaws of 1827, sec. XI).

14. This layout was continued when the Society moved to Burlington House in 1874, and it only came to an end in 1972 with the reconstruction of the meeting room there. For the contrast between the Royal Society and the Society of Antiquaries, see Needham and Webster (1905, pls. at pp. 235, 239).

15. The artist most commonly employed was George Scharf (b. 1788), who also worked as a theatrical scene painter. Scharf drew many of the illustrations for *TGS*, working from preliminary sketches supplied by the geologists. The blackboard was ordered by the Council in 1831 (GSL: BP, CM 1/3, p. 111).

recess might seem to have been related to the needs of summer fieldwork in geology. But in reality it simply conformed to a pattern already set by other learned societies, by the seasonal cycle of Parliament, the law courts, and the universities, and by the social life of the leisured classes. In fact the timing of the recess was rather poorly adapted to the needs of fieldwork, and the more active members often skipped the last few meetings, if other commitments allowed, in order to take advantage of good weather for fieldwork in late spring and early summer.

After the first few years the system had been adopted in the Geological Society—probably in deliberate contrast to the Royal Society—of electing a new president every two years, thus preventing any individual from gaining a permanent and potentially autocratic position. In the 1820s the anniversary business meeting was expanded to include a regular "anniversary address" by the president. From 1828 onward this address was printed in an annual report (together with reports on membership, finance, etc.); it grew into a general summary and assessment of the papers read during the previous twelve months, and it often expanded beyond the Society's own contribution to include a review of the current state of the science as a whole. The anniversary meetings thus provided an important occasion for successive presidents to give their own views on any long-continued theoretical debate, such as the Devonian controversy.

In the first decade or two, the leaders of the Geological Society insisted, as already mentioned, on a highly fact-oriented or "Baconian" approach to geology.[16] All matters of theory were rigorously excluded by those who held effective power in the Society. This was partly in reaction to earlier theoretical debates that seemed in retrospect to have been sterile, particularly the arguments between Neptunists and Vulcanists.[17] Partly it was a way of steering clear of socially and religiously divisive issues, such as those concerned with the relation between man, the Creator, and the natural world. And partly, as mentioned earlier, it followed naturally from the utilitarian goals of the early mineral resource center model for the Society. It has even been argued that the resultant inductivist methodology was the cause of the Society's rather unimpressive achievements in research in its earliest years.[18] But whatever the precise causes, the Society underwent a subtle change in the 1820s and got a new lease on life. There was a new generation of recruits, several of whom later became major participants in the Devonian controversy as well as in other important theoretical debates. The embargo on theorizing itself was relaxed gradually, as the original leaders of the Society lost power to newer members, and as it became clear that there were acceptable levels of theorizing that stopped short of the discredited older style of global cosmogonies.

16. A quotation from *Novum Organum* adorned the front of each issue of *TGS*. The term "Baconian," here and throughout this book, is used to denote what the philosophers and men of science of the early nineteenth century *attributed* to Francis Bacon, rather than whatever Bacon himself may have meant.

17. There is still no full account of these debates in the years around 1800, but see the brief summary in Greene (1982, chap. 1); for a classic treatment of some of their broader implications, see Gillispie (1951).

18. Laudan (1977).

One of the most far-reaching signs of this change was the adoption of a custom that was, and remained, virtually unique among British learned societies. When after its legal incorporation the Society drew up its byelaws, it concluded its regulations for meetings with the following provision:

> When the other business has been completed, the persons present shall be invited by the Chairman to deliver, from their places, their opinions on the communications which have been read, and on the specimens or drawings which have been exhibited at that meeting.[19]

On the face of it, this was no more than a decision to reintroduce into the meetings the element of free and informal discussion which, in the scientific dining club model, had been their original purpose. But in the learned society model that had long since become dominant, discussions were not so innocuous or unproblematical. For to allow discussion of a paper was to accept the possibility of alternative "opinions" on its subject, and that was to undermine the public image of natural science as straightforward and objective knowledge.

At the Royal Societies of both London and Edinburgh, a dignified silence greeted the delivery of papers; at the Linnean, discussion was further discouraged by not announcing before the meeting what papers were to be read; at the Astronomical, only factual comments were in practice permitted.[20] Even at the Geological Society, where informal discussion must have been facilitated by the parliamentary arrangement of the meeting room, a certain caution about the custom is apparent in the description of what were soon famous in English scientific circles as good healthy arguments. "I will not call them discussions—much less debates," said the president (Fitton) under whom the custom was started; but outsiders were more forthright. "Though I don't much care for geology," commented John Lockhart, editor of the influential Tory *Quarterly Review*, "I do like to see the fellows fight."[21] Charles Babbage, in his testy *Reflections on the Decline of Science in England* (1830), mentioned with approval how the Geological Society had "succeeded in a most difficult experiment" in allowing the free discussion of papers. A president of the Society in the early 1830s (Murchison) acknowledged even more positively how valuable the discussions were, as "the true safeguard of our scientific reputation." This implied that the authors of papers submitted to the "ordeal" of these discussions were thereby obliged to improve their argument and tighten their evidence.[22] But as already pointed out, to admit this was to abandon the "Baconian" ideal of indisputable factual knowledge that had dominated the earlier years of the Society, and implicitly to adopt an alternative that allowed an essential place for argument and interpretation, for rhetoric and persuasion, and above all, for *theory*.

Through its discussions the Geological Society found an institutional means for allowing the uninhibited expression of theoretical disagreements, and hence for allowing the development of a social role for the theorist in geology. Geological theorists were not, of course, to be a separate class from field geologists, but competent geologists of any kind were no longer to be

---

19. GSL, byelaws, sec. XII, clause 7 (adopted 1 May 1827).
20. Morrell and Thackray (1981, p. 50).
21. Fitton (1828, p. 61); Lockhart is quoted in Allen (1976, p. 70).
22. Babbage (1830, pp. 45–46); Murchison (1833a, p. 464).

debarred from openly theorizing about the significance of their observations.[23] The Society was able to reconcile this tacit acceptance of the value of "opinions" with its continuing public repudiation of theorizing, because its argumentative discussions took place in conditions of relative confidentiality, behind the closed doors of the Society's meeting room. Only the members and their guests could participate; the Society guarded jealously the privacy of what went on, forbidding any account of the discussions to be published. Only the anniversary addresses were exempt from running the gauntlet of argumentative discussion: only presidents were allowed to speak "six feet above contradiction." No record was kept of the discussions at ordinary meetings, even for the Society's own minute books: they were regarded as lying outside the official proceedings.[24] In this way, the Society could tolerate individual disagreements in the private and gentlemanly milieu of its own meetings, while maintaining toward the wider public its earlier politic stance of corporate theoretical neutrality.

What did get recorded in the Society's minute books was a summary of each paper read. The composition of these summaries was in principle the burdensome duty of the secretaries; in practice, they were sometimes written by the author, and if not, they were generally shown to the author for his approval. This was not unimportant, for from 1827 onward the summaries were printed and published. In that year the Society began to publish its *Proceedings* several times a year. Each issue contained summaries of all the papers read at one or more meetings, together with the names of new members and other information. One issue was devoted to the anniversary meeting, with the presidential address printed in full. To avoid postal costs the *Proceedings* were only delivered to members with a London address. But they reached a much wider audience than the Society's membership because the contents were regularly reprinted or abstracted in general cultural weeklies like the *Athenaeum,* in scientific monthlies like the *Philosophical Magazine,* and in foreign scientific journals. In this way the *Proceedings* became a highly effective medium for the rapid publication and diffusion of new geological research. Any author whose paper was read at the Society was assured of publication at least in this summary form.

After being read, a paper would be sent by the Council to a referee for an opinion on the desirability of publishing it in full in the *Transactions.* The referee was invariably a member of the informal inner circle of the Society and generally a member of the Council. The reports were often very brief, sometimes just to the effect that the summary already in the *Proceedings* was an adequate form of publication. Such a report would not necessarily be regarded by the author as any mark of rejection of his work, particularly if the paper had been brief or preliminary. An author might also ask to withdraw his paper—once read it was formally the Society's property—if he planned to

23. On the role of the theorist in geology (with particular reference to Charles Darwin), see Rudwick (1982b, pp. 194–96); the contrary case is argued by Herbert (1977, pp. 157–78). The implication of De la Beche's caricature (fig. 2.7), that theorists *were* a separate category, was polemical; Lyell was in fact a competent field geologist as well as a theorist.

24. As a result, the only surviving records of the discussions are in letters written afterward by those who had been present, usually to keep absent members informed about what had been said.

publish it elsewhere, for example, as part of a larger work. But if the paper was recommended for full publication, it would be voted on by the Council; if approved, it would take its place in the queue of papers waiting to be printed.

In 1822 the Society took control of the publication of its *Transactions,* the sales of which had languished. A new series was launched to give the journal a new public image, and the quality of illustration was improved by switching from engraving to the new technique of lithography, which was not only cheaper but also more effective for geology.[25] But the *Transactions* remained expensive; a part was issued on average less than once a year, and any given paper might have to wait a long time for publication. So by the time any important work appeared in this final form, it had often lost all novelty value and was more a monument to its author's reputation than a contribution to current scientific discussion. In practice, those involved in any debate such as the Devonian controversy paid much more attention to the summaries that were rapidly published in the *Proceedings.*

In parenthesis, it may be mentioned here that other periodicals, particularly the monthly *Philosophical Magazine,* were also used by leading members of the Geological Society as a medium for quick publication, but only in special circumstances.[26] A paper published rapidly outside the Society could serve to establish priority, particularly during the summer recess when there were no meetings; it could jump the queue of papers waiting to be published in the *Transactions* and avoid the hazards of refereeing, though the monthlies could not offer such fine facilities for illustration; and it could get into print the kind of material that was too personally vituperative to be acceptable to the custodians of the Society's gentlemanly standards.

To summarize this brief account, the Geological Society at the time of the Devonian controversy can be characterized as one of the liveliest and most innovative bodies within English scientific culture. It is not surprising that contemporary critics of that culture, such as Charles Babbage and William Swainson, specifically excluded the Geological Society from their strictures and praised its organization, procedures, and forms of publication.[27] For all its tendencies to cliquishness, its inner circle of highly active geologists really was active, not simply in running the Society but in promoting the growth of their science. To say this is not to subscribe to an idealized view of scientific activity: as the Devonian drama was to show, scientific growth may derive from, and even require, intense conflict and controversy.

2.3 The Société géologique

It is through no narrow nationalism that this account of the Devonian controversy will center on British geologists working mainly within the ambit of the Geological Society of London. It reflects, in the first place, the importance of that Society within the international community of geologists. This was

25. Rudwick (1976, pp. 156–57).
26. On the commercial scientific journals, see Brock (1980).
27. Babbage (1830, pp. 45–46); Swainson (1834, pp. 305–6, 429); Morrell (1976, p. 138).

even recognized by a French observer hostile to what he termed the "fausse civilisation" of England. When a geological society was started in Paris, not long before the beginning of the Devonian controversy, one of its founders reviewed the number of active geologists in Europe, country by country; he concluded a list totaling less than 400 for the whole of the Continent by adding rather grudgingly the members of the Geological Society of London, who at that time totaled more than 500.[28] The criteria for admission to the list were not stated and were certainly not uniform, but the figures are significant nevertheless as an indication of how the world of geology was widely perceived as being dominated by the English. A focus on the English scene is also justified more specifically as a genuine reflection of the way the Devonian controversy arose, developed, and was ultimately resolved. Nonetheless, it came to include the interpretation of rocks far from the shores of Britain, and many foreign geologists were drawn into the debate, at least in minor ways. In particular, the London society's sister body in Paris was an important if secondary arena for the controversy at certain phases in its development.

The Société géologique de France was founded in 1830, shortly before the July Revolution. It modeled itself closely after the London society, and in its own country it likewise became the most lively and successful of the first generation of societies devoted to specific branches of science.[29] Like many other French institutions, it was, despite its name, essentially metropolitan. It was run largely by an inner circle of active Parisian geologists, many of whom were professionals with salaried positions in the geological sciences at the École des Mines, the Muséum nationale d'Histoire naturelle, and other institutions. In effect, the Société complemented the Académie des Sciences, allowing for the discussion of geological matters in an environment less formal than that of the older and more official body.

The Société géologique was less than half the size of the London society—in 1836, for example, it had a total of 302 members, compared to the Geological Society's 810. As a mark of *égalité* it allowed no distinctions or different classes of membership. Foreigners could become members on the same terms as Frenchmen by paying the same entrance fee and subscription. In contrast to the limited number of foreign members in the London society, elected as a mark of distinction, the Paris society had a much larger proportion of foreigners—in the mid-1830s, almost one in three.[30] Most of the leading members of the London society had joined the Société géologique within a year of its foundation (for example, in order of election: Sedgwick, Greenough, De la Beche, Murchison, Buckland, and Lyell). Allowing for those more distinguished French geologists who were also foreign members in London, there was thus a significant overlap in membership between the two societies, particularly among more active geologists, as well as an important link between the French society and the less organized groups of geologists else-

28. Ami Boué (b. 1794), in *BSGF* 1:14. The total has been calculated from the higher figures, where a spread was given; the (approximately) 110 French members of the early SGF itself have been added. Boué gave the figure for the London society as only 300, so that the disparity with the Continent was in fact even more striking than he conceded.

29. Société géologique (1880, p. xix); Fox (1980, pp. 266–67). There is no adequate modern history of SGF; most of the information in the following summary is derived from the early volumes of *BSGF*.

30. Société géologique (1880, pp. xxiv, lii).

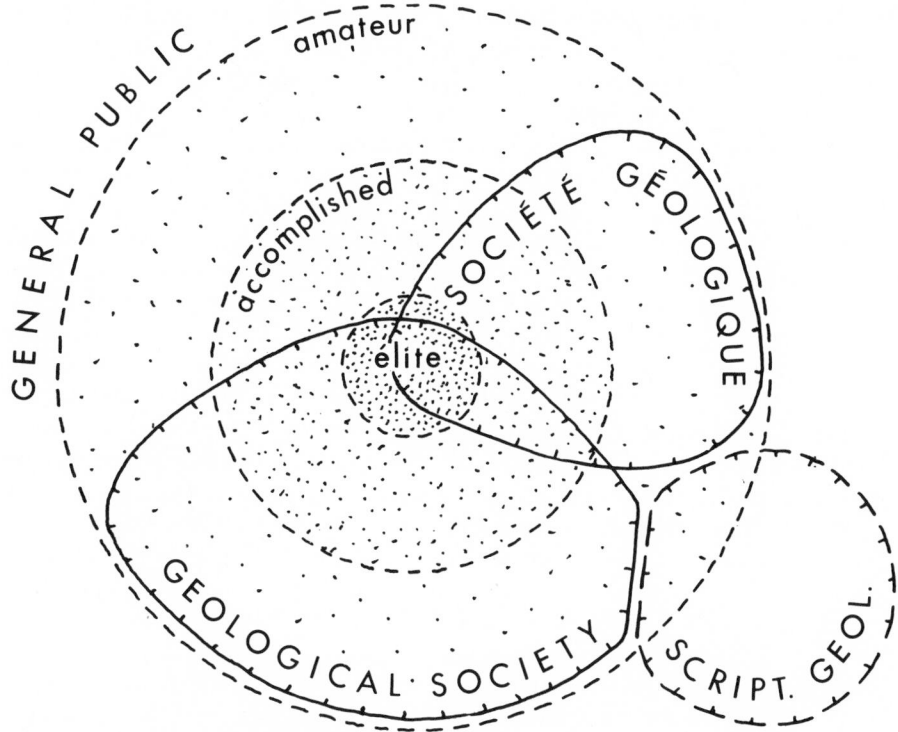

Fig. 2.3. The worldwide community of active geologists at the time of the Devonian controversy, drawn as a Venn diagram to show the overlapping membership of the Geological Society of London and the Société géologique de France (continuous boundaries). The outer dashed circle represents the limit of those whom contemporaries regarded as at least modestly active and competent in geology; beyond was the general public. The "scriptural geologists" are shown as almost completely outside that limit; the small overlap allows for the few who contributed factual material that the "philosophical" geologists found acceptable. The size of the envelopes gives a *very* rough and partly subjective impression of the relative numbers involved. The stippled zones of "elite," "accomplished," and "amateur" geologists are explained in § 15.4.

where in Europe. Between them, the two overlapping societies probably accounted for at least two-thirds of the worldwide total of those whom contemporaries regarded as active geologists (fig. 2.3).

The Société géologique met on alternate Mondays from November to July. Almost from the start, some discussion of the papers was not merely permitted but also recorded and published. But this may have inhibited the liveliness of the discussions compared to those in London; certainly it limited the vigor of what was printed, not least because members were in the habit of editing their remarks after the meetings.[31] Summaries of the papers and of the discussions were published in a *Bulletin*, but since this was issued only twice a year it lacked some of the topical immediacy of the London society's *Proceedings*.

The Société géologique did however initiate one important institutional feature that was never copied in London. This was at one and the same

31. See Rozet's complaint at SGF meeting on 5 February 1838 (*BSGF* 9:148).

time a means of recognizing and encouraging the work of its members in the provinces, and an imaginative innovation in relation to the intrinsic character of the science. Every September it held a *réunion extraordinaire,* an extended meeting in some provincial town, partly to attract and encourage the local geologists but also to allow for about a week of collective fieldwork.[32] The active inner circle of the Société would be conducted around some area of geological interest and importance by both Parisian and local experts, and they would discuss its significance both on the spot and after dinner in the evenings. This arrangement began modestly with a trip to Beauvais, not far from Paris, in 1831. The following year the Société went further afield, to Caen, and then to Clermont and Strasbourg. From 1835—in the years of the Devonian controversy—the field meetings were held successively at Mézières (including a trip into Belgium), Autun, Alençon, Porrentruy (just over the Swiss frontier), Boulogne, Grenoble, Angers, Aix-en-Provence, and Poitiers (see fig. 3.6). Some of these meetings were attended by English and other foreign geologists and allowed for on-the-spot discussion of the international implications of the rocks being studied.

2.4   The British Association

It is at least plausible to infer that the reason why the Geological Society never copied the Parisian society's annual meeting in the provinces was that similar provincial meetings began in Britain almost as soon as the Société géologique was founded. Such meetings were not formally any part of the London society's activities, yet they virtually were so in practice. The meetings were those of the British Association for the Advancement of Science. After the Geological Society, the geological Section of the British Association was the second most important institutional arena in the Devonian controversy. What it lacked in the frequency of its meetings was compensated by the more public circumstances in which the controversy was regularly given an airing (fig. 2.4).[33]

The founding of the British Association, only three years before the controversy began, was primarily the result of provincial initiatives. Originally it represented the claims of provincial scientific culture over and against those of the metropolis and the ancient universities.[34] It made a rather shaky start in 1831 with a meeting in York—the traditional cultural center of the north of England—in the face of apathy, skepticism, and even open disapproval from the metropolitan scientific elite (in this context the leading men of science at Oxford and Cambridge must count as "metropolitan"). As with the Geological Society a quarter-century before, the founders and early supporters of the British Association had in mind several disparate and partly incompatible

32. Société géologique (1880, p. xlv).
33. Fig. 2.4 is reproduced from Heywood (1843).
34. This section is based chiefly on Morrell and Thackray (1981), the fullest account of the foundation and early years of BAAS; see also Cannon (1978), Morrell (1971), and Orange (1971, 1972, 1975).

Fig. 2.4. Murchison delivering an evening lecture on the geology of Russia to a general audience at the British Association meeting in Manchester in June 1842. Note the large proportion of women in the audience. Note also the many traverse and columnar sections in use as visual aids.

aims: to create a nongovernmental body for national science policy *avant la lettre;* to enlist the knowledge and enthusiasm of local amateurs or "cultivators" of science in grand collaborative projects of "Baconian" science; and to raise the prestige and power of science and its practitioners in the eyes of the educated public and the government. From the Gesellschaft Deutscher Naturforscher und Ärzte (founded in 1822) they borrowed the idea of a peripatetic annual meeting, adapting it from the patchwork of independent German-speaking states to the network of provincial cities in Britain.

After the initial meeting, however, the leaders of the Association recognized the political urgency of first getting the support of men of science at the ancient English universities. In the absence of any encouraging signs from the more scientific of the two, the Association chose Oxford for its second meeting, but went to Cambridge the following year. At both meetings the leading resident geologists (Buckland, Sedgwick) played major roles in getting the Association the support of the metropolitan men of science and in giving it academic respectability. When the Devonian controversy erupted at the end of 1834, the Association had recently met in Edinburgh, the center of Scottish intellectual life; and it was due to visit the other non-English capital city, Dublin, the following year. Only from 1836 onward did it begin to visit the truly provincial towns and cities: during the period of the Devonian controversy it went in turn to Bristol, Liverpool, Newcastle, Birmingham, Glasgow, Plymouth, Manchester, and Cork (see fig. 3.5).

During this period the Association prospered and, despite some criticism, it became a prominent feature of British scientific life. At the same time, however, it came to be run by the same scientific elite that dominated the

London scene. The provincial groups that had been most active at its foundation were quickly swamped by the gentlemen of science based in London and the ancient universities: most of them moderate reformists (Whigs or Peelite conservatives) in politics, liberal Anglicans in religion. These men of science—at the Cambridge meeting in 1833, one of them (Whewell) coined the term "scientist" in its modern anglophone sense—quickly took over the effective power in the running of the Association and molded it to serve their interests. To some extent, therefore, the annual meetings of the Association became extensions of those of the London scientific societies. Since it normally met for a week during the late summer, its gatherings did not clash with those of the London societies, and indeed it gave the metropolitan men of science an occasion for an annual excursion to the provinces. On the other hand, there was one obvious and important difference: the Association's meetings were also attended by large numbers of people (including women) who did *not* belong to the metropolitan elite, particularly those from the region around the particular town or city being visited. In other words, the audience at the Association's meetings was much wider and more heterogeneous than at those of the London societies, although many of the chief performers were the same.

After the inaugural meeting, ideals of the unity of science retreated before practical necessity, and four specialist Sections were set up, again following the example of the German Gesellschaft.[35] In addition to the plenary sessions of the Association, which were directed primarily toward the general public, the Sections had their own sessions which were intended more for the active practitioners of a particular branch of science. The Sections met simultaneously (by 1833 there were six of them), so that participants were more or less obliged to specialize along the divisions imposed by the classification that the leaders of the Association had adopted. Each Section had its own president, vice-presidents, secretaries, and committee for each annual meeting; these appointments were soon de facto in the hands of the Association's central Council and were therefore controlled by the same coterie of mainly metropolitan scientists.

The committees of the Sections had the power to choose the papers to be read at each Section's meetings. This meant that the elite of specialists in each science could not only effectively regulate the degree of active participation by provincial members outside their circle, but they could also promote their own scientific causes by giving their own papers a prominent place in the Section's proceedings. This was important, because the Section meetings became highly effective arenas for the announcement of new results and for combating opposed opinions. The papers themselves were soon being published in summary form, often within a few days of being read, in two rival cultural weeklies (the *Athenaeum* and the *Literary Gazette),* not to mention local newspapers.[36] Furthermore, the discussions that followed the reading of papers were also reported in considerable detail, so that the arguments

35. The GDNA had first split into sections at the Berlin meeting in 1828, which also completed its transformation from a radical body suspected of subversion into a pillar of political respectability; see Pfetsch and von Gizycki (1975, p. 113).
36. On the cultural role of the weekly *Athenaeum*, see Marchand (1941) and Altick (1957, pp. 318–47),

and disagreements among the elite, and their treatment of provincial participants, were open for all to read. Such proceedings had enough entertainment
value, in some cases, to attract large audiences of nonspecialists; in the opinion
of some male participants the serious tone was still further jeopardized by
the admission of women into the Section meetings (unofficially from 1835,
officially from 1839).

These tensions were particularly acute in the Section that became an
important arena in the Devonian controversy. The flourishing condition and
respectable status of geology, and the early involvement of many leading
geologists in the affairs of the Association, ensured that geology (with physical
geography as an appendage) was assigned its own Section from the start.
Although it was lower in the implicit hierarchy of the Sections than those
devoted to the physical and chemical sciences, it was one of the most popular,
and its sessions attracted some of the largest audiences. In practice, the geological Section (or "Section C," as it became officially in 1835) was run by
much the same inner circle as the Geological Society (fig. 2.5; compare fig.
2.2).[37] They, in effect, determined the succession of the officers to run the
proceedings from year to year; they decided whose papers should be read,
on what day, and in what order. The Section became a kind of annual out-of-
town meeting of the Geological Society, providing an agreeable venue for the
leading figures to meet in the course of their respective fieldwork expeditions

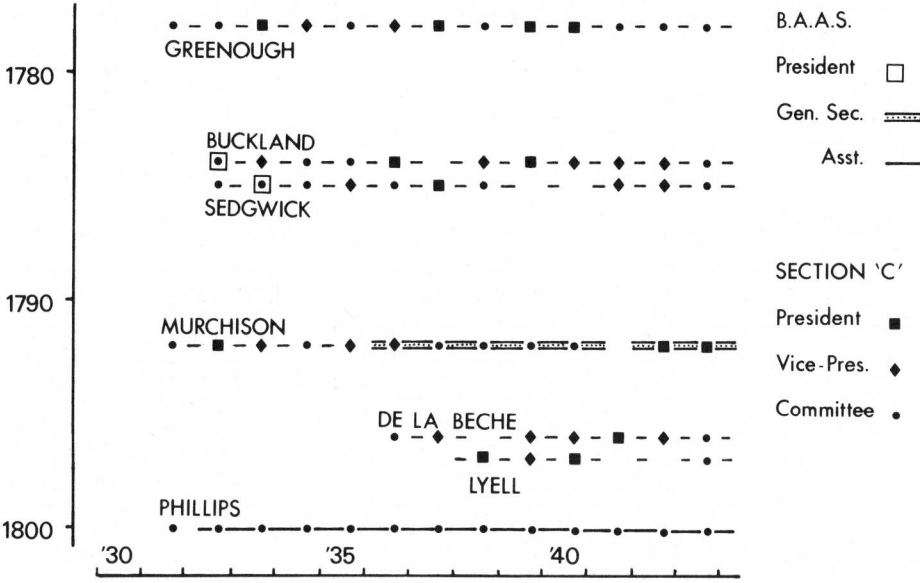

Fig. 2.5.   Officers in the geological Section ("C") of the British Association for the
Advancement of Science, 1831–43. This diagram shows how the geologists who became major figures in the Devonian controversy provided the Section with most of its
presidents and many of its other officers (Section C usually had two co-presidents,
one for physical geography). Individuals are arranged in order of age (year of birth at
left); compare fig. 2.2.

37. Fig. 2.5 is based on data in annual reports of BAAS and newspaper reports in *Ath.* and
LG.

before the Society's own meetings resumed in London in November. Some of them (e.g., Murchison) soon made a habit of presenting their latest results before the Section, however hastily assembled and provisionally expressed, with the promise of a more definitive presentation before the Society in the following months.

As already pointed out in more general terms, however, there was one crucial difference between the two arenas of geological debate. The audience at the geological Section of the British Association was significantly different from that at the Geological Society. It is not the case that at the Society the audience was exclusively one of specialists or active participants in the science, for its meetings were attended by many members (and their guests) who had little substantive knowledge or experience of geology. But these nonspecialists were nevertheless *gentlemen,* whose respectability had been vouched for, on the basis of personal knowledge, by existing members of the Society; and any arguments that broke out after the reading of papers were confined to the audience in the meeting room and not reported in the press. By contrast, the nonspecialists who formed the bulk of the audience at a Section meeting were likely to be in much larger numbers and were much more mixed in social status; furthermore, the arguments among the performers were disseminated afterward in published form. It was this contrast that was to be regarded as highly significant by some of those involved in the Devonian controversy. Was it right, they argued, for controversial matters of specialized science to be discussed before a large audience of mixed social composition and low average scientific competence?

2.5   The Art of Letter Writing

The major English participants in the Devonian controversy, along with other leading geologists, met each other quite frequently at the formal meetings of the Geological Society and its Council, over the convivial dinners of its Club, and at the regular soirées held by its more affluent presidents (Greenough, Murchison). Many were also members of the Athenaeum (founded in 1824), the London social club particularly favored by men of science and literature.[38] They knew each other well enough to be on quite informal terms even when they criticized each other's work and distrusted each other's motives. To discuss geological matters less formally than in a meeting, they would often invite each other to breakfast or dinner at home. This was easy to do because most of them lived within walking distance, or at least within a short cab ride, in the residential areas of central London (fig. 2.6).[39] Alternatively, they would

38. Cowell (1975).

39. Fig. 2.6 is based on a contemporary street map (Outhett 1840) and on addresses in GSL membership lists and correspondence of the 1830s and early 1840s. Since the purpose of the map is not antiquarian, no attempt is made to mark the exact position of particular private houses, but only their general situation. If an individual moved during the years of the Devonian controversy, only the later residence is shown. (The Linnean and Horticultural Societies have been misplaced in error: the former should be between "Darwin" and "Lyell," the latter between "Athenaeum" and "Sedgwick,")

Fig. 2.6. Map of part of London around 1840, showing places of work or residence of some of the participants in the Devonian controversy and locations of some scientific and other institutions. The map is designed to illustrate the small scale of scientific London and the ease with which men of science could meet at the various scientific societies and in each other's houses. Names in parentheses denote habitual lodgings or places of employment of out-of-town participants.

arrange to meet privately at the Geological Society earlier in the day before a scheduled meeting of the Society. Members living in the provinces would try to stay in London for at least a couple of nights in order to spend a whole day at the Society, meeting other geologists, looking at the latest additions to the library and the museum, and inspecting the diagrams and specimens displayed in connection with the papers to be read in the evening.

It can hardly be doubted that some of the most important arguments in the Devonian controversy, and on other similar issues, must have taken place as private conversations on occasions such as these. Except when some summary was reported later in writing by one of the participants, no trace of conversations has of course survived. Fortunately for the historian, however, the efficacy of informal conversations as a medium of exchange during the Devonian controversy was severely limited. Several of the major participants and many of the minor ones lived outside London. They were not able to attend the Geological Society's meetings as frequently as they would have liked, owing to the constraints of academic, ecclesiastical, and other professional duties (e.g., Buckland, Sedgwick). At least one (De la Beche), who had fallen on hard times financially, could not often afford the expensive fare to come to London by coach from the far southwest of England.[40] Such constraints, professional or financial, combined with the scattered distribution of geologists around England, forced them to exchange information and opinions and to carry on their arguments by means of frequent correspondence. Those resident in London were more fortunate in being able to meet more easily; but even they frequently had to use letterwriting to communicate with each other, since of course they had no telephones.

These gentlemen of science lived in a world, or more precisely in a social stratum within that world, in which spontaneous and fluent letterwriting was a routine accomplishment. Whether dashed off in haste or composed with deliberate care, their letters were almost invariably expressive, lucid, and revealing. Letters within London could be delivered by servant or messenger, or sent by prepaid twopenny post.[41] But for letters passing through the public mails beyond London there were further incentives that favored both concise writing and a full use of the space available. With the high cost of postage (often a shilling or more) falling on the recipient (unless the sender could get a friendly Member of Parliament to use his "franking" privilege to send it free) the writer felt an obligation to give his correspondent good value for his money, to fill the sheet with interesting and worthwhile material, but not to extend on to a second sheet (incurring still higher costs) unless it was really necessary. All this changed with the introduction of the cheap prepaid penny post throughout Britain in 1840; as soon as a low and uniform cost was incurred by the sender, the informative quality of long-distance scientific letterwriting fell off abruptly.

Even before then, however, the Post Office gave good value in terms of reliability and speed. For example, two close collaborators in the Devonian controversy (Sedgwick, Murchison) were able at one point to exchange letters between Norwich and London every day throughout the week, using the fast overnight mail coaches (§8.2). Letters from Cornwall, at the furthest tip of southwest England, or for example from Edinburgh, would reach London in two days.[42] Letters from Continental cities such as Paris and Bonn would take only a little longer, owing to the quite recent introduction of reliable steam-

40. One provincial member (Williams), coming to London on business and combining it with a Geological Society meeting, noted in his expense account that the one-way fare from Bridgwater in Somerset alone amounted to two pounds, ten shillings (SAS: DW, box 1, notebook 9).

41. See Robinson (1948, chap. 15).

42. The mail-coach system was at its fastest and most efficient in the late 1830s, just before it began to be superseded by the new railways: Robinson (1948, chap. 17). Even while it was still wholly dependent on horses, the postal system in Britain was faster and more reliable than it is in the 1980s.

ships across the English Channel; from Berlin might take a week, via a ship from Hamburg; but even from far St Petersburg a letter would reach London in about two weeks by Baltic and North Sea steamers. Such speed and reliability gave the exchange of scientific correspondence an immediacy and vitality that it had never had in earlier generations and, arguably, that it has never had since. That the historian of early nineteenth-century science can rely so much on the evidence of letters is not just a happy accident, the fortunate legacy of a culture that tended to hoard its manuscript papers; it is a genuine reflection of the centrality of correspondence during that period as a medium of scientific exchange.

2.6   The Romance of Fieldwork

The Devonian controversy, like other major issues in early nineteenth-century geology, was argued over in such public (or semipublic) arenas as the Geological Society and the British Association. The arguments continued more privately in conversation and correspondence. But all parties bowed to the official image of "Baconian" science, at least to the extent that they acknowledged publicly that fieldwork was the primary locus of encounter between the geologist and the phenomena of his science, and therefore an indispensable part of his activities (fig. 2.7).[43]

Fieldwork was generally a pursuit for the summer months, when the weather was more likely to make it agreeable to sit in a quarry hammering out fossils, to take a hired fishingboat close to a rocky shore to study otherwise inaccessible cliffs, or to walk long distances across country with a clear view of the topography. For fieldwork was mainly a pedestrian activity. Walking twenty miles or more a day, and being in the field from first light to dusk, was not thought unusual, at least by the English.[44] Geologists might use the efficient network of stagecoaches—expensive and uncomfortable though they often were—to reach a town that was situated at a convenient starting point for their work. The more wealthy might then hire a ponytrap to take their baggage and their specimens from one village inn to the next. But rock exposures were not always, or even usually, situated along the roadside; the greater part of each day in the field was generally spent on foot, walking across country with perhaps a hired pony or mule to carry the burden of specimens as they were collected.

43. Fig. 2.7 is reproduced from the first of a series of undated sketches (BGS: GSM 1/123) drawn in criticism of the theoretical pretensions of the first volume (1830) of Lyell's *Principles of Geology;* De la Beche lithographed and circulated the last of the series of sketches. The whole series is reproduced and analyzed in Rudwick (1975b). To give Lyell's book the Huttonian title was a polemical characterization.
44. It is a telling mark of national differences that the two young French geologists (Dufrénoy and Élie de Beaumont) who surveyed their country on government service were each recommended by their superior for the award of a *croix d'honneur,* on the grounds that their fieldwork had been arduous and entirely on foot: Brochant de Villiers, official report on 1830 fieldwork, dated January 1831 (copy in EMP: LEB, dossier I, 46).

Fig. 2.7.   De la Beche's caricature, drawn in about 1831, showing the plain "Baconian" field geologist—the traditional ideal within the Geological Society—confronted by the questionable arguments of a theoretical geologist. The theorist (probably Lyell) stands on ground marked "Theory" and conceals behind his back his unacknowledged "Theory of the Earth" (i.e., probably the *Principles of Geology*). He offers the field geologist (probably De la Beche himself) the tinted spectacles of theoretical supposition, through which to view the phenomena of the science. Note the polemical contrast between the outdoor clothes of the field geologist, with his hammer and collecting bag, and the elegant tailcoat and barrister's wig of the theorist, better suited to courtroom advocacy than empirical fieldwork.

Working on foot, wearing outdoor clothing, and being frequently soaked to the skin, geologists could find that their gentlemanly status went unrecognized by their social equals, particularly since at a casual glance they appeared to be engaged in that most menial of occupations, stonebreaking. One eminent English geologist (Sedgwick) relished the story of how, when hammering in the field one day, a passing lady gave him a shilling; that evening, transformed in clothing and unrecognized, he met her at dinner in the gentleman's house where he was staying, and he amused the other guests—if not the lady—by displaying the shilling and telling of its charitable donation![45] But such social inversion brought more important rewards. In the presence of quarrymen and miners, geologists remained conscious of their gentlemanly status. They often used their money to buy important specimens, and they would hire manual laborers for the heavier and therefore "commoner" work

45. Sedgwick's story is in Clark and Hughes (1890, 2:573–74); no date is given for the incident.

that was sometimes required.[46] Yet their relative disguise, and their evident readiness to share the hardships of outdoor life, seem to have enabled them to elicit valuable local information from the lower classes much more easily than if they had just descended from a carriage wearing the normal clothes of a gentleman. Conversely, the wearing of formal attire while "stonebreaking," which later became de rigueur among those employed on the Geological Survey in Britain, created a social discrepancy that must have puzzled the idle bystander (fig. 2.8).[47]

A hammer was the most conspicuous item of equipment that a geologist carried in the field; it was used not only for collecting samples of rock and extracting fossils but also for revealing the character of a rock exposure beneath its deceptive, weathered or lichen-covered surface. A magnetic compass, a simple gadget (clinometer) for measuring the tilt or angle of "dip" of strata,

Fig. 2.8. A governmental geologist at work in the field in gentlemanly dress: an illustration from the first regular memoir (1846) of the Geological Survey in Britain. The geologist is hammering an exposure of a major unconformity: younger and horizontal strata overlie and truncate older and vertical strata; the geologist is standing on a joint plane in the older rocks, at right angles to the (vertical) bedding planes (see § 3.3).

46. For example, Sedgwick paid twelve shillings for two trilobites that he felt might be important for the Devonian controversy: Sedgwick, notes for 12 July 1838 (SMC: AS, notebook XXXII, p. 1). De la Beche asked his superior for permission to employ a Cornish miner to help with the "commoner" manual tasks on his official survey (De la Beche to Colby, 18 September 1835 [NMW: TS, II, p. 138]).

47. Fig. 2.8 is reproduced from an engraving in De la Beche (1846, p. 248), based on a drawing by, and perhaps of, De la Beche himself. It shows New Red Sandstone strata lying unconformably on Coal Measures at Portishead in south Wales.

a little bottle of acid for detecting limestone (a drop would make a limey rock fizz), and a low-power handlens were important accessories. The best and largest-scale topographical map available formed the indispensable base for marking the position and character of rock outcrops and other geological features, and for sketching a provisional outline of a geological map. More detailed observations would be recorded in a notebook, while specimens were collected in some kind of bag or satchel. These few items were about all that even the most eminent geologist required in the field (see fig. 2.7, 2.8).

It is therefore no wonder that geology was such a popular science as an amateur pursuit. For local amateurs (here the term is fully appropriate) were able to contribute far more substantially to geology than to most other sciences. Not only did they need no elaborate apparatus or equipment; their local situation gave them a positive advantage over the metropolitan or university geologists who might occasionally visit their home areas. Within those areas, amateur geologists had the *time* that was needed to find fossils by patient hammering in local quarries, cliffs, or cuttings—for fossils were rarely to be found without that time-consuming work—and to build up a collection. In addition, as local gentry they soon became known to be willing to pay for any choice specimens that quarrymen and other laborers might come across from time to time. Their own reasons for amassing collections were often more aesthetic than strictly scientific, but provided they took care to record the localities where their specimens were found, they could have the added satisfaction of having their collections used and valued by some of the leading men of science in the country.[48]

So central to geology was the practice of fieldwork that geologists rarely kept any continuous record of their research except when they were in the field. Their notebooks lack the long-term continuity of the daily research recorded by experimental scientists in their laboratory notes. But within the more limited periods of their fieldwork, geologists' notes often provide remarkably detailed insights into both the routine procedures and the immediate conclusions of their research. While in the field during the day, they generally wrote quite brief and factual notes, often scrawled almost illegibly in pencil, with roughly sketched diagrams to match. At the end of the day, they would go over these notes, often making them more legible in ink and clarifying the diagrams (see fig. 7.2). At the same time they might add new and more explicitly interpretative notes, based on reflection on what they had seen during the day. They might also draw new and neater diagrams and even color them to make their meaning clearer, using watercolors taken with them for the purpose (see fig. 12.3). These notes and diagrams would later form the basis for what they wrote as draft papers and discussed with their colleagues.

In an obvious sense, the primacy of fieldwork reflected a recognition that the empirical locus of a science like geology was bound to lie "in the field." Even more than, say, botany, geology embodied an intrinsic spatial dimension; its practitioners would get nowhere, in all senses of that phrase, if they remained "closet philosophers." But the practice of fieldwork came to bear a far greater weight of significance in geology than a purely rational assessment of the subject matter required. It was loaded with sentiments that

48. On the role of the amateurs in British natural history generally, see Allen (1976).

united elements of romanticism and tacitly pantheistic religion with those of robust, manly Christianity and the gentleman's love of the countryside and its sporting pursuits. Fieldwork was a kind of "liminoid" pilgrimage away from effete urban luxury into a closer communing with rural nature; it was also initiation and ordeal. It was the mark of the true geologist; its sometimes arduous nature was the test of his apprenticeship and the badge of his continuing membership in the "brethren of the hammer."[49] Such sentiments gave the pursuit of geology in the nineteenth century affective overtones, and a consequent popularity and glamor, that it had never had before and that it was never to have again.

The foregoing sketch of the institutions of geology in the early nineteenth century provides the Devonian controversy with a social context. But a "con-text" has no significance without a "text." The final stage in setting the scene for a narrative of the controversy is therefore to outline the intellectual setting in which the argument took place.

49. On the elements of romanticism in gentlemanly geology, see Porter (1978). On the extension of van Gennep's classic concept of the "liminal" to cover a wider range of "liminoid" situations, see Turner (1974, chap. 5); his analysis of pilgrimage has suggestive parallels with geological fieldwork as it was practiced in the early nineteenth century.

Chapter Three

UNRAVELING
THE EARTH

3.1 Creation and Causation

The Devonian controversy took place within the collective research tradition that dominated early nineteenth-century geology. That tradition later came to be termed *stratigraphical* geology. Its primary intellectual goal was the delineation of a "succession" or sequence of distinctive major groups of *strata* which could be recognized throughout the world. Before outlining the procedures of stratigraphical geology, however, two common misapprehensions must be cleared out of the way.

In research on the sequence of the strata, temporal modes of analysis were usually subordinate to structural ones.[1] Nonetheless, all geologists were well aware that the structural sequence of the strata also represented a temporal sequence of past events. All geologists in the 1830s tacitly assumed that these events had taken place on a timescale so long that it was literally unimaginable, a timescale that dwarfed human lives and even the totality of human history. What is more, by the 1830s this vast timescale had already been routine and taken for granted for at least a couple of decades. No geologist of any nationality whose work was taken seriously by other geologists advocated a timescale confined within the limits of a literalistic exegesis of *Genesis*. This can hardly be emphasized too strongly, since the impression is still

---

1. For the "structural" style in geology, and its contrast with the causal and temporal goals of the "abstract" style, see Rudwick (1982a). The significance of "historicism" in geology is stressed by Oldroyd (1979), though without adequate recognition that it was rarely the dominant component of geological practice.

widespread among twentieth-century geologists and historians that the early nineteenth century was a period of conflict between geology and *Genesis*.[2]

The partial truth behind that erroneous impression is that there was indeed a heterogeneous group of writers who, with varying degrees of stridency, did claim the primacy of literalistic biblical exegesis, and who maintained that the record of the rocks could be compressed within the very short timescale that they claimed to extract from the biblical texts. But these so-called Mosaic or *scriptural geologists* had much the same social and intellectual relation to "philosophical" (or scientific) geologists as their indirect descendants, the twentieth-century creationists.[3] They were an important irritant and a serious disturbing factor in the scientific geologists' campaign to establish and maintain their own public image as a source of reliable and authoritative knowledge. For the scriptural geologists claimed to be an *alternative* source of knowledge, and a better one at that. The scientific geologists feared with good reason that the general public might be persuaded to accept the pronouncements of their rivals rather than their own; but for the ordinary business of discussion among themselves, they could and did assume broad agreement about the timescale of the history of the earth.

It is very difficult, however, to determine in quantitative terms what the scientific geologists of the 1830s had in the back of their minds, for that is precisely where it generally remained. They were very reluctant to hazard a guess at the magnitude of the timescale, not for fear of social disapproval, still less of persecution, but simply because even the most tentative estimates were open to criticism as groundless speculations. Nonetheless, a tacit consensus can be detected, not least from the iconography of theoretical diagrams, in favor of the view that the maximum thicknesses of the successive groups of strata were a roughly reliable guide to the relative duration of the successive periods in which the strata had accumulated.[4] With that as a clue, it is at once apparent, from both verbal and visual expressions of opinion, that the timescale represented by the whole sequence of strata dwarfed the period represented by the puzzling "Diluvial" deposits and the "Recent" or "Alluvial" sediments that were believed to represent the whole history of mankind (see fig. 3.4). Even if a fairly restrictive estimate were adopted for the latter, as it often was, the timescale represented by the strata as a whole would still be vast beyond imaginative comprehension.

The consensus among scientific geologists on this point was total and explicit in its rejection of the few thousand years posited by the scriptural geologists. It was equally total, though largely implicit, in its acceptance of a timescale that must have run at least into several hundred thousand years, if not into a few million, judging by the criterion just outlined. The consensus only broke down when some such vast timescale was put to work in the

2. Gillispie's pioneer book *Genesis and Geology* (1951) has a thesis more subtle than its catchy title; it is more concerned with the retreat from providentialist interpretations of the natural world in the early nineteenth century than with attempts to reconcile geology with the Bible,

3. The scriptural geologists still await their historian; for a preliminary survey of their work, see Millhauser (1954) and Gillispie (1951). The term "Mosaic" refers to their focus on the Creation narratives in the first book of the Pentateuch, traditionally the work of Moses himself. The term "philosophical" was still current in its older sense, being (in this context) roughly equivalent to the modern term "scientific."

4. One of the rare explicit discussions of this relationship from the period of the Devonian controversy is in Phillips (1837–39, 1:8–18).

explanation of specific features in geology; for example, when the total sequence of the strata was interpreted as an extremely fragmentary record of
the history of the earth and of life on its surface, in contrast to the usual view
that the known record was already fairly complete. In such conflicts, those
geologists (e.g., Lyell) who made the fullest explanatory use of time might
use the allegedly "inadequate" timescale of their opponents as a rhetorical
weapon, for it served to discredit them by implying they had some affinity
with the scriptural geologists.[5] But in fact the geologists so criticized would
protest—and there is no reason to doubt their sincerity—that they had no
reluctance about using a timescale even of "millions of millions of years,"
provided that it was adduced on good grounds.[6] In view of this virtual consensus on the vast timescale appropriate to geological thinking, the work of
the scientific geologists can be described and analyzed without further reference to the scriptural geologists, who remained socially and intellectually
marginal (see fig. 2.3).

This is not to say, however, that the scientific geologists were antire-
ligious secularists or agnostics *avant la lettre*. On the contrary, many of the
most prominent scientific geologists were men of acknowledged personal
piety, and some held ecclesiastical positions, the duties of which they fulfilled
conscientiously (e.g., Buckland, Conybeare, Sedgwick). But such men were
theological liberals, who were well aware of the critical hermeneutics being
developed by German biblical scholars at this period. They had not hitched
their religious beliefs onto literalistic modes of biblical exegesis; indeed, they
were among the most vehement critics of the scriptural geologists.[7] In their
view, the literalism of the scriptural geologists was just as outdated and unscientific (in the broad Continental sense of the word "science") as the deviant
"geology" that those writers proposed.

However important scriptural geology may have been for the external
public relations of the scientific geologists of the 1830s, its claims posed no
internal threat to the peace of mind of those with religious commitments.
They were not averse, and some (e.g., Buckland) were positively eager, to use
their science to impress upon the wider public an enlarged sense of divine
design in the world, thereby developing the long tradition of natural theology
into forms adapted to the latest scientific knowledge.[8] Furthermore, they had
in no way abandoned the traditional theistic interpretation of the natural world
as the product of divine activity; they had no hesitation about describing that
activity as "creation," even if the meaning of that term remained difficult to
specify precisely in scientific terms. In particular, although the term "creation"
was commonly used in referring to the origins of new organic species, it was
not incompatible with the belief that the creative action might have been
mediated by natural ("secondary") causes. Theories of "transmutation," as
organic evolution was generally termed, were only unacceptable if —as was

5. On Lyell's use of time as a rhetorical weapon in geological debate, see Rudwick (1970a).

6. Buckland (1836, 1:21); compare Conybeare's "quadrillions of years" in a letter to Lyell in
1841, printed in Rudwick (1967).

7. See, e.g., Sedgwick's attack on Ure (1829) in his 1830 presidential address to GSL (*PGS*
1:207–10). Sedgwick was particularly indignant because Andrew Ure, unlike other authors in
the genre, was a member of GSL. Conybeare was by far the most competent theologian among
the geologists: see, e.g., his Bampton lectures at Oxford on patristic theology (Conybeare 1839).

8. Buckland's Bridgewater treatise on *Geology and Mineralogy* (1836) was widely praised as
an outstanding example of such work.

notoriously the case with Lamarck's—they were taken to imply that *no* creative activity lay behind the natural process.

In practice, however, such interpretations remained an optional extra for the scientific geologists of the early nineteenth century. Those who were religious believers sensed no incompatibility or even strain between their scientific and their religious activities; they could get on with their everyday scientific research without constant or anxious concern for its religious implications. For this reason, a characteristic part of that scientific research—such as the Devonian controversy—can be fully described and analyzed without constant attention to its effects on the religious beliefs of its participants, because those effects were in practice indirect and marginal.

A second misapprehension about the geology of the 1830s, indirectly linked to the issue of religious beliefs, remains to be cleared away. Just as the ordinary business of stratigraphical geology could proceed among "philosophical" geologists without constant concern for the religious implications of geology, so it was only marginally affected by controversial questions about the causation of geological processes. Stratigraphical geology was not the sole collective research enterprise among geologists in the 1830s, but it was certainly the dominant one. Other branches of geology, particularly mineralogy and what in these years came to be termed palaeontology, were treated in practice as important but ancillary branches of the science. They were regarded as semiautonomous subjects, having links with chemistry and natural history respectively, but they were valued chiefly for their potential contributions to an understanding of the record of the rocks. Apart from such ancillary subjects, however, there was also at least one other collective enterprise of strictly geological research which was only loosely linked to the business of stratigraphical geology. This was dubbed at the time "dynamical geology" or *geological dynamics*.[9] It was concerned with geological events and processes in a causal rather than a structural or historical perspective. Causal analysis was central to dynamical geology, whereas for stratigraphical geology it was peripheral and—at least in everyday practice—largely dispensable.

Most geologists in the 1830s were *not* centrally concerned with causal questions at all, or at least not in the stratigraphical research that for most of them was central to their scientific lives. Causal problems such as the formation of the puzzling Diluvial deposits, the excavation of valleys, the elevation and subsidence of landmasses, and the action of volcanos and earthquakes were all problems of a quite different kind from the equally puzzling problems raised by stratigraphical research. The problems of geological dynamics were perceived as neither more nor less fundamental than those of stratigraphical geology: they were just different, and the linkage between the two enterprises was loose and weak. Many geologists contributed to both enterprises at various times, or even at the same time, but they treated them as essentially separate realms.[10] This is important to emphasize because there is a widespread impression among modern geologists and historians that the

9. Whewell (1837, 3: book 18, esp. p. 488).

10. For example, when Sedgwick and Murchison first undertook fieldwork in Devonshire in pursuit of the problems raised by the Devon controversy (§7.2), they also studied a raised beach there, which was relevant to the parallel but separate problem of recent crustal elevation; they published a short paper on it afterward, but it had no relation to the Devon problem.

geology of this period was focused primarily on fundamental conflicts over questions of causal interpretation. It is imagined that geologists were split into opposed parties of uniformitarians and catastrophists, of progressives and reactionaries, of enlightened scientists and bigoted obscurantists. This is not the place to debunk this historical myth, for such it is.[11] All that is needed here is to emphasize that the problems raised by research into geological dynamics did not in fact split geologists into any two sharply divided parties. In any case, those causal problems were only marginally relevant to what most geologists regarded as the central business of geology, namely, the unraveling of the record of the rocks and the decipherment of the history of the earth and of life.

3.2   Sections and Sequences

The procedures of stratigraphical geology first included a description of the strata or "beds" of rock exposed in, say, some coastal cliffs, and their grouping into a sequence of more or less distinct local *formations*, each characterized by some distinctive rock type (e.g., sandstone, shale, limestone). In fact, coastal cliffs provided an untypically easy example. More commonly it was a matter of piecing together the fragmentary evidence of artificial *exposures* in quarries, cuttings, and mines, and of natural exposures of rock in inland areas. The geologist would extrapolate specific formations between these scattered points by tracing the corresponding characteristic topography, soil, and vegetation across the country. This kind of survey could yield in the first place a two-dimensional *geological map* of the distribution of formations, as shown by their areas of observed and inferred *outcrop* at the surface of the land.

A geological survey could also yield a more theory-loaded visual expression, namely a *traverse section* (sometimes confusingly termed a "horizontal section"). A traverse section was (and still is) a visual representation of the structure of the superficial part of the earth's crust, as it might be seen if the crust could be cut open in a vertical slice running across country on a particular traverse (fig. 3.1).[12] In other words, it represented a hypothetical and artificial equivalent of a line of natural coastal cliffs, and indeed it was generally drawn with the same visual conventions as a stylized offshore view of such cliffs. But even where it could be based on the relatively continuous exposures provided by, say, a cutting for a canal (or later, for a railway), a traverse section was still intrinsically interpretative because it extrapolated observed exposures of strata downward into a more general visual statement about the local structure of the earth's crust.[13]

This extrapolation could be made with some confidence, however, because formations of strata were found to maintain constant relations to each other and fairly constant thicknesses, at least within any limited local area.

11. A pioneer attempt at demythologizing was by Hooykaas (1959), and more specifically his 1970 essay. Lyell's role in constructing the myth in the first place is discussed by Porter (1976).

12. Fig. 3.1 is traced from De la Beche (1830, pl. 5, fig. 2), which De la Beche redrew from Buckland and Conybeare (1822, pl. 32 no. 1); the southern end, not included in De la Beche's section, has been redrawn here from the latter source.

13. Rudwick (1976, pp. 164–72).

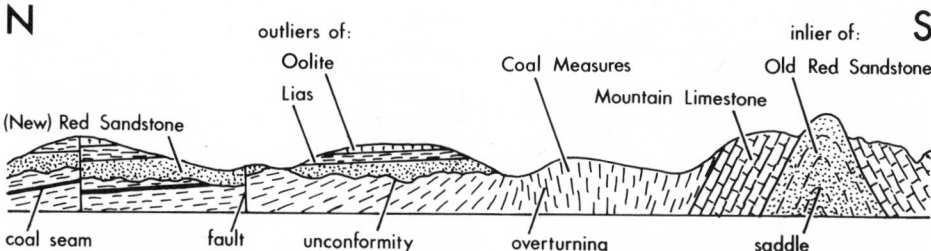

Fig. 3.1. A traverse (horizontal) section, in the style of the 1830s, to show various structural features. The section, redrawn from a book by Henry De la Beche (1830), runs about 8 miles (13 km) from the southern edge of the Bristol coalfield to the Mendip Hills in Somerset (see fig. 3.5); the vertical scale is exaggerated for clarity. In this and all other sections in this book (unless otherwise stated), conventional signs are used to represent dominant rock types: stippling for sandstones; parallel lines for shales; "bricks" for limestones.

The procedures for extracting a sequence of formations from a survey of an area were based on the rather obvious *principle of superposition,* namely that the sequence must be constructed from observable exposures showing which formation was lying on top of which. But although it was acknowledged as self-evident that the lower formation must have been deposited *before* the upper, the structural language of "upper" and "lower" was used much more commonly than the overtly temporal language of "younger" and "older."

The sequence of formations was most easily demonstrable in areas where the strata had retained an almost horizontal position but had been excavated subsequently into relatively deep valleys. The two areas which provided the most influential early exemplars were both of this kind: namely the strata around Bath studied by the land surveyor William Smith (b. 1769), and those around Paris surveyed by the mineralogist Alexandre Brongniart (b. 1770), assisted by his anatomist colleague Georges Cuvier (b. 1769).[14] If the strata were tilted with a gentle *dip*—whether this was regarded as original or attributed to subsequent crustal movements—the task was almost equally simple. A traverse section would normally be made along some line following the direction of the dip, and hence at right angles to the direction of horizontality of the tilted plane of the strata—the latter direction being known as the *strike* of the strata (like dip, a traditional miners' term). A traverse section, in conjunction with a geological map, could then be "read" as a visual summary of the three-dimensional structure of the area—part observed, part inferred.

A traverse section could serve in turn as the basis for the construction of a *columnar section* (sometimes confusingly termed a "vertical section"). A columnar section was (and still is) a visual representation of the sequence of formations in a given area, abstracted from the particulars of dip and surface topography in which they were observed. The observed outcrops were reconstituted as a column that depicted primarily the sequence and thicknesses of the formations, together with conventional signs or colors to indicate char-

14. For Smith's early work, see J. Eyles (1969) and Torrens (1979); for Brongniart and Cuvier, see de Launay (1940) and Coleman (1964).

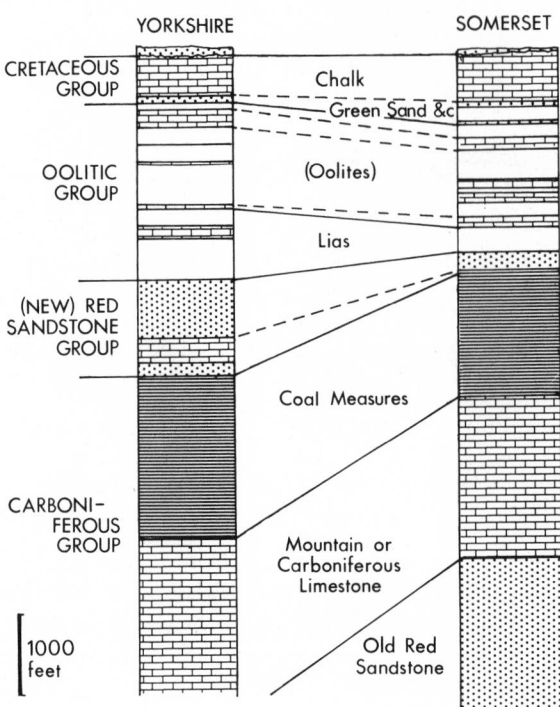

Fig. 3.2. Two columnar (vertical) sections, in the style of the 1830s, redrawn from a book by De la Beche (1830), showing the *correlation* of equivalent formations between Yorkshire and Somerset.

acteristics such as their rock types and fossils (fig 3.2).[15] Thicknesses could be measured directly in cliffs or quarries, or failing that, they could be computed from the observed dip of the strata and the breadth of outcrop of the formation, using geometrical techniques borrowed from the practice of mineral surveying. Just as traverse sections resembled offshore views of coastal cliffs, so columnar sections resembled the "logs" of boreholes or mineshafts.

Maps, traverse sections, and columnar sections were the indispensable "visual language" of a cognitive project that was profoundly visual from beginning to end.[16] This visual orientation ranged from the field geologist's need for a good "eye" for topography, soil, and vegetation as well as rock types and fossils, through his sketched maps and sections made in the field, to the completed maps and sections which he used as a means of visual communication with others. The steps by which maps and sections became a generally accepted visual language among geologists need not be traced here; it is sufficient to note that their use was well established by the time the Devonian controversy began. Maps and sections can therefore be used in this account of the controversy, not only as primary documents but also—without any anachronism—as interpretative tools for historical analysis.

15. Fig. 3.2 is traced from De la Beche (1830, pl. 1). The correlation lines connecting the two columns are implicit in the original labeling of the sections and are here made explicit according to a visual convention that developed somewhat later. Only the formations mentioned in the text are marked by name.

16. Rudwick (1976).

A geological survey of a limited area of relatively simple structure (e. g., gently tilted strata) could yield in this way a fairly unproblematic interpretation of a sequence of local formations. Articles or books containing such essentially *local* descriptions of the strata thus became the most characteristic product of the expanding enterprise of stratigraphical geology. Although the procedures and techniques involved reach back into the eighteenth century, the most influential exemplars were those already mentioned, by Cuvier and Brongniart (1811) and Smith (1815); but is was only around 1820 that the enterprise became manifest in increasing numbers of standardized local "memoirs." Thereafter the cognitive ambitions of this research were swiftly widened in two ways. First, geologists learned how to cope with areas of less simple structure; second, they began to have increasing success with extending their procedures to deal with wider regions.

3.3   Structures and Correlations

The techniques that had first been used in areas of very simple structure were found to be applicable also in areas where the strata were more highly tilted, folded, or otherwise disturbed. Procedures for detecting and describing such structures were developed by a series of precedents. Starting with simpler or better-exposed examples, geologists learned gradually to recognize more ambiguous or less well-exposed examples, and their confidence in the reality of these inferred structures grew accordingly.

Folds or flexures of the strata came to be recognized not only on a small scale, where they could be directly observed (e.g., in a cliff or quarry face: see frontispiece), but also on a much larger scale, where they could only be detected by careful mapping. If the strata were folded into an archlike form, with the convex side upward, they were said to form a *saddle* (see fig. 3.1); the converse form, with the concave side upward, was termed a *trough*. These were appropriate informal terms for geologists to use in a period of horse-drawn transport (around the 1830s and 1840s, they were gradually replaced by the more learned terms *anticlinal* and *synclinal*, respectively). When mapped along the axis as well as across it, a saddle might prove to have the three-dimensional form of a more or less elliptical *dome,* and a trough that of a *basin.*

In an area of flat-lying or gently dipping strata, a patch of an upper formation isolated on a hilltop was termed an *outlier,* while an isolated outcrop of a lower formation, exposed at the bottom of a valley, was termed an *inlier* (see fig. 3.1). Outliers and inliers would also be formed at the centers of basins and domes, respectively, even if the topography were flat. Although similar in appearance on a map, they were exact opposites in terms of solid geometry, and they differed in one decisive practical respect. If the formation concerned were economically valuable (e.g., coal strata), discovering by mapping that it was an outlier meant in effect circumscribing the limits of its exploitation. But discovering that it was an inlier implied that it might extend indefinitely beyond the limits of its outcrop, concealed by the surrounding higher strata but potentially accessible by sinking boreholes and mine shafts through them (see fig. 3.1). In ways such as this, the accurate mapping of formations and the interpretation of their structural relations had practical implications of the highest importance.

In addition to recognizing folds in the strata, geologists also learned not to be disconcerted by abrupt fractures, "dislocations," or *faults* cutting across the strata, even if these displaced the local sequence vertically or horizontally by hundreds of feet, or juxtaposed completely different formations (fig. 3.1). Faults had long been familiar on a small scale in the practice of mining, but it was the surveying work of geologists that demonstrated the much larger scale on which fractures of the earth's crust could occur.

Geologists also learned to recognize and interpret cases of "nonconformable" junctions between different formations. Such *unconformities*—the term was becoming established around the 1830s—were indicated where a lower set of strata was truncated at an angle and overlain by an upper set lying in a different orientation (fig. 3.1). This could be interpreted in terms of a distinctive temporal sequence of events. An earlier set of strata had first been deposited more or less horizontally, then tilted by some kind of crustal movement. The edges of the strata had then been truncated by some kind of erosion. At a later period a newer set of strata had been deposited on this eroded older surface, to be perhaps tilted in their turn by further crustal movement, and truncated finally by whatever erosion had produced the present surface topography. The interpretation was not often so overtly temporal as this summary suggests; more commonly an unconformity was treated simply in structural terms, as a sign that certain formations were missing in that particular area from a fuller sequence that might be found elsewhere.

As geologists gradually learned to detect and unravel fold and fault structures of more complex form, so they learned to recognize unconformities in more subtle forms. The easiest unconformities to recognize had been those where the contrast in the orientation of the strata, above and below the unconformity, was very great: for example, where a series of horizontal strata overlay and truncated a series of vertical strata (see fig. 2.8).[17] If the change of orientation was only slight, the unconformity might scarcely be noticeable in, say, a single quarry face, and might be indistinguishable from an abrupt change of rock type between two perfectly conformable formations. But if the formations were unconformable, even slightly, that fact would become detectable as the outcrops were traced across country; for then the formation immediately overlying the unconformity would be found to rest on successively lower or higher formations of the older sequence (see fig. 6.2). But the very subtlety of the evidence for such an unconformity, and its dependence on the competence of the mapping, could leave plenty of room for argument. Yet such cases were important because unconformities, however slight, were believed to represent gaps in the local sequence that might be filled elsewhere by many different formations, totaling hundreds or even thousands of feet in thickness. Conversely, it was taken for granted that any such major gap in the sequence *would* leave a detectable trace in the form of an unconformity, however slight. It seemed inconceivable that the deposition of strata could

17. The celebrated unconformities described and illustrated half a century earlier by the Scottish natural philosopher James Hutton (1726–97) in his *Theory of the Earth* (1795), and used by him as evidence of a succession of "former worlds," had also been of this kind.

have resumed, after some vast period of time, in precise conformity with much older strata, except perhaps locally and then fortuitously. Conversely, any claim that there was *no* unconformity between adjacent formations was greatly strengthened if a gradual transition or *passage* could be demonstrated between them, for example, from shale to sandstone by way of sandy shale and shaly sandstone.

By the 1830s the detection and analysis of folds, faults, and unconformities had become largely a matter of routine among geologists. Only at the edge of contemporary practice was there doubt and ambiguity. Faced with an exposure of vertical strata, for example, how could the geologist know which way to "read" the sequence? And if strata could even be *overturned* beyond the vertical, how could that fact be known, without subverting the fundamental principle of superposition on which the whole enterprise rested? There was no known criterion by which the original orientation of strata could be determined, or in other words no independent way of telling which side of any given stratum had faced upward when first deposited. So the correct direction in which to "read" a sequence of vertical or putatively overturned strata could only be determined by prior knowledge of the correct sequence derived from some less problematical area where the strata were less strongly folded (see fig. 3.1; fig. 6.4).

A much more common problem, particularly among the older rocks, was to determine the true orientation of the strata in the first place. In many formations this was no problem at all: the layers or strata were easily distinguished by changes of color or rock type, and accentuated by parallel cracks running across the face of the quarry or cliff. Other sets of cracks or *joints* might traverse the strata, often at right angles to the stratification or bedding of the rock; but generally there was no difficulty in distinguishing joints from the true bedding (see fig. 2.8). Among older rocks, however, it was not always so easy, and much depended on judgment and experience.

A particular difficulty arose with slates. In everyday or quarrymen's usage, a slate was any rock that could be split easily into fairly thin sheets. Some "slates" were just "thin-bedded" strata which split along what were clearly the original layers of the sediment.[18] Other "slates" were quite different, for by the 1830s it was recognized that the planes along which they split might be quite different in orientation from the original bedding. The latter might only be detectable with difficulty by looking closely for telltale bands of different color or more sandy layers running obliquely across the more conspicuous *slaty cleavage* (fig. 3.3).[19] Slaty cleavage was evidently a feature that had somehow been superimposed on shaly strata at a period much later than their deposition. But unless it was strongly developed, its detection required skill and experience. In a varied sequence, the slaty cleavage was more conspicuous in some strata (particularly the fine-grained) than in others; in some borderline cases, therefore, the very presence or absence of slaty cleavage was a matter of debate, even among competent geologists. Those

---

18. For example, the Stonesfield slate in England, used for roofing, and the Solnhofen slate in Germany, used as a lithographic stone; both were famous for their fossils and were assigned to the Oolitic group of Secondary formations.

19. Fig. 3.3 is reproduced from Sedgwick (1835a, pl. 47, figs. 2, 4, 5) and shows some of the older Greywacke strata of north Wales.

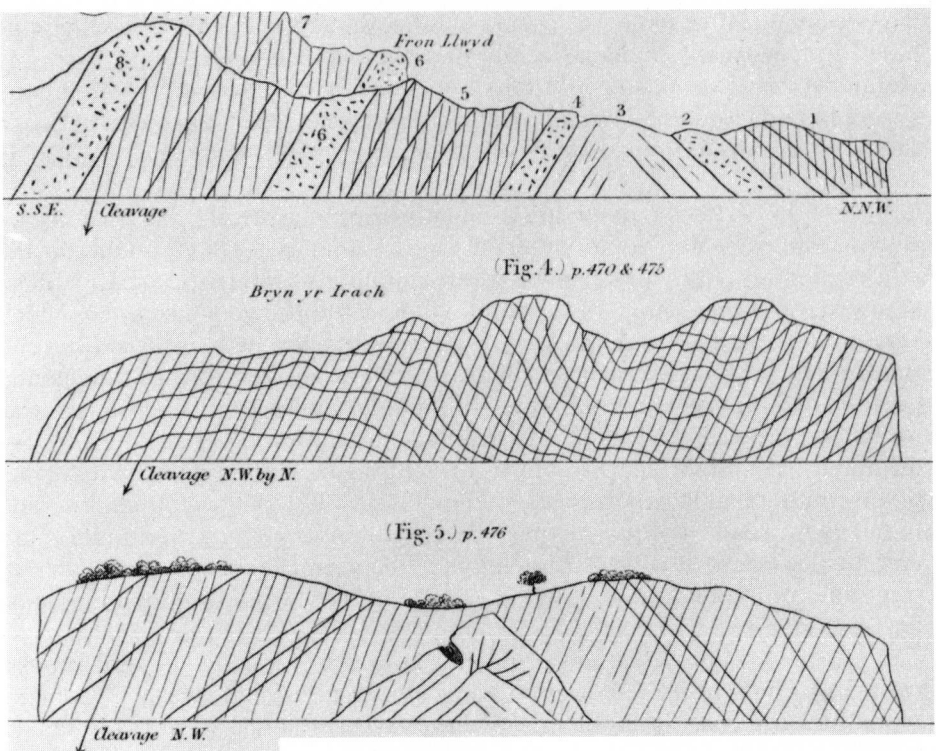

Fig. 3.3. Traverse sections of highly folded strata affected and partially obscured by slaty cleavage: an illustration from Sedgwick's classic memoir on rock structures (1835). Note that the coarser sandy strata (stippled in top section) are not affected.

with less experience of the problem were quite likely to mistake slaty cleavage for true bedding, and hence to derive quite spurious conclusions about the structure of a whole region and about the sequence of its formations.

An awareness of such deceptive features as jointing and slaty cleavage made leading geologists understandably skeptical about work among the older rocks by those who might be less competent than themselves. The difficulties of practical work in stratigraphical geology—particularly among the older rocks—were such as to put a premium on wide and long experience, and to make some kind of apprenticeship "in the field," however informal, almost a necessity. Only by the development of practical competence in fieldwork could a geologist hope to make any but the most low-level contribution to the science.

The various procedures of stratigraphical geology did not require commitment to any precisely specified theories about the *causal* origins of all these structures. For example, the tracing of folded and faulted strata did not commit the geologist to any particular theory of crustal movement—the nature of the forces responsible, or whether the movements had been rapid or not. Nor did the recognition and interpretation of unconformities imply any particular conception of the magnitude of the inferred time gap in the sequence of formations. Research in stratigraphical geology could therefore proceed without being disturbed by doubts or disputes on such matters of causal theory.

Indeed, as already mentioned, the standard style of explanation was far more structural than causal, or even temporal. The great virtue of stratigraphical geology, in the eyes of many geologists, was precisely its apparent detachment from theorizing of any kind, and they were often not fully aware of the necessarily theory-laden character of even their most routine procedures.

The second way in which the tradition of stratigraphical geology expanded its ambitions was in the lateral extension of the sequence of formations from one region to another. In areas of simple structure, where the strata were horizontal or gently tilted, it was relatively easy to trace the formations across country. It was Smith's great achievement to do this for much of England, using, for example, the almost continuous outcrop of the exceptionally distinctive Chalk as a kind of "marker" formation to match up the sequence on, say, the south coast with that in the Midlands and then with that on the northeast coast of Yorkshire. Yet even Smith ran into difficulties when he found in certain areas some distinctive formations that were wholly missing from the sequence elsewhere.[20] This highlighted what became the central problem of stratigraphical geology as geologists sought to expand it from the local toward the programmatically global.

The problem was that of *correlation*, a term used to denote the matching of the sequence of formations in one area with that in another (see fig. 3.2). The goal of constructing a single sequence that would have worldwide validity had in fact been present implicitly in much earlier descriptions of strata; the general working assumption—it hardly merited the status of a theory—was that there was indeed such a universal sequence to be found. But discrepancies and problems soon multiplied when geologists attempted to correlate their own local sequences with each other's, or with the one that had gained some status as a "standard" sequence; namely the one constructed or at least propagated by the famous Saxon mining geologist, Abraham Werner (1749–1817), mainly on the basis of strata in central Europe. For example, a distinctive formation of red sandstone, known by a traditional German miners' term as *Rothe todte liegende*, was at first correlated with a formation of similar appearance in Britain; but the latter was then found to lie below the even more distinctive Coal strata, as well as above them as in the German states. This is where the principle of superposition became an effective working maxim: obviously a correct correlation between two regions had to avoid such inconsistencies of sequence. This particular problem was solved by the recognition that in Britain there were two superficially similar formations: an upper one, which became known as *New Red Sandstone* and which was equated with the German formation, separated by the Coal strata from a quite distinct lower formation, not present in Germany, which became known as *Old Red Sandstone* (see figs. 3.1, 3.2).

This problem—and there were many similar ones—focused the attention of geologists on the need to clarify the criteria on which correlations were to be made, particularly when comparing distant regions or in cases where no physically continuous "marker" formations were available. Smith in England, and Brongniart and Cuvier in their joint work around Paris, found that a close study of the enclosed fossils could often help to trace certain specific

20. Laudan (1976).

formations across country within a given region. Furthermore, they found that fossils could often help to discriminate between otherwise rather similar formations.[21] This use of fossils was not entirely original, since it had long been conventional to characterize a formation in a general way in terms of its fossils as well as its rock type. But the newer work did demonstrate the heuristic value of fossils in stratigraphical geology with much greater force and precision. As a matter of working practice, rather than as a corollary of any clearly articulated theory, geologists from the 1820s onward paid increasingly close attention to fossils and accorded them greater weight in proposing correlations between the sequences of formations in different areas. Nonetheless, the formations themselves remained the primary units of description. They were necessarily defined in the first place in terms of characteristic rock types: otherwise they could not have been used in the fundamental practice of geological surveying on which the whole enterprise rested.

In practice, therefore, the criterion of rock type or *mineral character* was not an alternative to the criterion of *characteristic fossils*, for the two were used to complement each other. It was not so much a question of conflicting theories, but more a matter of differing emphases in geological practice. Stratigraphical geology developed with relatively little debate on the level of causal explanatory theories, not only because of the widespread suspicion about theorizing in general, but also because it was easy to give theoretical grounds for justifying *either* of the alternative emphases in geological practice. The recurrence of a given rock type at different positions in the total sequence of formations could be translated into causal terms by inferring that specific types of depositional environments had recurred at different periods. But the possibility that a given assemblage of fossils might likewise recur could be explained similarly in terms of the probable recurrence of specific ecological environments. Where there was an argument about stratigraphical correlations, therefore, it tended in practice to turn on the balance of plausibility.

Stratigraphical geology, in summary, was a highly *practical* enterprise in every sense of the word. It was based on a subtle and largely tacit body of practical maxims, rules, and precedents, the effective application of which could only be learned through practical experience itself.

3.4   Defining a Global Sequence

By the time the Devonian controversy broke out at the end of 1834, research in stratigraphical geology had already been remarkably successful in plotting the main outlines of a classification of the sequence of formations, at least throughout western Europe. There were some indications that it would turn out to have worldwide validity.

Stratigraphical geology inherited a framework of classification from the more scattered work of earlier generations. In the course of the eighteenth century it had become customary to distinguish three main categories of strata or *sedimentary* rocks, which were termed Primary, Secondary, and Tertiary

---

21. J. Eyles (1985) gives convincing if circumstantial evidence that Brongniart learned of Smith's techniques while on a visit to London in 1802, during the brief peace of Amiens.

(volcanic and other *igneous* rocks formed a category apart and need not be considered further here). The Primary, Secondary, and Tertiary rocks were distinguished initially—and not always explicitly—by a variety of criteria, of which superposition or relative age was only one. The *Primary* or "Primitive" rocks (e.g., granite and gneiss) generally formed the cores of mountain ranges; they were usually crystalline; they were either not noticeably stratified at all, or else their strata were highly folded or crumpled; and they contained no fossils. The *Secondary* rocks were of varied types (e.g., sandstones, shales, and limestones) and well differentiated into local sequences of stratified formations; they were often widespread on the flanks or foothills of mountain ranges, and were often moderately tilted or folded; and many formations (particularly the limestones) contained abundant fossils. The *Tertiary* rocks were generally found on low ground in restricted areas; they were usually horizontal strata of unconsolidated materials (e. g., loose sands and gravels), and thus not "rocks" at all in the everyday sense; and most of the fossils they contained were similar to living organisms.

A fourth category was later intercalated between the Primary and Secondary, and termed appropriately *Transition* rocks (in French, *terrain de transition;* in German, *Übergangsgebirge).* They appeared to be transitional in relative position and relative age, and in topographical position, rock type, stratification, and fossils (fig. 3.4).[22] One particular rock type, a hard, dark,

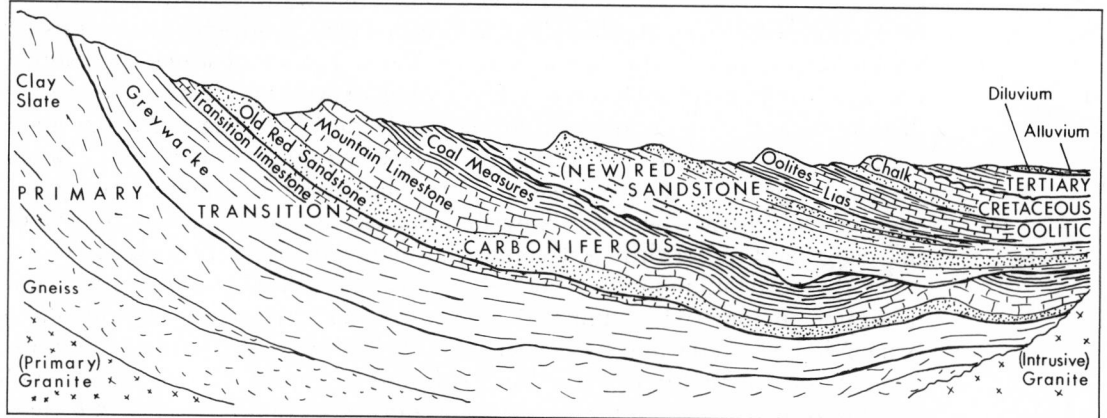

Fig. 3.4.   Part of Buckland's "ideal section" of the earth's crust from his Bridgewater treatise (1836), to show the relations of Primary, Transition, Secondary, and Tertiary formations (the Secondary, as customarily defined, ranged from the Carboniferous to the Cretaceous, inclusive). The diagram also shows incidentally Buckland's conception of the relative thicknesses of the main groups of formations, and hence implicitly of the relative timescale of the earth's history. He has often been classed erroneously with the "scriptural geologists," but the diagram implies that, like all scientific or "philosophical" geologists, he ascribed a vast age to the prehuman (i.e., pre-"Alluvium") deposits. Note that Buckland depicted granite *both* as the oldest Primary rock (left) *and* as an igneous rock that had intruded much younger strata (right).

22. Fig. 3.4 is traced from Buckland (1836, 2: pl. 1), simplified by the omission of all igneous intrusions, volcanic rocks, and faults. The plate was engraved earlier in the 1830s, before the first attempts to describe the Transition formations with greater precision. Unlike most other authors at this period, Buckland extended the term "Transition" upward to include the Mountain or Carboniferous Limestone; the more common usage is shown here.

sandy or slaty rock known by an old German miners' term as *Grauwacke,* was so characteristic and common in the Transition group that the term Grauwacke (or its half-anglicized version, Greywacke) was often used as a virtual synonym for Transition.[23] Fossils in the Transition or Greywacke strata were rare and poorly preserved, except at a few localities, and were generally of organisms quite unknown in the living state.

The defining characteristics of these four main categories of strata underwent gradual modification in the course of more detailed research in particular regions, but the terms themselves remained in use. The criterion of relative structural position—and hence implicitly of relative age—came only slowly to have preeminence in geological practice. It was only in the 1820's that the concept of structural position became clearly distinguished from simple topographical position. This was partly a result of the discovery that in mountain ranges some of the rocks at very high altitudes—which previously would have been regarded as Primary—contained fossils characteristic of Secondary or even Tertiary formations.[24] It therefore became clear that, if superposition were respected as the supreme criterion, then fossils were the most powerful ancillary criterion for the identification of the correct place of formations in the classification.

The most striking successes of the newer stratigraphical geology were achieved among the Secondary strata. The Secondary sequences that had been described in various local areas began to be linked up by correlation, especially after Smith's map of England and Wales (1815)—the first to cover such a wide area with an explicit emphasis on the sequence of strata—had shown the heuristic value of such surveys. Over much of northwest Europe the highly distinctive and very widespread formation of the Chalk came to act as a valuable "marker" for the top of the Secondary sequence (see fig. 3.2). Below the Chalk the sequence was less uniform, but several distinctive formations, such as Oolite, Lias, and the New Red Sandstone mentioned earlier, could be recognized under different vernacular names over a very wide area.

As more and more local sequences were described in increasing detail, it became customary to define distinctive major groups or series of formations that could be traced from one region to another, or identified in distant regions, even if their constituent formations differed in detail. By the 1830s such *groups, terrains,* or *Gebirge*—the terms most commonly used in the three main scientific languages—were beginning to gain general acceptance as a useful level of classification, intermediate between the very broad categories such as Transition and Secondary, and the often variable and rather local formations. At the top of the Secondary sequence, for example, a *Cretaceous* group was defined to include not only the distinctive Chalk itself but also a series of closely associated lower formations that were less easy to correlate in detail. At the bottom of the Secondary sequence, below the New Red Sandstone, a *Carboniferous* group was defined, likewise containing a variety

23. The spelling "Greywacke" is used throughout this book, except in quotations and in direct reference to German sources. To avoid confusion, "Greywacke" with an initial capital will be used to denote certain formations assigned to a Transition age, while "greywacke" without a capital will denote a specific rock type, whatever its age; likewise with Chalk and chalk, Culm and culm, etc. Early nineteenth-century geologists were less consistent.

24. A memoir by Alexandre Brongniart on some of the Alpine regions (1823) was decisive in this respect.

of formations (see fig. 3.2). At the top of the Carboniferous group was the formation known by the old miners' term *Coal Measures (terrain houiller; Steinkohlengebirge)*, which contained most of the coal seams that were so important for the growing industrialization of the West European economies. Underlying this was a quite different formation known indifferently as *Mountain Limestone* or *Carboniferous Limestone*. At the base of the Carboniferous group there was in some regions a third major formation, the Old Red Sandstone mentioned earlier.

Of all the major groups, the Carboniferous received particular attention, owing to the supreme economic importance of the Coal Measures. Not all workable coal seams were confined to that formation, but most of the highly productive ones were found there. National geological surveys, whether publicly or privately financed, went to exceptional lengths to delineate as accurately as possible the outcrops of Coal Measures strata; fortunately, these were generally quite distinctive even without their coal seams. The mapping of adjacent areas as either younger or older than the Coal Measures was almost equally important, for areas with younger strata at the surface might have concealed coal seams at greater depth (see fig. 3.1). Britain was found to be exceptionally well endowed with coalfields in the Coal Measures. In the 1830s British mines accounted for more than half (57%) of the total world output of coal (fig. 3.5).[25] Two rich coalfields on the continent gave Prussia and the newly independent Belgium a lesser but significant boost to their incipient industrialization (together, 23% of world output). By contrast, France had only a few small patches of *terrain houiller* within her post-Napoleonic frontiers, apart from a concealed extension of the Belgian coalfield, and hence a much smaller output of coal (14% of world output) in relation to the large area of the country (fig. 3.6).[26]

The strata below the Secondary sequence posed greater problems. Below the lowest formation of the Carboniferous group (i.e., usually either Old Red Sandstone or Mountain Limestone) there was in many regions a confused and confusing mass of Transition or Greywacke rocks. These were often strongly folded and faulted; they were not generally differentiated into distinctive formations that could easily be traced across country and used to decipher the true sequence; and they rarely contained more than a few obscure and poorly preserved fossils. Some geologists doubted whether it would *ever* be possible to reduce the Transition rocks to the kind of orderly sequence that was becoming clearer every year in the Secondary strata. This was a major challenge to the expanding program of stratigraphical geology, and it formed the intellectual arena for the Devonian controversy. The problem was that of penetrating into the Transition rocks with the methods and procedures that had been proving so spectacularly successful in the Secondary strata. The

25. Fig. 3.5 is traced from Phillips (1838a), the "second edition" of Douglas and Edmonds (1950).

26. The percentages are derived from figures given in a contemporary French geological textbook (Burat 1834, p. 269) rather than from more conventional sources for economic history, because in the present context the way *geologists* perceived this disparity is of primary importance. The total output was given as 133 million *quintaux métriques* (i.e., 13.3 million metric tons) *per annum*. Fig. 3.6 is based on the index map to Dufrénoy and Élie de Beaumont (1841a). The symbols are used here for outcrops too small to be shown to scale. The anthracite or culm in Devon was not marked on the French map but is added for comparison.

Fig. 3.5.   Part of John Phillips's small-scale geological map of Britain (1838), redrawn to show only the outcrops of the Coal Measures (black), and the areas of still older rocks (stippled). This map is also used to show the locations of the annual meetings of the British Association for the Advancement of Science (to 1842), and some other place names mentioned in the text.

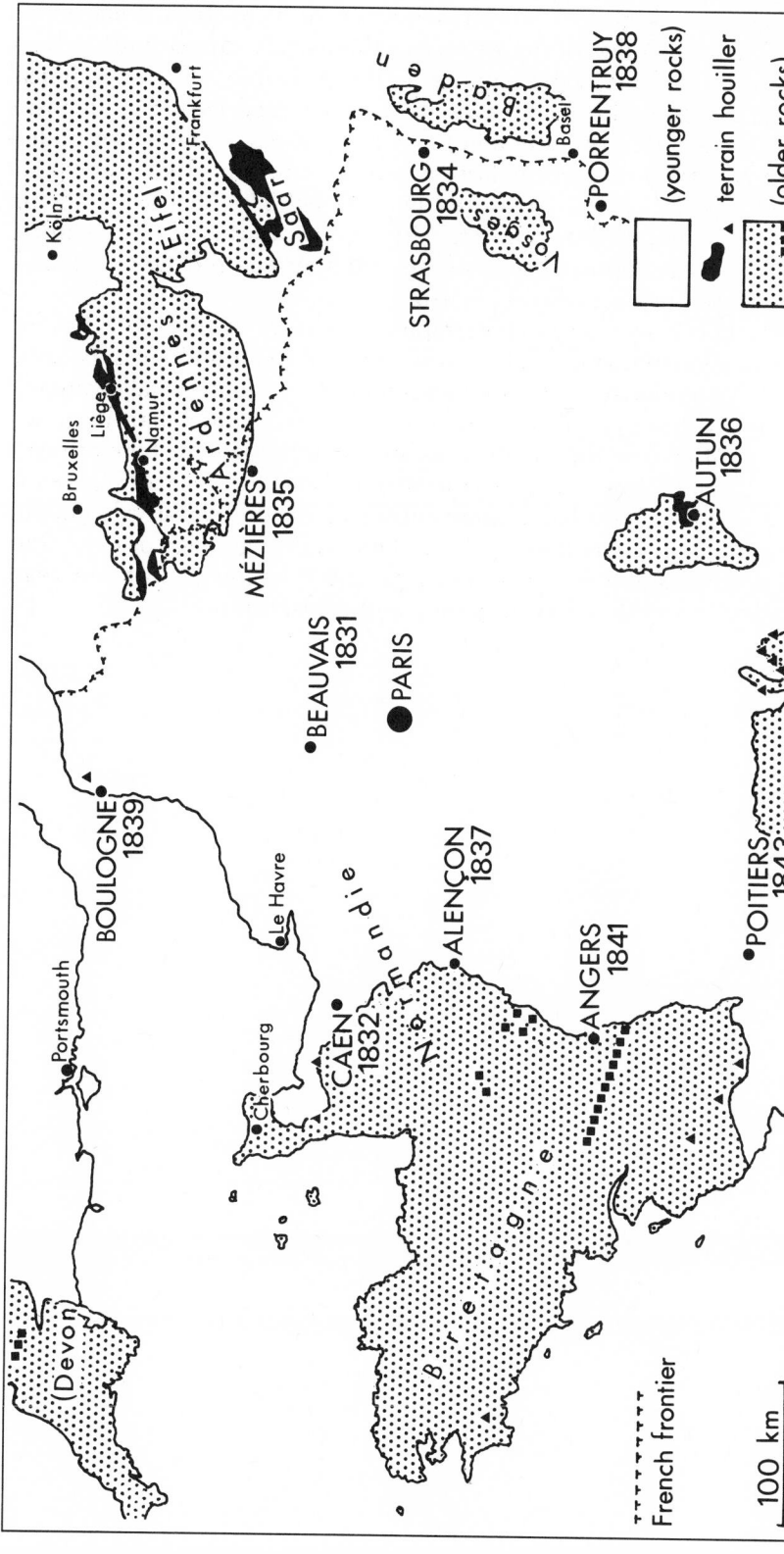

Fig. 3.6. Part of the great official geological map of France published in 1841 but completed several years earlier by Dufrénoy and Élie de Beaumont, redrawn to show only the outcrops of the Coal Measures or *terrain houiller* (black areas and black triangles) and the areas of still older rocks (stippled) with occurrences of anthracite in the *terrain de transition* (black squares). This map is also used to show the locations of the annual field meetings of the Société géologique de France (to 1843), and some other place names mentioned in the text.

prizes at stake were high. An orderly sequence of Transition strata would carry a detailed knowledge of the history of the earth's surface further back than ever before, lending further support to the principles of stratigraphical geology generally. In addition, the Greywacke contained important deposits of coal (or its harder variety, anthracite) in some regions, and a better understanding of it was therefore particularly urgent for countries like France that were poor in deposits of Coal Measures age (see fig. 3.6). Last but not least, it was generally anticipated that the Transition rocks would contain whatever traces remained of the earliest history of life, and hence carry scientific knowledge back to some record of Creation itself.

The foregoing account of stratigraphical geology in the 1830s has deliberately been given with minimal reference to individual geologists, because it is intended to summarize tacit geological practice and consensual geological knowledge at the time of the Devonian controversy, rather than to give a historical account of how the practice became standardized or how the consensus was reached. Now, however, the way is clear for the participants in the controversy to be introduced more fully and for the narrative to begin. But it begins far from the particularities of Devonshire geology or even from the general problems of the Transition strata, and it begins almost four years before the controversy flared up in public and explicit form.

Part Two

THE PLOT UNFOLDS

Chapter Four

PRELUDE: UNRAVELING THE GREYWACKE

4.1 A Father Figure for Geology (1831)

Late in 1828, as he lay dying, the chemist and mineralogist William Wollaston (1766–1828) had made his last will and testament, disposing of a fortune partly derived from his success in making platinum a commercially workable material. Among other legacies, he had given the Geological Society the most generous donation it had yet received, namely £1000 of government stock. The Society had resolved to use the interest to establish a gold medal in the donor's memory, to be awarded for outstanding geological research.[1] Early in 1831 the Council of the Society decided to give the first Wollaston medal to William Smith, for being "a great original discoverer in English geology, and especially for his having been the first in this country to discover and to teach the identification of strata, and their succession, by means of their imbedded fossils."[2]

The decision was neither straightforward nor uncontroversial. Smith, as a professional land surveyor and no gentleman, had not been welcomed into the bosom of the Society when it was founded, and he might not have felt comfortable if he had. Furthermore, the first president had sought to mobilize the Society's membership in a collective project of constructing a theory-free geological map of the country. This was incompatible with Smith's

1. *PGS* 1:110, 175.
2. Council minutes, 11 January 1831 (GSL: BP, CM 1/3, pp. 82–83). The phrase "in this country" was a later addition, probably inserted at the insistence of some member of the Council who felt that otherwise the priority claim being made for Smith would be indefensibly chauvinistic.

one-man project, which was parallel in intent but far from theory-free. A mutual antipathy between Smith and the leaders of the Society had deepened in those early years into rivalry and hostility.[3] The proposal that Smith be awarded the first Wollaston medal was thus pregnant with symbolic significance and political intent. By honoring the doyen of the professional surveyors, the Society would publicly make amends for its earlier exclusion of him, and hence implicitly acknowledge the contribution that his social group could make to geology. But at the same time, by praising Smith specifically for his use of fossils, the Society would publicly endorse the validity of that criterion in the correlation of strata and the making of geological maps, and hence implicitly turn its back on the skepticism about Smith's work that its early leaders had expressed.

Of these two aspects of the award, the social was relatively straightforward. In a period of increasing tension and conflict between the classes of English society, such a symbolic act of reconciliation could have value for the public relations of the Geological Society, to put it at its lowest. This was probably the point that appealed particularly to the then president of the Society, who chaired the Council meeting at which the award was decided, and who announced the award with an effusive speech at the anniversary meeting a month later. The president was Adam Sedgwick (b. 1785), the 46-year-old professor of geology at Cambridge and a prominent member of the inner circle of the Society (fig. 4.1).[4]

Sedgwick was the son of a country clergyman in Yorkshire, and he had earned his university education the economical way, like Isaac Newton a century and a half before, as a "sizar" at Trinity College, Cambridge.[5] He had done well in the strongly mathematical curriculum, was elected a Fellow of his college and in due course was ordained. In 1818, when the long-established Chair in geology at Cambridge became vacant, he had competed for it. Although his qualifications were even more meager than those of his opponent, he had been elected Woodwardian Professor and thereby acquired a small income of £200 a year. Unlike most of his predecessors, he had taken to his adopted science with zest and energy; he had at once begun an annual series of lectures during the winter—though they were extracurricular—and a parallel series of lengthy field expeditions every summer. He had joined the Geological Society soon after his appointment at Cambridge, and it remained his major scientific affiliation and gave him his main outlet for publication. He had earned respect in the 1820s for his stratigraphical research on some of the English Secondary strata, in which he had paid due attention to fossils without giving them any special emphasis. His real love, however, was for the upland regions, such as the Cumbrian or Lake District hills not far from

3. Laudan (1976).

4. *PGS* 1:270–79. The medal was not ready in time for the meeting and was not presented to Smith until the Oxford meeting of BAAS in 1832. Fig. 4.1 is reproduced from an engraving by Samuel Cousins after a painting by Thomas Phillips (NPG: 33648).

5. The late-Victorian biography of Sedgwick (Clark and Hughes 1890) is still indispensable; Speakman (1982) adds useful material on his Yorkshire background. This and all subsequent biographical sketches in this book are based in part on standard sources, such as the *Dictionary of Scientific Biography*, Poggendorff's *Biographisch-Literarisches Handwörterbuch*, and the relevant national biographical dictionaries; other sources are only cited where they supplement these in important ways.

Fig. 4.1.   Adam Sedgwick in 1833, from a portrait by Thomas Phillips.

Fig. 4.2.   George Greenough in the 1830's.

his family home (at Dent; see fig. 3.5), where fossils were rare and of little use.

The hard rocks of such areas seemed to Sedgwick's contemporaries to match the toughness of his own character. For Adam, or "the first of men," as they called him jokingly, was a bluff, plain-speaking Yorkshireman, a man of straightforward moral principles and genuine piety. Sedgwick's rural upbringing had given him a healthy respect for the yeomen farmers who were his boyhood neighbors; although he was no political leveler, the proposal to honor Smith would have appealed to him on social grounds. There is also no reason to doubt the sincerity of his public praise for the surveyor's pioneer work in English geology. But like many of his colleagues at the Geological Society, Sedgwick probably had his reservations about Smith's simplistic use of fossils.

That scientific aspect of the award was in fact far more controversial than the social. In particular, the terms of the award would have been seen as a clear rebuff to the Society's first president, George Bellas Greenough (b. 1778), whose map, as already mentioned, had been constructed on quite different principles than Smith's. Greenough was not only one of the last of the founding members to remain active in geology, he was also still a force to be reckoned with in the Society's internal politics in the early 1830s (fig. 4.2). The son of a lawyer but orphaned at an early age, his gentlemanly style of life had been secured when a grandparent left him a substantial fortune, derived from a lifetime's successful but less than gentlemanly business as an apothecary. Greenough had been sent to the great Prussian university of Göttingen to complete a legal education, but instead he fell under the spell of the natural historian Johann Friedrich Blumenbach (b. 1752), who had fired him with enthusiasm for the study of the earth.[6] Göttingen had also given him a cultured, cosmopolitan outlook that contrasted sharply with the philistine chauvinism of many of his English contemporaries.

Greenough had returned to England a wealthy and cultivated young bachelor. He sat for several years in the unreformed House of Commons, and later built himself a grand Italianate villa on the edge of the Regent's Park in London (see fig. 2.6). But unlike other fashionable young men about town, he had decided to devote himself to science in general and to geology and physical geography in particular. He had been one of the chief instigators of the Geological Society; despite his youth—he was still under thirty at the time—he had become its first president; and he had steered it uncompromisingly through its early crisis with the Royal Society (§2.2).[7] Two decades later he still regarded the Geological Society as his creation, and with some justification.

In his earliest fieldwork around the turn of the century, Greenough had tried to evaluate some of the then contested theories of Neptunism, Vulcanism, and Plutonism in geology, but he had concluded that none was satisfactory.[8] In apparent reaction to all such global theorizing, he had then turned to an extreme "Baconian" empiricism, and his forceful views had made this the dominant approach in the early Geological Society. In his only book, entitled, significantly, *A Critical Examination of the First Principles of Geology* (1819), he had argued not only that all theorizing in geology was premature, but that even the basic concepts of the science needed to be purged of all unacknowledged theoretical overtones. What was of value in geology, he maintained, was the diligent accumulation of unadorned factual observations. He had tried to compile his geological map of England and Wales in conformity with this "atomized empiricism."[9] The map was produced in direct competition with Smith's, but Greenough considered that his was the more soundly based because it did not depend on Smith's questionable assumptions about characteristic fossils. In practice both maps depicted the outcrops of formations recognized in the first instance by their rock types.[10] But Greenough regarded fossils as just one empirical character among others; in some instances it might prove useful for correlation, but it was not a criterion that could be relied on uncritically. His and Smith's mutual animosity was therefore due in part to a substantive conflict of scientific opinion, though of course it was heightened by the social tension between the wealthy gentleman and the self-made surveyor.

Their hostility was later aggravated by the way the publication of Greenough's map (1820) made Smith's map (1815) virtually obsolete, drying up its sales. For although Greenough incorporated much of Smith's local detail—with scant acknowledgment—his map also summarized in visual form a large

6. Greenough is an unjustly neglected figure, and there is no modern biographical study of him: see, however, Rudwick (1962) and the introduction to Golden (1981). Fig. 4.2 is reproduced from an undated portrait in Woodward (1907: 12).

7. Rudwick (1963).

8. Rudwick (1962).

9. His "atomized" concept of facts is reflected in the many large volumes of manuscript notes that he compiled, most of which contain little more than lists of unconnected instances of some phenomenon or physical feature. (Many of these volumes are now in UCL: GBG.)

10. Laudan (1976, 1977).

mass of other research, both published and unpublished, by the members of the Geological Society and many local informants. Its depiction of the outcrops of formations and their boundaries was much more accurate than Smith's, particularly outside the areas of Secondary strata which had been central to Smith's work. It remained the indispensable starting point for attempts by others to map local areas in greater detail; by 1831, Greenough, now a mature 53-year-old, had been engaged for several years in revising and improving it for a second edition. So he would naturally have felt that the citation for the award to Smith gave his rival altogether too much credit, and thereby detracted from his own achievement. In any case the decision would have been seen within the Society as being, at the very least, a signal victory for those who were turning Smith into a figurehead for their own scientific ambitions.

There is much to suggest, in fact, that the real driving force behind the award was not Sedgwick, the current president, but Roderick Impey Murchison (b.1792), an ambitious 39-year-old geologist who was due to take over from Sedgwick after the meeting at which the award was to be announced (fig. 4.3).[11] Murchison, who was Sedgwick's junior by seven years, came from the Scottish landed gentry, but his father had died when he was still young. He was given a military education (at Great Marlow), but his military career was less brilliant than he liked to recall in later life: he was briefly on active service during the Peninsular War, but at the height of the hostilities with France he was stuck with a posting in Ireland, far from the fighting. Yet army life seems to have suited, or fostered, his exceptionally pugnacious temperament, and in later life his language was peppered with military metaphors almost to the point of self-parody.[12] When peace finally returned, Murchison had resigned from the army, married a young heiress, and settled down to life as a country gentleman. But his extravagant expenditure—not least on hunting—had later obliged him to move to less expensive surroundings in London, although, thanks to his wife's family, he was able to retain a gentlemanly style of life.[13] Encouraged by the chemist—and geologist—Humphry Davy, and by his highly intelligent wife Charlotte, Murchison had then decided to adopt science as a full-time pursuit and, in effect, as an alternative route to social respectability. He had quickly chosen geology as a branch of science that could incorporate outdoor exercise and be combined conveniently with hunting foxes and shooting pheasants. He had joined the Geological Society in 1825 and became one of its secretaries only a year later.

Within geology, Murchison had soon focused on stratigraphical research, where his lack of formal scientific training was not the handicap it would have been in the study of minerals or fossils, and where his love of the countryside could find full expression in fieldwork. Sedgwick had become informally his teacher and mentor; Murchison learned the practice of fieldwork in the course of their joint expeditions, as far afield as the Scottish Highlands and the Alps. But Murchison's research had begun to diverge from Sedgwick's, not in the sense of disagreement but rather as a matter of different

11. The Victorian biography of Murchison (Geikie 1875) is too laudatory to be of much value. Though not a biography, Secord (1986) is a fine modern account of one aspect of his work. Fig. 4.3 is reproduced from a sketch by William Drummond, published June 1836 (BML: 1865.6.10.1219).
12. Secord (1982).
13. Secord (1986) makes explicit what Murchison's biographer hinted at too discreetly.

emphasis.[14] While Sedgwick remained primarily concerned with unraveling the structure of a region, Murchison was more interested in establishing the sequence of strata and formations. Where Sedgwick's field notebooks were packed with observations of strike and dip, and notes on inferred folds and faults, Murchison's were full of sketched traverse sections and notes on the constituent formations and their fossils. While Sedgwick was drawn toward the tough and complex Primary and Transition rocks of the upland areas, Murchison was attracted by the clearer sequences and richer fossil contents of the Secondary formations.

Murchison's orientation had been decisively confirmed by one of his earliest pieces of solo fieldwork, in which he had set out to determine the true stratigraphical identity of an anomalous patch of coal-bearing strata at Brora, not far from his birthplace in the eastern Highlands. On his way north in 1826 he had stopped to meet William Smith at Scarborough on the Yorkshire coast. Smith and his young nephew and assistant (John Phillips) had shown Murchison a formation that, guided by its position in the whole sequence, they correlated with the Oolitic strata further south, although it contrasted strongly with the usual rock types of that group and even resembled the Coal Measures. On studying the Brora coalfield, Murchison concluded that Yorkshire held the key to the Scottish anomaly. The fossil plants associated with the Brora coal were the same as those of the unusual "Oolitic" strata he had seen in Yorkshire and were quite different from those of the much older, true Coal Measures. In other words, this example had convinced Murchison that fossils were far more reliable than rock types in determining the correct position of strata in newly studied or isolated areas.[15] The same rock types (in this case, coal seams and associated sediments) might recur in quite different parts of the whole sequence, but he believed that a close attention to the fossils could be relied upon to distinguish between different instances.

This example not only convinced Murchison of the heuristic value of Smith's emphasis on fossils in stratigraphical research, but it also suggested to him a potent if unconventional ally in his ambitious ascent to fame as a geologist. As already mentioned, it was probably Murchison who chiefly urged Smith's claims to the first Wollaston medal early in 1831. Certainly it was Murchison who lavished further praise on him later that year, when the British Association assembled in Yorkshire, where Smith had settled. At this inaugural meeting, where Murchison was one of the few metropolitan gentlemen of science present to give the new body unstinted support, he publicly dubbed Smith the "father" of English geology.[16] This was partly to make him serve as a symbol of social solidarity for the Association, just as he had been for the Geological Society. But the act had social meaning beyond that. The anthropological overtones of the epithet "father," though doubtless not in Murchison's mind, were nonetheless appropriate. Smith was being made into a symbol for a new and more strongly fossil-based style of geology. It hardly mattered that Smith's own early use of fossils had been simplistic and far from unproblematic. For those like Murchison who were promoting his claims in his

14. Secord (1986).

15. Murchison (1827).

16. Morrell and Thackray (1981, p. 89). At the end of 1831 Murchison was also the instigator of one of the Association's first political lobbies which successfully won for Smith a small government pension: ibid., pp. 325–26.

Fig. 4.3.   Roderick Murchison in 1836, from a sketch by William Drummond.

Fig. 4.4.   Henry De la Beche in 1842, from a sketch by William Brockedon.

old age, historical accuracy was less important than present usefulness.[17] As a symbolic or totemic figure, he could be used as a powerful rhetorical weapon in the struggle to establish the ascendancy of a geology based on the primacy of "characteristic fossils." This was a struggle in which Murchison saw himself ranged against an older generation led by Greenough. But Murchison had yet to find a distinctive place for himself in this contest for power within the social world of geology.

4.2   Sedgwick's Greywacke Grindstone (1831)

Any geologist in 1831 who was keeping abreast with the science would have been well aware that the Transition or Greywacke strata were a promising and underdeveloped field for research. The consensual view of the place of those strata within the record of the rocks is reflected and summarized in the introductory *Geological Manual* published that year by Henry Thomas De la Beche (b. 1796), a respected 35-year-old geologist who was then serving as a secretary of the Geological Society under Murchison's presidency (fig. 4.4).

De la Beche was no Frenchman, but the grandson of a plain Thomas Beach, who had given his name a Norman flavor, perhaps to reinforce his social status as the owner of a plantation in the West Indies. De la Beche had grown up in England, however, after his father's death; living at Lyme Regis

17. Laudan (1976).

in Dorset, one of the richest areas for fossils in the country, his later interest in geology is unsurprising. Like Murchison, who was only four years his senior, he had been given a military education at Great Marlow, but there the parallel, and De la Beche's military career, had ended abruptly. He had been expelled for insubordinate behavior that the authorities attributed to the "dangerous spirit of Jacobinism," but in later life he successfully concealed this early disgrace.[18] One positive legacy of his military training was that it stimulated or improved his abilities as a draughtsman, which he later put to good use in his geological research and not least in scientific caricature (see figs. 2.7, 5.4).

When De la Beche had come of age and gained legal control of the family estate in Jamaica, he became, like Greenough, a wealthy young man about town with an income of some £3000 a year; unlike Greenough, he had married. He had joined the Geological Society in 1817, at the age of only 21, and the Royal Society two years later; it is clear that he had decided, like Greenough earlier and Murchison several years later, to devote his time and money to science in general and geology in particular. In the 1820s he had published several competent papers on stratigraphical geology and also a volume of translated selections from the Paris *Annales des Mines*, which reflected his interest in the practical uses of geology and its international dimension. But his early sense of social exclusion had been aggravated when his wife applied for legal separation and went to live with another man. A subsequent period of ill health, perhaps not unconnected with his personal troubles, had at least given him a reason to travel extensively on the Continent and had broadened his experience of its geology.

The success of De la Beche's *Geological Manual* reflects not only its popularity with the general public but also its acceptability among his fellow-geologists.[19] As in most such compendia of this period, De la Beche described the main groups of formations in inverse chronological order, moving from the relatively clear upper (and younger) parts of the sequence to the more obscure lower (and older) parts. So after summarizing the strata of the "Carboniferous Group," including the Old Red Sandstone as its lowest formation, De la Beche dealt at some length with the "Grauwacke Group." He gave a long list of fossils known from the Greywacke strata and commented that at that period there was evidently "not that poverty of organic structure which once was supposed."[20] On the contrary, there was a full range of animal fossils, including even some traces of fish, and also quite a variety of plant fossils. The very earliest traces of life were, he believed, in still older rocks, which, he suggested, dated from a time when organisms might already have been quite abundant but mostly soft-bodied and therefore not preserved as fossils. The oldest strata of all, on the other hand, contained no traces whatever of any fossils, and De la Beche suggested that they might predate the first origins

18. McCartney (1977) gives valuable new biographical information based on NMW: DLB. Fig. 4.4 is reproduced from a sketch dated 22 April 1842, by William Brockedon (NPG: 2515/94).

19. Within two years of its first publication the *Manual* was already into an enlarged third edition; American and German editions had also appeared, and a French edition was in preparation (De la Beche 1833a, preface).

20. Ibid., p. 452. De la Beche retained the original German spelling "Grauwacke," probably as a deliberate gesture of internationalism, when most other English geologists were shifting to "Greywacke."

of life.[21] These very oldest "stratified rocks" such as gneiss and schist, and the granites that often underlay them, remained confused and obscure in De la Beche's summary. But the Greywacke itself was treated as a part of the total sequence that was almost as "normal" as the Carboniferous strata above it.

As a record of the period during which a moderately diverse fauna and flora first became discernible, the Greywacke took its place within a history of the earth that displayed a broadly directional development toward the present.[22] In this respect De la Beche's treatment of the Greywacke would have commanded general assent from other geologists throughout Europe and beyond. What was signally missing, however, from De la Beche's *Manual*—or any other similar compilation of the same date—was any description of a sequence of formations *within* the "Grauwacke Group" comparable to those that by then were well established among the Secondary strata. The reason was simple. Virtually no such sequence, even of purely local validity, had yet been described from any region of Transition rocks. Geological maps that attempted to generalize current knowledge of the distribution of rocks over wide areas, such as Greenough's for England and Wales (1820) or Friedrich Hoffmann's for northwest Germany (1830), depicted vast tracts of undifferentiated Greywacke. The only local detail that was commonly shown within those tracts was an indication of the outcrops of bands of "Transition limestone" or *Übergangskalkstein* (see fig. 4.10). Such limestones were potentially important as the source of most of the best preserved fossils that were known from the Transition strata, but even they seemed to be confined to the upper part of that sequence (see fig. 3.4).

To find a well-defined sequence of Transition formations was just what Sedgwick set out to do in the autumn of 1831, when he started fieldwork in the largest area of Greywacke rocks in Britain (south of Scotland), namely, north and central Wales. His purpose was quite specific. The standard general summary of the British strata had been published nearly a decade earlier by William Conybeare (b. 1787), who had become interested in geology as a young don at Oxford and was an influential early member of the Geological Society. Conybeare in turn had revised an earlier work by William Phillips (1775–1828), a London publisher and one of the founders of the Geological Society, and their joint *Outline of the Geology of England and Wales* (1822) had quickly become a classic work of reference. But they called their book "volume I" and had taken their description of the British formations—in the customary retrospective order—only as far down as the base of the "Carboniferous Group," that is, as far as the base of the Old Red Sandstone. After Phillips's death, Sedgwick had promised Conybeare, who was now a country clergyman in south Wales (at Sully, near Cardiff), to help him write a sequel on the Greywacke and still older rocks. But before he could fulfill this obligation he had to extend his earlier study of the rocks of Cumbria to those of Wales and other regions.

21. Ibid., sec. X, "Lowest Fossiliferous Rocks," and sec. XI, "Inferior Stratified and Non-fossiliferous Rocks." De la Beche did not term the latter "Primary" or "Primitive," though other writers did. The "Unstratified Rocks" (sec. XII) were dealt with last of all, though De la Beche acknowledged that granites were *not* necessarily the oldest rocks, and that at least *some* were of igneous origin (see fig. 3.4).
22. On this "directionalist synthesis" in geology around 1830, see Rudwick (1971).

Sedgwick therefore began work in Wales with the specific aim of tracing the whole sequence of strata below the Old Red Sandstone. He had as his companion Charles Darwin (b. 1809), whom he intended to teach the rudiments of geological field practice before that young product of his university sailed for South America as an unofficial gentlemanly naturalist on the *Beagle*[23] Sedgwick could have started at either the bottom or the top of the Welsh sequence. The rocks in the northwest, on the island of Anglesey, were regarded as Primary, so that the oldest Greywacke rocks could also be expected in that vicinity. At the other end of the scale, it was equally well known that in parts of the Welsh Borderland the Old Red Sandstone passed down without a break into the youngest of the Transition or Greywacke strata. The latter point had been noted explicitly a full decade earlier in a widely cited article by Sedgwick's counterpart at Oxford, William Buckland (b. 1784); Buckland and his friend Conybeare had later substantiated it in the course of a long joint memoir on the coalfields of the west of England.[24]

Like Sedgwick, only a year his junior, Buckland (fig. 4.5) was the son of a country clergyman, had earned his university education economically—in his case through a "closed" Devonshire scholarship to Exeter College, Oxford—and had been elected a Fellow of his college and later ordained. With a growing interest in geology, Buckland had joined the still youthful Geological Society in 1813 and was appointed Reader in Mineralogy at Oxford, beginning an annual course of lectures the following year. In 1818 he had successfully petitioned the Prince Regent, through the Tory Prime Minister, Lord Liverpool, to be given the additional title of Reader in Geology. This had given public recognition to the way he had expanded his course to cover the whole scope of the new science of geology, including especially the study of fossils.[25] Like Sedgwick's lectures at Cambridge, Buckland's were strictly extracurricular and attracted more dons than students. But they had been the most popular scientific lectures in Oxford in the 1820s; they were highly influential in establishing the academic respectability of geology in a university that was generally indifferent to science when not actually hostile; and they had caught the attention and enthusiasm of several undergraduates who became some of the most active geologists of the next generation. When Buckland was tempted to accept a better-paid ecclesiastical position elsewhere, and might have been lost to science altogether, he had been kept in Oxford by being appointed a Canon of Christ Church, again through Liverpool's patronage. This had given him a substantial income of £1000 a year and had enabled him to marry, and his geological work had continued unabated.

Buckland's inaugural lecture at Oxford in 1819 had seemed to mark an attempted rehabilitation of the use of geology to support the historicity of *Genesis*. But Buckland was no scriptural geologist; the "Deluge" he had inferred from his celebrated research on cave faunas, although identified at first as that recorded in *Genesis*, was far from literalistic in character. His early

---

23. Barrett (1974).

24. Buckland (1821, p. 465); Buckland and Conybeare (1822, especially pp. 283–84). The conformable junction was described on the northern edges of the Forest of Dean and Bristol coalfields (at May Hill and Tortworth, respectively).

25. Edmonds (1979) gives useful new information on Buckland's early life; Edmonds and Douglas (1976), on his teaching. See also Rupke (1983). Fig. 4.5 is reproduced from a sketch by William Brockedon, dated 7 February 1838 (NPG: 2515/87).

emphasis on the geological traces of the Deluge was partly an attempt to reassure his Oxford colleagues that the pursuit of geological science would not subvert Christian faith and piety, and that it was therefore a proper subject to be taught at the university that was the intellectual center of established religion in England.[26] Most of Buckland's earlier research, however, had been straightforward stratigraphical work on particular areas or formations; like Sedgwick, he accepted the value of fossils for correlation, though without any special emphasis.

The article by Buckland mentioned earlier had been firmly in that uncontroversial tradition: it was one of the first thoroughgoing attempts to correlate the whole British sequence with the one known on the Continent. In that international context, Buckland's report of a conformable passage from the lowest Secondary strata (i.e., the Old Red Sandstone) down into the Transition or Greywacke rocks in the Welsh Borderland was of clear importance, because on the Continent—and indeed in Britain too—that position was more usually marked by a gap in the sequence and a major unconformity. Buckland would therefore have awaited with interest the results of Sedgwick's projected study of the whole Greywacke sequence in Wales.

In the event, Sedgwick started his fieldwork along the north coast, probably in the hope that in a single traverse he might get a first rough outline of the entire sequence, from the Carboniferous strata near the border with England to the Primary rocks of Anglesey. He found, however, that the conformable passage that Buckland had commented on further south did not extend to this area, and that instead there was the more usual unconformity below the Carboniferous. Since the top of the sequence was evidently missing in north Wales, he therefore began instead at the bottom. Taking the Primary rocks of Anglesey as his base, he concentrated first on the older Greywacke strata around Snowdon (see fig. 3.5). The structure of the region and the sequence of the strata proved to be complex and difficult to decipher. There were few distinctive formations and few fossils, and the structure and sequence were confused by an almost ubiquitous slaty cleavage. It was, he recalled afterward, "like rubbing yourself against a grinding stone."[27]

4.3 Murchison Gets out of Lyell's Shadow (1831–33)

While Sedgwick was struggling with heavy autumn rain in Wales, Murchison was rounding off a period of no less than four months in the field, which he had only interrupted briefly for the inaugural meeting of the British Association in York.[28] Although it was Sedgwick who in effect had given him his practical training in fieldwork, Murchison's theoretical ideas were influenced

26. Buckland (1820, 1823). Rupke (1983) puts the cave research into its Oxford context. Its respectability in contemporary scientific eyes is indicated by the fact that it gained for Buckland the Royal Society's Copley Medal.

27. Secord (1986); Sedgwick to Murchison, 20 October 1831, quoted in Clark and Hughes (1890, 1:382); Lyell, journal entry for 17 January 1832 (recording Sedgwick's remarks), quoted in K. Lyell (1881, 1:366).

28. The following interpretation of Murchison's 1831 fieldwork is based on Secord (1986), to be amplified in a separate article.

Fig. 4.5.   William Buckland in 1838, from a sketch by William Brockedon.

Fig. 4.6.   Charles Lyell in 1836, from a portrait by J. M. Wright.

more by Charles Lyell (b. 1797), a prominent London geologist five years his junior (fig. 4.6). One major purpose of Murchison's fieldwork in 1831 was to explore the validity of a theory that Lyell had begun to work out three years earlier, while they were traveling together in France and Italy. This was that there was a direct causal connection between volcanic and other igneous activity, on the one hand, and the elevation and folding of strata, on the other.[29] Hoping to test the idea on a variety of regions, Murchison began his season's fieldwork with a brief three-week tour along the edge of the Greywacke area in the Welsh Borderland. He ended the season with a study of the Tertiary ("Crag") strata in East Anglia, again following up Lyell's earlier work on the Tertiary of Italy and Sicily. As his young fellow-Scot James Forbes (b. 1809) had noted perceptively after meeting him earlier in the year, "[Murchison] is a diligent observer . . . but does not appear possessed of much originality."[30]

That at least could not be said of Lyell, who had struck the same commentator as "a remarkable man," but also as one "sanguine in his opinions, hasty in his generalizations," and reluctant to accept the reality of any phenomenon unless he had a theoretical explanation for it.[31] Like Murchison, Lyell came from the Scottish country gentry, but he had grown up in the south

29. For a narrative of this earlier fieldwork, see Wilson (1972, pp. 183–223). For some of its implications, see Rudwick (1969).

30. J.D. Forbes, journal of a visit to London in the spring of 1831 (SAU: JDF, boxes 1/2, notebook 10). I am indebted to Dr Secord for drawing my attention to Forbes's fascinating vignettes of the leading English geologists.

31. Ibid.

of England. As a schoolboy he had developed a taste for natural history but turned to geology while an undergraduate at Buckland's college at Oxford, attending Buckland's lectures three years running.[32] On graduating he had moved to London to study law, but although he later practiced for a time as a barrister, his interests were moving more and more toward geology. He was soon accepted into the inner circle of the Geological Society; having established contact with the leading geologists in Paris, he had been made its foreign secretary. The Society remained his main scientific affiliation; with a fine sense of where his career interests lay, he avoided taking on similar duties in the other scientific bodies to which he belonged.[33]

Unlike Greenough, Murchison, and De la Beche, Lyell needed to supplement a small allowance from his father with income from other sources. He had tried to make his interest in science pay, initially by writing general articles for the influential *Quarterly Review*. Then, in the spring of 1831, he had got himself appointed professor of geology at the recently established King's College in London; but the lectures there, like those by Buckland and Sedgwick at the older universities, were extracurricular, so Lyell could expect little income from fees.[34] Authorship remained a more hopeful source of income. So popular and fashionable was geology that there seemed to be an insatiable demand for elementary books on the science, so Lyell had decided to try his luck in that market. But unlike De la Beche's *Manual*, for example, Lyell's book had turned in the course of writing into a massive work that was far more than a mere introduction. The first volume (1830) of what this self-confident 33-year-old termed the *Principles of Geology* had sold surprisingly well, and his publisher, John Murray, paid him handsomely. It had also earned him respect from his fellow-geologists, though they had not accepted his conclusions uncritically.

Lyell's *Principles* had expanded into a major treatise primarily under the impact of a powerful but controversial theoretical vision. This had required him to approach geology from a novel direction and to reinterpret much that other geologists regarded as familiar and well understood. His major expedition to France and Italy in 1828–29 (initially in the company of Murchison) had confirmed and enlarged his conviction that a more thorough study of geological "causes now in operation," such as volcanos, earthquakes, erosion, and sedimentation, was a prerequisite for any real progress in geology. It was important, in his view, because these present processes could be shown to be "true causes" (*verae causae*) adequate to explain most if not all the records of the past history of the earth. Other geologists did not deny this in principle, but they did question the range of applicability of Lyell's kind of interpretation, and as mentioned earlier (§3.1), most of them were more concerned with stratigraphical geology than with causal analysis or geological dynamics.[35] Lyell had gained a few enthusiastic supporters and many more who admired

32. The Victorian biography of Lyell (K. Lyell 1881) is still indispensable; Wilson (1972) gives a supplementary chronicle of his life to 1841. Fig. 4.6 is reproduced from a lithograph by W. Drummond after a painting by J. M. Wright (BML: 1865.6.10.1215).

33. Morrell (1976).

34. Rudwick (1975a). On scientific lecturing in London generally, see Hays (1983).

35. Laudan (1982); Rudwick (1971).

some aspects of his work, but others were critical of his high-level theorizing and of what they regarded as his special pleading (see fig. 2.7).

Such criticism came not only from the antitheoretical Greenough, whom Lyell regarded as a die-hard reactionary in geology, but also from Sedgwick, whose penetrating comments were highly unwelcome to him.

Murchison had attached himself firmly to Lyell in this continuing debate, at least to the extent that he hoped to develop some of Lyell's ideas on British soil in the course of his fieldwork in 1831. But he did have other less controversial objectives as well. One was to help revise Greenough's general map for its forthcoming new edition; like other members of the Geological Society, he regarded that project as still being a collective endeavor, in practice if not formally. One of the important geological boundaries that was known to be inaccurate on the original map was that between the Old Red Sandstone and the Greywacke along the borders of Wales. So since he was visiting that area anyway for a different (and Lyellian) purpose, Murchison took the opportunity to locate the boundary, finding it first near Llandeilo and later on the river Wye near Builth (see fig. 3.5).

There is no evidence to suggest that at the time he was surprised in any way to find that at both localities the junction between the two formations was conformable, for as already mentioned, that fact was well known from Buckland's work. In the event, he did more than just trace the course of that junction northward from where he had first located it. When he got to Shropshire, near the end of his brief tour of the Borderland, he was shown the local Transition strata by Thomas Lewis (b. 1801), the curate of Aymestrey, who, having attended Sedgwick's lectures while he was at Cambridge, had become a keen local geologist. Here to the west of Ludlow, Lewis showed Murchison a sequence of limestones and shales, lying conformably below the Old Red Sandstone, and almost as unaltered, flat-lying, and full of fossils as the Secondary formations that Smith had first explored in the south of England (fig. 4.7).[36]

It was probably during the following winter, while reflecting on this remarkably Secondary-like sequence and looking over the specimens he had collected, that Murchison made a major decision about the direction of his research. Instead of returning to East Anglia in the summer of 1832 as he had planned, to work further on the Tertiary strata, he went back to the Welsh Borderland and began a systematic study of the Transition formations. In retrospect he constructed a romantic myth around what he claimed had been a "Eureka! moment" in 1831, when near Builth he had found the conformable junction between the Old Red Sandstone and the Transition strata. He forgot or concealed the fact that this conformable relation was already well known, that the river Wye near Builth was not even the first place he himself had seen the junction, and that at the time he had paid no special attention to it. He tried to eliminate Lewis's priority by claiming that, even before seeing the local geologist's home ground, he himself had not only found the junction but had realized its significance. His mythical moment of discovery became

---

36. Fig. 4.7 is reproduced from Murchison (1839, 1:195, pl.), engraved from a sketch by Lady Harriet Clive. On Lewis and his relation to Murchison, see Thackray (1979).

Fig. 4.7.   Ludlow Castle and the surrounding countryside: a sketch by one of Murchison's aristocratic acquaintances in the Welsh Borderland. The gently dipping strata in the foreground lay conformably below the Old Red Sandstone, and belonged to what Murchison in 1835 termed the Ludlow division at the top of the Silurian system.

the retrospective starting point for a research project he had not in fact envisaged at the time. His reasons for embarking on it, however, are plain enough. There were evident scientific rewards to be gained from a successful deciphering of the Transition strata, and the realization that those strata were exceptionally unaltered and full of fossils in the Borderland made it an opportunity too good to miss. By contrast, his main projects in 1831, in dynamical geology and Tertiary stratigraphy, would only have led him to follow in Lyell's footsteps and to remain in his shadow. In exploring the sequence of Transition formations downward from the Old Red Sandstone, Murchison found a project that might make him independently a leading man of science. It was also a project that gave him a distinctive personal stake in the furtherance of fossil-based methods in geology, and hence ranged him even more clearly behind his adopted father figure, William Smith.

Having once found a worthy and promising project for his research, Murchison applied himself with great energy to its development. In 1832 he spent his entire field season in the Borderland, exploring further the formations that evidently represented the upper part of the Greywacke, but that were fortunately so much easier to work on than Sedgwick's "crusty old rocks" to the northwest. Sedgwick may have envied his colleague's simpler task, but he had no reason to feel competitive. On the contrary, he probably welcomed Murchison's unexpected decision to make a thorough study of the upper part of the Greywacke sequence, for this division of labor could only ease the burden of his own obligation to Conybeare. Sedgwick and Murchison both felt they had made good progress that summer. Both took it for granted that their sequences, although probably overlapping a little, were in the lower and upper parts of the Greywacke, respectively, and that they would be able to link them up with no great difficulty in due course.

Murchison, always more ready than Sedgwick to put pen to paper, first announced his preliminary results early in 1833. In his presidential address to the Geological Society, he emphasized that the Old Red Sandstone was as conformable to the Transition or "fossiliferous grauwacke" formations below it as it was to the Mountain Limestone above, and that the older strata contained a rich and largely undescribed fossil fauna. He added generously that the value of his own work would be enhanced when it could be linked with Sedgwick's.[37] A few weeks later he presented a paper giving his own results in greater detail. This was a general summary of all the rocks in the parts of the Borderland he had studied. He estimated the "vast formation" of the Old Red Sandstone at more than 4000 feet in thickness, and then described an even thicker sequence of six formations in the Transition rocks beneath. He emphasized that he had distinguished these formations "by the evidences of fossils and the order of superposition."[38] His hearers would have been in no doubt about the significance of that pregnant phrase. Murchison gave public notice, in effect, that he intended to apply Smith's methods to strata that had hitherto resisted such treatment, namely, the Transition or Greywacke.

4.4   Weaver and the Irish Question (1832–33)

Murchison's conception of the importance of his new research project, and his ambitions for it, went far beyond the delineation of a sequence of local formations in the Transition or Greywacke. Though not stated explicitly, neither in his presidential address nor in his subsequent paper, the vast thickness of the Old Red Sandstone and the distinctive set of fossils in the upper Greywacke formations had an ulterior significance. Both served to accentuate the separation of the economically valuable Coal Measures from any older formations. For Murchison was well aware that most of his fellow geologists believed that deposits of coal, or its harder variety, anthracite, were not uncommon among the Greywacke strata of other regions. In his *Manual*, for example, De la Beche had mentioned that seams of anthracite had long been known in the Greywacke of north Devonshire, and that an even clearer example had been reported the previous year from Ireland. He also mentioned a recent paper by the American geologist Amos Eaton (b. 1776), summarizing the North American coal deposits. He accepted Eaton's report that, while many might be equivalent to the Coal Measures of Europe, some at least were definitely in the Greywacke. De la Beche maintained that all these anthracites, and the plant fossils associated with them, indicated that the plant life of the

37. Murchison (1833a, p. 449), read at GSL on 15 February.
38. Murchison (1833b, p. 475), read at GSL on 27 March and 17 April.

Greywacke period must have been broadly similar to that of the subsequent Carboniferous period.[39]

Such reports of coal or anthracite deposits in Greywacke strata were flies in Murchison's ointment. They were incompatible with his emerging conception of the Greywacke as a set of formations long predating any flora that could have provided the debris to produce such deposits. Of the two reports that De la Beche cited from the British Isles, the one from Ireland had seemed to him the more striking, because it had been studied recently and in detail by a respected geologist. The geologist was Thomas Weaver (b. 1773), who had trained at Werner's Bergakademie in Freiberg and had been the first editor of Werner's work in English, and who had since worked for the government in Dublin, exploring Irish mineral resources. In a paper read at the Geological Society in 1830, Weaver had claimed that in southwest Ireland (province of Munster) most of the economically important seams of coal were clearly in Transition strata and not in Coal Measures, because they lay under, not over, the distinctive Mountain Limestone. Furthermore, he had reported finding fossil plants in close association with the coal, of the same kind as those of the Coal Measures ("Equiseta and Calamites").[40] As Sedgwick recalled in his presidential address a few months later, Weaver's paper had been "heard with no little surprise by many members," among whom Murchison was almost certainly one. But Sedgwick himself had regarded its conclusions as an unsurprising confirmation of what was already known from other regions outside Britain.[41] De la Beche, as mentioned earlier, was of the same opinion, but he had been sufficiently aware of the importance of the report to check the matter privately with Weaver before sending his *Manual* to the printer.[42]

The potentially controversial character of Weaver's paper is also reflected in its subsequent fate, which was unprecedented and wrapped in mystery. It had been refereed in the usual way soon after it was read and recommended for full publication in the *Transactions*.[43] But it was not in fact printed, and no further action was taken for more than a year and a half. Weaver had gone to Mexico to do some more surveying and could not easily be consulted, but that is no adequate explanation for the mysterious delay, nor for the fact that a new referee was then brought in to give a second opinion. A second time the paper was approved for publication, and at the end of 1832 it was at last sent to the printer.[44] Only five weeks later, however, at a meeting chaired by Murchison as president, the referee "made a communication to the Council, which induced them to suspend the printing of the paper until they could communicate with Mr. Weaver."[45]

39. De la Beche (1831, sec. IX); Eaton (1831).

40. Weaver (1830), read at GSL on 4 June. The fossil plants were interpreted by botanists as the remains of giant horsetails (see fig. 5.3).

41. Sedgwick (1831, pp. 281–83), read at GSL on 18 February.

42. Weaver to De la Beche, 23 March and 23 April 1831 (NMW: DLB).

43. Council of GSL, 18 June and 5 October 1830 (GSL: BP, CM 1/3, pp. 53, 65); Weaver to De la Beche, 12 April 1831 (NMW: DLB). This referee was probably Greenough, whose undated report (GSL: BP, COM P4/2, p. 245) made recommendations that match the Council's subsequent decisions about the paper.

44. Council of GSL, 21 November and 19 December 1832 (GSL: BP CM 1/3, pp. 182, 185).

45. Council of GSL, 23 January 1833 (GSL: BP, CM 1/3, p. 204).

What the unnamed referee had written is not recorded, nor why he had so abruptly changed his mind. But Sedgwick was apparently recalling that second referee when he mentioned much later that the one who had "knocked over Weaver's case" had been the leading Dublin geologist Richard Griffith (b. 1784).[46] Griffith had trained as a mining engineer and had held a variety of posts in Ireland—Weaver had been his unsuccessful rival for one of them. At Greenough's suggestion, Griffith had embarked on an unofficial geological survey of Ireland, which by 1832 he was carrying on under the cloak of directing the government's survey of the rateable (i.e., taxable) value of Irish landed property.[47] In the course of this work, he had been over the part of Munster where Weaver had worked and might have been in a position to issue a caveat about his rival's earlier conclusions. But it is unlikely that he would have reversed his favorable report on Weaver's paper at just this moment, without being prompted from elsewhere.

Sedgwick later reminded De la Beche that Weaver's paper had been suppressed because "there were unbelievers in the Council of the Geol. Soc."[48] Since De la Beche himself had been at the meeting, Sedgwick felt it unnecessary to identify whom he meant. But he recalled that what the "unbelievers" had doubted was not the report of coal as such in the Transition strata of Ireland, but rather Weaver's claim that the fossil plants associated with it were of species known from the much younger Coal Measures. It was probably no coincidence that Weaver's paper was thus stopped in its tracks only three weeks before Murchison's presidential address, in which he announced the first results of his research in the Welsh Borderland. Certainly Weaver's claim, if it had appeared in the *Transactions* shortly afterward, would have been in embarrassing contrast to Murchison's insistence on the distinctive character of the Greywacke strata he had been studying, and of their fossils. If such strata elsewhere contained not only coal seams but even the same *species* of plant fossils as the true Carboniferous coal strata, Murchison's discoveries would be reduced to the status of local curiosities.

For Murchison, in striking contrast to Weaver, had found no such traces of plants, let alone coaly deposits, in the Transition strata of the Welsh Borderland. Far from treating his own area as an exception in this respect, he made it at once the rule. This became explicit, at least in private, only a few weeks after he read his first report to the Geological Society. When a Radnorshire physician sent him a local press cutting about recent prospecting for coal in his county, Murchison wrote back in high indignation at the "gross state of ignorance" of the mineral surveyor who had been responsible. The surveyor deserved "severe Castigation" for ignoring "the whole of the geological facts" that Murchison himself had determined:

> In not one of these [Greywacke formations] is there a trace of coal nor of vegetable substance out of which Nature might have elaborated that mineral, altho' the strata *abound* in Organic remains. These organic remains are the medals by which we read off the age of these deposits,

46. Sedgwick to Murchison, 3 February 1837 (GSL: RIM, S11/119).
47. Herries-Davies (1980).
48. Sedgwick to De la Beche, 17 December 1834 (NMW: DLB).

and they prove to us to be of Species which are never found in the Carboniferous Series.[49]

If the absence of any trace of coal were a general feature of the Greywacke or Transition strata, and not just peculiar to the Welsh Borderland, it would put a scientifically authenticated limit to the current scramble to prospect for new coal deposits. It would focus attention on areas where such prospecting had at least some chance of success, and prevent landowners from wasting their money by putting down boreholes in areas where there was no chance at all. But the use of this science-based criterion would depend on the ability of prospectors to recognize when they were in fact among the Transition strata and not among Carboniferous or still younger rocks. Murchison was convinced that this could be achieved by following the principles of the "father" of English geology. The "medals" of nature—Smith himself had often used the traditional archaeological metaphor—were an infallible guide to the true position of strata. Murchison's claim that he had discovered a distinctive fossil fauna in the upper Greywacke strata was therefore not only an important extension to the history of life that was being pieced together by geologists. It was also a claim with major economic implications. Furthermore, if it proved valid, it would certainly consolidate the authority of geological gentlemen, like himself, over and above that of scientifically ignorant mineral surveyors. Smith's principles were thus to be deployed against Smith's own social equals.

The other leading figure whose recent work would have qualified him to be an "unbeliever" in the Weaver affair was Lyell, who at this time was preparing the third and culminating volume (1833) of his *Principles of Geology*. In that work he put forward a theoretical model of great sophistication to account for constant but piecemeal change in the composition of the faunas and floras at the earth's surface, in correlation with its ever-changing physical environments. He was sufficiently confident of the general validity of this model to anticipate that it would yield a method of giving quantitative, and perhaps even absolute, dates to geological formations.[50] It followed from the model that while particular species might have had long ranges in time, an assemblage of many species would not remain unchanged for long; two formations deposited at widely separated periods would be expected to have few if any species in common. Lyell would therefore have been highly skeptical of Weaver's reports that the Transition strata of Ireland contained Coal Measures plants, and he would have been likely to attribute it to some mistake in the geological interpretation of the area, if not in the identification of the fossils.

Whatever the truth behind the suppression of Weaver's paper, the underlying issue is clear. Murchison and Lyell were both staking their future

49. Murchison to Davies, 14 June 1833 (NMW: bound with Davies's copy of Murchison 1839). Radnorshire lies north of Builth (see fig. 3.5); Aaron Wall Davies (b. *c.* 1794) was a physician in Presteign. Murchison seems to have been criticizing prospectors for coal in the older strata of the Borderland even on his first field trip there two years earlier: Murchison, notes dated 14 July 1831 (GSL: RIM, N58, pp. 26–30). I am indebted to Dr Secord for drawing my attention to both these important sources.

50. Lyell (1833); Rudwick (1970a, 1978). The model posited the piecemeal "creation" and extinction of separate species, *not* the transmutation (in modern terms, evolution) of one into another.

reputations in science on the outcome of their respective major research projects; both projects, for different reasons, were faced with an awkward major anomaly if Weaver's report from Ireland were authentic. The two London geologists were making a defensible act of judgment in doubting that report, since the technical difficulties of working with the Greywacke strata were such as to make a mistake on Weaver's part a not unlikely explanation of the anomaly. What Sedgwick found indefensible was not the skepticism of the "unbelievers" but their action in suppressing Weaver's work. As he commented to De la Beche, "the paper *ought* to have been printed & if there had been any mistake the blame wd. have fallen on the author & not on the Socy."[51] This was a fair comment, for the Geological Society, like other learned bodies, had from the beginning taken care to dissociate itself publicly from the opinions of its authors. Every issue of the *Transactions* bore a notice to this effect, which was an overt expression of the Society's officially nontheoretical policy. But on a matter with such far-reaching economic implications, there were evidently some members who felt that the usual disclaimer was not enough to protect the Society's public reputation as the fount and guardian of reliable knowledge of British geology.

4.5   The Proprietor of the Welsh Borderland (Summer 1833 to October 1834)

The suppression of Weaver's paper proved ineffective anyway. His claim, if not the full evidence for it, already stood on public record in the Geological Society's *Proceedings*. When later in 1833 Murchison read the new and enlarged edition of De la Beche's *Manual*, which he doubtless did with great attention, he would have found that the anomaly of the alleged Greywacke coal was proving to be hydra-headed. Far from expunging Weaver's report on the grounds of its dubious authenticity, De la Beche had defiantly added to it a whole series of other reports of anthracitic coal in the Greywacke of Normandy, Brittany, the Vosges, and Baden (see fig. 3.6). Above all, De la Beche issued an implicit challenge to Murchison's claims. Anthracite had long been known in the Greywacke of north Devon and was presumably of plant origin: hence "there seems no good reason," he pointed out, "why grauwacke should not contain its coal-beds as well as other great deposits."[52] As was appropriate between gentlemen, however, De la Beche's challenge remained discreetly veiled.

So did Murchison's response. Reading in De la Beche's book that the French edition had not been ready in time for its extra material to be included, Murchison hastened to write to him about it. "I am highly desirous," he told him, with at least an appearance of polite esteem, "that everything in your excellent manual should appear on the Continent in the best possible form." Not knowing he was already too late, he explained to De la Beche what he hoped to achieve. "It is not merely the Upper Greywacke the relations of

---

51. Sedwick to De la Beche, 17 December 1834 (NMW: DLB).
52. De la Beche (1833a, p. 427).

which I am desirous of illustrating, but the whole mass of strata from the Mountain Limestone downward, which part of the Series has in my conscience no development in any part of the World equal to that seen in *my* regions."[53] His research was so far confined to the Welsh Borderland. But the implicit claim was that *his* newly characterized series of Greywacke formations was potentially of much wider validity. The proprietorial attitude was quite as marked as that of any of his prospective patrons. For a group of local landowning aristocracy and gentry, headed by Lord Clive, the Earl of Powis, had now urged him to write a book on what he told them was a region of great scientific significance.[54] This enlarged Murchison's stake in his research in a new and welcome direction, for it promised him social rewards in addition to those he hoped to earn from his fellow gentlemen of science.

Early in 1834 he revealed to those gentlemen what he had told De la Beche in advance. In papers read to the Geological Society at two successive meetings he outlined his conclusions, first on the Old Red Sandstone and then on the Greywacke formations (fig. 4.8, B).[55] Below a residual capping of Mountain Limestone, the Old Red Sandstone—its estimated thickness now more than doubled to no less than 10,000 feet—was divided into three parts, including a middle part with fossil fish. Murchison insisted on the virtual absence of any plant fossils, "as demonstrating the hopelessness of ever finding any workable quantity of coal in the old red sandstone of this part of the kingdom."[56] Despite that final qualification, he clearly meant the claim to apply to the Old Red Sandstone generally and, a fortiori, to the Greywacke rocks below it. The lowest part of the Old Red Sandstone, according to Murchison, contained a few fossil shells and formed "beds of passage" with complete conformity downward into the "upper grauwacke series," which in this revision was divided into four parts (fig. 4.8, B, I–IV). Among this varied sequence of shales and sandstones were two major limestones, of which the "Wenlock and Dudley limestone" was already celebrated among collectors for its spectacular fossils. Murchison's sequence was closed off at the bottom by a "vast system" of strata without fossils (fig. 4.8, B, V), and thereby implicitly separated from Sedgwick's still lower sequence in northwest Wales.[57]

Murchison's research was duly summarized by Greenough a month later, when as president he reviewed the past year's work in his anniversary address. Greenough made no reference to its possible wider significance in the search for coal, for that had hardly yet been made explicit. But he welcomed Murchison's achievement, and with characteristic internationalism, he expressed the hope that it would provide the "clue" by which the equivalent rocks throughout Europe and North America could be rescued from being a "mere scene of confusion."[58] Murchison could hardly have asked for more generous treatment by the president. Later in the year, and more predictably, he was publicly praised by Lyell for having taken the "first great step" in

---

53. Murchison to De la Beche, 12 December 1833 (NMW: DLB).
54. Thackray (1978b, p. 63).
55. Columns (A) and (B) in fig. 4.8 are based on Murchison (1834c) and Secord (1986); for column (C), see fig. 3.2, Somerset column; column (D) is based on De la Beche's letters to Sedgwick in November 1834, and De la Beche (1834b).
56. Murchison (1834a, p. 12), read at GSL on 8 January.
57. Murchison (1834b), read at GSL on 22 January.
58. Greenough (1834, pp. 50–51), read at GSL on 21 February.

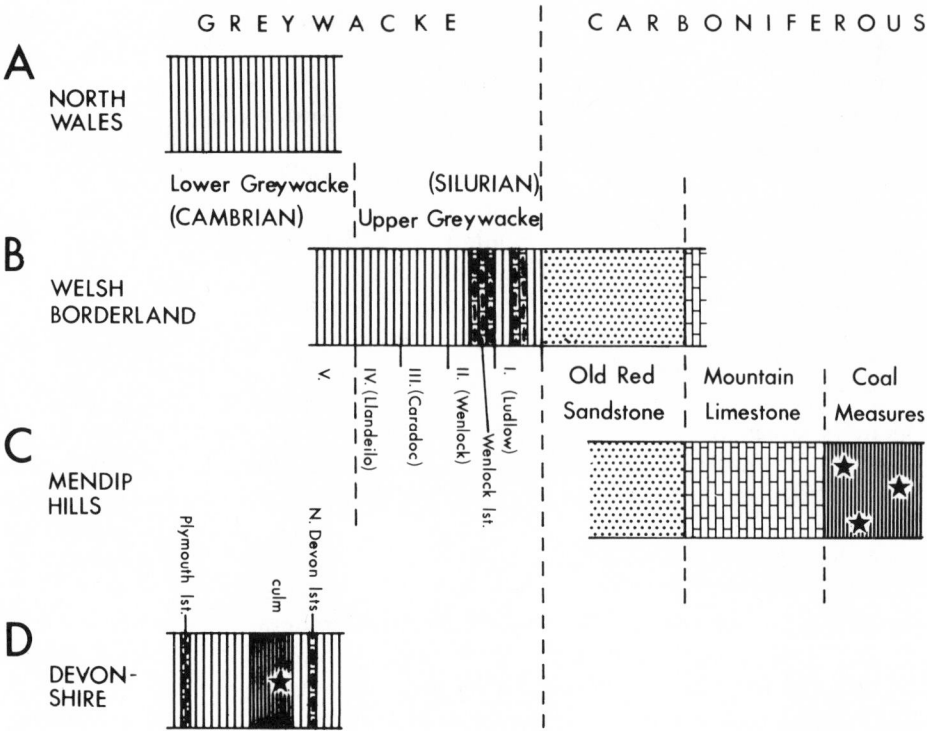

Fig. 4.8.   Diagrammatic columnar sections to show correlations of Greywacke and Carboniferous strata in 1834. The correlation between (A) and (B) is that agreed between Sedgwick and Murchison in the summer of 1834 and accepted by other geologists; the correlation between (B) and (C) was consensual and unproblematic. The correlation between (D) and the others became highly controversial as soon as De la Beche proposed it in public in December 1834; note the huge discrepancy in the position of the "coal plants" (shown by star symbols) in (C) and (D). This and all similar columnar sections later in this book are drawn on their side in order to emphasize that they are historical interpretations of early nineteenth-century opinions about sequences that also represented the flow of *time* (from left to right). The scale is roughly proportional to contemporary estimates of the thickness of the formations; Murchison estimated the Old Red Sandstone in (B) at 10,000 feet. Names in parentheses are those proposed by Murchison and Sedgwick later, in the summer of 1835.

deciphering the older strata, and his results were duly incorporated into the first full revision of Lyell's *Principles of Geology*.[59]

Before that appeared, however, Murchison took steps to ensure that his achievement would also be properly recognized in the first lecture course that Lyell's successor was due to give at King's College, London. Lyell had given up his professorship after only two years because he reckoned it did not yield enough income to justify the effort, and he now hoped to earn more from authorship.[60] A less wealthy geologist, however, had been happy enough

59. Lyell (1834, 1: preface, dated 20 May 1834; 4:314, table II). This so-called third edition contained what was only the second edition of the stratigraphical part of the *Principles*.
60. Rudwick (1975a).

to take over from him (fig. 4.9). John Phillips (b. 1800), the son of an excise officer, had, like Greenough, been orphaned at an early age. He was brought up in Yorkshire by his uncle, William Smith. Although he had no formal education beyond the age of fifteen, Smith trained him as a surveyor, and Phillips had acted for several years as his assistant. In particular he had helped with the cataloging of the fossils that Smith had used in his mapping, and this had laid the foundations for Phillips's later research on fossils and fossil-based geology. In 1825, thanks to the patronage of William Vernon Harcourt (b. 1789), a son of the then Archbishop of York and an influential local man of science, Phillips had been appointed Keeper of the museum of the Yorkshire Philosophical Society (founded in 1822), which was rapidly becoming one of the most important of English provincial scientific bodies. He supplemented the small income of this post in the city of York (initially only £60 a year) by building up a successful circuit as a scientific lecturer throughout the north of England.[61]

When the British Association held its inaugural meeting in York in 1831, Harcourt was its main driving force at the higher social levels, but Phillips had been largely responsible for its organizational success. The following year, when it became clear that the Association was a viable body, Harcourt had got Phillips a part-time salaried post as its assistant general secretary; at £100 a year, this doubled his regular income. Working closely with the two honorary general secretaries, he came into personal contact with all the leading men of science in Britain. He was soon on friendly terms with most of

Fig. 4.9.   John Phillips in 1840, from a portrait by Alexander Craig.

61. Edmonds (1975a); Morrell and Thackray (1981, pp. 439–44). John Phillips was no relation of Conybeare's earlier collaborator William Phillips. Fig. 4.9 is reproduced from a painting by Alexander Craig of Glasgow (NPG: 23461; present location of original unknown); other paintings in the same series are dated 1840, the year BAAS met in Glasgow.

the inner circle of the Geological Society, and despite his rather humble origins, he was treated by them as almost—if not quite—their social equal. When Lyell resigned his professorship at King's College, the influence of the London geologists ensured that it was Phillips who was appointed to succeed him. But by then he had established his provincial base so securely that he chose to remain in York, visiting London only to give the required annual lecture course.[62]

Like Buckland's and Sedgwick's courses at the ancient universities, Phillips's could be expected to draw an audience more of established men of science than of mere students, so it would be an important occasion for influencing the opinions of the metropolitan scientific elite. In contrast to most of his contemporaries, Phillips organized his review of stratigraphical geology not from top to bottom but along true chronological lines, beginning with the oldest strata. When Murchison sent him a summary of his own latest results, Phillips told him how he would incorporate them in his lectures. He would deal first with the Primary rocks; then with Sedgwick's older Greywacke rocks in Cumbria and with what Phillips evidently regarded as their equivalents, the "Devonshire Tracts"; and "finally your Shropshire country to make the climax & fix the classification of old rocks for the future."[63] Murchison would have been gratified to hear this, for Phillips's plan reflected exactly his own view of the place of his research in the larger scheme of things.

A high priority for both Murchison and Sedgwick was now to link their respective sequences more securely into a single order of successive formations. In the summer of 1834 they worked together in the field, covering wide areas of Wales and the Borderland. In particular, they traversed from Murchison's area near Welshpool to Sedgwick's area near Bala (see fig. 3.5). They concluded to their own satisfaction that Sedgwick's sequence was indeed lower than Murchison's, with only a minor overlap between them (fig. 4.8, A and B).[64] Murchison now felt confident enough of the distinctiveness of his "upper grauwacke series" to issue a prospectus for his projected book and to call for subscribers.[65] He gave a brief progress report at the British Association meeting in Edinburgh in September, and published a fuller account in the scientific monthly edited in that city by the doyen of Scottish geologists, Robert Jameson (b. 1774).[66]

At three points this article went beyond the papers Murchison had read to the Geological Society nine months earlier. First, as a direct result of their joint fieldwork, he now explicitly located Sedgwick's sequence below the lowest formation he had himself defined, effectively insulating his own sequence from any older formations that Sedgwick might in due course describe. Second, he stated unambiguously that there were no remains of land plants

62. Murchison had earlier hoped to get Phillips a similar position at the London University (later, University College London). Phillips did give one course of lectures there in 1831, but no permanent post had materialized (Edmonds 1975b).

63. Phillips to Murchison, 10 April 1834 (GSL: RIM, P14/9). On Lyell's lectures the previous year, see Rudwick (1975a).

64. Secord (1986).

65. Thackray (1978b, p. 64).

66. Murchison's paper (1835a) to BAAS was reported in *Ath.*, no. 360, p. 698, and in *LG*, no. 922, p. 637 (both for 20 September 1834). The article (Murchison 1834c) was in the October issue of *ENPJ*; Jameson, who had studied with Werner, was the professor responsible for geological teaching in Edinburgh.

below the Mountain Limestone, and no traces even of coaly material below the Old Red Sandstone. This made public his claim to have found a reliable if negative guide in the search for coal, and implied his rejection of all the contrary cases that De la Beche had cited (see fig. 4.8, columns C and D). And third, he now emphasized explicitly—also for the first time in public— the value of establishing *"types"* for major parts of the total sequence of stata, from the well-known Secondary formations downward. It was now clear to other geologists that he was proposing his own Borderland sequence as the type for the upper part of the Greywacke sequence, and hence that he believed it would be found valid far beyond that region.

4.6   De la Beche and the Devonshire Survey (1834)

When Phillips quite casually equated the "Devonshire Tracts" with Sedgwick's older Greywacke strata in Cumbria (and by implication with those in north Wales too), he was merely reflecting a general opinion about the geology of southwest England. It was as "Grauwacke" that much of that region had been colored on Greenough's great map (1820), which, by integrating its author's fieldwork with that of many colleagues and correspondents, had long since become the indispensable basis for more detailed research (fig. 4.10).[67]

Greenough had also marked the string of well-known granite areas, from the bleak uplands of Dartmoor westward through Cornwall to Land's End. These and the schists and slates surrounding them, for which Greenough used the miners' term "Killas," were of great interest because of the valuable mineral veins that traversed them and made the region one of the most important sources in the world for nonferrous metals. The format of a map allowed Greenough to be noncommittal on many matters of theory. But he had probably believed, in common with most other geologists, that the granites were true Primary rocks, forming the core of the region with the Killas and Greywacke overlying them. Certainly this was the view adopted by Sedgwick at that time: in the paper that emerged from his earliest season of fieldwork, he had referred unambiguously to the "primitive ridge of Devonshire and Cornwall," and claimed explicitly that "granite is the fundamental rock of the whole region." But while most geologists in the early 1820s would have agreed with Sedgwick that the facts were "at variance with the Huttonian hypothesis," a decade later he and they had changed their minds.[68] By that time James Hutton's old "Plutonic theory" of the origin of granites as hot fluid intrusions was enjoying a revival, not least through Lyell's persuasive efforts. This implied that in Devon and Cornwall the apparent division of the older strata into Killas and Greywacke might be due to nothing more than what Lyell termed the *metamorphic* alteration of the Greywacke where it was near the granite.[69]

67. Fig. 4.10 is redrawn from Greenough (1820), but without distinguishing the Killas from the Greywacke; the bands of culm and limestone were marked by Greenough as "carbonaceous slate" and "calcareous slate," respectively.

68. Sedgwick (1820, title and pp. 96, 115).

69. On Hutton (1795) and his "Plutonic theory," see, e.g., Dott (1969) and Gerstner (1971). On metamorphism and its role in Lyell's geological theorizing, see Rudwick (1970a).

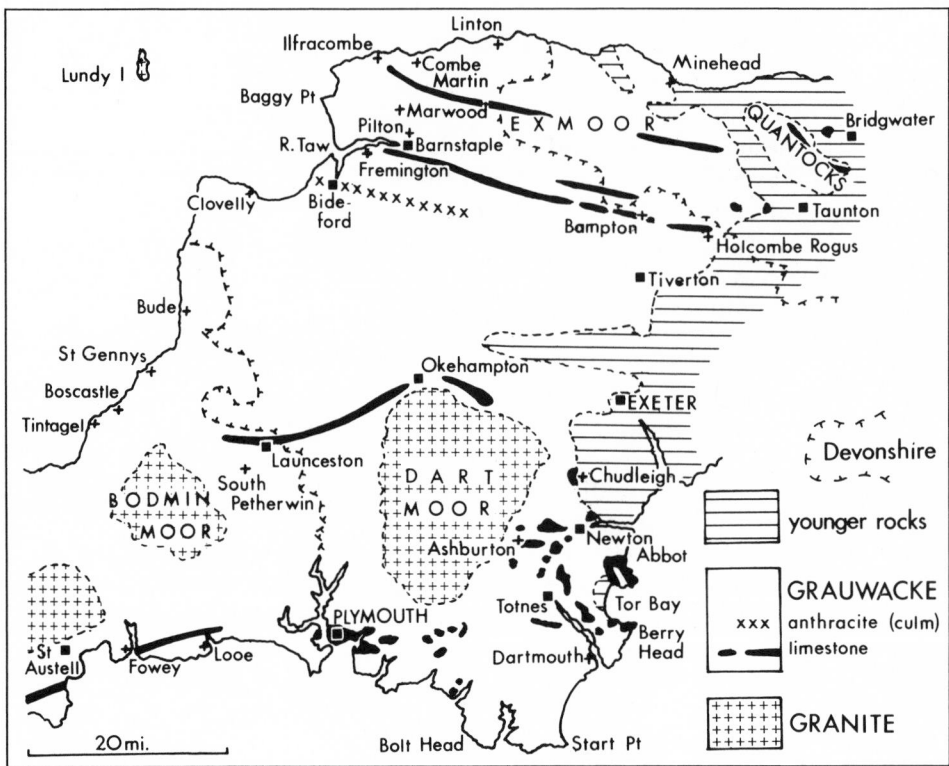

Fig. 4.10.   The geology of Devonshire and adjacent areas, redrawn from Greenough's map (1820), to show the areas of "Grauwacke, Killas" and other older (pre-New Red Sandstone) rocks; all mineral veins, etc., omitted. This map is also used to show place names mentioned in the subsequent text.

The swing of opinion on this issue was displayed vividly at the British Association's meeting in Edinburgh in September 1834, shortly before Murchison revealed his latest conclusions on the Welsh Borderland. Henry Boase (b. 1799), a physician in Penzance and a keen Cornish geologist, had just published an ambitious *Treatise on Primary Geology*, which the widely traveled Lyell later characterized bluntly as "an attack on the Plutonic theory by one who was never out of Cornwall."[70] But at the meeting, when Boase read a paper summarizing his conclusions, it was not Lyell but Sedgwick who acted as spokesman for the metropolitan gentlemen of science. The provincial geologist was no match for what one reporter correctly perceived as "the phalanx arrayed against him".[71] The attack was not however an arbitrary exercise of cultural firepower. Boase showed himself quite unaware of the technical problems, such as those posed by jointing and slaty cleavage, that lay in wait for any geologist bold or rash enough to tackle the older rocks (§3.3).

As for the geology of Devon and Cornwall, the outcome of Boase's public defeat at Edinburgh was to make it clear that the granites and their

70. Boase (1834); Lyell to Fleming, 7 January 1835 (printed in K. Lyell 1881, 1:444–46).

71. *LG*, no. 922 (20 September 1834), p. 635. See also *LG*, no. 921 (13 September 1834), p. 632; *Ath.* no. 360 (20 September 1834), p. 698; and Morrell and Thackray (1981, p. 461).

surrounding rocks would have to be treated as a separate problem from that of unraveling a sequence of formations in the Greywacke. If the granites were intrusive, and if the Killas rocks were the products of metamorphism, southwest England would have no "Primary Geology" at all![72]

At the opposite end of the scale, there were known to be large areas of Secondary strata in the east of the region. But if the focus of attention were on the Greywacke, these strata—from the New Red Sandstone upward—could simply be ignored, because they overlay the older rocks with an unconformity so gross that it could rarely be a source of confusion. Ignoring, then, this concealing cover of Secondary strata to the east, a large part of the rest of southwest England had been colored by Greenough as Greywacke, or more precisely, as "Grauwacke, Culm &c." The term *culm* was used by miners for any impure coaly rock; Greenough's adoption of the word expressed his conviction that the seams of anthracitic coal in this region were an integral part of the Greywacke. In this opinion he was followed without hesitation by De la Beche and by most other geologists, except Murchison. Most of these seams of culm were much too thin and impure to be of any commercial value; the few exceptions were worked here and there along a narrow east-west strip in north Devon (fig. 4.10).

Greenough had also marked within the Greywacke of Devon and Cornwall four discontinuous bands of thin, impure limestones, quarried locally for lime burning (fig. 4.10). But in south Devon, particularly around Plymouth and Tor Bay, there were scattered patches of much more massive limestone with quite abundant fossils. The best known of these was at Plymouth. Its fossils had been described to the Geological Society by Richard Hennah (b. 1756), an elderly but keen fossil collector who was chaplain to the military garrison and had the limestone literally on his doorstep.[73] Greenough had colored this and the other limestone patches in south Devon as integral parts of the Greywacke, implying that they were roughly equivalent to the "Transition Limestone" of other regions (e. g., the well-known limestone of Wenlock and Dudley in the region that Murchison was later to make his own). Some years later, in a paper on southeast Devon, De la Beche had assigned the patches around Tor Bay to the Mountain Limestone, maintaining that they formed outliers overlying the Greywacke unconformably. But Greenough was unconvinced. He doubted the field evidence for an unconformity and objected that the fossils (at least those from the limestone at Newton Abbot) were "either such as are found no where else, or such as are common both to the transition limestones and to the mountain [limestone]."[74] No one knowing that Greenough was using such arguments could fairly accuse him of ignoring fossil evidence in geology. But equally it was clear that the fossils of the south Devon limestones were somewhat peculiar. Until they were studied more

72. Except perhaps for two small areas on the south coast, namely the schists around Bolt head and Start Point in Devon, and the serpentine rocks of The Lizard in Cornwall.

73. Hennah (1817, 1828, 1830). His locally published book (Hennah, n.d.) had rather crude lithographic illustrations of some of the fossils, but the specimens were not identified by name.

74. Greenough to De la Beche, 28 July 1829 (NMW: DLB), reacting to De la Beche's paper (1829) read at GSL on 16 November 1827. Smith (1815) had also colored the south Devon limestone patches as Mountain Limestone, but he did not mark the other lines of limestone nor the line of culm.

closely they could hardly count as evidence for the correlation of Devonshire rocks with any others.

All in all, Devon and Cornwall remained an underdeveloped region for geology, in spite of all the interest in the mining areas. It was known to be one of the largest regions of Greywacke rocks in Britain, but it had yet to be tackled systematically. A systematic survey, however, was just what De la Beche was beginning to carry out. In 1831 the income he received from his Jamaican estate had declined suddenly and disastrously. This had imperiled his whole career as a gentleman of science, and he resigned from being a secretary of the Geological Society—a decision that was minuted tactfully as being "on account of urgent occupations."[75] His *Manual*, as one of the best-selling introductory books on the science, must have helped his financial situation at least a little. But he had also tried to increase his longer-term income by taking the first steps toward getting his research subsidized by government money.

He had already taken advantage of the existence of large-scale topographical maps of Devonshire—the Ordnance Trigonometrical Survey had begun its work in southwest England—to make his own geological map of a part of that county for his own interest. But in 1832, using Sedgwick as an intermediary, he had submitted to the Board of Ordnance a proposal that he should be paid a fee of £300 for completing a geological survey of virtually the whole of Devon and some adjacent areas by coloring eight sheets of the Survey's map. Lord Grey's Whig administration was sympathetic, and De la Beche had been appointed to the ad hoc position of geological surveyor within the Ordnance Survey. He was expected, however, to publish at his own risk whatever memoir or sections he might deem necessary to explain the map itself; he was paid no expenses and no salary, and he had to wait until each sheet was complete before getting even a proportion of his fee.[76] Indeed, all that the Board of Ordnance had undertaken to do at government expense, apart from paying De la Beche his fee, was to engrave inconspicuous geological boundaries and other symbols onto their uncolored topographical maps. De la Beche's geological maps were thus not to be published formally or officially by the Survey; he was left to make private financial arrangements with James Gardner, the Survey's mapseller in London, for the making and selling of special copies at a higher price, with the geology displayed in color added by hand.[77]

This ad hoc and niggardly arrangement was in striking contrast to the generous government support already being given to an official geological survey of France. What made the contrast even more galling for De la Beche was that the inspiration for the French survey had come from Britain, indeed from Greenough. It was the publication of Greenough's map that had prompted the senior Parisian geologist, André Brochant de Villiers (b. 1772), to suggest

75. Council of GSL, 16 November 1831 (GSL: BP, CM 1/3, p. 136).

76. For this summary I am greatly indebted to Dr McCartney, who has kindly allowed me to read part of his unpublished work on the early history of the Geological Survey. On the Ordnance Trigonometrical Survey, see Seymour (1981).

77. Colby to De la Beche, 9 April 1832 (NMW: DLB). This arrangement explains the uncertainty surrounding the exact dates at which De la Beche's sheets became de facto available to other geologists.

a geological survey to the French government. That proposal had been approved, and in 1823 Brochant and two younger colleagues had spent some six months in Britain—of course at government expense—to study British geology and Greenough's methods. In 1825 they had begun an intensive program of geological mapping; by 1834 they had covered the whole vast area of France, and their map was almost complete, apart from the engraving.[78]

The geological mapping of France was intended primarily to survey the mineral resources, particularly coal, which might enable the country to compete better with Britain in the new industrial economy. It is therefore not surprising that the government's instructions for the survey had focused explicitly on the known regions of older rocks (see fig. 3.6). The work had been divided simply between the two younger surveyors; Pierre Dufrénoy (b. 1792) was to take the western regions, Léonce Élie de Beaumont (b. 1798) the eastern, while Brochant supervised them both and later accompanied them on joint expeditions.[79] Of the three, Élie de Beaumont had become the English geologists' chief contact with the French, as well as being one of the most respected geologists on the Continent. In addition to his more technical teaching at the École de Mines, his annual course of lectures at the Collège de France gave him an influential platform as a leading interpreter of geology in France.

While the French geologists conducted their summer fieldwork with full government support, spending the winters comfortably in Paris studying their specimens and discussing their progress with each other, De la Beche had begun his operations in Devonshire single-handed. His urgent need to earn his fee obliged him to do fieldwork not just in the pleasant weather of the summer season but throughout the year, and he was unable to afford frequent coach trips to London. He remained a member of the Council of the Geological Society, but could rarely attend the meetings. He felt deeply the isolation that his financial situation forced on him, because it was in such contrast to the freedom enjoyed not only by the other leading members of the Society but also until then by himself. In Greenough, however, De la Beche found a staunch ally. They shared not only a suspicion of speculation and a respect for what they regarded as plain observational facts, but also an appreciation of the value and importance of geological mapping. Once De la Beche had been commissioned to survey Devon, Greenough adopted the role of his patron and protector. While Thomas Colby (b. 1784), the director of the Trigonometrical Survey, was De la Beche's superior and acted as his formal contact with officialdom, Greenough acted unofficially as his London spokesman at the Geological Society. The newly impoverished De la Beche was understandably grateful for the patronage of the older and wealthier man, and his attitude to Greenough bordered on the deferential.

De la Beche later recalled that by the spring of 1834 he had completed the geological coloring of the southern half of his area, which he had begun

---

78. Brochant de Villiers (1835). Brochant was also the editor of the French edition of De la Beche's *Manual.* De la Beche had sent a copy of his first sheet to Paris, as a sample of what he hoped to do, even before his official appointment (Élie de Beaumont to De la Beche, 8 April 1832 [NMW: DLB]).

79. Bequey to Brochant de Villiers, 29 June 1825; Bequey to Elie de Beaumont, 15 July 1825 (EMP: LEB, dossier IV, 8; dossier I, 20); Brochant de Villiers (1835).

even before his appointment.[80] As he began work on the northeasterly area, he wrote from Bridgwater in Somerset to tell Greenough of his progress and mentioned that he was looking out for the "upper Grauwacke" formations that Murchison had recently described.[81] No positive report followed that summer, but this was in no way surprising. It merely confirmed the general opinion that the "Devonshire Tracts," as Phillips had called them, were composed of *older* Greywacke. In other words, if the area was ripe for annexation by anyone, it was expected to be Sedgwick, not Murchison. Yet that inference left the anthracite in the Greywacke of north Devon as a glaring anomaly, at least for Murchison. The stage was therefore set for a major argument, the repercussions of which were bound to reach far beyond southwest England.

80. De la Beche to Greenough, 21 January 1835 (ULC: GBG). He submitted sheet 22, covering part of southeast Devon, with his proposal: De la Beche to Colby, 9 May 1832 (NMW: TS, I, pp. 61–65).

81. De la Beche to Greenough, 30 March 1834 (ULC: GBG).

Chapter Five

THE CONTROVERSY GOES PUBLIC

5.1 De la Beche's "Interesting" Fossil Plants (October and November 1834)

When in the summer of 1834 De la Beche moved his survey westward into north Devon, he was well aware that he would at last have a chance to study at firsthand the culm or anthracite there, which had already figured in his gentlemanly shadowboxing with Murchison. Conversely, Murchison was well aware that the Devonshire case threatened his whole ambitious project, unless some grounds could be found for doubting its authenticity. Weaver's Irish case might be safely out of sight, if not out of mind.[1] But if the Devonshire case were confirmed, Murchison's position would certainly lose plausibility, for De la Beche's competence as a field geologist would not be so easy to impugn as Weaver's had been. Above all, Devonshire was not on the other side of the Atlantic, or even of the English Channel or the Irish Sea. Of all the alleged instances of coal in the Greywacke, it was by far the nearest to the Welsh Borderland, over which Murchison had now in effect claimed exclusive geological rights.

On the west coast of Devon, near Bideford, De la Beche duly found not only the culm or anthracite itself, which was mined there on a small scale for local use, but also some fossil plants in strata closely associated with it

---

1. That the Irish anomaly was still on Murchison's mind is suggested by the fact that he borrowed the whole file on Weaver's paper from the Geological Society: Lonsdale to Murchison, 11 July 1834 (GSL: RIM, L13/9).

(fig. 5.1).[2] De la Beche sent a box of his Culm fossils to the Geological Society, asking William Lonsdale (b. 1794), its curator and librarian, to get them authoritatively identified. Lonsdale was the Society's key administrator, handling its day-to-day business on behalf of the officers and Council, and acting informally as an information exchange for all the leading members. Like Murchison, he had served in the Napoleonic wars, but afterward he retired to his native Bath as an ex-officer and got a curatorial post in the local museum. When in 1829 he was offered a similar position at the Geological Society, he had welcomed it as a promotion, because although he was paid only £125 a year it did at least put him at the center of British geological life. After his move to London he had no time—and too poor health—to do much further fieldwork, but his museum responsibilities had enabled him to become a leading authority on fossils, particularly on fossil corals.

Specialization in palaeontology was already such, however, that De la Beche needed to have his fossil plants identified by a more appropriate expert. He therefore asked Lonsdale to send them on to John Lindley (b. 1799), who was Lonsdale's counterpart at the Horticultural Society, the professor of botany at the London University, and co-author of a massive new work on fossil botany.[3] De la Beche explained that he was sending a short report on the fossils to Greenough as president, to be read to the Society once the specimens had been identified. Fully aware of the effect his report would have, he told Lonsdale that "as the whole [plant-bearing Culm formation] forms an undoubted part of the Grauwacke series, the specimens, as I need not tell you, acquire considerable interest."[4] The understatement was impressive. Three weeks later he sent Sedgwick a summary of his results, for Sedgwick to incorporate into his and Conybeare's proposed book on all the older rocks of Britain (§4.2). "I always like speaking to eye when I can," he wrote, and did so by sketching a traverse section from Dartmoor northward to the Bristol Channel. Although he called it "a most abominable piece of exaggeration," his section was in fact drawn with characteristic caution, and Sedgwick had to mark it with the bolder extrapolations implicit in De la Beche's verbal explanation in order to see its full implications (fig. 5.2).[5]

De la Beche reported in effect that the whole of Devon was quite simply a sequence of strata ranging from the oldest near the south coast to the youngest near the north coast. He referred to the strata that included the Plymouth limestone of south Devon as "the lowest part of the grauwacke series" in the region. These strata (fig. 5.2, d) had been intruded by the Dartmoor granite. North of Dartmoor they were succeeded by "a great mass of arenaceous grauwacke" showing strong, small-scale contortions in the cliffs on the west coast

2. Fig. 5.1 is reproduced from part of sheet 26 of the Ordnance Survey's topographical map (scale of original: one inch to one mile) colored geologically by De la Beche, probably in 1834 after his first fieldwork there. His sketch of the cliffs was drawn in the margin of this map (BGS: map coll.).

3. Lindley and Hutton (1831–37); the work was dedicated to Murchison, who had first suggested it. William Hutton (b. 1798) was an insurance agent in Newcastle, and a keen collector of the fossils from the coalfield there.

4. De la Beche to Lonsdale, 11 October 1834 (GSL: BP, LR 1/90). For convenience, the formation containing the seams of culm or anthracite will henceforth be termed *Culm*, with an initial capital.

5. Fig. 5.2 is reproduced from De la Beche to Sedgwick, 5 November 1834 (ULC: AS, IA, 123).

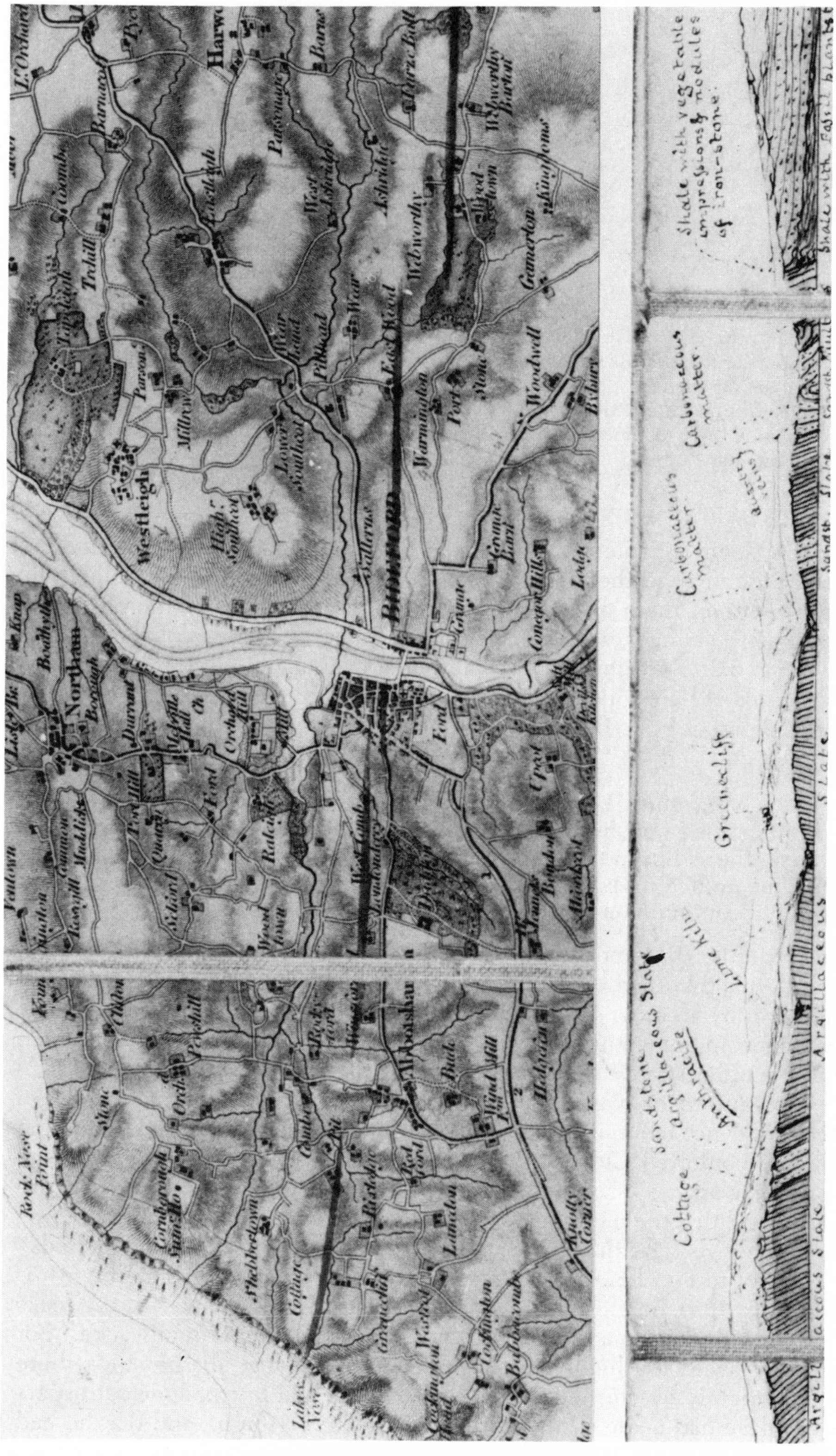

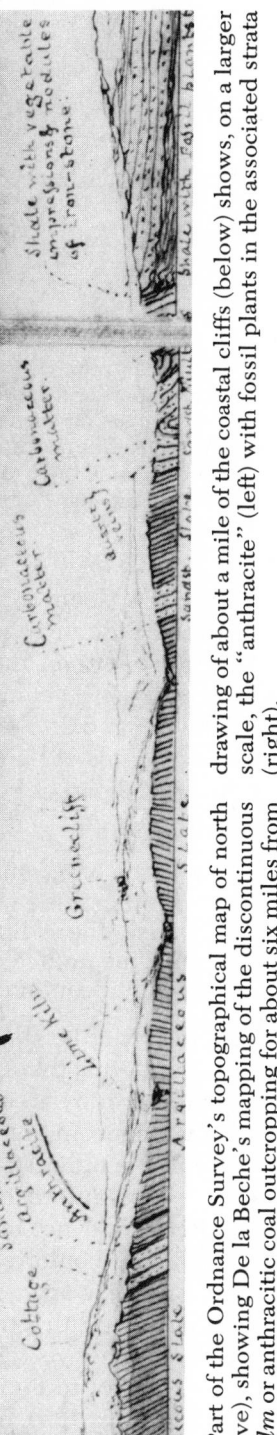

Fig. 5.1.   Part of the Ordnance Survey's topographical map of north Devon (above), showing De la Beche's mapping of the discontinuous seams of *culm* or anthracitic coal outcropping for about six miles from the coast eastward through Bideford, in an area of Greywacke. His drawing of about a mile of the coastal cliffs (below) shows, on a larger scale, the "anthracite" (left) with fossil plants in the associated strata (right).

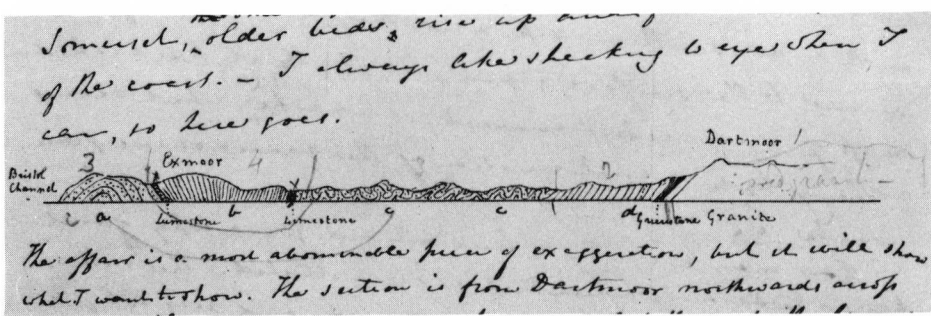

Fig. 5.2.   De la Beche's traverse section of Devonshire, from Dartmoor (at right) northward to the Bristol Channel (left), as sketched for Sedgwick in November 1834. The numbers and extrapolated structural lines were added by Sedgwick in pencil after receiving the letter. The position of the anthracite and fossil plants of Bideford, as reported to Sedgwick a little later, was in the upper (left) part of (c). Note that De la Beche equated the two bands of limestone running across north Devon on Greenough's map (see fig. 4.10).

(fig. 5.2, *c*; see frontispiece). This was in fact the formation that included the culm of Bideford and the fossil plants De la Beche had collected, though he did not mention them at this point. Still further north, and therefore—he inferred—still younger, was "a mass of rocks composed chiefly of argillaceous slates" (fig. 5.2, *b*), forming the upland area of Exmoor, with one of Greenough's bands of limestone near its southern margin (see fig. 4.10). Now came the crucial part of De la Beche's report, which in effect determined the whole interpretation:

> As I went north I expected to find some [still] newer beds come in, but in this I was disappointed, for the whole [series] got turned up with a southern dip, and the limestone again came to the surface, and beneath it the arched mass of arenaceous grauwacke (*a*), which seems nothing but a continuation of the arenaceous grauwacke *c c*.

In other words, De la Beche inferred—as Sedgwick duly sketched in on the diagram—that the strata *b* were folded into a tight U-shaped trough, succeeded to the north by an open archlike saddle. The strata *b* were thus inferentially the youngest in the whole sequence, and the anthracitic culm was well down within the older strata *c* (equivalent to *a*).

This inference was probably based purely on structural evidence; De la Beche did not mention fossils once in this report to Sedgwick. Crossing the highly contorted Culm strata in central Devon, he probably made the reasonable working assumption that the overall direction of dip was unchanged, and that the more or less vertical strata were continuing to become progressively younger the further he traversed northward. The "newer beds" he expected to find before he reached the north coast were probably Murchison's; but if so, he must have been dismayed to find instead that the rocks were folded there into a saddle with no sign of the "upper grauwacke." For in order to avoid the highly implausible inference that all the strata there were completely overturned, he must then have had to conclude retrospectively that he had made a mistake further back (i.e., south), and that he had passed over a tight trough structure without noticing it.[6] This interpretation would then have seemed to gain support from the most northerly of Greenough's four bands of limestone, for De la Beche inferred that this was the *same* as the band he had crossed further south, repeated by the fold he had found it necessary to postulate.

Given this interpretation of the structure of Devon, the wider implications followed directly. There must be an immensely thick sequence of conformable strata in Devon, the oldest being in the south (including the Plymouth limestone) and the youngest in the inferred trough running through Exmoor. Like Greenough before him, De la Beche assigned the whole sequence to the Greywacke. Since it did not include any strata resembling Murchison's "upper grauwacke" of the Welsh Borderland, he concluded that it must be wholly older or lower in the total sequence, and hence roughly equivalent to Sedgwick's lower Greywacke in north Wales (see fig. 4.8, D).

Later in November, in a further letter to Sedgwick, De la Beche repeated his inference that there was a trough structure in north Devon in what he termed the "Devonian grauwacke." He was here using the word "Devonian" in an everyday sense—the only sense it possessed.

The adjective was used (as it still is) simply to denote a native inhabitant of Devonshire or any feature of that county. De la Beche called its Greywacke rocks "Devonian" in just the same way as Sedgwick might have called the rocks of his own research areas Cumbrian Greywacke or Welsh Greywacke. But De la Beche now added, with a deceptively casual "by the way," that he had found fossil plants close to the seams of culm and had sent some of them to the Geological Society. Since the fine cliff sections (see fig. 5.1) left "no doubt " that they came from the Greywacke, he pointed out again that "these plants are interesting."[7] His report that "ferns and Calamites" were plentiful indicates that, as a competent field geologist, he did not need to wait for Lindley's report to know that these fossils were at least *roughly* similar to those of the much younger Coal Measures. But he would not have found this at all surprising, since he had already expressed in print his belief that the flora of the Greywacke period was probably quite similar to that of the Carboniferous (§4.4). But he was well aware that, at least in Murchison's eyes, his report would be found not merely "interesting" but disturbingly anomalous.

When De la Beche sent off that second letter to Sedgwick, he could not have known that Lonsdale was sending him Lindley's report on the fossils by the same day's post. De la Beche recalled later that when it reached him in Devon, he felt more than a little surprise. For Lindley had identified a series of plant species *all* of which were well known from the Coal Measures of the rest of England and Wales (fig. 5.3).[8] Where De la Beche expected a rough affinity, Lindley pronounced an exact identity. Lindley complained of

6. The way the strata are shown on the section suggests that he had noticed no positive evidence for a trough: compare the sketched course of the strata in the inferred trough with that in the saddle further north.

7. De la Beche to Sedgwick, 22 November 1834 (ULC: AS, IA, 124).

8. Lindley to Lonsdale, copied in Lonsdale to De la Beche, 21 November 1834 (NMW: DLB), and *PGS* 2:106n. Lindley identified species of *Pecopteris, Sphenopteris, Calamites, Asterophyllites, Cyperites,* and *Lepidophyllum;* Fig. 5.3 is reproduced from illustrations of much better preserved specimens of some of the same species, from the Coal Measures, in Lindley and Hutton (1831–37, pls. 79, 153, 156).

A

B

C

Fig. 5.3.   Some fossil plants from the Coal
Measures, as illustrated in the monograph
by Lindley and Hutton (1831–37); the
same species of *Pecopteris* (A), *Sphenopteris* (B), and *Calamites* (C) were identified by Lindley in November 1834 among
the poorly preserved specimens sent by
De la Beche from the Greywacke of Devonshire.

being sent such "wretched morsels," and said he had identified De la Beche's fragmentary specimens only "as far as I can make them out." Yet such was Lindley's authority in his own specialized field that no mere geologist would have dared to cite the poor preservation of the Devon plants as grounds for doubting his identifications. Now it was not only Murchison and Lyell who would have to face up to an anomaly, but De la Beche too. The fossil plants of Bideford were proving more "interesting" than he had bargained for.

5.2   Preconceived Opinions versus Facts (December 1834)

A few days later, at the next meeting of the Geological Society, there was an unusually good attendance: eighteen members introduced a total of no fewer than twenty-seven guests, and doubtless many other members also attended.[9] De la Beche himself was not among them: as he had explained to Sedgwick, he had no money to spare for the coach fare to London. Many in the audience may have come primarily to hear the second part of a long paper by the young American geologist Henry Darwin Rogers (b. 1809) on the remote central and western areas of North America. But they may also have been drawn by rumors that the reading of De la Beche's letter, which was to conclude the meeting, would provoke a good argument, despite its deceptively factual title "On the anthracite found near Biddeford in North Devon."[10] In his letter, De la Beche claimed publicly, as he had already claimed privately to both Lonsdale and Sedgwick, that the anthracite and fossil plants—which were displayed in the customary way on the table—definitely came from an integral part of the Greywacke, and he now located them about two-thirds of the way up the whole sequence. He reinforced this conclusion by stating that the relevant strata, like the Greywacke elsewhere in Devon, had been affected by slaty cleavage, and he had sent specimens to demonstrate this. Despite a "very diligent search," he had failed to find any trace of the "interesting beds of the upper grauwacke noticed by Mr. Murchison," and he clearly implied that the whole of the Greywacke sequence in Devon must therefore be still older (fig. 4.8, D).

As Greenough reported to De la Beche the next day, the reading of this short note gave rise to "a very animated discussion."[11] The understatement matched De la Beche's prediction that his note would be found "interesting." In fact, as Greenough's account made clear, the discussion was so animated that the guests may well have felt that the Society's celebrated custom of unbridled argument had got out of hand. For, as De la Beche heard soon afterward from Edward Turner (b. 1798), one of the Society's secretaries and the professor of chemistry at the London University, "your discovery of Coal and its fossils in the Greywacke was last Wednesday turned topsy turvy with-

9. The GSL attendance books for this period have been lost, but the names of guests and of those who introduced them are recorded in GSL: BP, OM.

10. Rogers (1834); De la Beche (1834b), read at GSL on 3 December (GSL: BP, OM 1/7, pp. 89–99). "Biddeford" was an alternative spelling. Although a foreigner, Rogers had been elected an ordinary member of GSL during his visit to London in 1833: see his own account, in E. Rogers (1896, 1:104–8).

11. Greenough to De la Beche, 4 December 1834 (NMW: DLB).

out scruple."[12] The course of this important discussion can be reconstructed in some detail from these two accounts.

The opening was described vividly by Greenough: "Murchison led the attack, and expressed his astonishment that so experienced a geologist [as De la Beche] should have fallen into so great a mistake—as to fancy that the specimens on the table had anything to do with transition rocks." But as Turner made clear, the reason for Murchison's startling rejection of De la Beche's report was that the plant fossils from Devon were said by Lindley to be identical to those of "the ordinary coal fields." Although Murchison admitted he had never been to north Devon, he now felt so confident about the application of Smithian principles to the older strata that he insisted that De la Beche's rocks must really be Coal Measures and not Greywacke or Transition at all. There could hardly have been a more direct affront to an experienced field geologist than for Murchison to assert in this way that De la Beche had completely mistaken the character of the rocks he had studied.

By his own account, Turner then adopted the role of an impartial arbitrator and asked Murchison the pertinent question. Did he suppose that the culm and its plants came from a hitherto unrecognized outlier or isolated patch of Coal Measures overlying genuine Greywacke rocks, or that the *whole* sequence of so-called Greywacke in Devon was really Carboniferous in age? Murchison replied that he thought the former explanation more likely. This implied a still greater piece of incompetence on De la Beche's part, not to have recognized what would surely be a major unconformity between the coal strata and the far older Greywacke (see fig. 4.8, C and D).

"Greenough then defended you as well as he could," Turner told De la Beche, perhaps implying that it did not carry much conviction. According to Greenough himself, he pointed out that Murchison had "confessedly never seen the country," whereas De la Beche had studied it carefully; and that the rocks were well exposed. In a characteristic reference to the Society's "Baconian" traditions, he added that "it was not our practice to give credit to the results of abstract reasoning when opposed to actual observation."

According to Turner, "Lyell then spoke, more guardedly than Murchison, but inclined to the same side." But Greenough reported more colorfully that Lyell, "being equally ignorant of the country," had supported Murchison "with so much spirit as to excite the cheers of several of his hearers." The content of Lyell's speech was not recorded in either account, but almost certainly it was based on the improbability of a flora wholly composed of Coal Measures species being found in the far older strata of the Greywacke.

Turner then returned to the fray, and "said as much as I could for an absent man." He argued that a geologist as experienced as De la Beche, with good coastal cliffs to study, could not possibly have overlooked a major unconformity as Murchison had suggested. He adopted instead the second of the two explanations he had proposed earlier, inferring that De la Beche could only have "erred, along with the President and all other Geologists who had seen North Devon, in mistaking for Greywacke [what were really] rocks of the Carboniferous series." This implied that no crass mistake had been made about the structure of the region, but only a more pardonable error about the

12. Turner to De la Beche, 7 December 1834 (NMW: DLB).

dating of *all* the rocks of north Devon. This brought into the open the underlying issue of the relative reliability of fossils and rock types in a case such as this where those two criteria for dating were in conflict. Murchison had unhesitatingly given priority to the Coal Measures character of the fossils. Turner, on the other hand, accepted that De la Beche's rock specimens looked just like ordinary Greywacke, though he felt he had to admit "the possibility of deception by mere mineralogical character."

Murchison, however, did not accept the conciliatory solution that Turner had proposed. According to Greenough, he insisted again that De la Beche had "made a mistake, confounding things essentially different, and supposing the Coal plants to be regularly interstratified with the Greywacke when in point of fact they formed a separate series resting on its edges." Greenough strongly denied this supposition of a major unconformity and claimed that the Culm strata were perfectly conformable with the rest—"as far as I could rely on my own observations made some years since but yet perfectly present in my mind." He implicitly added his support to Turner's suggestion, for he recalled—"with a view to pacify my opponents"—an analogous instance where he and Buckland had long mistaken the true age of some of the Coal Measures in Pembrokeshire (in southwest Wales), because they were so much like Greywacke in appearance. But he wiped out the force of this eirenical concession by citing one of the analogous cases of anthracite in the Greywacke of Normandy (in the Bocage, département of Calvados), thus giving his opponents an unwelcome reminder that the Devonshire case was not unique. He also repeated his argument that experienced observers such as De la Beche, and Weaver in Ireland, were unlikely to have made the same error, "if the error were one as clear as my opponents, reasoning wholly upon what ought to be, not what was, assumed."

Not unnaturally, this barbed insinuation of apriorism provoked Murchison once more. According to Greenough, he spoke with indignation.

> Could it have been imagined that such a line of argument should have been taken by the President? With great respect to the chair he differed from him toto coelo—was it to be endured that at this time of day and in this room they should still hear mineralogical characters relied on as tests of geological position? Was grauwacke to be called grauwacke because it had the external characters of grauwacke?

Whether or not the phrasing was reported accurately, this captures what was now emerging as the real issue as seen by Murchison and probably by Lyell too. The reliance on rock type or "mineralogical character" above fossil content as a criterion of relative age, implicitly by De la Beche and explicitly by Greenough, was branded by Murchison as an outmoded throwback to the older Wernerian "geognosy." He presented the issue polemically as a crusade by the rising generation—including of course himself—against diehards like Greenough; and De la Beche, though younger than Murchison, had joined the wrong side.

As his trump card, Murchison cited an even younger but already distinguished man of science, Louis Agassiz (b. 1807), the professor of natural history at Neuchâtel and Cuvier's effective heir in the study of living and fossil fish. Two meetings before, Agassiz had summarized for the Society his vast current work on the classification of fish and their geological history, for which he had been awarded the second Wollaston medal earlier in the year. He had mentioned that some well-known fossil fish found in an ancient-looking black slate in the Swiss Alps were of species known elsewhere from the Chalk. As Greenough reported, perhaps more pungently than Murchison himself, Agassiz had shown them all that "black was white" and that "the Glarus slate was good honest chalk." In other words, it was claimed that in this spectacular case the true age was indicated by the fossils, not the rock type.[13] The intended moral was clear: the Devon culm was to be dated by its plants, as the Alpine slate was by its fish, and the ancient Greywacke appearance of the rocks was to be discounted in both cases equally.

"So ended the conflict," Greenough reported; "Murchison declared he would never use the word Grauwacke again so long as he lived, and departed." For in Murchison's view the word could not be used *both* as a name for a particular rock type *and* as a name for all rocks of a certain age. Turner told De la Beche that "the Society obviously leaned to the side espoused by Murchison"—an impression probably heightened by what sounds like a well-timed exit by that actor.

Turner himself found it incredible that De la Beche should be mistaken, but expected him if necessary to make "a great correction" to his mapping, "losing your own case in the interests of Science and truth." He saw clearly that the argument hinged on Murchison's firm conviction that fossil species characteristic of the Coal Measures could not possibly be found in rocks even older than the upper Greywacke strata. "It will be an important fact if you can prove that they can," Turner added with the customary understatement. Greenough likewise told De la Beche to "fortify yourself against objections of every kind," and urged him to send a more complete collection of specimens, so that his "discovery"—in which Greenough fully believed—would be "beyond the reach of suspicion." He suggested special attention to the conformable relation between the plant-bearing Culm strata and those to the north and south, and he urged De la Beche to collect the fossils from the nearby limestone strata: "if you establish that all the strata are conformable, these fossils would decide the question." With regard to the value of fossils, Greenough was once again far from being the reactionary diehard that Murchison had insinuated. But, as always, Greenough was most concerned for the reputation of his institutional baby: "all this was very amusing," he admitted, "but such discussions I am sure will not tend to the benefit of science or the credit of the society."

The controversy was now out in the open, at least among members of the Geological Society and their guests. De la Beche had already cited the culm of Devonshire as important evidence for the reality of land plants and coal deposits in the ancient Greywacke strata. If the case were genuine, De la Beche's fieldwork would be vindicated, but both Murchison's and Lyell's theory-based conclusions would be seriously undermined, albeit for different reasons. Far more was at stake than the geology of one corner of an English

13. Agassiz (1834b), read at GSL on 5 November. The case of the Glarus fish had already been published in German (Agassiz 1834a) and was probably mentioned in the discussion at GSL. Agassiz was in Britain to study public and private collections for his monograph on fossil fish (Agassiz 1833–43), and had also attended the Edinburgh meeting of BAAS (see Andrews 1982).

county. The argument impinged on the highest level of theoretical interpretation of the history of the earth, on the most fundamental methods to be used in deciphering that history, and on the most important single contribution that geology was expected to make to economic progress.

5.3   De la Beche Springs to His Own Defense (December 1834)

When he received Greenough's and Turner's reports of the meeting, De la Beche was naturally incensed that his competence had been impugned by two geologists who had never seen the area in question. He wrote a stiff rejoinder at once, to be read out at the next meeting; but fearing he had been too harsh, he wrote a more personal letter the following day, telling Greenough that anything he found "offensive or ungentlemanlike" should be cut out.[14] "If I could afford it, I would run up to town for the next Geol. meeting and give my objectors that good humoured trimming which I think they deserve." But afford it he could not. "The present state of West Indian property has nearly smashed me," he explained; his fee for the Devon survey was not even covering his expenses, and he feared he would soon have to give up geology and return to Jamaica to live on his estate. Lord Melbourne's Whig administration had just collapsed, and with it De la Beche's best chance of getting some sinecure that would enable him to continue in a life of science. The only prospect that remained for him in England was for his surveying work to be put on a more permanent footing. He must have known that in any such proposal the Board of Ordnance was certain to ask the president of the Geological Society for an expert opinion, and Greenough's presidency still had more than two months to run. The pitiable story of creeping poverty that De la Beche related was thus not without ulterior significance. But the same situation also lent urgency to his self-defense. If those attending the previous meeting had included men in public life, as well they might, the rumor would quickly spread in Whitehall and Westminster that public money was being spent on an incompetent geologist who could not even recognize the supremely important Coal Measures when he saw them. It was therefore imperative for De la Beche to retrieve his reputation without delay from the doubts that his critics had cast on it.

Having taken steps to ensure that his rejoinder would be read at the next meeting of the Society, De la Beche turned to the most influential figure who was not yet embroiled in the controversy. Sedgwick, as the professor of geology at Cambridge, was also certain to be asked his opinion by the Board of Ordnance; besides, that opinion carried great weight among the geologists themselves. De la Beche therefore lost no time before writing to tell Sedgwick how his brief report from Devon had caused "a most singular exhibition" at the Geological Society.

> Murchison and Lyell, who confessedly never saw a square yard of the country, attacked me most fiercely, particularly the latter, declaring their

---

14. De la Beche to Greenough, 9 and 10 December 1834 (ULC: GBG, 45, 46).

Fig. 5.4. "Preconceived Opinions *v[ersus]* Facts": De la Beche's caricature of the scene at the Geological Society when his report of fossil plants in the Greywacke of Devonshire was criticized by Murchison and Lyell:

> [De la Beche:] This, Gentlemen, is my *Nose*.
> [His Critics:] My dear Fellow!—your account of yourself generally may be very well, but as we have classed you, *before we saw you*, among men *without noses*, you *cannot possibly have a nose*.

This was sent to Sedgwick, and a similar version to Greenough, in December 1834.

perfect conviction that I had made a gross mistake as to the geological position of the beds whence the plants were derived, &c. &c. &c. Now as I had toiled day after day, for months in the district, examining every hole and cranny in it, this was a pretty good go of preconceived opinions against facts, which are so plain that the merest infant in geology could make no mistake.[15]

To illustrate this highly charged account of what had happened at the meeting, De la Beche drew a caricatured version of the scene. Beneath the surface, the caricature was no light-hearted joke, nor was it just for Sedgwick's benefit. It was loaded with rhetorical significance, and De la Beche ensured that it would gain wide circulation among those whose judgment counted by sending redrawn versions of it to Greenough and Turner (fig. 5.4).[16]

15. De la Beche to Sedgwick, 11 December 1834 (ULC: AS, IA, 125).

16. Fig. 5.4 is reproduced from a drawing in ibid. The second version is in De la Beche to Greenough, 14 December 1834 (ULC: GBG, 47), which also mentions the copy to Turner; it is entitled "Abstract Notions v. Facts," differs slightly in the legend, and shows the geologists observing the nose through *tinted* glasses (compare fig. 2.7). De la Beche told Greenough he was thinking of lithographing the drawing, as he had on earlier occasions (Rudwick 1975b; McCartney 1977), to give it still wider circulation, but there is no evidence that he did so.

The title of the caricature, "Preconceived Opinions *versus* Facts," took up one of the themes in Greenough's contribution to the meeting, and it was calculated to make both him and Sedgwick spring to the defense of the methodology they cherished in their science. De la Beche depicted himself demonstrating his manifest possession of a nose, just as he claimed to have merely reported the simple existence of fossil plants in the rocks of north Devon. By contrast, he showed his critics flatly denying that plain fact of anatomy, just as they had rejected his geological report out of hand. Their reason in the caricature was that, *"before we saw you,"* they had classed him among "men *without noses,"* just as (in De la Beche's interpretation of events) they had insisted—before going to Devon to look for themselves—that true Greywacke could not possibly contain the plants De la Beche claimed to have found in it.

As on an earlier occasion (see fig. 2.7), the significance of the caricature extended to the visual design itself. De la Beche depicted himself, the honest working geologist, dressed in a practical topcoat suitable for fieldwork in the often inclement weather of Devonshire. His critics, by contrast, were all shown wearing elegant tailcoats more suitable for a convivial evening at the Geological Society Club. Of course both Murchison and Lyell were highly competent field geologists, but De la Beche's point was that they had pronounced on his observations before doing their own fieldwork *in Devon*. Most significantly, however, the spectacles that De la Beche himself wore in real life (fig. 4.4) were transferred to his critics. They were shown holding spectacles *through* which they were viewing the disputed fact of De la Beche's nose, just as he alleged that it was the distorting or selective effects of their *theoretical* convictions that led them to deny the validity of his factual observations. In every detail his caricature was thus carefully contrived to bring the influential Sedgwick on to his side and to detach him from Murchison, to stiffen Greenough's resolve as his London spokesman and protector, and to strengthen collective opinion at the Geological Society in his favor before the controversy went any further, as he knew it was bound to do.

A few days later, at the next meeting of the Society, De la Beche's formal letter was duly read out.[17] It opened with a disclaimer: he had merely sent a few specimens for the Society's museum, which he thought "might eventually prove interesting, little dreaming that I should bring upon myself any serious attack by so doing." As before, this was highly disingenuous, because he knew his fossils would be found "interesting" immediately, precisely because they were good evidence bearing on a disputed point of high theoretical and practical significance. His main defense, however, as in his letter to Sedgwick, lay in his emphasis on his own command of the clear and unambiguous field evidence, and he protested that "some little deference" was therefore due to his conclusions.

> It appears, however, that this will not do:—the plants are found to be of the same species with some discovered in the coal measures; *therefore,* say those who object to my statement, I *could not* have got them from the grauwacke. Let us hope that the day is past when preconceived opinions are to be set up, as good as arguments, against facts; because

17. De la Beche to Greenough, 9 December 1834, read at GSL on 17 December. (Here and throughout this book, archival details are not repeated at second and subsequent citations in the same chapter.)

if they are, let *that fact* at least be clearly understood, and let us be consistent, and no longer boast of our adherence to the Baconian philosophy.

With that ringing declamation, as with the caricature he had circulated to illustrate it, De la Beche clearly hoped to range on to his own side all those who stood by the Society's traditional approach to the science.

De la Beche denied that his report on the Devon fossils had been influenced by any such lapse of scientific method: "the evidence of my senses compels me to see that the beds whence I took the fossil plants form part the grauwacke," he insisted, in good "Baconian" style; "therefore I conclude that the plants, whatever they may resemble, are plants of the grauwacke." He claimed that although he had been struck by the Coal Measures appearance of the Devon fossils, "my *preconceived opinions* led me to consider that there *must* be some difference," and that Lindley's report had therefore surprised him. This was probably not disingenuous, for he knew that Murchison claimed that the *other* (animal) fossils of the Greywacke were substantially different from those of the Carboniferous formations. His discovery had therefore faced him with an anomaly of his own, although a less serious one than that of his critics.

As usual, he took the opportunity to point out that his report was far from unique, if only his critics would look beyond the shores of Britain. He recalled how Élie de Beaumont had been met with a similar incredulity when he first announced an analogous but even more surprising discovery, namely that in the French Alps some characteristic Oolitic (or more precisely, Liassic) fossil belemnites occurred in strata sandwiched among some containing fossil plants of Coal Measures appearance. Brochant had gone there to check the authenticity of the report for himself; and Adolphe Brongniart (b. 1801), the son of Cuvier's collaborator and already a leading specialist on fossil plants, had confirmed that the plants were indeed of Coal Measures species. The locality (near Moutiers, département of Savoie) had become famous—indeed notorious—as soon as Élie de Beaumont's report was published in 1828; for the evidence of the belemnites conflicted seriously with that of the plants.[18] This suggested that "characteristic fossils" could *not* always be relied on to give unambiguous indications of the true relative ages of formations.

This was what made the Alpine analogy such a powerful argument in De la Beche's counterattack on the Devon issue. For, as he pointedly emphasized in his letter to the Geological Society, those who had been skeptical about Élie de Beaumont's report had "one by one dropped off when they and others took the trouble to look at the, now, well known localities themselves." The moral of the story was obvious. De la Beche therefore concluded his letter by issuing a stiffly formal challenge to his critics.

> Either that they should forthwith proceed to examine the district in question, I engaging to furnish, for their assistance, any documents relating to it that I may possess, they engaging to give in an immediate statement of what they have seen to the Geol. Soc., or that they should

18. Élie de Beaumont (1828); Adolphe Brongniart (1828a). The younger Brongniart had begun work on a massive monograph on fossil plants (Adolphe Brongniart 1828b, 1828–37). Belemnites were interpreted as the internal skeletons of cephalopod mollusks and were known *only* from the Oolitic and Cretaceous groups.

now print their objections to my statement in some periodical work, and I will engage to reply to such objections.[19]

The gauntlet was down. De la Beche challenged his critics either to what he was confident, would have to be a retraction within the gentlemanly confines of the Society, or to open conflict in the more exposed arena of the scientific press.

At this moment, however, the imminent battle was abruptly called off. As Murchison reported afterward to De la Beche, "the President after reading out in a most *impressive* manner your rejoinder, interdicted all observation thereon." Greenough had earlier expressed his fear that the conflict might damage the Society's reputation, and probably he also considered it inappropriate to discuss a communication that contained no geological "facts" that had not already been reported.[20] But he cannot have been unaware that the prevention of discussion would also leave his protégé with the last public word on the matter, at least for this occasion. Whatever his reasons, Greenough did not escape criticism for his action, particularly from William Fitton (b. 1780), a wealthy London physician and prominent member of the Society, under whose presidency the custom of allowing discussion had been instituted and who therefore regarded himself as the custodian of the Society's gentlemanly moral standards. As Murchison told De la Beche, "our impartial Censor, Fitton, gave *us all* (from the President downwards) a solemn flagellation. The President catching it for his interdict & the stoppage of free discussion, your opponents for having been too *vifs*."[21] After the meeting, Greenough took Murchison home in his carriage and tried to dissuade him from writing to De la Beche about what had happened, on the grounds that it "was unnecessary as the thing was all over."[22] But on an issue of such fundamental importance to the science, even the president of the Geological Society could not halt the process of debate. The controversy was certainly not "all over."

5.4   Restoring a Gentlemanly Civility (December 1834 and January 1835)

As soon as he heard how his original note on the Devon plant fossils had been received, De la Beche had taken action to ensure Greenough's continued support and to try to win Sedgwick on to his side. Greenough had now indeed defended him at the Geological Society, though with the questionable use of presidential censorship. Sedgwick could not attend that second meeting, but wrote the same day from Norwich. Only the previous month, Lord Brougham, on almost his last day in office as Lord Chancellor, had offered Sedgwick a prebendal stall at the cathedral there. Sedgwick had accepted with alacrity,

19. De la Beche to Greenough, 9 December 1834.
20. Probably this argument was also used to justify the omission of any mention of De la Beche's letter in the printed record of the meeting (*PGS* 2:109–12).
21. Murchison to De la Beche, n.d., pmk 19 December 1934 (NMW: DLB). In 1830 Fitton had played a major role in the famous unsuccessful agitation to get a man of science (Herschel) elected president of the Royal Society, rather than the royal Duke of Sussex: see MacLeod (1983).
22. Murchison to De la Beche, n.d., pmk 19 December 1834.

for the duties of his new position—taking services and preaching sermons—
only required his presence in Norwich for two winter months each year and
could be combined with his college and university positions at Cambridge.
More important, this piece of Whig patronage for science offered a stipend
of £600 a year, thus quadrupling Sedgwick's regular income. Although not
matching Buckland's, and without the convenience of having the ecclesiastical
position in his university town, it did give Sedgwick for the first time quite
substantial means to support a life of active scientific research.

Sedgwick told De la Beche he had been "delighted" to hear of the
discovery of fossil plants in Devon. "I *never* had the *shadow of a doubt* of
their being in the grauwacke," he told him, and added, "neither was I very
much astonished when I learnt that your plants were identical with true coal
measures species."[23] De la Beche could now evidently count on the influential
support of the geologist with the widest experience of the Greywacke rocks
of Britain. Sedgwick was well aware that the crucial issue concerned the
proper use of fossils for establishing the relative order of formations. "The
fossiollogists [*sic*] are going mad," he exclaimed to De la Beche; "their art is
only subsidiary." He did not deny the value of fossils, but insisted that they
could never safely be given priority: "the *foundation of our conclusions*, as
far as they are geological, must ever rest on the evidence of superposition
given by actual sections."

This was exactly De la Beche's view: it was the evidence of the fine
coastal cliffs in Devon that had made him confident that, whatever the fossil
plants might turn out to be, they certainly came from an integral part of the
Greywacke sequence (see fig. 5.1). The chief "fossiollogist" whom Sedgwick
had in mind was almost certainly Lyell, who had just been awarded the Royal
Society's coveted Royal Medal, not for the geological "principles" that figured
in the title of his monumental work, but for his concrete results in piecing
together by purely fossil evidence the sequence of the Tertiary deposits scattered throughout Europe. But that work had depended crucially on a theory-
based conception of piecemeal faunal change, and could not be verified independently by the fundamental criterion of visible superposition.[24] Sedgwick
had urged Murchison's merits for the award, but he had to concede that his
candidate's research, unlike Lyell's, was not yet fully published.[25] Now, to
Sedgwick's dismay, Murchison seemed to be following Lyell in putting fossil
evidence into first place, even against the apparently clear evidence of superposition in the cliffs of Devonshire.

For the time being, however, Sedgwick saw himself principally in the
role of a peacemaker. Above all, he was concerned to repair De la Beche's
relationship with his collaborator Murchison. "By your *brusque* manner of
attacking his positions, now and then," he told him, "you may have set up his
back a little." The comment was accurate and the military metaphor appropriate. When De la Beche—like Weaver before him—claimed to have found

23. Sedgwick to De la Beche, 17 December 1834 (NMW: DLB).

24. Greenough to De la Beche, 4 December 1834 (NMW: DLB). For De la Beche's reaction
to the award, see De la Beche to Greenough, 10 December 1834 (ULC: GBG). For Lyell's
interpretation of the Tertiary, see Rudwick (1978).

25. Sedgwick to Greenough, 25 November 1834 (ULC: GBG, 1534); Sedgwick to Murchison,
n.d., pmk 23 December 1834 (ULC: AS, IC, 12).

fossil plants of Coal Measures species in lower Greywacke strata, one of Murchison's strategic strong points had indeed been threatened.

> He has, as you know, examined an astonishing fine series of upper fossiliferous Grauwacke; and finding no coal plants in it, resolved in his own mind that the said plants could not be found any where else. I have once or twice argued this very point with him, when speaking of Weaver's paper. This accts. well enough for his reluctance to admit your facts.[26]

For facts they were, in Sedgwick's view, and facts they would remain. De la Beche's caricature had duly resonated with Sedgwick's own feelings, and he concluded in characteristic style: "the more opposition the better—facts must prevail over mere opinions." Assuring De la Beche that there was no conspiracy against him, he added an implicit warning against Greenough's mischief making: "remember that a story always improved in travelling & the account of the attack made on you was probably more piquant than the reality."[27]

As already mentioned, Murchison ignored Greenough's attempt to silence him and wrote to De la Beche immediately after the second formal letter had been read. "I should not have slept soundly without it," he told him; he protested that he had said "nothing which exceeded the limits *of fair discussion*." But he insisted that the original report from Devon had been no mere "little notice," as De la Beche had disingenuously claimed, but "a most important novelty in English Geology," and he feared that what he bluntly termed De la Beche's "*Solecism*" would soon become generally accepted "if sanctioned by your name & reputation."[28] In this way Murchison neatly shifted the burden of proof from himself to his opponent, conveniently glossing over the fact that it was his own position, not De la Beche's, that was the real novelty.

Murchison had probably been shown one of the copies of De la Beche's caricature; certainly he responded, like Sedgwick, in the language of facts and opinions. But this time De la Beche's joke proved a boomerang; for Murchison, unlike Sedgwick, was quite prepared to insist on the essential place of theoretical interpretation in geology.

> I am as far as ever from *being able* to agree with you *in opinion*. What an impertinent fellow, you have almost a right to exclaim! He has not seen an inch of the Coast, & he contests a matter of fact & of personal observation. I *do not* in the slightest degree contest the facts, & the correctness of your observation of the physical features of N. Devon, but I still cling to the opinion that the *phenomena may* be explained in a *different manner*; & why do I so doggedly adhere to this belief?

---

26. Sedgwick to De la Beche, 17 December 1834. The comment supports the inference that Murchison had been one of the "unbelievers" who had blocked the publication of Weaver's paper nearly two years before (§4.4).

27. Ibid. De la Beche feared a conspiracy to undermine his reputation as a competent geologist because Greenough had told him that, in addition to the attack on his report of plants in the Devon Greywacke, his earlier paper on the Tor Bay area (1828) had been criticized at another recent GSL meeting. Sedgwick denied that there had been any such attack: ibid.; Greenough to De la Beche, 4 December 1834 (NMW: DLB); Austen (1834), read at GSL on 19 November.

28. Murchison to De la Beche, n.d. (NMW: DLB). This was almost certainly the letter referred to in a covering note, Murchison [to De la Beche], n.d., pmk 19 December 1834.

> Because it is a sequitur of the establishment of your inferences, that strata can no longer be identified by their fossil remains.[29]

This brought Murchison's theoretical commitment into the open: Smith's method of identifying formations by their characteristic fossils was at stake. De la Beche, "though so deeply imbued with these sound principles of our science," appeared to be "all at once launched into this old Wernerian heresy that mineralogical characters are to establish the age of a formation." With this comment Murchison cleverly flattered De la Beche while implying that he would not wish to be associated with the "reactionary" faction in the geological community. Surely, he suggested, De la Beche could not mean to claim that his Devon rocks were of Greywacke age simply because they were of greywacke rock type: "I am quite [sure] you will be the last person to set up *that character alone* as a *proof* of the age of the beds." By contrast, Murchison located himself firmly in the forefront of progressive research on this issue: "ever since I have hammered a rock the results arising from identification of deposits by zoological and vegetable evidences have been gaining ground from long continued & patient observation."

De la Beche's deceptively factual report from Devon had implicitly questioned the validity of that approach. For the fossil plants he had collected were, on his interpretation, separated from the identical species in the true Coal Measures by a vast thickness of intervening formations: the Mountain Limestone, the Old Red Sandstone, and Murchison's own upper Greywacke formations (see fig. 4.8, C and D). The implausiblity of De la Beche's claim would be apparent, Murchison maintained, if one made a comparable leap in the other direction, from the Coal Measures upward in the sequence to the far younger Cretaceous strata, where the fossil flora (in the "Green Sand" below the Chalk) was known to be radically different (see fig. 3.2). In the face of such evidence for the overriding validity of the fossil criterion, Murchison told De la Beche he would "doubt & *continue to doubt* any such isolated cases as that at Biddeford," relegating them to the status of minor anomalies. For he himself claimed to have found just the evidence that made the Devon report so implausible.

> I happen to have worked out patiently *the very descending series* which *does* separate the Coal Measures from the Older Greywacke rocks, & I know & have *proved*, that *all the classes of organic* remains become more and more divergent in *generic* & *specific* characters from those of the Carboniferous Series in proportion as we descend, and that with the finest possible sections the work of 4 years has not produced a fragment of any vegetable approaching to those of the Carboniferous era.[30]

Murchison made no reference to De la Beche's public challenge that he (and Lyell) should go to Devonshire to look for themselves at the disputed strata. Instead, following the maxim that attack is the best form of defense, he issued his own challenge in return.

> *Whenever* you produce the orthoceratites, trilobites, leptaena &c. of the *true* transition series, in *direct association* with these Coal plants (*all*

29. Murchison to De la Beche, n.d. (see note 28).
30. Ibid.

of which can be paralleled in the true Coal fields) *then,* but not till then can I be convinced . . . I shall seriously be very much obliged to you to point out the instances where such proofs *as I require* can be obtained.[31]

Such direct evidence was indeed just what De la Beche had not produced; for he had not yet followed Greenough's advice to collect the animal fossils from the limestones near the Culm strata and to see whether they resembled those of Murchison's upper Greywacke formations.

Murchison ended his long letter on a conciliatory note by thanking De la Beche for sending him a copy of his newly published *Researches in Theoretical Geology,* calling it "one of the cleverest & best written works we have ever had." Yet he got its title wrong—misnaming it, by a significant slip, merely "Theoretical Sketches"—and he had evidently not read it carefully enough to notice De la Beche's trenchant criticism of the identification of strata by fossils when that principle was divorced from the more fundamental evidence of superposition.[32]

De la Beche responded to Murchison's letter at once, doubtless with some relief that his opponent seemed not to be concerned to damage his reputation after all. Reconciliation was the order of the day: "when I read your straight-forward, manly, and, I may add, friendly letter," he told him, "I fairly wished all grauwacke, fossil plants, and geological discussions at the devil, since they could cause even the appearance of sharp shooting between us."[33] He explicitly accepted the validity of Murchison's upper Greywacke sequence and its subdivisions with their characteristic fossils, though only on the *local* scale of the region in which Murchison had done his fieldwork. Implicitly he denied that the same formations and the same fossils would necessarily be found in other regions such as Devon, let alone still further afield. He claimed again that his report from Devon had been correct, telling Murchison that Sedgwick agreed with him. He also mentioned the usual Continental and American analogues; while conceding that their existence could not prove the Devon case, he insisted that "it takes away all *a priori* argument against the possibility of the thing,—nay even its probability."

Just as Murchison had sought to calm the situation by praising De la Beche's work, so De la Beche now reciprocated with some carefully worded praise for Murchison's "patient, long-continued, and unwearied personal examination" of the Welsh Borderland and the important results it had yielded. "They will eventually bring you more credit," he assured him, "than if you had been employed for the last four years in constructing a geological theory, which, however brilliant it may appear at the moment, might run the chance, like others, of being blown to pieces by the discovery of twenty, or even two, new facts." With this studied phrasing, De la Beche drew a veiled contrast between Murchison's work, with its undoubted empirical value at least within its own region, and Lyell's *Principles,* which was, in De la Beche's opinion,

---

31. Ibid. Orthoceratites were cephalopod mollusks with straight shells; trilobites were considered to be extinct relatives of the crustaceans; *Leptaena* was a brachiopod shell characteristic of Murchison's upper Greywacke (see fig. 10.3).
32. Ibid.; De la Beche (1834a, chap. XI, XII).
33. De la Beche to Murchison, 21 December 1834 (GSL: RIM, D12/2).

just such a precariously hypothetical work.[34] The comment was thus a covert bid to win Murchison on to his own side by detaching him from his alliance with Lyell.

De la Beche had distinguished sharply between his two chief opponents. He had told Sedgwick he thought Murchison "an honest fellow" who at least showed him no personal dislike to his face, whereas Lyell "takes no pains to conceal his." He added, "I *do not* suspect a wilful attack from one quarter but I do from the other."[35] In his reply, Sedgwick hastened to assure him that in his opinion neither of the London geologists bore him any personal ill will. But he too tacitly distinguished between them: "our friend Murchison is sometimes a little touchy," he admitted, whereas Lyell was "extremely sensitive to any remarks on his published opinions." He recalled how his own critical remarks on Lyell's *Principles*, in his presidential address to the Geological Society three years before, had provoked Lyell to write him "one of the most angry letters I ever read," so that all Sedgwick's tact and good humor had been needed to repair the relationship.[36] Though he did not make the point explicitly, Sedgwick must have been well aware that De la Beche's *Researches* had embodied an even more weighty and sustained critique of the *Principles*, so that Lyell's reported antagonism toward its author could have come as no surprise.

Murchison had now at least gone through the motions of reconciliation. Lyell, whom Sedgwick had characterized as "an upright man of Gentlemanlike feelings," took longer to get around to any similar gesture. But after the New Year holiday he too wrote to assure De la Beche that there had been no conspiracy against him.[37] Like Murchison, he was careful to praise De la Beche's *Researches*, despite its criticism of his own theoretical approach to geology. In a somewhat patronizing manner, he compared it favorably with Boase's *Primary Geology* (§4.6), which he called "a work after Greenough's own heart." The Cornish geologist's book was, he wrote, a work "endeavouring to unsettle every received doctrine, to pull down as much as possible & build up as little." This comment was probably calculated to detach De la Beche from his London champion, just as De la Beche had covertly tried to detach Murchison from Lyell himself. Lyell denied that De la Beche had any cause for complaint about the way his work had been treated at the Geological Society, but he did not enlarge on the matter, because he said defiantly that he would "show fight" when they next met in London.

De la Beche, however, could not afford to go to London and may never have heard Lyell's promised "true account" of the meetings there. "Poked out in this non-scientific corner of our Isle," he complained to Greenough soon afterward, "I know little of what is passing in the scientific world."[38] He told Greenough that he had heard from both Murchison and Lyell about their

34. See De la Beche (1834a) and his earlier caricatures criticizing the *Principles* (Rudwick 1975b).

35. De la Beche to Sedgwick, 11 December 1834.

36. Sedgwick to De la Beche, 17 December 1834, referring to Sedgwick (1831, pp. 298–307).

37. Lyell to De la Beche, 2 January 1835 (NMW: DLB). De la Beche's reply, when he found the letter on returning from fieldwork, was cordial; he warned Lyell that Boase's work was unreliable, but otherwise added nothing of substance to the debate: De la Beche to Lyell, 24 January 1835 (APS: CL).

38. De la Beche to Greenough, 13 January 1835 (ULC: GBG).

"flourish of adverse trumpets" in the Devon affair, and that Murchison had explained his views in a "straight-forward handsome way." But both had continued to insist that their objections were to De la Beche's interpretation, not to his facts. "This appears to me, as the cockneys would say—'Werry hodd'," he commented; "I don't very well see how I could draw conclusions from facts of which I knew nothing, and consequently I don't see that there were any conclusions to object to." As before, this was calculated to appeal to Greenough's predilection for unvarnished facts. De la Beche omitted to mention that he had known his fossil plants would create a stir even before Lindley reported that they were of Coal Measures species.

"Unless fired at again," he told Greenough, "I shall trouble my head no further about the matter." But it was now clear that Murchison and Lyell had no intention of letting the matter rest. Behind an appearance of restoring a gentlemanly civility, neither of his opponents had made any real concession of substance to De la Beche. On a issue of such importance, renewed exchanges of "fire" were sooner or later inevitable.

5.5   Marshalling Support behind the Scenes (December 1834 and January 1835)

Meanwhile Murchison had not been idle in marshalling support for his own side of the argument. After De la Beche's first note on the Devon coal plants had been read at the Geological Society, Murchison wrote at once to Buckland, evidently to get a second opinion on the "facts" from a geologist who had seen the area at firsthand. Buckland told him that he well remembered the Bideford culm from his own fieldwork there more than twenty years before, and that he suspected it was "only a well baked form of the Welsh coal" on the other side of the Bristol Channel. By itself, that remark might have meant that Buckland agreed with Murchison that the Devon Culm strata were really just Coal Measures in a slightly altered form. But since he added that "there are Transition lime Stone beds not very far off which will help your identifications," it is clear that he did *not* agree with Murchison at all.[39] Evidently he regarded the whole set of strata in north Devon as Transition or Greywacke in age, culm and all, and like Greenough he expected the fossils in the nearby limestones to confirm that inference.

Having failed to get Buckland's support, Murchison tried next to win Sedgwick's. He too had seen the disputed area at firsthand, though likewise many years before. But Sedgwick had already ranged himself decisively on De la Beche's side in the argument, and he had little patience with either the tone or the content of Murchison's continuing attack on De la Beche's work.

39. Buckland to Murchison, 5 December 1834 (GSL: RIM, B34/9); Murchison's "long epistle," to which this was a reply, has not been located. Buckland later refereed De la Beche's original note and gave his opinion that as a preliminary report it did not warrant fuller publication in *TGS*: Council of GSL, 17 December 1834 (GSL: BP, CM 1/4, p. 4); Buckland to GSL, 13 January 1835 (GSL: BP, COM P4/2, p. 44).

De la Beche announces *a fact*. This may be right or wrong. But why say that this was throwing down *the gauntlet*, or why argue with temper about the matter? In my own mind I have not the shadow of a doubt that he is right because I have crossed the country, tho' I have not examined the spot. Had you found a coal formation in your [upper Greywacke] groups, with plants differing specifically from those of the true carboniferous series, I would have understood what you write, but at present I do not. I must however off to my ride.[40]

Sedgwick's daily exercise cut short his critique, but his point is clear enough. Murchison had no good reason to deny the possibility of Coal Measures plants being found in rocks as ancient as the lower Greywacke, unless he first found in his own (upper Greywacke) formations some positive evidence of what the plant world had really been like at that period. Until he found such fossils, Murchison's argument depended perilously, in Sedgwick's opinion, on purely negative evidence.

By contrast with this rebuff, De la Beche had already got Sedgwick's full support. But he had also taken steps to reinforce his position in a way that Murchison did not attempt. Even if he had agreed that his report from Devon was, as Murchison put it, a "novelty in English geology," De la Beche would have regarded that status as irrelevant in view of the many foreign analogues to his discovery of coal plants in the Greywacke of Devon. So as soon as he had written off to Greenough and Sedgwick, De la Beche asked Élie de Beaumont to confirm that he was still confident of the validity of the cases he had reported from the Greywacke of France. His letter reached Paris in time for Élie de Beaumont to report the Devon discovery at the meeting of the Société géologique only three days later. He added on his own account that there were indeed analogous cases in northwest France, where coal seams in the *terrain de transition* were associated with fossil plants, of which some at least were referable to species of the *terrain houiller* or Coal Measures (fig. 5.5).[41] At the end of the week he sent De la Beche a long letter in which he reported on the meeting and set out the French evidence in some detail.[42] He explained that he had discussed the question at length in his lectures at the Collège de France in February 1833. No skepticism had been expressed by the other geologists present, and that summer he and Dufrénoy had carefully checked the point in the field.

Élie de Beaumont told De la Beche that in the older rocks of northwest France he and his colleague distinguished three main *terrains*, separated by major unconformities and as distinct from each other as the Oolite, Chalk, and Tertiary among the younger strata (fig. 5.6). The oldest, or *terrain de transition ancien*, comprised gneisses, schists, and greywacke. Then came an intermediate group, the *terrain de transition supérieur*; as he had mentioned

40. Sedgwick to Murchison, 23 December 1834 (ULC: AS, IC, 12). Murchison's letter, to which this was a reply, has not been located.

41. Élie de Beaumont at SGF, 15 December 1834 (*BSGF* 6: 90). Fig. 5.5 is reproduced from Dufrénoy and Élie de Beaumont (1841b, pp. 222, 230). The mine was at Haye-Longue, upstream from Chalonne (département of Maine et Loire).

42. Élie de Beaumont to De la Beche, 20 December 1834 (NMW: DLB). As Élie de Beaumont mentioned, his conclusions had already been published in Burat's *Traité de Géognosie* (1834, pp. 255–59). They were repeated four months later in a letter to Murchison, from which fig. 5.6 has been traced: Élie de Beaumont to Murchison, 19 April 1835 (GSL: RIM, E4/5). See also Dufrénoy at SGF, 20 April 1835 (*BSGF* 6:238–39).

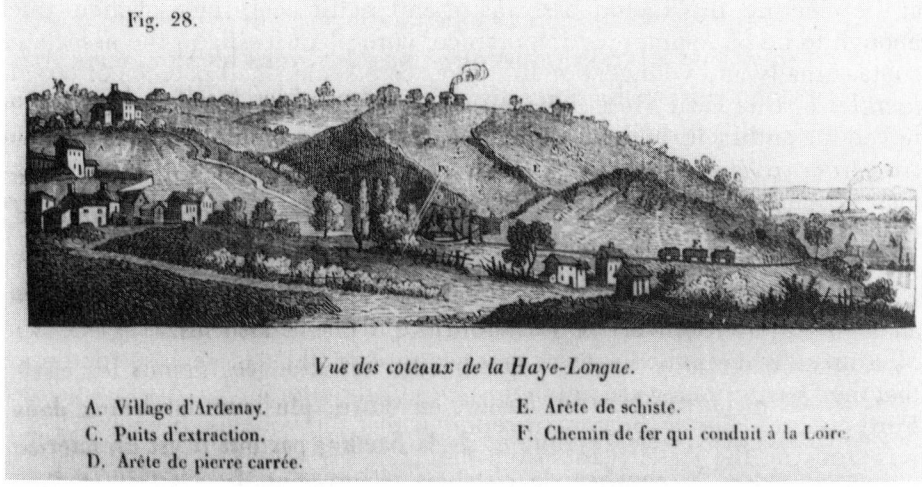

Fig. 28.

Vue des coteaux de la Haye-Longue.

A. Village d'Ardenay.
C. Puits d'extraction.
D. Arête de pierre carrée.

E. Arête de schiste.
F. Chemin de fer qui conduit à la Loire.

A

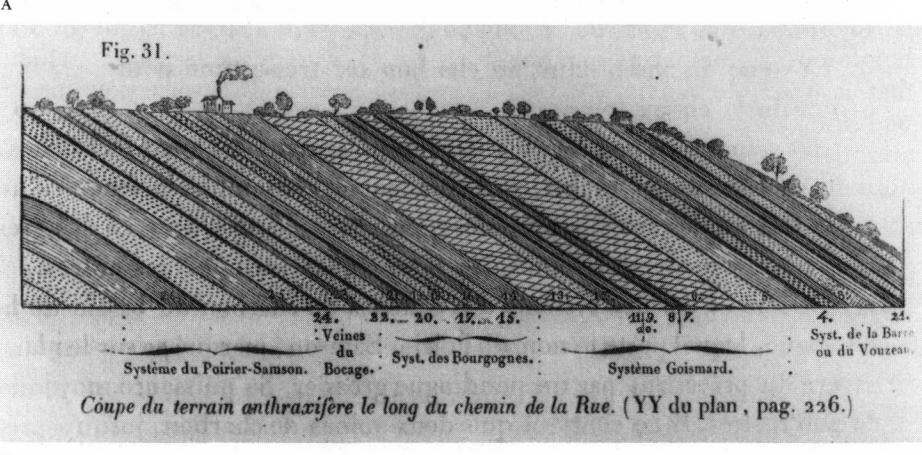

Fig. 31.

Coupe du terrain anthraxifère le long du chemin de la Rue. (YY du plan, pag. 226.)

B

Fig. 5.5. A mine producing anthracitic coal from the *terrain de transition supérieur* or upper Greywacke, near Angers in northwest France (A), and a section of the strata (B): illustrations from the memoir published in explanation of the official geological map of France (1841). Note the railway taking the coal from the mine for shipment down the Loire.

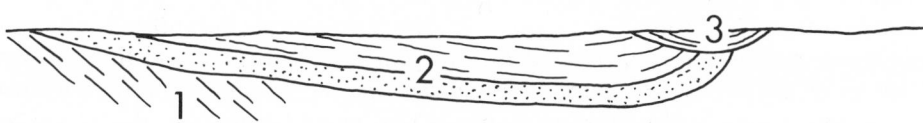

Fig. 5.6. Élie de Beaumont's diagrammatic traverse section showing his interpretation in 1834–35 of the geological structure of the older rocks in northwest France. The *terrain de transition supérieur* (2), with anthracitic coal seams and fossil plants, is separated by major unconformities from the *terrain de transition ancien* (1) below, and from small outliers of *terrain houiller* or true Coal Measures (3) above.

at the meeting, this included seams of anthracitic coal, some of them thick enough to be of commercial importance, with plant fossils in the associated strata. Finally and youngest of all, there were some small patches of *terrain houiller* or true Coal Measures (see fig. 3.6). Élie de Beaumont said he believed the anthracite and plant fossils were mainly from the upper part of the *transition supérieur*, and therefore relatively near the Carboniferous in the total sequence. Nevertheless, they were demonstrably *not* part of the true Coal Measures, since they were separated from them by a major unconformity.

This report was of great importance to De la Beche because, as the French geologist pointed out, the *transition supérieur* seemed to be closely analogous to the Culm strata in Devonshire, while the existence of true Coal Measures higher in the French sequence supported De la Beche's contention that the Devon strata were not themselves Carboniferous but distinctly older. Furthermore, Élie de Beaumont reported that the younger Brongniart, who was studying the fossil plants from the Transition strata in various parts of the Continent, considered provisionally that of twenty-five plant species so far identified from northwest France only four were also known in the Carboniferous. Commenting on Brongniart's data, Élie de Beaumont concluded that between the two floras "il y a d'une part assez de différence pour montre que ce sont *deux formations distinctes*, et d'autre [part] assez de ressemblance pour que la nature des plantes que vous avez trouvés avec les anthracites du nord de Devonshire ne doive pas paraître un fait singulier."

The month that had had passed since the reading of De la Beche's first report had been busy with maneuvers behind the scenes. There is no mistaking who had been the more successful protagonist. Murchison had failed to get Buckland's support and had been rebuffed by Sedgwick. De la Beche, by contrast, had the weighty support of both Sedgwick and Élie de Beaumont, the acknowledged authorities on the Greywacke or Transition rocks of Britain and France, respectively. Above all, he had the assurance that the French reports could be relied upon as genuine analogues to his Devonshire case.

5.6   Weaver Recants His Heresy (January and February 1835)

A few days before he received this welcome news from Paris, the first hint had reached De la Beche that he might need all the reinforcement he could get from the Continent. For when Murchison wrote to defend his rejection of De la Beche's report, he told him in passing, "I understand that Weaver has quite abandoned his case."[43] What lay behind Weaver's change of mind is as mysterious as the fate of his paper two years before. On that occasion his report of coal in the Greywacke of Ireland was suppressed very shortly before Murchison announced the first results of his own research, in relation to which Weaver's report would have been an embarrassing anomaly (§4.4). It seems equally unlikely that it was pure coincidence that Weaver retracted his claim so promptly after De la Beche's work had presented Murchison with

43. Murchison to De la Beche, n.d. (see note 28).

another awkward anomaly of the same kind.[44] There must be a strong suspicion that Murchison, in order to undermine the plausibility of De la Beche's report, put pressure on Weaver to make a public statement without delay on that aspect of his Irish work that most closely paralleled De la Beche's conclusions.

Whatever exactly had happened behind the scenes, Weaver wrote from Cork early in the new year to send the Geological Society a box of the Irish "Coal Plants," and he told Lonsdale that "the analogy to those of other coal measures series will strike you." He also sent a formal letter to the president, to accompany the specimens, and called it significantly his "recantation concerning the Munster *transition* coal."[45] In his letter, which was duly read out at the meeting of the Society the following week, Weaver reported that further fieldwork had shown him that there was generally "a well-characterized formation of old red sandstone" sandwiched between the true Transition rocks and the limestone underlying the coal seams. He no longer regarded that limestone as "Transition " on account of its trilobites and other fossils, but as Carboniferous, so that the sequence now seemed perfectly normal: Transition, Old Red Sandstone, Mountain Limestone, Coal Measures. He concluded that "the anomaly disappears."[46]

Whether or not there was any discussion of Weaver's letter, its significance for the Devon case would have been clear to all who had followed the affair to this point. Certainly Murchison set it firmly in that context when he wrote to Sedgwick the next day.[47] Adopting Weaver's own epithet, he termed it a "public recantation," a phrase clearly linked to what he had called the "old Wernerian heresy." Yet in fact Weaver's retraction was *not* the result of substituting fossil-based identifications for Wernerian criteria of rock type. On the contrary, Weaver implied that he had been misled, at least in part, by the Transition appearance of the *fossils* in the disputed limestone formation, and conversely that he had corrected the mistake by recognizing by its *rock type* a formation he had earlier overlooked, namely the Old Red Sandstone. In other words, however much Murchison may have been delighted to see the overthrow of one of the anomalies to his grand theoretical scheme, he could not properly attribute it to the superior heuristic power of the use of fossils in geology.

Weaver's Irish case might now be safely eliminated, but the Devon and Continental cases remained. Murchison reserved judgment on Devon. He accepted Sedgwick's reproof about his pugnacious attitude toward De la Beche, but assured him that "a friendly interchange of opinions" had eliminated any acrimony. "I see the force of your position quoad De la Beche," he wrote, "& shall hold my tongue *till I have seen the place*." Clearly he believed that a firsthand study would eliminate De la Beche's troublesome claim about Devon, just as Weaver's new fieldwork had now removed the Irish anomaly. "As for

44. Only two weeks before De la Beche's report was read, the Council of GSL had reviewed the backlog of unpublished papers and had noted Weaver's as "not ready; deferred by consent of the author and in his hands": Council of GSL, 19 November 1834 (GSL: BP, CM 1/3, pp. 318–19). There seems to have been no expectation that Weaver would so soon bring the matter back into discussion.

45. Weaver to Lonsdale, 12 January 1835 (GSL: BP, LR1/156).

46. Weaver (1835), read at GSL on 21 January. A revised version went ahead for full publication in *TGS*: Weaver to Lonsdale, 7, 12, and 24 April 1835 (GSL: BP, LR1/209, 210, 223); Weaver (1838a).

47. Murchison to Sedgwick, 22 January 1835 (defective transcript in ULC: AS, III.D.7).

the cases on the Continent," he added, "I do not believe there is *one where* the *same species* of coal [? plants] are in ancient [? Greywacke] rocks."[48] The comment suggests that he feared the foreign cases might be harder to crack than the British, and that the Greywacke on the Continent might contain *some* fossil plants, although not exactly the same assemblage as in the Coal Measures. Murchison had probably not yet heard the news, but his was just what Élie de Beaumont had reported to De la Beche a month earlier (§5.5).

Lyell was more hopeful. He told Murchison he would note Weaver's "recantation"—he too used the word with its overtones of heresy—in his next edition of the *Principles of Geology*.[49] The footnote he then wrote was a little gem of concealed rhetoric. He noted that Murchison had found no trace of plants in the Transition strata, whereas "MM. Élie de Beaumont, Virlet and De la Beche" claimed they had; but that Weaver had recently changed sides on this point.[50] While avoiding any overt claim that De la Beche and the French geologists were mistaken (the uninitiated reader might have got the impression that De la Beche too was a Frenchman!), Lyell implied that they like Weaver might fall into line in due course and assign their plant-bearing strata to the Carboniferous. The whole series of anomalies would then be eliminated, and order restored.

Those on the other side of the controversy, however, distinguished sharply between the two cases and declined to regard Weaver's "recantation" as undermining the Devon report at all. "That seems to have been a queer mistake of Weaver's," De la Beche remarked to Greenough, "however he has handsomely come forward to avow it."[51] Buckland told Murchison likewise that he was not surprised at Weaver's "concessions," but that he regarded De la Beche's report as quite another matter.[52] He recalled again how he had studied the Bideford locality more than twenty years before: "I then was of the same Opinion that De la Beche is now—and have since had no facts before me to affect my opinion one way or the other." He saw no reason to doubt that the Devon case was analogous to all those on the Continent, and wanted to know whether and why Murchison still objected to it.

Being assured of Lyell's continuing support, Murchison now felt he could answer Buckland more defiantly, despite Sedgwick's criticisms. He protested that, while having "no unwillingness to swallow any geological monstrosity" if the evidence were adequate, he remained "quite incapable of digesting this anomaly" of Coal Measures plants in the Greywacke.[53] "*Four years ago* & before I began to study these rocks," he maintained, "I should have thought it all very proper." But his work on the upper Greywacke formations in the Welsh Borderland, "to the tune of *20,000* feet of thickness," had convinced him of "the impossibility of the same terrestrial plants having continued to exist through all the *ages*" which must have elapsed between the deposition of De la Beche's alleged lower Greywacke and that of the true

48. Ibid.
49. Lyell to Murchison, 5 February [1835] (GSL: RIM, L17/17).
50. Lyell (1835, 1:196n).
51. De la Beche to Greenough, 21 January 1835.
52. Buckland to Murchison, 6 February 1835 (DRO: WB, F.219).
53. Murchison to Buckland, n.d. [? February 1835], wmk 1834 (UMO: WB, box B.T.). The opening of this long letter is missing; its contents suggest strongly that it is the reply to Buckland's letter of 6 February 1835.

Coal Measures (see fig. 4.8). For the *animal* life that had existed over the same vast period was recorded in "a beautiful & complete series of fossils changing proportionately & regularly" as they were traced downward through his upper Greywacke formations, and it was reasonable to expect the flora to have altered likewise in the course of time. De la Beche's "unexpected Solecism" was thus not only "avowedly an exception to every rule in England"; more seriously, Murchison claimed, it "violates all the laws which we may suppose can have regulated animal & vegetable distribution" during the history of the earth. He therefore warned Buckland that even *"the possibility of such a case as that of De la Beche"* should not be mentioned in his forthcoming Bridgewater treatise except with the utmost caution. "Where is the grand Greywacke Coal Series of Ireland?" he asked Buckland rhetorically, confident that "the little coal hole of Biddeford must follow the fate of the great Irish Coal hole into which Weaver tumbled."

More positively, Murchison tried to answer the criticism that Sedgwick and others had made against his "tenacity" in the matter, namely, that "as I have no *land* plants in my Series below the coal [measures] I have no right to contest the possibility of the miraculous Biddeford forest." To meet this objection, Murchison abruptly reversed his earlier emphasis on the scarcity of plant fossils even in the Old Red Sandstone (§4.5), and claimed that this oldest of Carboniferous formations "does contain *innumerable* indica of vegetable matter of terrestrial origin, *none of which have the least resemblance* to the Coal Measure plants." Yet since he admitted that these fossils had seemed too poorly preserved to be worth collecting, and that Lindley had advised him to record them in his book merely as unidentifiable terrestrial plants, it is clear that Murchison was having to scrape the bottom of the empirical barrel in order to find any fossil evidence to support his assertions about plant life before the Coal Measures period.[54]

Evidently both De la Beche and Buckland had now lost confidence in Weaver's credibility, at least on such an important matter of interpretation. But De la Beche's own reputation would be harder for his opponents to dent, particularly since his opinion on the Culm strata of Devon was now supported by all the other geologists who had studied the area for themselves. Furthermore, De la Beche had gone back to Bideford, since hearing of Weaver's retraction, to check on the position of the culm. "All right and tight," he reported to Greenough, "in the grauwacke to all intents and purposes."[55] The argument was not likely to be resolved as easily as Murchison and Lyell imagined.

A month after Weaver's retraction had been read at the Geological Society, its members assembled to hear Greenough speak for the last time as president. The customary ballot for new officers was a mere formality. It was a foregone conclusion that the next president would be Lyell; among other changes, De la Beche was to take over from him as foreign secretary, a function he could perform adequately even from the depths of Devonshire. Before handing over the presidency, however, Greenough had one last chance to comment without opposition, at least in public, on contentious matters such as the Devon controversy.

54. Ibid.
55. De la Beche to Greenough, 29 January 1835 (ULC: GBG).

Greenough accepted Weaver's "error" without comment, as if it did not greatly affect the argument. He noted that the age of the American coal and anthracite deposits was still uncertain: he conceded that the great Pennsylvania coalfield, although "said to occur in the higher beds of grauwacke," might turn out to be Carboniferous in age, but he considered that other American deposits (those of Rhode Island and Massachussetts) were in genuine Greywacke or even among Primary rocks. This implied that De la Beche's report was in no way peculiar or unique. When he dealt explicitly with that report, Greenough noted with fine understatement that it had "given rise to a good deal of discussion," and he defended it vigorously. He pointed out that clear coastal sections of the strata were available, and he cited Smith's geological map as testimony to the Greywacke age of the strata. Neither Smith nor any other geologist, he emphasized, had ever expressed in print any suspicion that the Culm might be as young as the Coal Measures; "nor am I aware," he added pointedly, "that such suspicion is entertained even now by any one who has seen them in situ."[56] This cleverly brought Smith—the father figure of the fossil-based geologists like Murchison—on to his own side, while at the same time alluding to the fact that those who had criticized De la Beche's report had not yet been to Devon to see the rocks for themselves.

Greenough claimed that a geologist as experienced as De la Beche was unlikely to have been mistaken, and that anyway there were similar cases known on the Continent. But his strongest comment was methodological. He criticized the assumption that the ranges of fossils in the sequence of formations were already precisely known, "unless indeed we choose to suppose, amid all the obscurity that surrounds us, that our knowledge has already reached a maximum, and that nothing more can ever be visible than that which we have been accustomed to see." That arrogant assumption, Greenough implied, was what really underlay Murchison's and Lyell's criticisms of De la Beche's report, and such apriorism struck at the foundations of the whole empirical edifice of geological science.

Greenough used his final moments of authority as president to eulogize De la Beche as a true empirical geologist, and his *Researches* as a model that the Society should emulate.

> Unshackled by authority, unenslaved by preconceived opinions, unseduced by the love of novelty, free from all vanity of authorship, concise, methodical, exercising his judgment continually, his fancy seldom, the author may not obtain that popularity which with less merit he might have easily commanded; but such a work cannot fail to be appreciated here.[57]

Uncovering the allusions that would have been clear enough to his audience, Greenough presented De la Beche as the very opposite of all that Lyell stood for. De la Beche's book might not win as much popular success or so many editions as Lyell's *Principles*, to which Greenough had earlier referred rather grudgingly, but he implied that it was the sounder and more scientific of the two. "You will find an ample guarantee for more brilliant anticipations of success," he concluded, "in the youth, the spirit, the abilities and the character

56. Greenough (1835, pp. 155, 159, 163–65), read at GSL on 20 February.
57. Ibid, p. 173.

of my successor."[58] But in fact it must have been with great unease that he handed over effective power in the Society to Lyell with the Devon controversy unresolved.

5.7   Phillips Interprets the Anomaly (February to April 1835)

De la Beche, backed by Greenough, had taken his stand in the Devon affair on the plain "fact" that the disputed strata were an integral part of the Greywacke, whatever the plant fossils might turn out to be. De la Beche admitted that he had been surprised when Lindley identified *all* of them as Coal Measures species; he had only been expecting a general affinity between the two floras, not an exact identity (§5.3). Both Sedgwick and Élie de Beaumont had told De la Beche they were not surprised by Lindley's report, but neither had suggested *why* a Greywacke flora should turn out to be identical to the well-known flora of the far younger Coal Measures. It was left to Phillips to supply a possible explanation for what was, on De la Beche's side, a puzzling anomaly, if not one of the same magnitude as his opponents faced.

Phillips had just sent De la Beche a copy of his newly published *Guide to Geology*. A week after Greenough's address, De la Beche returned the compliment with copies of his own latest works; he praised Phillips's little book and urged him to write a similar introduction to the study of fossils, which he knew was the younger geologist's strongest claim to fame.[59] He admitted however that he was less confident than Phillips about the reliability of fossils in correlation, and he mentioned specifically Élie de Beaumont's discovery of Oolitic (Liassic) belemnites and Coal Measures plants in closely associated strata in the French Alps (§5.3). "Judging only from organic remains," he commented, this would imply the absurdity of "a rock both lias and coal measures at the same time." Since he told Phillips his skepticism was directed at "the geological value of terrestrial vegetation," it is clear that, like Élie de Beaumont himself, he had resolved that anomaly by discounting the evidence of the plants while accepting that of the belemnites. It was in this context that De la Beche referred to the "curious matter" of his recent discovery in Devon. He had of course treated his own anomaly in the same way as Élie de Beaumont's, again discounting the evidence of the plants while accepting in this case the clear Greywacke position of the disputed strata. He told Phillips there had been "some skirmishing" about his discovery when it was reported at the Geological Society; he added, with a characteristic tilt at his critics, that "the fact is as stated, as can readily be seen by those who will take the trouble to see."

A month passed before Phillips got round to a reply, but when he did so he ranged himself enthusiastically behind De la Beche.[60] Somewhat deferentially he remarked on the "very unexpected coincidence of thought and feeling" that he had noticed between De la Beche's writing and his own

---

58. Ibid, p. 175.
59. De la Beche to Phillips, 27 February 1835 (UMO: JP, 1835/8); Phillips (1834).
60. Phillips to De la Beche, 1 April 1835 (NMW: DLB).

ideas. He attributed this to the fact that their respective views had been *"found in the impressive communion with nature which many of our book men have not enjoyed."* If Phillips had been shown a copy of De la Beche's caricature (fig. 5.4), as he may well have been, he would surely have relished its polemical contrast between the field geologist and his indoor critics. He claimed that only his own similar openness to empirical experience had saved him from "a blind adherence to some overstrained system" of theory. As one example of his own changes of mind, he cited his earlier belief "that every Rock of the oolitic Suite had its own 'peculiar' fossils." It may seem surprising that the nephew and pupil of William Smith should have dismissed as a "fantasy" Smith's own view of the various formations of the Oolitic sequence, for these were the supreme exemplar of the use of characteristic fossils. But Phillips was no uncritical adherent of his uncle's pioneer work, and his comments on De la Beche's Devon report show how far his conception of the relation between fossils and strata had diverged from Smith's.

Phillips told De la Beche that he found the discovery of "plants of the Carboniferous era in Grauwacke" very interesting but not surprising, and for three reasons. First, he thought "the Rate of Change of Organic forms" might be different for terrestrial organisms, implying that their fossil record might be, as it were, out of phase with the more complete record of marine life. Second, he referred to Élie de Beaumont's Alpine discovery as evidence that "certain districts might really enjoy an immunity from those revolutions of organic life which happened to others." Like De la Beche, Phillips evidently trusted the belemnites rather than the plants in this case. But his comment added a *causal* explanation for the anomaly, since it suggested that in the region that had since become the Alps the Coal Measures flora might have survived much longer than elsewhere, as a result of local ecological circumstances. A similar explanation would of course be directly applicable to De la Beche's Devon discovery. Phillips's third reason for finding that discovery unsurprising was in effect a generalization of the specific case.

> Though not a common doctrine I am quite prepared to find the "ages" of particular races of organic beings (especially terrestrial) not exactly synchronous in all districts; because the forms of organic life are probably *not* functions of the *time* but of the physical conditions, which, no doubt, in general did vary with the time.

What is striking about all three arguments is that they stem from a conception of fossils as the remains of truly living organisms, not merely as characteristic objects found in certain strata. As had already become clear from the first volume of his *Illustrations of the Geology of Yorkshire* (1829), dealing with the younger Secondary strata of that county, Phillips's approach was far more biological or, more precisely, ecological, than that of any of the other geologists who had so far entered the controversy. It was nearest to Lyell's in its awareness of the subtle interactions in space and time between organisms and their environments. Yet Phillips realized that even Lyell's assumption of a uniform rate of organic change was itself questionable, and that a fully ecological point of view would have to allow for the possibility that fossils might characterize particular environments far more than particular periods. Applying these general conceptions to the explanation of the Devon problem, it was quite possible in Phillips's view that a flora of so-called Coal Measures species might have flourished locally in specific ecological conditions during the Greywacke period, long before it was able to spread widely and become the "characteristic" flora of the Coal Measures period. Phillips was thus pointing toward a possible causal explanation that reached far beyond De la Beche's and Greenough's dogged insistence on the mere empirical "facts" of the case.

This put a valuable new argument into De la Beche's hands, and it may well have strengthened his resolve to criticize the dogmatic use of characteristic fossils as forcefully as possible in the volume on *Geology* that he was just completing for the *How to Observe* series. In that introductory book he termed the notion of characteristic fossils a "hasty generalization," which was "mischievous" if upheld in the face of contrary facts, and anyway "utterly at variance" with what was known of the present geographical distribution of organisms.[61] But Phillips's remarks on the Devon problem ended with a comment that indicated how his support might turn out to be two-edged.

> The first account I had of your discovery was that a great many of our [Yorkshire] *mountain Limestone fossils* were in your grauwacke; this was a novelty to me, yet I expect something of this kind will at last turn out to be true. There are so many Novelties in my list of the fossils of the Yorkshire Limestones that it ought not to startle us if some of them should be repeated in other rocks.

Who had told him this is not clear, but the report evidently referred to the *animal* fossils found in the strata associated with the Culm, not to the plant fossils that had caused the controversy. Phillips seems to have considered that this could be assimilated with the same explanation: no one could claim to know with any certainty the geological ranges of the animal fossils that he was busy describing from the Mountain Limestone, and some of them might well extend down into the Greywacke. But Murchison and Lyell, if and when they heard about it, would have seen at once that the report could be turned to their own advantage. For it enlarged the apparently "Carboniferous" aspect of the Devon strata beyond just the plant fossils, and thereby expanded the scale of what they already regarded as a massive anomaly. It would have increased the suspicion that De la Beche had somehow made a major error in his mapping work for the Ordnance Survey.

5.8   De la Beche's Livelihood Secured (May to August 1835)

Such suspicions were just what De la Beche least wanted to reach the ears of the Board of Ordnance. In anticipation of Greenough's anniversary address, De la Beche had written to tell him of the progress of his survey and had asked him to use the occasion to urge its continuation, adding in confidence that he wanted to move into Cornwall "in the first instance."[62] Greenough had duly reported that all but one of De la Beche's eight sheets were finished, and the first four already published. Directing his remarks at more influential

---

61. De la Beche (1835, pp. 17–18, 23, 220, 238).
62. De la Beche to Greenough, 21 January 1835.

ears than those of his fellow geologists, he had added the hope "that this work so admirably begun may not be suffered to terminate here."[63]

By May, De la Beche was able to report to the Board that the mapping he had been commissioned to do was now complete, and that he was willing to continue the survey elsewhere.[64] The Board referred his proposal to what might be termed, *avant la lettre*, a governmental advisory committee on science policy. This consisted of the President of the Geological Society, the Reader in Geology at Oxford, and the Woodwardian Professor of Geology at Cambridge: in effect, De la Beche's proposal was referred to Lyell, Buckland, and Sedgwick.[65] The questions put to this ad hoc committee were: Had the geological map of Devon been well executed? Should the work be extended under De la Beche's direction? What was the best mode of organizing it? What was the likely expense?

This impinged on the Devon debate in one very important respect: any formal statement of opinion on the quality of De la Beche's work would have to take into account a matter that had been the subject of heated argument in the previous months. De la Beche was evidently confident of Sedgwick's support, having written to him beforehand to ask him to use his personal influence with the Marquis of Lansdowne (b. 1780), Lord President of the Council in Melbourne's new administration and a prominent Whig supporter of science. He could also have been confident of Buckland's support, for in arranging a meeting of the committee Buckland commented quite casually to Sedgwick that "we all know the Perfection of the Work," while he told the secretary of the Board of his "very high opinion" of it.[66] Both Sedgwick and Buckland had already expressed the view that De la Beche's report of "coal" plants in the Greywacke of Devon was well founded or, at the very least, plausible. Lyell's attitude was quite another matter, but it would have been difficult for him to oppose the joint judgment of his two more senior colleagues.

In the event, the committee reported that De la Beche's mapping was "the result of great labour combined with great skill" and that it was unequaled in Europe for its detailed accuracy.[67] Appealing skillfully to the Board's patriotism, they recommended the continuation of the survey so that it could produce a map comparable to the one that had almost been completed in France. Appealing equally to their practical concerns, the three geologists also emphasized the economic value of the survey, not least in preventing the wastage of vast sums of money on the fruitless search for coal. They recommended the establishment of a new geological branch within the Ordnance Survey, with De la Beche at its head and some assistants at his disposal. In

63. Greenough (1835, p. 154), read at GSL on 20 February. He had made similar comments the previous year: Greenough (1834, p. 51), read at GSL on 21 February.

64. De la Beche to Ordnance, 25 May 1835 (NMW: TS, II, p. 93).

65. Board of Ordance, minute of 25 May 1835 (NMW: TS, II, p. 94). The request was first sent in error to Greenough instead of Lyell: Ordnance to Greenough, 25 May 1835 (ULC: GBG); Ordnance to Sedgwick, 25 May 1835 (ULC: AS, IB, 15). Like the Royal Society (see Hall 1981), the Geological Society was used not infrequently by the government as a source of scientific advice.

66. De la Beche to Sedgwick, 13 May 1835; Buckland to Sedgwick, 20 May 1835 (ULC: AS, IB, 913); Buckland to Byham (Ordnance), 30 May 1835 (NMW: TS, II, p. 96).

67. Buckland (with Sedgwick and Lyell) to Ordnance, 12 June 1835 (NMW: TS, II, pp. 98–102).

place of his fee of £300 for some three years' work, they proposed a salary of £500 a year, with £1000 a year for expenses. De la Beche could hardly have asked for a more generous proposal.

The committee's report was duly accepted by the Board of Ordnance and passed on to the Treasury for financial approval. In the course of protracted negotiations, the proposed generous funding for expenses was cut out, and the Treasury asked with characteristic caution how long it would take to complete a geological survey of the rest of England.[68] On the official level, De la Beche maintained that no answer could be given, since it would depend on the assistance he was given; unofficially, he pressed home his advantage by telling Francis Baring, the Secretary to the Treasury, that he estimated it would take only ten years if he had two or three assistants, but twenty-one if he had to do the work alone.[69] He waited impatiently for final approval of his new appointment, hoping it would come through in time for him to begin work in Cornwall while the summer weather still allowed the use of a boat to study the coastal cliffs. He had to ask Sedgwick once more to use his personal influence behind the scenes, not only with Lansdowne but also with Thomas Spring-Rice (b.1790), the Chancellor of the Exchequer.[70]

Murchison for his part could now at least rest assured that De la Beche's activities would not threaten his own publication plans. The very day after the committee of three met, Buckland wrote to tell him that the proposed official survey would not get round to his Welsh districts for several years, so that Murchison's "Alarms" about his priority there were unfounded.[71] However skeptical Murchison had been about De la Beche's work, he could now afford to get on with his own research and shelve his doubts until he could go and study Devon for himself.

Despite those doubts, and Lyell's too, De la Beche's reputation was now as high as ever in official quarters. His proposal for the establishment of a public collection of specimens to illustrate the practical value of geology was sympathetically received by Spring-Rice; by August he had been allotted some space just off Whitehall for a small Museum of Economic Geology (see fig. 2.6).[72] Since the new museum came under the jurisdiction not of the Board of Ordnance but of the Department of Woods and Forests, the risks of De la Beche's nascent career in governmental science were now neatly spread; any enemies he might have would find it more difficult to dislodge him from two positions than from one. Despite bureaucratic delays over the establishment of the new geological survey, De la Beche was clearly set on a course of expansion, and his employment in science looked more secure.

Half a year after it first erupted, the Devon controversy had reached, if not a stalemate, then at least a tacit truce. Until it was resolved, however,

---

68. Board of Ordnance, minute of 15 June 1835; Treasury to Ordnance, 30 June 1835 (NMW: TS, II, pp. 104–6).

69. De la Beche to Ordnance, 4 July 1835; De la Beche to Baring, 9 July 1835 (NMW: TS, II, pp. 113, 119–20).

70. De la Beche to Sedgwick, 3 July 1935 (ULC: AS, IB, 27).

71. Buckland to Murchison, 12 June 1835 (DRO: WB, F.221).

72. De la Beche to Spring-Rice, 13 July 1835 (NMW: TS, II, p. 121); McCartney (1977, pp. 34–35).

it remained not only a massive anomaly in the path of both Murchison and Lyell, but also a serious if dormant threat to De la Beche's public reputation and career. Beyond the personal level, it remained one of the most important unsolved problems not only for geological theory but also for the science's prospective contribution to the industrial economy.

Chapter Six

TAMING THE TRANSITION

6.1 Murchison Defines the Silurian System (June and July 1835)

When, some six months earlier, Murchison made his dramatic exit from the meeting at which De la Beche's original report from Devon was read to the Geological Society, he had—according to Greenough's account—sworn never to use the term Greywacke again (§5.2). For he saw clearly that at the very least its ambiguity was aggravating the disagreement. De la Beche seemed to him to be identifying the disputed strata in Devon as Greywacke in age, and hence pre-Carboniferous, merely on the grounds that the rocks were of a greywacke appearance or rock type. Soon after that meeting Murchison had therefore suggested to several other geologists that the term Greywacke should be replaced with its virtual synonym Transition, in order to obviate the ambiguity once and for all.

Buckland had approved, asking if Murchison proposed to divide the Transition into upper and lower parts, as the French geologists had done, in order to accommodate, respectively, his own and Sedgwick's formations.[1] Both Sedgwick and Phillips, however, had criticized the proposed term for being too imprecise. "We *must* if possible get rid of the word *transition*," Sedgwick had insisted, adding more specifically, "you have no good base line, and an immense thickness of older *transition rocks* are below your series."[2]

1. Buckland to Murchison, 5 December 1834 (GSL: RIM, B34/9).
2. Sedgwick to Murchison, n.d., pmk 23 December 1834 (ULC: AS, IC, 12); Phillips to Murchison, 12 December 1834 (GSL: RIM, P14/10).

Murchison had duly taken the hint to keep off Sedgwick's territory, but in the following months he had come to feel the need to characterize his Borderland formations in some more distinctive way than as "Upper Transition" or the now unacceptable "Upper Grauwacke." He therefore decided to revive the memory of the Silures, the warlike British tribe that had fought the Romans in part of what was now the Welsh Borderland, and to use their name to characterize both his region and his formations.[3] The Silures' historic resistance to a foreign power made them an appropriate symbol for Murchison's own bid to forestall any possible foreign competition, by coining an unambiguously British term for his former "Upper Grauwacke" strata.

Murchison told Buckland he had decided to use the term "Silurian System" for all the formations between the Old Red Sandstone and the Greywacke "properly so called." Claiming that Greenough and De la Beche already approved of the new term, he urged Buckland to adopt it in his "Magnum Opus" and to drop the term Transition altogether. Asserting that the "simplicity of the project & the absence of all theory" should make it generally acceptable, he explained the Silurian by an idealized section through the Borderland (fig. 6.1).[4] This showed his thick sequence of formations below the Old Red Sandstone, dipping as gently and regularly as Smith's Secondary formations in central England (see fig. 4.7). The four subdivisions he had already outlined in public (§4.5) were now named after appropriate localities: in order from top to bottom, *Ludlow, Wenlock, Caradoc* and *Llandeilo* (see fig. 4.8, B, I–IV; also fig. 3.5). The Greywacke was newly defined as excluding all these formations and was implicitly left for Sedgwick's strata; Murchison's were shown as sharply separated from them by a major unconformity.

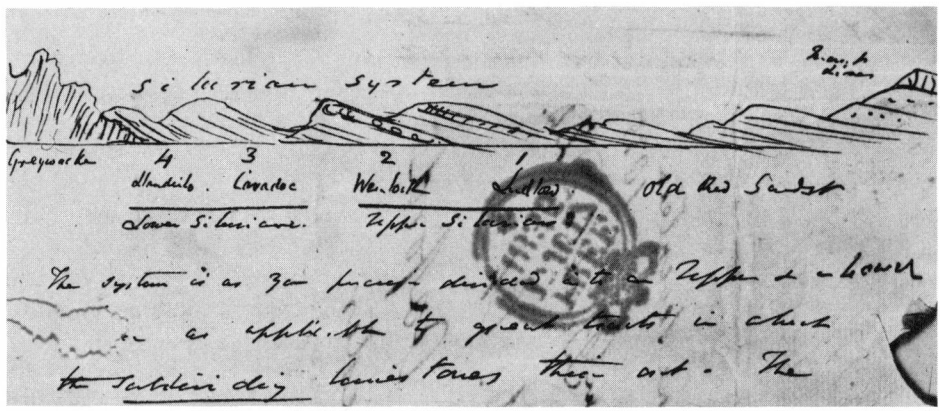

Fig. 6.1.   Murchison's idealized section of the Welsh Borderland, showing his definition of the Silurian System and its four subdivisions. The lowest (4, Llandeilo) rests unconformably on a redefined Greywacke (left), and is followed by two middle formations (3, Caradoc; 2, Wenlock). The uppermost (1, Ludlow) is overlain conformably by thick Old Red Sandstone and some residual Mountain Limestone (right). This sketch was sent to Buckland in June 1835, and a woodcut based on it was published shortly afterward.

3. He had to strain the historical evidence to make the territory of the Silures extend as far north as Shropshire, the "type" area for most of his formations: Secord (1982).
4. Fig. 6.1 is reproduced from Murchison to Buckland, 17 June 1835 (UMO: WB, box B.T.).

The most surprising aspect of Murchison's proposal was his use of the word "system" to denote what other geologists would have termed a group, series, or *terrain*. In English, the word "system" was generally used only in a loose and colloquial sense, to refer to a *local* area of some series of strata with a particular direction of strike. But in French the word *système* had come to have a much more theory-loaded meaning, as a result of its use in Élie de Beaumont's influential but controversial theory of crustal movement.[5] On that theory, *all* areas of strata that had been elevated or folded during a particular *révolution du globe* were expected to have the same orientation, wherever they were found; the word *système* therefore referred as much to the episode of crustal activity as to the strata that had been affected by it. When Élie de Beaumont told Murchison he approved of his "système silurien," he probably interpreted it in terms of his own theory.[6] Conversely, English geologists might have felt confused by its ambiguity: was it a descriptive and local term or a putatively general but theory-loaded one? Certainly Buckland avoided using the word "system" in his reply; he simply told Murchison he approved of the "Silurian Series" as a name for the formations in his "Silurian Region."[7]

To ensure his priority, Murchison wrote an article "On the Silurian System of Rocks" for rapid publication in the *Philosophical Magazine*.[8] This public definition of the Silurian and its subdivisions gave Murchison's formations a set of proprietary brand names, as it were, but otherwise it added nothing of substance to what he had published the previous autumn (§4.5). The most pertinent comment came from Élie de Beaumont. Writing to Sedgwick, he praised Murchison's work for the way it helped "de *secondariser* le groupe Barbare des terrains de transition."[9] The neologism and the metaphor were apt: the Silurian was seen as having tamed the barbarous confusion of at least the upper part of the Transition into as much order and simplicity as the Secondary formations.

Murchison's work was bound to focus attention once again, however, on the magnitude of the discrepancy between the allegedly plantless Silurian system and the reports of plant fossils in the Greywacke of other regions, most notably De la Beche's from Devonshire. Collecting the latest information for his forthcoming Bridgewater treatise, Buckland was openly incredulous at Murchison's continuing claim that the Silurian strata were totally lacking in plant fossils: "if De la Beche has found them in the Grauwacke of Biddeford, they must have existed during all your Silurian Period and we should expect at least a few Traces of them."[10] Murchison replied however that although he and his "aides de Camp"—the local geologists of "Siluria"—had searched for plant fossils, they had found only a few obscure traces in the Old Red Sandstone and none at all in the Silurian strata.

> De la Beche's theory is that the same vascular Cryptogamic vegetation was in existence during the whole Greywacke periods. If so it is as you

5. Élie de Beaumont (1829–30); Greene (1982, chap. 3).
6. Élie de Beaumont to Murchison, 17 June 1835 (GSL: RIM, E4/9).
7. Buckland to Murchison, 19 June 1835 (DRO: WB, F.216).
8. Murchison (1835b), published in July issue of *PMJS*. This contained a woodcut based on the section he had sketched for Buckland (fig. 6.1).
9. Élie de Beaumont to Sedgwick, 26 July 1835 (ULC: AS, IF, 36); Élie de Beaumont to Murchison, 14 June 1835 (GSL: RIM, E4/8).
10. Buckland to Murchison, 19 June 1835.

observe most remarkable that not a vestige of it should have found its way into these deposits (many of which have all the characters of *shallow sea* and shore accumulations) which I have so studiously examined and which are *so well elaborated*. Of course I cannot go against *his facts* but I shall stick to my own text.[11]

Like De la Beche, Murchison was here claiming credibility for his report on the basis of both the extent and thoroughness of his own fieldwork and the clarity of the exposures. Yet the absence of fossil plants remained undeniably negative evidence.

Murchison's insistence on rejecting the possibility of any trace of plant remains in his Silurian strata is all the more striking when compared with his treatment of the parallel problem of fossil fish. He told Buckland to be "very cautious in asserting upon mere negative evidence the non existence of fishes in the Lower Silurian System." The Old Red Sandstone, he insisted, was "proved by my researches to be a Piscina throughout." Agassiz had already begun to analyze and describe those fish, not only from the Welsh Borderland but also those in the much richer collections from the Old Red Sandstone of Scotland. Murchison had already found traces of this striking and largely new assemblage of fossil fish in the uppermost Silurian; he evidently expected that with more intensive collecting the record would be extended still further down the sequence.[12] So what he anticipated for the record of fish was just what he emphatically denied for the record of land plants; the value of negative evidence was renounced in the one case but insisted on in the other.

The only plausible explanation for this striking inconsistency is that the two groups of fossil remains played contrasting roles in Murchison's wider ambitions for his Silurian rocks. The fish served to cement the link between Silurian and Old Red Sandstone, to underline the continuity that was manifest in their conformable relation, and to enrich the range of fossils that demonstrated the fullness of the living world even in the Silurian period. A record of land plants, if found, could in principle have served precisely the same functions in the characterization of his new system. But for Murchison to have admitted even the possibility of finding land plant fossils in Silurian strata would have been to eliminate at a stroke all his ambitions for making his system a decisive terminus in the hunt for coal.

6.2   The Dublin Festival of Science (July and August 1835)

With his stake in the Silurian system established on public record, Murchison began his summer's fieldwork, tracing the outcrop of his Silurian rocks westward from "Siluria" into southwest Wales. Although he had not forgotten how De la Beche had challenged him to go to Devon, he had to secure the Silurian first. He had hoped that Sedgwick would join him in Pembrokeshire before they both crossed to Dublin for the British Association meeting. In the event,

11. Murchison to Buckland, 24 June 1835 (GSL: WB, M9). The "vascular Cryptogamic" plants were the tree ferns and giant horsetails and club mosses of the Coal Measures (see fig. 5.3).
12. Ibid. On the Scottish fish, see Andrews (1982).

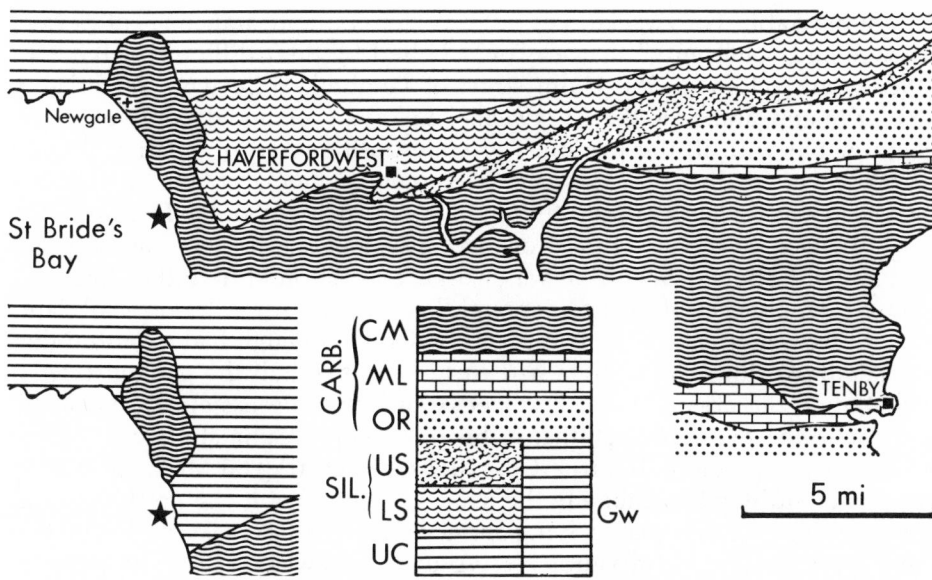

Fig. 6.2. Murchison's interpretation of part of Pembrokeshire, redrawn from the map published in his *Silurian System* (1839) but based on his fieldwork in 1835 (for the location of St. Bride's Bay, see fig. 3.5). The Culm-like Coal Measures (CM), folded in a narrow east-west trough, overlie the Mountain Limestone (ML) conformably in the east, but become unconformable to the west and rest successively on Old Red Sandstone (OR), Upper and Lower Silurian (US, LS), and Upper Cambrian (UC). The inset shows a small part of De la Beche's earlier map of the same area (1826), showing the Coal Measures resting on undifferentiated Greywacke (Gw). The stars mark the point in the coastal cliffs where plant-bearing strata were attributed to the Coal Measures and the Greywacke, respectively: Murchison claimed that this showed how De la Beche might have made an analogous mistake in Devonshire.

he had to work on his own, but he did find a possible explanation of the mistake he was convinced De la Beche had made in Devonshire.[13]

In St. Bride's Bay on the west coast of Pembrokeshire, Murchison sketched a cliff section that he captioned "Greywacke of de la Beche." He noted that the strata were "undistinguishable from many strata of the Silurian Series," yet they contained "coal plants." By itself, this might have been taken as confirmation of the stand De la Beche had taken over the disputed strata and plant fossils of Devon. But Murchison believed he had found the vital clue to De la Beche's mistake. For here in Pembrokeshire these "Culm Measures" passed conformably down into what he identified as "Millstone Grit"— the well-known sandstone that formed the normal intermediate formation between Coal Measures and Mountain Limestone. In Murchison's view this was "totally unlike any thing in the Transition Series," yet De la Beche had colored it as Greywacke on his map of Pembrokeshire some years before (fig. 6.2).[14] Traced further north (near Newgale), the same Culm rocks lay directly

13. Murchison's plan to get Sedgwick to join him in Pembrokeshire is mentioned in Murchison to Buckland, 2 July 1835 (GSL: WB, M1).

14. Murchison, notebook "July 1835 Vol. 19" (GSL: RIM, N74, p. 86), notes after dated entry for 20 July. Fig. 6.2 is traced from the main map in Murchison (1839); the inset is from De la Beche (1826, pl. 1), to which Murchison's notes refer.

on unmistakable Greywacke, yet with a "general appearance . . . of perfect passage and conformity" that was only belied by their different strike.[15] In other words, although the Culm strata of Pembrokeshire were remarkably similar in appearance to the Greywacke, and even seemed deceptively to be conformable with it, the evidence of superposition and orientation proved that they were really Coal Measures. The potential anomaly of the fossil plants then evaporated, and so, apparently, might the anomaly of those in Devonshire.

In fact, however, Murchison was privately scoring too easy a point against De la Beche. While he was working in the field he either forgot or ignored the concession that Greenough had made during the original "animated discussion" of the Devon plant fossils. Greenough had admitted that the Pembrokeshire Culm strata had earlier deceived all the geologists who had studied the area (§5.2). A few weeks after that discussion, De la Beche had confirmed to Greenough that in his opinion the Culm strata in Pembrokeshire were definitely the true Coal Measures, resting on Mountain Limestone and Old Red Sandstone in the usual way; but that to the north the Coal strata became unconformable, lying on successively older rocks, so that locally there were "true coal measures *resting* upon grauwacke." The situation in Pembrokeshire was therefore no proper analogy for Murchison to cite against him; it was, De la Beche claimed, "altogether a different affair from the Bideford culm—the position of which *in* the grauwacke is as plain as a pikestaff."[16] The only mistake that Murchison could properly hold against De la Beche was one quite local detail. In one part of St. Bride's Bay, where the easily recognizable Mountain Limestone had been cut out by the unconformity below the Culm strata, De la Beche had marked as Greywacke a small area of rocks that Murchison identified as Carboniferous (namely, Millstone Grit). But for Murchison even that small mistake was significant, for it suggested that his opponent might have made a similar misidentification on a much larger scale in Devonshire.

It was not only Murchison who used the summer season to attack indirectly—but at the empirical level—the anomaly that De la Beche's Devon report had posed. Just as Murchison nibbled at its edges by trying to find a similarly deceptive instance in Pembrokeshire, so Lyell crossed to Paris determined to scrutinize critically the supporting analogy that De la Beche had cited from the French Alps. He had many other concerns more pressing for his current research, particularly those connected with his work on the geologically recent elevation of land masses and the mode of formation of volcanos. But that only makes it the more striking that the very first item on his agenda for his work in Paris read: "Belemnites found by [Élie] De Beaumont in Alpine coal-plant lias: see them."[17] Élie de Beaumont was also first on his list of people to see in Paris. Although this was mainly in connection with the elevation problem, Lyell did make notes on the French Alpine localities and may have considered visiting them.[18]

---

15. Murchison, notebook 19, p. 96.

16. De la Beche to Greenough, 21 January 1835 (ULC: GBG, 49).

17. Lyell, notes after dated entry for 8 July 1835 (KHK: CL, notebook 57, p. 81): agenda headed "Paris" and probably written when he arrived there.

18. Ibid., p. 75; notebook 58, p. 11. Lyell's stay in Paris, and his subsequent fieldwork in Switzerland, are described in letters printed in Clarke and Hughes (1890, 1:450–57); see also Wilson (1972, pp. 413–20).

In the event, however, Lyell went instead to the Swiss Alps. His study of the specimens in Paris, combined with his knowledge of the highly folded and disturbed character of the Alpine rocks, may have convinced him that the anomaly was sufficiently dubious to be left on one side, at least for the time being. Nonetheless, the priority he assigned to the problem beforehand indicates the importance he attached to such a glaring exception to his increasingly confident interpretation of the history of life. The reports from the French Alps might be as questionable as Weaver's Irish case had proved to be. But De la Beche's report from Devonshire, not to mention Élie de Beaumont's instances of Greywacke coal on French soil, was still outstanding.

While Lyell and his wife were touring Switzerland, De la Beche was confined to Cornwall, using the calm summer weather to study the cliffs— and asking authority to commandeer Coastguard boats for the purpose in order to undercut the exorbitant rates charged by local fishermen.[19] With those two exceptions, most of the leading English geologists converged on Dublin for the meeting of the British Association. Sedgwick took the Sunday services on board the steamship that had been put at the Association's disposal to bring participants over from England; no one present could doubt the positive alliance of science with religion that the leaders of the Association wished to foster. Its intended role as an agent of social reconciliation was equally clear: Protestants and Catholics cooperated in Dublin with unaccustomed cordiality, and Trinity College gave honorary degrees not only to several distinguished gentlemen of science—and to the foreigner Agassiz—but also to the humble William Smith, the "respected father of geology" whose education had fallen far short of any university. All in all, it was a festive occasion.[20]

The geologists were allotted one of the largest meeting rooms, and as expected their Section was one of the most popular. As a tactful concession to the home team, the presidency had been assigned to Griffith, who exhibited a draft copy of his map of Ireland, colored to match Greenough's.[21] Murchison and Sedgwick held their joint paper in reserve until the last day, so that it would come as the climax of the week's work. The Section's committee, dominated by the metropolitan gentlemen, arranged for a paper by a mere provincial, which impinged directly on the Devon controversy, to be read the previous day, perhaps anticipating that it could be dismissed as summarily as Boase's had been at the Edinburgh meeting the previous year (§4.6). This time the provincial was David Williams (b. 1792), rector of a country parish (Bleadon) at the seaward end of the Mendip Hills in Somerset. Williams had joined the Geological Society in 1828, but had submitted his first paper in 1834—on fossil mastodon bones in limestone caves in the Mendips—with all the deference of the beginner: "I did not write it for critical and practising geologists," he had told Lonsdale.[22] With further fieldwork his confidence had

---

19. De la Beche to Colby, 2 and 3 August 1835 (NMW: TS, II, pp. 128–29). He had duly got his first assistant, Henry McLauchlan (b. 1791): Colby to Robe, 18 July 1835 (NMW: TS, II, p. 127).

20. Morrell and Thackray (1981, pp. 182–85); *Ath.*, nos. 407–8 (15 and 22 August 1835), pp. 617, 642; *LG*, no. 969 (15 August 1835), pp. 513–14.

21. Hardy (1835, pp. 17–19); *Ath.*, no. 407, p. 619; *LG*, no. 969, p. 514.

22. Williams to Lonsdale, 7 April 1834 (GSL: BP, LR1/32); Williams (1834), read at GSL on 9 April.

grown, however, and he had begun to explore and map the older rocks of west Somerset and north Devon, though apparently without any communication with De la Beche.

Williams reached Dublin full of enthusiasm. The British Association, he believed, "will prevent an author from being unjustly swamped by a villainous criticism or being meretriciously installed in a seat of Fame he does not deserve—in fact a man's calibre will be fairly ascertained and appreciated."[23] He was indeed just the sort of provincial man of science that the leaders of the British Association hoped to attract; but since they also intended to keep them in their proper place, it remained to be seen whether they would treat him as fairly as he anticipated.

In what one reporter found a "striking paper," Williams supported the absent De la Beche. He claimed that the fossil plants of north Devon were indistinguishable from those of Pembrokeshire, yet the culm of Devon was "decidedly transition" whereas that of Pembrokeshire belonged to the true Coal Measures. He concluded bluntly that the case showed "at least an exception to the law, that strata may be identified by their imbedded organic remains." Not content with that provocative criticism—made quite probably in the presence of Smith himself—he also attacked two other "popular or fashionable theories," namely those of the gradual cooling of the Earth and of the evolving composition of its atmosphere.[24] The comments proved him well read in the current geological literature, but were hardly likely to endear him to those who reserved to themselves the authority to pronounce on such high-level matters.

Publicly, however, the leading geologists held their fire. Sedgwick, having perhaps heard from Phillips about the explanations he had put forward to De la Beche earlier in the summer (§5.7), suggested that the anomalous plant fossils from Devon might be explained in terms of a differential rate of change between the land flora and the marine fauna, or by their incomplete preservation. Murchison tactfully thanked Williams for his paper, but stressed that De la Beche's interpretation of Devon needed better proof.[25] Privately, they were less tolerant of the upstart provincial: reporting afterward on how the meeting had gone, Sedgwick told Lyell with unusual vehemence that Williams's paper had been "ignorant and impudent."[26]

The following day, Sedgwick and Murchison duly read their joint paper on the Transition formations of Wales. Murchison repeated the Silurian subdivisions he had published the previous month and stressed how he had now traced the Upper and Lower Silurian no less than 120 miles, from Shropshire to Pembrokeshire.[27] For the older formations, Sedgwick had rejected Élie de Beaumont's proposed term "Hercynian," Murchison's "Snowdonian" or "Dimetian," and his own "Cumbrian." But he did follow the principle behind all those suggestions and finally chose another geographical term, "Cambrian,"

23. Williams, notebook for "Decr. 1834 to Augt. 1835," undated note headed "British Association" (SAS: DW, box I, notebook 7).

24. Hardy (1835, p. 83); *Ath.*, no. 408, p. 646; *LG*, no. 970, p. 535; Williams (1836).

25. Hardy (1835, p. 83).

26. Sedgwick to Lyell, 20 September 1835 (K. Lyell 1881, 1:448).

27. Hardy (1835, pp. 107–8); Sedgwick and Murchison (1836).

after the Roman name for Wales itself.[28] He outlined three subdivisions of the Cambrian—upper, middle, and lower—lying conformably below his colleague's Lower Silurian strata.

In the discussion, Murchison again relegated the term Greywacke to the older strata, firmly excluding the Silurian. Sedgwick noted the gradual disappearance of fossils as his Cambrian strata were traced downward, thereby implying that they contained the very beginnings of the fossil record, if not of life itself.[29] He also stressed how even at that unimaginably remote period, events had clearly been under the "usual causal laws" of nature: even at the beginnings of the record, there was no place for chaos. Greenough cautioned that the authors' subdivisions might only be of local validity, but he predicted that the broader systems themselves would in due course be traced on the Continent and perhaps even to America. Warming to the infectiously amicable atmosphere, Murchison responded with praise for Greenough's mapping. Phillips, notwithstanding his uncle's presence in Dublin and perhaps in the meeting room itself, stressed that all such groups of formations needed to be identified by their distinctive assemblages, not by single "characteristic" fossils. And in a gloss on what Sedgwick had said, he stressed that even the oldest fossils showed no sign of any "inferior structure," thereby implicitly closing the door to any dangerously Lamarckian interpretation of the record.[30]

In summary, the final session had established a clear consensus among the leading geologists. The ancient Transition strata, right back to the beginnings of the fossil record, could indeed be tamed into order and simplicity by the judicious application of fossil-based methods. A reliable start had been made on British soil with the delineation of the Cambrian and Silurian (see fig. 4.8, A and B). It remained to fit into this scheme the troublesome anomaly of De la Beche's report from Devon and to explore how widely afield the new systems could be recognized.

6.3 Anglo-Belgian Correlations (September 1835)

The Dublin meeting over, the leading gentlemen of science dispersed. Sedgwick and Murchison stayed in Ireland: they joined Agassiz and Phillips at a country house party, where they admired the fine fossil collection of the young Viscount Cole (b. 1807), and later they did some fieldwork in the north. Greenough and Buckland, on the other hand, crossed to France and joined the summer field meeting of the Société géologique, fully intending to compare

28. Élie de Beaumont to Murchison, 27 June 1835 (GSL: RIM, E4/9); Murchison to Buckland, 2 July 1835 (GSL: WB, M11); Buckland to Murchison, 17 July 1835 (NMW: DLB, copy); Sedgwick to Murchison, 25 July 1835 (GSL: RIM, S11/90). Élie de Beaumont's suggestion was derived from the Harz mountains, the classic area for the concept of *Grauwacke*. Buckland urged the choice of an unambiguously British term to forestall foreign rivals.

29. The Lower Cambrian was said to contain no fossils at all, and the other two divisions very few. In Sedgwick's major paper on the Cumbrian region in the north of England, which after lengthy revision was at last in press, he emphasized that the absence of fossils in the lowest Greywacke—in Cumbria, Wales, Devon, and Cornwall alike—could *not* be attributed along Lyellian lines to their destruction by metamorphism: Sedgwick (1835b, p. 66), read at GSL on 5 January 1831, published 18 November 1835.

30. Hardy (1835, pp. 108–9).

the newly clarified British sequence of Transition strata with one of the finest sequences on the Continent. The French geologists met this year at Mézières, not far from the Belgian frontier and the edge of the Ardennes massif (see fig. 3.6). The party of twenty-five included six foreigners, four of them British. The first few days were spent studying the Secondary strata around Mézières. For the second half of the meeting the geologists traversed down the valley of the Meuse through the uplands of Transition rocks, past the frontier town of Givet into Belgium, and all the way to the mining and industrial center of Namur and beyond.[31] To interpret what they saw, they had two leading Belgian geologists to guide them.

The older of the two was Jean-Baptiste d'Omalius d'Halloy (b. 1783). As a young man, d'Omalius had worked as a geologist for the French government, and after the fall of Napoleon had served as governor of his native province of Namur. When in 1830 Belgium won its independence from the northern Netherlands, he lost that administrative position, retired to his country château in the Condroz south of Namur, and devoted himself to science.[32] This enforced early retirement had given him time to complete his *Éléments de Géologie* (1831); he had summarized his mature conclusions by distinguishing three major *terrains* of older strata in Belgium, correlating them provisionally with strata elsewhere (fig. 6.3, A).[33] The *terrain houiller* was manifestly equivalent to the Coal Measures of England; it formed the great Belgian coalfield running through Namur and Liège (fig. 3.6), the basis for the new kingdom's growing industrial strength. Underlying these coal strata was the *terrain anthraxifère*, a varied series of sandstones, shales, and limestones, which d'Omalius equated collectively with the Mountain Limestone and Old Red Sandstone of the English sequence.[34] Lowest of all, and forming the uplands of the Ardennes, was the *terrain ardoisier*, which d'Omalius identified as the German-speaking geologists' *Grauwacke*.[35]

The star at the Mézières meeting, however, was not d'Omalius but his brilliant young protégé André-Hubert Dumont (b. 1809). Dumont had originally followed his father's vocation, qualifying at Liège as a mining surveyor. Having become interested in geological questions beyond the immediate demands of mining, he had entered the competition set by the Académie des Sciences in Brussels for an essay on the geology of his native province. In 1830 he had won the prize with a massive memoir; d'Omalius, who had been one of the judges, had urged him to study at the university. Even before graduating, Dumont had been elected a *correspondant* of the Académie and had been encouraged to extend his mapping to cover the whole country. By the time he came to the meeting at Mézières, plans were in hand for his

31. Report of *reúnion extraordinaire* of SGF, 1–10 September 1835 (*BSGF* 6:323–57).
32. Fourmarier (1968).
33. D'Omalius (1831, chaps. 18–20); borrowing from d'Omalius (1828), Rozet (1830), and Dumont (1832). See below.
34. The correlation had been accepted by Conybeare (1833) in his general report on geology for the British Association, when he proposed the term "anthraxiferous" for the strata between the Coal Measures and the Greywacke; however the term was not widely adopted in England. The word *anthraxifère* referred to the characteristic black limestones of Belgium, some of them quarried for ornamental "marble," which were believed to owe their color to carbonaceous material.
35. D'Omalius noted, however, that in German usage *Grauwacke* extended upward as far as the productive coal strata and therefore included his *terrain anthraxifère*.

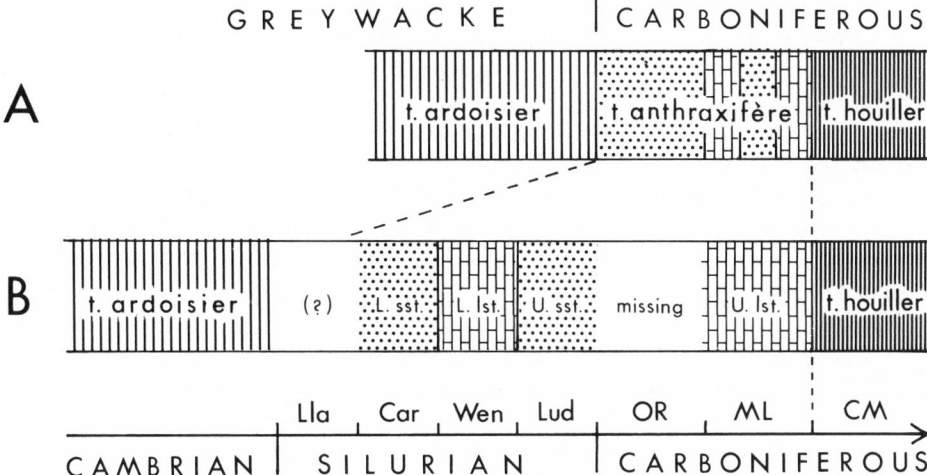

Fig. 6.3. Correlations between the Belgian and British sequences of older strata. (A): the correlation proposed by d'Omalius in 1831. (B): the correlation proposed by Buckland during the field meeting of the Société géologique in September 1835, incorporating Murchison's and Sedgwick's work in Britain and Dumont's in Belgium, and involving a radical reinterpretation of the *terrain anthraxifère*. The British sequence is reduced for simplicity to a linear timescale, as classified in 1831 (top) and 1835 (bottom). Lla[ndeilo], Car[adoc], Wen[lock], and Lud[low] are Murchison's four divisions of the Silurian (see fig. 6.1). Dumont later modified Buckland's correlation by inferring that the upper sandstone group (U. sst.) probably included the equivalents of the Old Red Sandstone (OR) as well as the Ludlow.

nomination—at the age of only 26—to a new chair in mineralogy and geology at Liège.[36]

Dumont's memoir on the province of Liège (1832) had included not only a detailed geological map but also a series of bold traverse sections, in which the surface geology was interpreted in terms of a series of huge folds, some of them overturned. For the Mézières meeting he prepared a similar section along the line to be traversed by the party through the province of Namur, showing a similar series of folds, for which the evidence was to be seen along the sides of the trenchlike valley of the Meuse (fig. 6.4).[37] Dumont had presented a copy of his memoir to the Geological Society in 1833, but there is no evidence that the leading English geologists took any notice of it: Buckland recalled much later that they had been skeptical about the reality of such enormous fold structures when put forward by a young and unknown author. Dumont brought another copy to Mézières to present to the Société géologique, and he was elected a member on the spot; this time, however, seeing was believing—since perception was guided by so convincing an interpreter.[38]

36. Thoreau (1968); Renier (1949).

37. Fig. 6.4 is traced from *BSGF* 6:pl. 3, fig. 1, but omitting most of the *terrain ardoisier* at the ends of the section. Published with the report of the meeting, this section was almost certainly available to the participants in the field, if only in some provisional form.

38. Annual report of GSL for 1833 (*PGS* 2:31); *BSGF* 6:324. Buckland's later comment was made at the GSL anniversary in 1840 (*PGS*, 3:206–07).

Dumont's unraveling of the structure was based on detailed mapping and a careful attention to the complex solid geometry of the folded strata. But the detection of the folds also required the identification of a series of distinctive formations, which Dumont grouped into four subdivisions of d'Omalius's *terrain anthraxifère*. He had given lists of their fossils in his memoir, but in practice he recognized them more by their rock types, as they reappeared again and again when traced across the axes of the folds. These subdivisions or *systèmes*—Dumont used the word idiosyncratically—included two groups dominated by limestones (*système calcaire supérieure* and *inférieure*) alternating with two that were mainly sandstones and shales (*système quartzo-schistose supérieure* and *inférieure*).[39]

Important though Dumont's structural conclusions were for current debates on the mechanisms of folding and elevation, it was mainly his sequence of formations that captured the attention of the geologists as they traversed down the Meuse valley, carefully collecting fossils on the way.[40] At the conclusion of their fieldwork, an evening session at Namur was devoted to a discussion of possible correlations between the Belgian sequence they had just seen and the British; in line with Murchison's ambitions, the latter now seemed set to become a standard for the whole of Europe. It was Buckland who told the Société of Sedgwick's new "Cambrian System" and Murchison's "Silurian," and who boldly proposed a provisional correlation between the two sequences as a whole (fig. 6.3, B).

The top and bottom of the correlation were hardly contentious. Like d'Omalius, Buckland correlated the *terrain houiller* with the Coal Measures and the *terrain ardoisier* with the Greywacke—the latter now defined more precisely as the Cambrian and equated not only with north Wales but also with Cornish rocks and fossils (at Tintagel). As already mentioned, d'Omalius had assigned the whole of the intervening *terrain anthraxifère* to the Carboniferous, remarking of its lowest group, "c'est l'*Old Red Sandstone* des Anglais" (fig. 6.3, A).[41] Buckland proposed a drastic revision. He correlated Dumont's upper limestone group with the Mountain Limestone, regarded the Old Red Sandstone as missing altogether in Belgium, and equated Dumont's remaining three *systèmes* with three of Murchison's divisions of the Silurian (fig. 6.3, B).[42]

The subsequent discussion showed that these correlations were based substantially on fossil evidence as well as that of perceived similarities of rock type. In opening the session, Constant Prévost (b. 1787), the leading Parisian geologist in the party, had already noted the similarity between Dumont's lower limestone group (e.g., at Givet: fig. 6.4, 3) and the south Devon limestones (e.g., at Plymouth).[43] Buckland incorporated this suggestion into his scheme, but further equated the Devon rocks with those of Murchison's Wenlock group (the Dudley limestone). Greenough agreed with Buckland's

39. In place of these cumbersome terms, the following account will refer to Dumont's *systèmes* as the upper and lower limestone groups and the upper and lower sandstone groups.

40. *BSGF* 6:345–53.

41. D'Omalius (1831, p. 318); cf. Rozet (1830, p. 146). D'Omalius had referred to Dumont's memoir (1832), although it was at that time still unpublished.

42. *BSGF* 6:354.

43. *BSGF* 6:353. Prévost had studied the geology of Devon in Lyell's company in 1824: Wilson (1972, pp. 126–28).

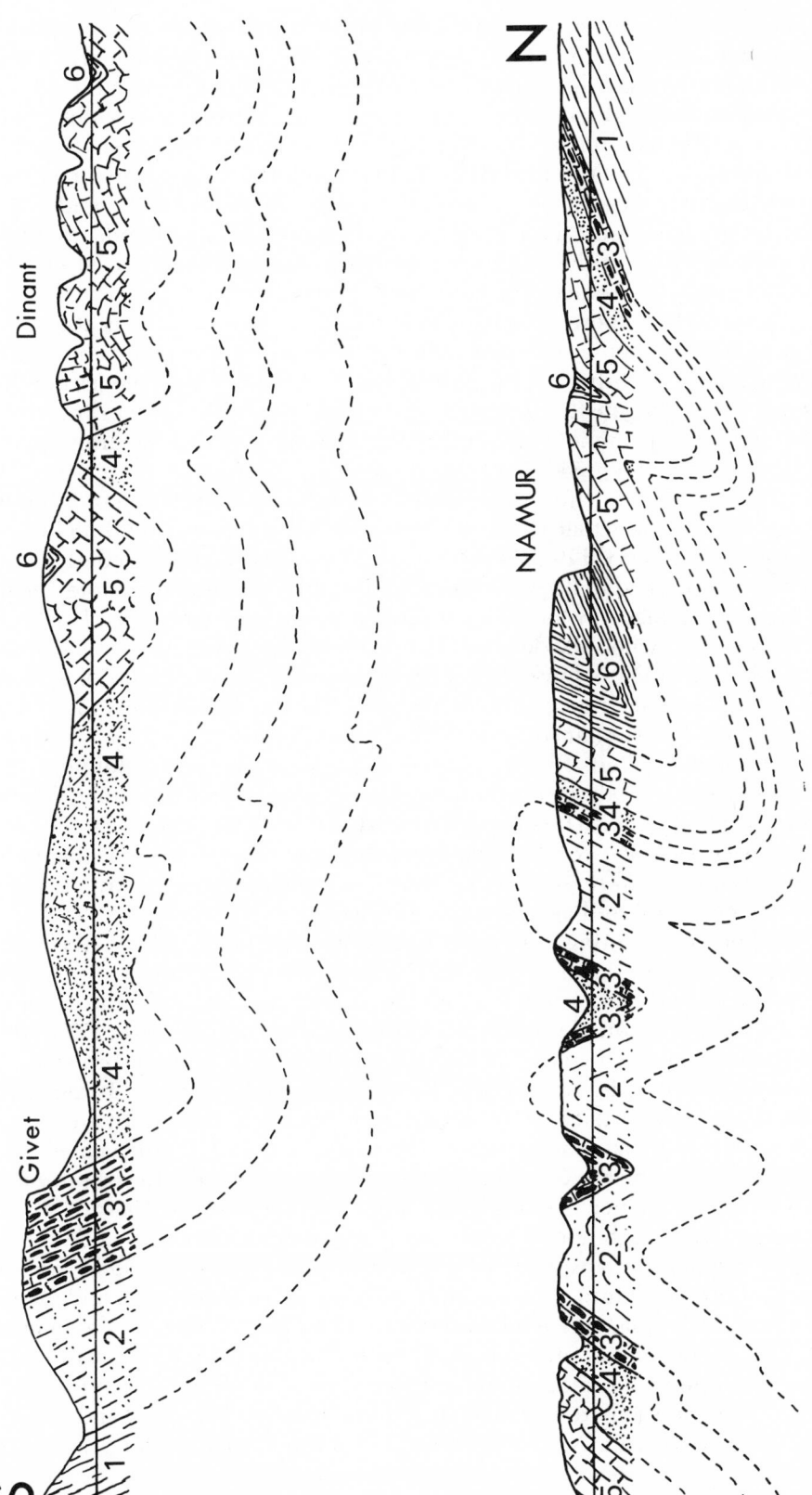

Fig. 6.4.   Part of Dumont's traverse section across the northern part of the Belgian massif, showing the huge folds inferred along the line of the Meuse from Givet northward to beyond Namur: redrawn from the section prepared for the field meeting of the Société géologique in September 1835. (1): *terrain ardoisier*; (2–5): *terrain anthraxifère* (lower and upper sandstone and limestone groups); (6): *terrain houiller* or coal strata. Note that the sequence is constant but the thicknesses highly variable. The section is about 30 km long; the datum line represents the level of the Meuse; the extrapolations of the folded strata are Dumont's.

scheme generally, but expressed doubts on this particular point, since the Belgian limestones were notably lacking in the trilobites (particularly *Calymene*, the prized "Dudley locust" of fossil collectors) that characterized the true Wenlock strata. Buckland said he had collected several species of corals in the lower Belgian limestones (around Givet), which he believed were also found at both Dudley and Plymouth. But Greenough was not convinced that the two English limestones were really similar in their fossils, and more seriously he considered that Dumont's two limestone groups did not differ from each other—in either rock type or fossils—as widely as the Mountain Limestone and the Wenlock (or Dudley) in England.

Nonetheless, there seems to have been a general consensus at Namur that Buckland's scheme was plausible in its broad outlines, and that some part of Murchison's Silurian was represented in Belgium by some part of the *terrain anthraxifère*, just as Sedgwick's Cambrian was by the *terrain ardoisier*. At the very least, Dumont's upper limestone group was now securely established as the direct equivalent of the Mountain Limestone in Britain. But, in effect, the price to be paid for Buckland's provisional recognition of Silurian strata in Belgium was that Dumont's lower sandstone group could no longer be equated with the Old Red Sandstone, as d'Omalius had proposed; that vast British formation had to be left in Buckland's table with the terse note "manque" against its place in the Belgian sequence, without further explanation.

The other major puzzle about the proposed correlations was that while the fossils of Dumont's lower limestone group were agreed to be like those of the south Devon limestones, neither the Belgian fossils nor those from Devon seemed as similar to those of the Upper Silurian (Wenlock) as Buckland's scheme implied. But in a sense this only exported to Belgium the existing uncertainty about the proper place of the Devon limestone fossils within what had hitherto been undifferentiated Greywacke (§4.6). Finally, the Belgian excursion showed the geologists no trace of coal or fossil plants in the *terrain anthraxifère*; to that extent, the sequence supported Murchison's generalizations and told against De la Beche's.

When the meeting was over, Buckland, Greenough, d'Omalius, and Prévost all traveled on to Bonn for the annual meeting of the Gesellschaft Deutscher Naturforscher und Ärzte. There they met not only the leading German-speaking men of science, including Leopold von Buch (b. 1774), the doyen of Prussian geologists, but also Élie de Beaumont and the Brongniarts *père et fils* from Paris. Among British geologists present were Turner and Lyell—the latter just back from Switzerland. Whatever the international aspirations of the British Association, a meeting like that at Dublin paled into insignificance beside the German *Versammlung* as a truly pan-European gathering of men of science. Among much other business in the *Mineralogische Abteilung*, Greenough hoped to get international agreement on standard coloring for geological maps; and Conybeare had urged him to get some agreed scheme of nomenclature for such units as groups and formations, and above all some agreed replacement for Greywacke. Lyell had his own battles to fight with Élie de Beaumont and von Buch on the contentious problems of elevation. But one of the most important items of news for the assembled geologists, as Turner duly reported to Murchison, was that Buckland and Greenough had arrived from Namur convinced of the presence of the Silurian in

Belgium. He added that they were leaving its detailed study to Murchison—whether this was with the agreement of the Belgians was left unclear![44] In any case, the field meeting of the Société géologique and the German *Versammlung* had clearly signaled the start of the extension of the Silurian to the Continent.

6.4 Murchison's Surreptitious Fieldwork (September and October 1835)

Meanwhile, Sedgwick and Murchison had returned from Ireland, but before going their separate ways they stayed a few days in Cheshire at the country seat of Cole's close friend Sir Philip Egerton (b. 1806), with whom Murchison shared an enthusiasm for both fossils and pheasants. While he was there, he reported to Buckland on what they had seen since leaving Dublin; but more important were his reflections on the fieldwork he had done before crossing to Ireland.

> From my work in Pembrokeshire I am more than ever disposed to doubt De la Beche's Bideford case of Coal plants in the Greywacke. Let this be *between ourselves*, but I have fully resolved to revisit Pembroke in about a fortnight and to cross over to Bideford to confirm or reject the belief with which I am now inpressed; for I have got the clue already to the mistake which I more than suspect our friend has fallen into. Sedgwick is *now* (having seen my sections) almost of my opinion and thinks De la Beche must be wrong. Of all this anon—I told you to be cautious about carrying the Flora of the Coal Measures down into the Greywacke. I am still of opinion that there is not one *authenticated* & *proven* case in Europe.[45]

This comment did more than underline Murchison's continuing skepticism about the Devon case and those on the Continent. For if Sedgwick had indeed been persuaded by the Pembrokeshire analogy, De la Beche would be deprived of his most powerful supporter.

Murchison proved not to be mistaken. A few days later, after Sedgwick reached his family home in Yorkshire, he wrote to Lyell with news of the Dublin meeting, mentioning how Williams in his "ignorant and impudent paper" had supported De la Beche's interpretation of the north Devon strata and their plant fossils. But the following comment revealed unequivocally how Sedgwick had indeed transferred his support.

> Murchison is going to have a brush at it [the Devon Culm], and I suspect strongly will succeed in turning it out of the older system, and putting it where it ought to be. I am the more inclined to this belief now, as he has proved De la Beche to be wrong in a part of Pembrokeshire

44. Sessions of *Mineralogische Abteilung* of GDNA, 19–25 September 1835, summarized in Harless and Nöggerath (1836, cols. 706–17) (Buckland's correlation is noted in col. 715); Turner to Murchison, 25 September 1835 (GSL: RIM, T13/1). On Greenough, see *BSGF* 6:357; Conybeare to Greenough, 2 September 1835 (ULC: GBG). On Lyell's "famous fight" over the theory of craters of elevation, see Wilson (1972, p. 420).

45. Murchison to Buckland, 11 September 1835 (GSL: WB, M12); the phrase "having seen my sections" refers to his field notes, since Sedgwick had not been to Pembrokeshire.

where he has coloured a part of the undoubted coal-measures as greywacke.[46]

With Sedgwick's defection from De la Beche's side, the balance of power in the controversy would be shifted drastically in Murchison's favor.

Meanwhile, Murchison had written to ask Conybeare to accompany him on his projected further trip to Pembrokeshire. Since Conybeare had studied the area many years before, his conversion would strengthen Murchison's case still further. Conybeare excused himself, however, on the reasonable grounds of his many family commitments; he claimed that, in any case, he could not add anything to what he had told De la Beche long before. With characteristic generosity, he took responsibility for whatever errors there might have been in De la Beche's paper, which after all had been read at the Geological Society more than twelve years earlier: "I am afraid I must claim the merit of all the blunders, but in our then indefinite ideas of grauwacke they were quite natural."[47] On the main issue, however, Conybeare was not yet convinced by Murchison: "your conjectures on the Bideford anthracite are very ingenious," he admitted, "but I can't get rid at once of my beliefs that it is true transition—I am delighted that you mean to examine it personally."

Although he could not persuade Conybeare to join him, Murchison did call in to see him and discuss the issues en route. He then spent three weeks in Pembrokeshire, primarily, of course, to improve his work on the Silurian strata. But he also confirmed—"to my complete satisfaction," as he put it— the "mistake" that De la Beche, following Conybeare, had made in St Bride's Bay (§6.2); and he attributed it to the ambiguous use of what he now called the "absurd" word Greywacke.[48] Although he knew that Conybeare had conceded the error, Murchison still hoped to use it to undermine the plausibility of De la Beche's interpretation of north Devon by urging the likelihood of an analogous mistake there.

Murchison invited Conybeare to join him on his trip across the Bristol Channel to north Devon, but again Conybeare demurred, mentioning this time the vile weather. He reiterated that it was he, not De la Beche, who had been responsible for the original error in Pembrokeshire: "I humbly acknowledge my fault," he pleaded with mock penitence, "pray don't whip me too hard this time." After his dose of face-to-face persuasion from Murchison, Conybeare, like Sedgwick, was now shifting his opinion on the main issue. "As to the Devon Anthracite," he told Murchison, "I shall be glad on the whole if you can succeed in establishing your views—this would assuredly simplify the matter." He was still puzzled that the Culm was so closely associated with "the neighbouring zones of old (greywacke) limestone," but he was confident that "at all counts discussion will establish the matter on *sure* grounds one way or the other."[49] Conybeare's shifting opinion might be more

46. Sedgwick to Lyell, 20 September 1835 (Clark and Hughes 1890, 1:446–48).

47. Conybeare to Murchison, "Saturday," pmk 19 September 1835 (GSL: RIM, C17/5), referring to De la Beche (1826), read at GSL on 16 May 1823.

48. Murchison, notebook "[from] 17 September 1835, vol. 21" undated notes, c. early October 1835 (GSL: RIM, N76, pp. 89–91). Lonsdale had criticized him for calling the term *Greywacke* "absurd": Lonsdale to Murchison, 24 July 1835 (GSL: RIM, L13/11).

49. Conybeare to Murchison, n.d., pmk [? 13] October 1835 (GSL: RIM, C17/6).

hesitant and less weighty than Sedgwick's, but it was nonetheless a significant indication of how the balance of plausibility was changing.

The heavy autumn rain to which Conybeare referred was perhaps the final factor in Murchison's decision not to cross to Devonshire after all. As he told Egerton, he had already "completed a very handsome tailpiece for my Silurians" by tracing their lower and upper divisions all the way to the furthest tip of south Wales. He had been "poaching to some purpose in De la Beche's country," as Egerton quipped in return, and had "found a greater *variety* of game there [i.e., in Pembrokeshire] than he ever dreamt of." Murchison's somewhat surreptitious plans to "invade" De la Beche's more recent "territory" of north Devon—revealed at least to Buckland, Conybeare, and Sedgwick, but obviously not intended for the ears of De la Beche himself—could safely be postponed, while during the winter months he got down to writing his "big book" on the geology of the region of "Siluria."[50]

6.5   The Greywacke Defended and Attacked (November 1835 to February 1836)

While the more affluent gentlemen of geology were touring Britain and the Continent, De la Beche had kept his nose to the Greywacke grindstone in Cornwall. When in the early summer he had applied for the continuation of his survey, he had told the Board of Ordnance that he had finished his mapping of Devon (§5.8). By the end of November the engraving and coloring of the map was also complete, and he asked the Ordnance to send presentation copies to public bodies in London (including the Geological Society), Edinburgh, Dublin, Paris, Berlin, and the United States.[51] The copies sent to Paris must have arrived only a few days after Élie de Beaumont's superior, Brochant de Villiers, presented the Académie des Sciences with an advance copy of their completed geological map of France.[52] Although this was on a much smaller scale than De la Beche's survey, the completed coverage of their whole country may have given the French geologists some satisfaction at seeing the superiority of their system of more generous state support. But at least a sample of De la Beche's finished work was now available in some of the main national centers of geological research.

Meanwhile he had been busy mapping the first of his Cornish sheets, covering much of the coast of north Cornwall to the west of his earlier survey. He seems to have had no inkling of Sedgwick's new-found skepticism about the Devon coal plants, for he made no reference to that contentious issue when he wrote to Sedgwick just before Christmas to tell him how he had

50. Murchison to [? Egerton], probably late October 1835 (Geikie 1875, 1:231–32): the printed extract reads like the letter to which Egerton to Murchison, "Sunday," pmk 1 November 1835 (GSL: RIM, E1/9), was the reply. See also, Murchison to Sedgwick, 30 October 1835 (ULC: AS, IIID, 8).

51. De la Beche to Colby, 30 November 1835 (NMW: TS, II, p. 141). The American copy was to go to Yale, doubtless on account of the presence there of Benjamin Silliman (b. 1771), the professor of chemistry and natural history, and the doyen of American geologists.

52. Brochant de Villiers (1835), read at Académie des Sciences on 30 November. The geological coloring was already complete on the copy that Brochant exhibited. Full publication was planned for the end of 1836 but was held up by delays in the engraving of the topographical features.

found quite abundant fossils in some of the older Greywacke in north Corn-wall.[53] He described these strata (in the Padstow area) as a westward extension of those long known elsewhere on the north Cornish coast (at Tintagel), which in turn could be traced eastward round the north sides of both the Bodmin Moor and Dartmoor granites. This letter was in fact intended to give Sedgwick a privileged preview of what De la Beche had also sent more formally to the president of the Geological Society.

De la Beche's letter was read (in his absence) at the first meeting in the new year. He used the occasion to reiterate implicitly his original view of the Devon problem. He made no mention of the plant fossils and anthracite of north Devon, but he referred unrepentantly to *all* the older strata of Devon, as well as those of Cornwall, as "Grauwacke." He stated that he had discovered some "natural divisions" among them, "founded upon marked characters." He implied that it might be possible to correlate them with Sedgwick's Cam-brian divisions in Wales, and he said he was "of opinion that the whole of the district is older than the Silurian formations of Mr. Murchison."[54] The other geologists concerned, whether or not they were present at the meeting, would have agreed with this as a generalization about much of the region. The crucially important exception was, of course, the plant-bearing Culm of north Devon; Murchison was now more than ever convinced that these strata were really Coal Measures in disguise. The limestones of south Devon seemed rather like those of the Silurian, although their fossils were somewhat anom-alous (§6.3). But for the rest of De la Beche's Greywacke, a Cambrian dating was perfectly acceptable to all. Sedgwick, indeed, told De la Beche he was "greatly delighted" at his detailed report from Cornwall, for it gave promise of a more adequate characterization of his Cambrian system in terms of its fossils and was likely to extend its validity beyond Wales.[55]

At the very next meeting of the Geological Society, Murchison had a chance to reply to De la Beche with an equally veiled reference to the Devon problem. His paper was primarily a summary of his recent work in Pembro-keshire; it described his successful recognition of Silurian strata there, and their conformable relation to the Old Red Sandstone and other Carboniferous strata above and to the Cambrian strata below. It was particularly important for Murchison that he had been able to identify the Silurian strata by recog-nizing characteristic Silurian fossils, although the rock types were not the distinctive ones he had described in the Welsh Borderland. For this dem-onstrated how fossils—always in conjunction with the observable sequence of the formations—were more reliable than rock types for the identification of the strata.

Murchison made this theoretical point with special reference to the term "grauwacke," which, as foreshadowed in his correspondence, he now publicly and emphatically rejected. He concluded scornfully that its tradi-tional sense—which De la Beche had used in his letter read at the previous meeting—was "no longer consistent with the advanced state of geological science." For he claimed to have found strata of greywacke rock type, within

53. De la Beche to Sedgwick, 20 December 1835 (ULC: AS, IB, 52) and 14 January 1836 (FMC: Perceval, L/106).
54. De la Beche (1836), dated 18 December 1835, read at GSL on 6 January 1836.
55. Sedgwick to De la Beche, 8 February 1836 (NMW: DLB).

the single county of Pembrokeshire, ranging in age from the Cambrian up to the Coal Measures. He therefore urged that the term should either be "rigidly restricted to some of the very oldest sedimentary deposits"—thus excluding his own Silurian strata—or preferably used simply to denote a specific rock type, examples of which, he claimed, "actually occur in strata formed in many successive epochs." This conclusion must have pleased the president, since such recurrences of supposedly unique rock types—attributed to the recurrence of similar depositional environments and postdepositional events—had been a leading theme in Lyell's own research.

Murchison's conclusion contained, however, an implicit criticism of De la Beche that was much more specific. He explained in some detail how the Coal Measures in Pembrokeshire were locally (in St Bride's Bay) in apparent conformity with the far older Greywacke strata, with which they had been confused—notably, of course, by De la Beche—because they were so similar in rock type (§6.2). In what was implicitly a calculated insult to De la Beche, Murchison maintained that "mistakes might easily result from such juxtapositions," since the two series of strata "might to an unpractised eye appear indistinguishable." But whereas the unpracticed eye of a De la Beche might be deceived, Murchison was not. "Even where the order of superposition is not to be detected," he insisted, "essential differences are invariably to be observed, in the coal shales never containing those organic animal remains which are so abundant in the Silurian system, whilst the latter never contains a single plant similar to those which abound in the former."[56] In other words the evidence of characteristic fossils would always lead the practiced eye of a Murchison to the correct conclusion.

Sedgwick strongly criticized Murchison afterward for rewriting the history of the Silurian research and making all earlier contributors—including himself—look like "asses."[57] But the relevance of Murchison's paper for the Devon controversy would have been apparent to all in his audience who had followed the debate to this point. If De la Beche had mistaken Coal Measures for Greywacke in Pembrokeshire, however locally, he might have made the same mistake on a much larger scale in Devonshire. Murchison's inference was clear: the Culm series of Devon, with its fossil plants of Coal Measures species, must really be Coal Measures strata with the same confusingly greywacke-like appearance as those in Pembrokeshire. He had now persuaded Sedgwick into this opinion (§6.4), but they had still not clinched the matter by studying Devon for themselves. This may explain why, a month later, Lyell was so carefully uncontroversial in his anniversary address. He reviewed the relevant work, aligned himself with Murchison in the rejection of the traditional usage of Greywacke, but otherwise avoided making any provocative remarks on the still simmering controversy.[58] The time for that would come in due course, once Murchison had cast his self-styled "practised eye" on the rocks of Devonshire itself.

---

56. Murchison (1836), read at GSL on 20 January (GSL: BP, OM 1/7, pp. 383–92). The Devon problem was certainly on his mind while preparing this paper: Lonsdale to Murchison, 26 December 1835 (GSL: RIM, L13/12).

57. Sedgwick to Murchison, 25 January 1836 (GSL: RIM, S11/96).

58. Lyell (1836), read at GSL on 19 February (GSL: BP, OM 1/8, pp. 48–96).

6.6   Successes for the Fossil Criterion
(February to April 1836)

In his anniversary address, Lyell mentioned that Murchison's book on the strata of "Siluria"—with what Murchison had called its "tailpiece" into Pembrokeshire—was almost complete. But it still needed the expert description of fossils, which Murchison had largely delegated to others. This was where his project was intimately linked with Phillips's forthcoming monograph on the fossils of the Mountain Limestone of Yorkshire. For Murchison's claims for the distinctive character of the Silurian fauna could only be established against the standard of the Carboniferous fauna that immediately succeeded it. "Do not fear any guess work as to Species in my Vol.," Phillips told him; even with the tactful omission of any reference to the Silurian fauna, he could state that in the Mountain Limestone "the multitude of species is enormous and nearly all new to books."[59]

Phillips was not merely being deferential or chauvinistic when he praised Murchison's forthcoming publication as "a national work" and "*a peculiarly English Book.*" For he saw clearly that, *if* Murchison's Silurian could be established as a system of general validity, its "type" region would have an importance beyond the shores of Britain, as a clear demonstration that "the laws of distribution of organic remains in relation to succession of deposits" applied with the same force to those ancient periods as to more recent times in the earth's history. But Phillips, like other geologists, considered that the wider validity of the Silurian was most seriously put in question by the anomaly of the Devon Culm. "Upon my life you astonish me quoad De la Beche's Grauwacke," he exclaimed, on hearing how Murchison proposed to resolve that anomaly. "You are cutting yourself out a great deal of further work. Can you avoid going into Devonshire to see how the thing stands there? into the south of Ireland? again to determine the limits of your classification."

Phillips's monograph on the Mountain Limestone was published a few weeks later. It was dedicated jointly to Sedgwick and to Harcourt, his patron in York, though Sedgwick was aggrieved at its lack of adequate acknowledgment of earlier research.[60] Besides a large number of other Yorkshiremen, the subscribers included all the leading British geologists and fossil specialists. Perhaps the book's most original feature was what Phillips termed his "method of variation." This was in fact less a method than a conclusion. His detailed study of the Mountain Limestone sequences all over the Pennines, most of them flat-lying, had enabled him to trace the remarkable *lateral variation* in the strata, when traced from Derbyshire in the south through Yorkshire to Northumberland in the north. The conclusion was relevant beyond Phillips's region and to formations other than the Mountain Limestone. It was a striking and persuasive reminder to other geologists that strata that were unquestionably contemporaneous could vary laterally, so radically that their equivalence

59. Phillips to Murchison, 2 February 1836 (GSL: RIM, P14/14).
60. Phillips (1836, introduction dated 1 March); Murchison and Sedgwick both had copies by early April. Sedgwick to Murchison, 7 April 1836 (GSL: RIM, S11/101).

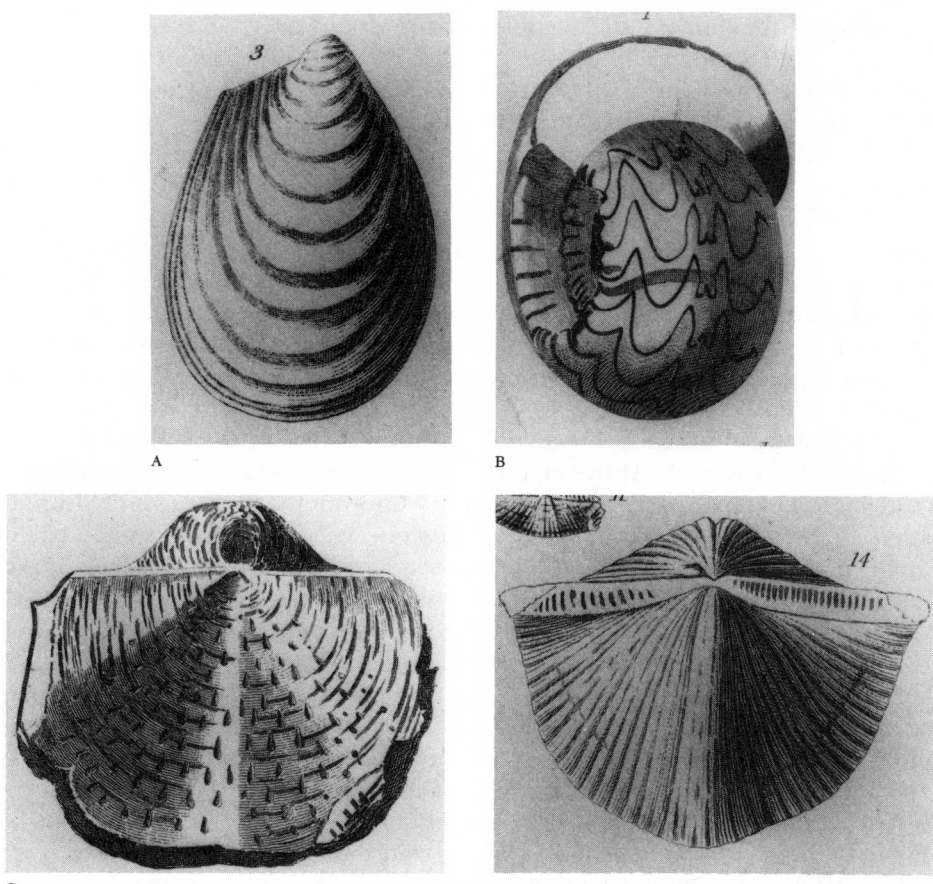

Fig. 6.5. Some fossils from the Mountain Limestone of Yorkshire, as illustrated in Phillips's monograph (1836) and cited subsequently in discussions about the age of the strata of Devon and Cornwall: (A), the bivalve mollusk *Posidonia*; (B), the cephalopod mollusk *Goniatites*: (C), the brachiopod *Productus*; (D), the brachiopod *Spirifer*.

would be hard to recognize if the intermediate areas had not been preserved. It was also a reminder that major formations such as the Mountain Limestone, however clear-cut they might be shown on geological maps and sections, were often in reality less sharply defined and might shade imperceptibly into the formations above and below. In particular, Phillips was at pains to emphasize the importance of the intermediate or *"transition group"*—using the word in a quite different sense from Transition—by which the Mountain Limestone passed gradually down into the Old Red Sandstone.[61]

Murchison was disappointed at the lack of what he regarded as real "geology," that is, of maps, traverse sections, and the like in Phillips's book. For it was devoted largely to the detailed description of the strata and even more to the enumeration and illustration of the Mountain Limestone fossils (fig. 6.5). But Murchison was realistic enough to recognize that its importance lay precisely in Phillips's command of the fossil evidence. He told Sedgwick

61. Phillips (1836, pp. 15–18, table on p. 11).

he believed that almost *all* Phillips's fossils would prove to be "specifically distinct" from his own "Silurians," though he conceded that Phillips's transition beds at the base of the Mountain Limestone "might contain a species or two like mine."[62] The comment shows how he both hoped and expected that the Silurian and Carboniferous would prove almost completely distinct in their fossil assemblages. The Old Red Sandstone, lying between Phillips's fossils and his own, was treated tacitly as of no importance in this context; although very thick, it had few fossils, and those few (mainly fossil fish) were peculiar in both senses of the word. If the fossil specialists who were studying the Silurian fossils confirmed Murchison's expectations, the Silurian could confidently be launched on the geological world as a "system" that, whatever the geographical scope of its validity, indubitably represented a distinct major epoch in the history of life.

Geological practice, however, was not waiting for the full publication of Murchison's results. At the meeting in Bonn the previous autumn, Greenough and Buckland had reported that they were convinced there were Silurian strata in Belgium (§6.3). One of their hearers, von Buch, had at once taken up the notion of the Silurian and Cambrian systems and used it in the following months to bring order to the varied Greywacke regions of central and eastern Europe. In March 1836 he summarized his provisional conclusions in a letter to Élie de Beaumont, which was read to a meeting of the Société géologique.[63]

Von Buch claimed to have distinguished the Cambrian from the Silurian on fossil grounds: for example, the Cambrian had many species of the brachiopod *Orthis*, but none of the *Spirifer* that was so abundant in the Silurian and later strata. He reported that the Cambrian was rare in central Europe (except in the Fichtelgebirge and near Prague), but that the strata near St Petersburg were positively a "modèle" for it; the Silurian was reported from Lithuania and the Baltic island of Gotland. With these reports, von Buch was applying Sedgwick's and Murchison's concepts in a way that raced ahead of their own research. However tentative his conclusions, he was identifying their systems by characteristic assemblages of fossils more boldly and over a wider terrain than even Murchison or Buckland. The Silurian and Cambrian systems, defined less than a year before, were already being used in an almost routine and unproblematic manner. The Transition was well on the way to being tamed.

---

62. Murchison to Sedgwick, 17 March 1836 (ULC: AS, IIID, 10). Fig. 6.5 is reproduced from Phillips (1836, pl. 6, fig. 3; pl. 20, fig. 1; pl. 8, fig. 20; pl. 9, fig. 14).

63. Von Buch (1836), read at SGF on 21 March; his inferences were based partly on published reports (e.g., Pander 1830). At the same meeting, Élie de Beaumont read extracts from a letter from Sedgwick, summarizing his Lower and Upper Cambrian *systèmes* in Cumbria and Wales and emphasizing their gradual transition into Murchison's Silurian: *BSGF* 7:152–55.

Chapter Seven

DEVONSHIRE
PUT TO THE TEST

7.1 Preparing an Invasion (February to June 1836)

When De la Beche sent Sedgwick a full summary of his first season's fieldwork in Cornwall, he apparently had no inkling that Sedgwick had become highly skeptical about the Culm strata in Devon (§6.4). Nor did he know that Sedgwick had begun to make his own inquiries about the disputed rocks and their fossils. Sedgwick had in fact asked one prominent local geologist, Major William Harding (b. 1792) of Ilfracombe, to tell him more about the Devon anthracite; Harding had promised to send him specimens, and duly did so some months later.[1] Another local informant, the physician Edward Moore of Plymouth, hearing indirectly of Sedgwick's new interest in Devonshire, had written to suggest that the Plymouth fossils must be of Upper Cambrian age; for he considered that they differed both from the older fossils of north Cornwall and north Wales (Tintagel and Snowdon, respectively) and from those of Murchison's Wenlock division of the Silurian.[2] At the last meeting of the Geological Society before the summer recess, yet another keen local geologist, the young country gentleman Robert Austen (b. 1808), presented a competent paper on the geology of his own corner of Devonshire, commenting on the abundant marine life of "the transition epoch" preserved in the local limestones. Austen's interest in geology had been aroused by Buckland's lectures at Oxford; in 1833, on marrying an heiress, he had resigned his Fellowship at Oriel

---

1. Harding to Sedgwick, 16 November 1835 and 1 June 1836 (ULC: AS, IB, 54; IC. 1).
2. Moore to Sedgwick, 10 January 1836 (ULC: AS, IB, 92).

College and had settled on an estate near Newton Abbot, where he had begun to amass one of the finest fossil collections in Devon.[3] Comments and contributions from amateurs such as these would have made Sedgwick realize that the local geologists of Devon were well worth cultivating, if only for their expert knowledge of the best fossil localities.

Early in February, Sedgwick suggested to Murchison a plan for joint fieldwork that summer, which would at last take them both into Devon. He proposed going in turn to west Somerset, south Devon, Cornwall, north Devon, the south of Ireland, and the west of Scotland: "I think this is a good plan, what say you to it?"[4] The primary purpose of this fieldwork was evidently to explore the extension of their respective Cambrian and Silurian strata into new areas, but in the course of their tour the disputed area of north Devon would be taken in their stride. Sedgwick did not propose going behind De la Beche's back, as Murchison had intended to do the previous autumn (§6.4): on the contrary, he suggested that they should call on De la Beche to discuss the issues while in Cornwall, *before* moving into north Devon (De la Beche's surveying duties were expected to prevent him from accompanying them there).

Murchison agreed with Sedgwick's proposal, and their plans for another major season of fieldwork went ahead. In June, shortly before they were due to set out, Murchison sent the first part of his big book to the printers.[5] Following normal printing practice, he would have known he could make substantial alterations in the proof stage. Nevertheless, his action does imply that he regarded his research on the Silurian strata (or more accurately, on the *region* of Siluria, i.e., the Welsh Borderland) as virtually complete. This is what Lyell had announced, doubtless with Murchison's approval, in his anniversary address earlier in the year (§6.6). It is therefore clear that Murchison did not expect to have to make any major revision of his Silurian work in the light of his forthcoming fieldwork. He was convinced that De la Beche was mistaken about the fossil plants of Devon, and that the Culm would prove to be merely a confusing variety of ordinary Coal Measures, like that in Pembrokeshire. But once that anomaly was safely eliminated, Murchison probably expected that most of the remaining older strata of Devon and Cornwall would turn out to be equivalent to the Cambrian of Wales, as De la Beche and most other geologists already assumed. He may well have hoped that he might also find *some* Silurian strata in Devon, which would extend the validity of the Silurian concept outside its original region, just as his work in Pembrokeshire had done. But it is unlikely that he expected any such finds to necessitate major alterations in his general classification of the Silurian system; at most an extra chapter, added to the book while it was in press, would be sufficient.

When Sedgwick and Murchison left London early in July, one geologist—not counting the local residents—had already preceded them into Devonshire. The irrepressible Williams was continuing his independent studies

3. Austen (1836), read at GSL on 25 May. His estate was at Ogwell, west of Newton Abbot, and close to quarries yielding some of the best fossils from the south Devon limestones and associated strata (see fig. 4.10).

4. Sedgwick to Murchison, 8 February 1836 (GSL: RIM, S11/98). No contemporary documents support Murchison's much later claim (Geikie 1875, 1:249–50) that he, and not Sedgwick, first suggested this joint fieldwork.

5. Thackray (1978b, p. 65).

of the geology of north Devon and west Somerset. "In my Geol. researches," he had noted some months earlier, "my only object is to accumulate materials for 'some wise master builder' hereafter to use in the structure of an edifice to Truth that shall be immortal—when once this is effected I am sure that by the wise and the good the labourer will also be remembered."[6] But his paper at the Dublin meeting had shown that he was by no means content merely to present others with factual building blocks, and that he had ambitions to comment on the highest issues of theory and method in geology (§6.2). Since this impinged on the competence of the wise master builders, the laborer could not in fact be so confident that his efforts would be appreciated.

After the Dublin meeting the previous summer, Williams had spent some time in north Devon, where he noted what he termed "Old Red Sandstone" alternating with the slaty rocks of "the upper series of the Grauwacke."[7] Had Murchison known what Williams was doing, he would have rejected that "mineralogical" use of the term Old Red Sandstone as vehemently as he had publicly rejected the term Greywacke. Yet the fact remained that Williams believed he had recognized rocks *like* the Old Red Sandstone among the varied strata that "the wise and the good" regarded as older Greywacke or Cambrian. Now Williams had preceded them into Devon. There he noted that some of the seams of culm (at South Molton) were closely similar to those of the Coal Measures in the Bristol coalfield near his home.[8] But in his opinion this did not make them genuine Coal Measures; on the contrary, he was convinced—like De la Beche—that they were an integral part of the Greywacke. This was just what Sedgwick and Murchison entered the region determined to disprove.

7.2   The Devonshire Campaign (July and August 1836)

Sedgwick and Murchison took the night coach from London to Bristol, and then went straight on to Bridgwater in Somerset to begin their fieldwork (their itinerary can be followed on fig. 4.10). For at least part of the time, they traveled with a hired pony trap or on horseback. Their notebooks indicate that they had with them not only a copy of Greenough's older geological map of the region (§4.6), but also a set of De la Beche's newly published large-scale maps. They were often highly critical of the latter: *"this portion of De la Beche's map wretchedly worked out,"* Murchison noted, for example, at one point (Combe Martin); *"half the Limestone* omitted—no angles of inclination given—& the strike not made apparent."[9] Yet there can be little doubt it was De la Beche's map that enabled them to cover so much ground so quickly and fruitfully, because for all its defects its large scale helped to direct them to the most significant localities—whether quarries or limeworks, road cuttings or coastal cliffs.

6. Williams, notebook for "September 1835 to March 1836," note between dated entries for 4 and 24 December 1835 (SAS: DW, box I, notebook 8, pp. 16–17).

7. Ibid., pp. 2–3, notes of September 1835.

8. Williams, notebook for "June 1836 to September 1836," notes before dated entry for 5 July 1836 (SAS: DW, box I, notebook 9).

9. Murchison, notebook "Devon with Sedgwick," undated note [July 1836] (SMC: Murchison notebook 22, p. 23).

Their very first quarry showed how far their field practice diverged from their public statements. "Limestone which de la Beche has mapped as Grauwacke Limestone, which Sedgwick thinks is Mountain [Limestone]," Murchison noted; "from Silurian I reject it indignantly. Unfortunate mass!"[10] From the beginning of their tour in southwest England, they made provisional identifications of the ages of the strata they studied simply on the basis of appearance or rock type, and *not*—or only secondarily—on the basis of fossils. This does not mean that they regarded fossils as unimportant; on the contrary, they evidently collected fossils wherever they could, and sometimes hired quarrymen to collect for them to save their own time. But such collections were made for future study by the relevant specialists; they themselves relied on very general impressions of the fossils they saw.

As suggested by Sedgwick in his original plan (§7.1), they first studied the Quantock Hills in Somerset, an isolated patch of older rocks that was nearer the Mendip Hills than any other such area in southwest England (see fig. 3.5; fig. 4.10). The Mendips were known to have a full sequence of normal Carboniferous rocks (fig. 3.2; fig. 4.8, C), whereas Sedgwick and Murchison suspected that Devonshire contained some of abnormal appearance, namely the Culm. Probably they hoped that the area would be as intermediate in its geology as it was in its geographical position. But in the event the Quantocks provided them with no such easy key to their main problems. It was perhaps for that reason that they then moved directly into north Devon instead of heading south as Sedgwick had first suggested, although as a result they would see the most controversial area *before* visiting De la Beche in Cornwall.

They worked their way along the north coast, studying the fine natural cliff sections. On his sketch of one cliff (near Minehead), Murchison marked the rocks "old Greywacke," having in private no compunction about using that term in the traditional sense he had so firmly rejected in public (§6.5).[11] Further west (at Linton), they evidently thought at first that the rocks were Lower Silurian; Murchison recorded at the end of the day how he "first perceived the real analogy to the Lowest Caradoc Sandstones or Lower Silurians from the abundance of Encrinite heads . . . fossils weather as in the Lower Silurians of Pembroke." But soon afterward he was identifying this thick "*arenaceous system*" as "very ancient Cambrians," purely on the basis of a resemblance in rock type to Cambrian strata in the Welsh Borderland and in southern Scotland.[12] At Ilfracombe they met Harding, Sedgwick's local informant in the previous months; they studied the nearby cliff sections in his company, dined at his house, and were doubtless told in detail where to go and what to look for.

They then turned south toward Barnstaple; traversing in the direction of the dip of the strata, they could now study the sequence more easily, and they reckoned they were seeing successively younger strata, all dipping steeply to the south. Before reaching Barnstaple, Murchison noted that they had found "the base of the true Silurian System," implying perhaps that the strata further

10. Ibid., p. 1, note dated 5 July 1836.
11. Ibid., p. 13.
12. Ibid. pp. 15, 16, 29; cf. Sedgwick, notebook "Devonshire 1836," note dated 8 July (SMC: AS, notebook 29, p. 3). "Encrinites" were fragments of the stalks of fossil crinoids (stalked echinoderms). The other strata referred to were termed "Purple Longmynd" ("V" on fig. 4.8, B) and "Lammermuir series," respectively.

north had seemed to them neither Cambrian nor "true" Silurian in character. The Silurian identification, made initially on grounds of rock type, was then confirmed on the outskirts of the town (at Pilton) where they discovered what Murchison regarded as "the Lower Silurian fossils." He termed the strata there "a very good representative of the Llandeilo Flags," even though they failed to find the fossil most characteristic of that lowest division of his Silurian in Wales (the large trilobite *Asaphus buchii*), and Murchison noted no other species by name.[13] On this traverse he and Sedgwick made no reference to the tight trough structure that De la Beche had in effect been forced to infer to the north of Barnstaple, which had made him interpret the Culm strata as a reemerging continuation of the older rocks to the north (§5.1). If Sedgwick remembered this from the sketch De la Beche had sent him two years earlier (fig. 5.2), he must have dismissed it as just another of De la Beche's errors.

Continuing south from Barnstaple, Sedgwick and Murchison soon reckoned they had entered the area of the contentious Culm strata. They had just concluded that the strata immediately to the north—and therefore in their opinion older—were Lower Silurian; and they suspected that, contrary to De la Beche's claim, the plant-bearing Culm rocks were true Coal Measures. They were therefore expecting to find evidence of a major unconformity between the two sets of strata. In the event, no such evidence was forthcoming—if they had found any it would certainly have figured prominently in their notes. But this was not too disturbing, since it could be attributed to the fact that the junction between the two sets of strata seemed to run somewhere under the Taw estuary, and was therefore not exposed.

As for the Culm strata themselves, Sedgwick noted a clear distinction between what he called the black "*Culm Limestones*" and associated shales in what they regarded as the lower part of the series, and the sandstones, shales, and thin seams of anthracite in the main body of the Culm. The former contained the distinctive fossil shell *Posidonia* (see fig. 6.5, A); the plant fossils that had caused the whole controversy were found in the latter. Neither geologist expressed any doubt about the age of the Culm proper: Sedgwick noted how even in small details the rocks were "precisely as the *coal measures.*" When they studied the coastal sections of these strata beyond Bideford, Murchison noted confidently, "N.B. No slaty cleavage in any beds of the Culm series."[14] This assertion was directly contrary to De la Beche's original report that the Culm was affected by cleavage just as much as the other strata, this being one of several features that it shared with the rest of the Greywacke (§5.2). Murchison now denied this and thereby reinforced his claim that the Culm was much younger. As he noted at the end of the day, "the ride [from Bideford to Clovelly] indeed convinced us that far as we went, 6 miles across the strata—all was in the same carbonaceous series—*wholly unlike* any thing in the Greywacke Series."[15]

It was probably at Bude—still in the middle of the Culm or "carbonaceous series"—that their path crossed with that of Williams. The encounter

---

13. Murchison, notebook 22, p. 31. *Asaphus buchii* was among the "marvellous things" Murchison had found in Wales the previous summer: Murchison to Sowerby, 20 July 1835 (RSL: JCS, 684, 24); Murchison (1839, pls. 24, 25).

14. Sedgwick, notebook 29, pp. 6–10 (13 July 1836); Murchison, notebook 22, p. 42.

15. Murchison, notebook 22, p. 50.

was not explicitly recorded at the time by any of them; although traveling geologists were hardly thick on the ground, Sedgwick and Murchison probably felt that this was one provincial amateur they could safely ignore. But Williams listed at this point a series of "reasons why this transition district is of *far higher antiquity* than Mendips & B[risto]l Coalfield," probably after hearing the other geologists express a contrary conclusion. His reasons were obscurely expressed, but were based negatively on the total absence of the Mountain Limestone in Devon, and positively on the similarities between the Devon rocks and those of Wales and Brittany.[16] Like De la Beche, Williams was thus convinced that *all* the north Devon strata were older than the Carboniferous; his reference to the "incalculably long period" of deposition of the Carboniferous strata indicates incidentally that this country parson was no scriptural geologist.

Sedgwick and Murchison, on the other hand, now felt sure not only that Williams was wrong to regard the Culm as Greywacke, but that De la Beche's interpretation was even more profoundly mistaken. The Culm was not well within the total Greywacke sequence, with still younger strata to the north; on the contrary, the Culm itself was uppermost, and the strata to the north were much older. Since they both assumed that the strata further south, in South Devon and most of Cornwall, were once again older Greywacke or Cambrian, this suggested that the Culm strata of central Devon were in the form of a huge trough. This inference could not be confirmed, however, as simply as in an area of gently folded Secondary strata. It was not just a matter of following the strata southward along the coast and watching them change gradually from a southerly dip through horizontality to a northerly dip. For the Culm strata were almost everywhere complicated by "extraordinary troughs and saddles," or in other words by strong small-scale folding which made the general regional structure difficult to infer from simple observation (see frontispiece and fig. 5.2, *c*).[17] Their inference could only be confirmed by recognizing the older rocks again as they emerged from beneath the Culm on the southern flank of the putative trough.

Their immediate task, therefore, was to locate this southern boundary of the Culm area. Bearing in mind their earlier confident claim that the Culm strata were quite distinct in appearance from the older strata, locating this boundary should have been no great problem. Some miles south of Bude, Sedgwick duly noted that "a northern dip begins to prevail at high angle—& finally the [Culm] system ends against the slates and flagstones [i. e., the older strata] at the mouth of St Gennis river."[18] At the end of the day, Murchison too recorded how they had worked their way along the coast, looking for the "long expected *junction*" between the Culm and the older strata. But they had become more and more puzzled as they followed the rocks southward. "From the carbonaceous appearance of the shales & the little difference of

16. Williams, notebook "June 1836 to September 1836," memorandum immediately after notes dated 16 July, at Bude (SAS: DW, box 1, notebook 9); their respective notebooks indicate that they were all in Bude on that day. Williams had studied the geology of Brittany in 1834 (ibid., notebook 5). His encounter with the other geologists was recalled explicitly in a letter written later in the year, though he placed it at Boscastle: Williams to Lonsdale, 25 November 1836 (GSL: BP, LR2/252).

17. Sedgwick, notebook 29, p. 11 (16 July 1836).

18. Ibid., pp. 12, 13 (16 July 1836).

lithological structure between the Sandstones on this coast of Cornwall & those of Bideford & Clovelly," Murchison recalled, "we still imagined we were *in* the deposits of the carboniferous age." But still no distinct junction appeared. Sedgwick thought at one point that he had found the black Culm limestone that marked the base of the Culm in the north; but when this proved a mistake, Murchison admitted that they became "quite *alarmed* at the thickness of our carbonaceous system." Eventually they realized what—so they believed—had happened. "Instead of an unconformable junction we found a gradual passage & became convinced that we had *PASSED* by many miles the limit, the Bude system being that of S. Wales Cambrian." Murchison noted that as a result they "resolved to work back to Bideford by other lines [i.e., traverses] to clear up this most obscure point."[19]

This was a serious admission: to term the matter "most obscure" was decidedly an understatement. Even with the best exposures that they could expect to find anywhere, namely the coastal cliffs, they had failed to detect any change of rock type (or for that matter, any fossil evidence) by which to distinguish the Culm, which they believed to be Coal Measures in age, from what they regarded as Cambrian strata. Even more damaging to their case was Murchison's use of the phrase "a gradual passage" to describe the relation between the two sets of strata. A gradual transition was precisely what De la Beche had claimed from the start; conversely, Murchison's argument, that De la Beche had made a gross mistake, depended on showing that there was an unconformity. Failing that, he needed at the very least to show that there was a deceptive "juxta-position" between the two formations, like the one he had found in Pembrokeshire (§6.2, §6.4). Such a junction they had signally failed to find.

They continued along the coast, still among what they regarded as Cambrian rocks, as far as the well-known fossil-bearing strata (at Tintagel) that De la Beche had mentioned in his report from Cornwall (§6.5). Following those strata inland past Launceston, they found a limestone in quarries south of the town (at South Petherwin) that reminded Sedgwick of limestones on the north coast of Devon (at Combe Martin) and in Cumbria (at Coniston).[20] This would have supported their inference that, however problematic the southern boundary of the putative Culm trough had proved to be on the coast, they were now in any case back on Cambrian ground.

From Launceston they traveled directly to Plymouth, where they met Hennah, the most notable local collector (§4.6), and were able to see the fossils he had found in the Plymouth limestone. After studying its associated strata in the field, Sedgwick concluded that "the thickness of this system over the Plymouth L.S. is enormous," or in other words that the Plymouth limestone, whatever its exact age, lay far down in the sequence of the Cambrian strata. From Plymouth they continued their southward traverse all the way to "the old chloritic system" or "ancient rock" of Bolt Head near the southernmost tip of Devonshire.[21]

---

19. Murchison, notebook 22, pp. 55–58. The quotation illustrates how Murchison, like Sedgwick and other geologists, still used the word "system" in its colloquial sense, to mean a formation or set of formations.

20. Sedgwick, notebook 29, pp. 14, 15 (18 July 1836).

21. Ibid., pp. 17–19 (19–21 July 1936).

To summarize their provisional interpretation of what had of course been only a rapid reconnaissance traverse from north to south: Sedgwick and Murchison were clearly agreed that the geological structure of Devonshire was basically quite simple. In central Devon there was a broad trough of Culm strata, inferentially of Coal Measures age. They had failed to find any direct evidence for the anticipated unconformity between the Culm and the older rocks to the north and south, but they were convinced that the latter did rise from beneath the Culm on both sides of the central trough. To the north, the strata were, in their opinion, mainly of Cambrian age, with a little Lower Silurian; to the south there was no sign of Silurian strata or fossils—they may have taken this as indirect evidence for an unconformity—and they attributed the very thick sequence of older strata wholly to the Cambrian, placing the well-known Plymouth limestone and its abundant fossils far down in the sequence (fig. 7.1, A). At the very base were the ancient schists of Bolt Head and Start Point.

Their next move was probably designed to confirm this provisional interpretation by traversing back toward the north by another route, although this involved abandoning Sedgwick's earlier suggestion that they should penetrate further into Cornwall and visit De la Beche. Although they did not meet Austen, they studied briefly the thick limestones he had recently described around Tor Bay and Newton Abbot (§7.1), and collected some of their fossil corals and trilobites. Passing this time to the east of the Dartmoor granite, through an inland area of sparse exposures, they failed to find any but the most obscure and ambiguous indications of an unconformity on the southern boundary of the Culm trough (near Chudleigh). They therefore

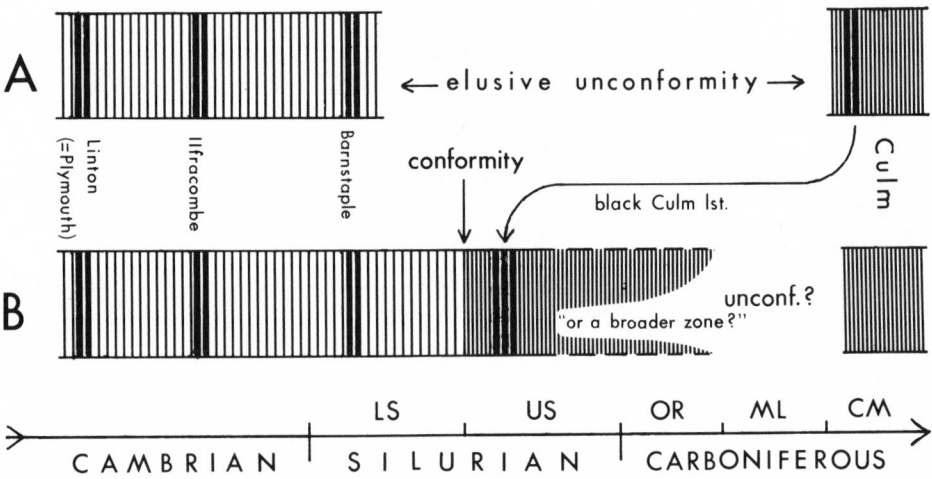

Fig. 7.1. Diagrammatic columnar sections to show (A) Sedgwick and Murchison's agreed interpretation of the correlation between Devonshire and their "standard" sequence in Wales and the Borderland, as formulated during their fieldwork in July 1836; and (B) the alternative correlation that Murchison considered briefly while in the Bampton area, in order to explain the lack of evidence for any unconformity below the Culm. In this and all subsequent diagrams depicting correlations of the Devon strata, the closely spaced shading represents the Culm, the broader shading the pre-Culm rocks, and the black stripes the various limestones.

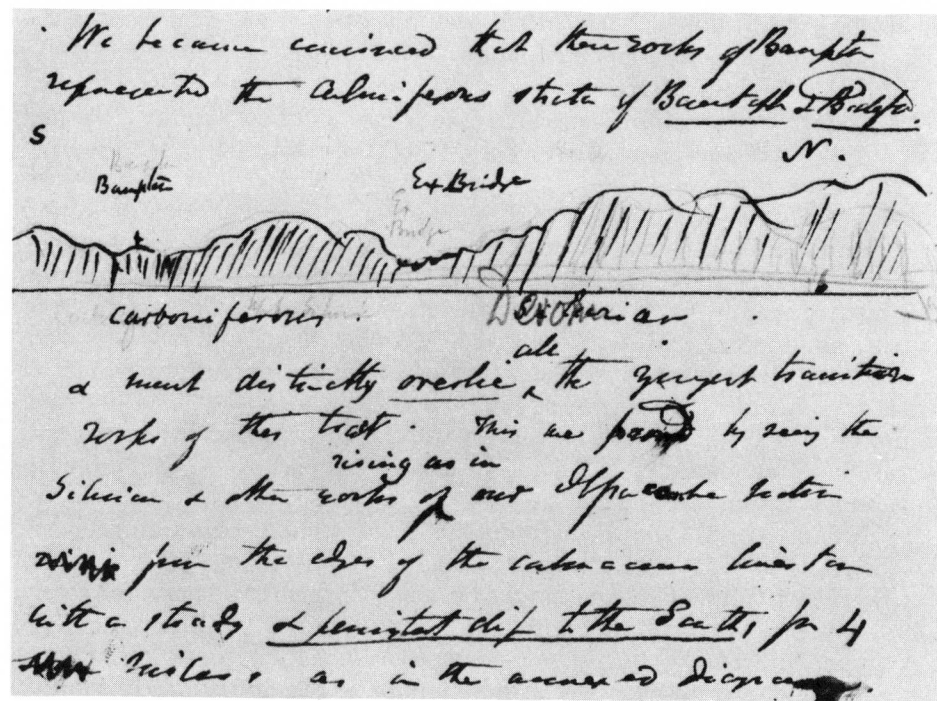

Fig. 7.2.   Murchison's section through Bampton and the south flank of Exmoor, sketched in the field in July 1836. Contrary to what he and Sedgwick anticipated, but in accordance with what De la Beche had maintained, Murchison conceded here that the Culm ("Carboniferous") passed down without a break into the older strata ("Silurian"). (Murchison altered the latter attribution to "Devonian" at a much later date.)

traveled rapidly through Exeter and Tiverton, back to the northern boundary. Around Bampton and to the east (at Holcombe Rogus) they duly found the black Culm limestone again; this time they had better luck with the exposures. Murchison noted at the end of the day that the Culm strata "most distinctly *overlie* all the youngest transition rocks of this tract," since they had found "the Silurian & other rocks" emerging from beneath them, the whole sequence having "a steady & *persistent dip to the south*" (fig. 7.2).[22]

Once again, as on their earlier and parallel traverse from Ilfracombe through Barnstaple, they had evidently rejected the trough that De la Beche had postulated through Exmoor. Not only had they found no direct evidence for such a structure in the dips of the strata; even more decisively, they had concluded that the two bands of limestone that De la Beche had equated when inferring his trough (see fig. 5.2) were really quite distinct. The limestone running out to the north coast at Ilfracombe they now regarded as Cambrian; the second of Greenough's four bands of limestone (see fig. 4.10), running through Bampton and south of Barnstaple, was the black Culm limestone, distinct in both rock type and fossils. With De la Beche's trough dismissed as an illusion based on the fallacious equation of those two distinct

---

22. Murchison, notebook 22, pp. 92, 93 (fig. 7.2 is reproduced from p. 93); Sedgwick, notebook 29, pp. 23, 24 (25 July 1836).

limestones, the Culm was now more securely established than ever as the uppermost and therefore youngest of the formations.

Sedgwick and Murchison had failed again, however, to find the expected sharp "juxtaposition" at its base, let alone an unconformity. "We had now pretty well convinced ourselves," Murchison admitted, "that although the carbonaceous [i. e., Culm] limestone was clearly *overlying* all the older transition rocks, still that it was so connected with, & so conformably united with the underlying fossiliferous beds of the Silurian & Cambrian Systems that it could not be separated from them."[23] This was a major admission, though only made privately. Even if De la Beche had been wrong about the structure and the sequence, he had consistently claimed what Murchison now admitted, namely, that there was no natural break whatever between the two sets of strata. Murchison realized this, as his following note shows. "The question then occurs, is it [the Culm limestone] = to the Upper Silurian?—or does it represent a broader zone? Is it to be considered as part of the Culm deposit or as merely an underlying calcareous mass of another epoch?"[24]

Although he still believed firmly that the main body of the Culm—with the fossil plants—was of Coal Measures age, he here raised the possibility that the black Culm *limestone* might be much older. If it were completely conformable to Lower Silurian strata, as it appeared to be, then perhaps it was Upper Silurian; while his second query—"does it represent a broader zone?"—suggested that it might also be equivalent to formations further up the sequence, such as the Old Red Sandstone (fig. 7.1, B). In that case the unconformity—or at least deceptive juxtaposition—that their general interpretation still required might perhaps be found where they had not hitherto thought of looking for it, namely, between the Culm limestone and the overlying main body of Culm. For they still needed to equate the latter with the Coal Measures, if the original anomaly posed by De la Beche's fossil plants were to be eliminated.

Anyway, the relations between the plant-bearing Culm, the Culm limestone, and the "Lower Silurian" strata obviously needed further attention. Sedgwick and Murchison therefore traveled westward along the line of this obscure but crucial boundary, all the way back to Bideford; but the paucity of their recorded notes suggests that they had no further success. Then, rather surprisingly, they cut south again right across the Culm area to the northern edge of Dartmoor (at Okehampton). Probably they wanted to check, first, whether the Culm was metamorphosed by, and therefore older than, the intrusive granite; and second, whether the limestone marked by Greenough along the edge of the granite (fig. 4.10) was black Culm limestone, which would help confirm their inference about the southern edge of the Culm trough. On both points they probably found positive indications, if not conclusive evidence.[25] Despite their total failure to find the kind of unconformable boundary to the Culm that they had expected, Murchison and Sedgwick were now more confident than ever that at least the main body of the Culm was indeed Carboniferous in age, that it formed a "basin" or trough occupying

23. Murchison, notebook 22, p. 97.
24. Ibid.
25. Sedgwick, notebook 29, pp. 25, 26 (28 July 1936). This interpretation of the detour is supported by the form of the map that Murchison drew soon afterward (see below).

the whole of central Devon, and that it rested on Lower Silurian strata on the north side, and perhaps directly on Cambrian on the south. This was a major reinterpretation and, in their own view, a drastic correction of the structure and dating proposed by De la Beche.

From the edge of Dartmoor they returned to Barnstaple, and then back to Ilfracombe, this time around the coast. Murchison recorded only that it was a "glorious day of discovery," a tantalizingly cryptic note which probably reflects their excitement at seeing the fine coastal cliffs (around Baggy and Morte Points), displaying the whole sequence of Lower Silurian and Cambrian strata more clearly than they had seen them before.[26] From Ilfracombe they crossed the Bristol Channel to Swansea, and then traveled westward to Pembrokeshire, where they studied particularly the "Culm measures" near Tenby. Almost certainly their aim was to check that these unquestioned Coal Measures strata were indeed indistinguishable from the Culm strata of Devon, while the latter were still fresh in their mind's eye, so that they could be used as an analogical argument to explain De la Beche's crucial mistake in Devon (§6.2, §6.4).[27] After all this joint fieldwork, they parted for a time to work on their respective strata elsewhere in south Wales. Later they met up again in north Devon, probably to make a similar check in reverse, namely, to confirm that the older Devon strata were indeed like those in Wales. Murchison soon packed his hammer and left for Bristol to help prepare for the meeting of the British Association, of which he was now one of the general secretaries. Before joining him there, Sedgwick used his remaining days to traverse south across Exmoor and to look again at the puzzling Culm limestones in the Bampton area.[28] Then he too got a coach to Bristol.

Their joint fieldwork in Devon and north Cornwall had lasted only about four weeks (not counting a period of about two weeks in Wales), but in that time they had worked intensively and to great effect. There is no evidence in their respective notebooks that their opinions on what they had seen differed in any major way. They had jointly created a radical new interpretation of the geology of Devonshire, which promised to eliminate once and for all the anomaly of Coal Measures plants in the Greywacke, at least on British soil. So far, however, they had convinced only themselves. It remained to convince others, and they arrived in Bristol determined to use the British Association meeting to do just that.

7.3   The Grand Attack at Bristol (August 1836)

The Marquis of Lansdowne, whom De la Beche had more than once asked Sedgwick to approach to press his case for the establishment of a governmental geological survey (§5.8), should have presided over the British Association meeting at Bristol. But almost at the last moment the sudden critical illness of his son obliged him to withdraw; he was replaced by a much less influential

26. Murchison, notebook 22, p. 113 (29 July 1836).
27. Murchison, notebook "No. 2, 1836 July [Vol.] 23" (GSL: RIM, N77, pp. 1–8); Sedgwick, notebook 29, pp. 27, 28.
28. Sedgwick, notebook 29, pp. 45–50 (18–19 August 1836).

nobleman, the Marquis of Northampton (b. 1790), who was however a respectable man of science, active particularly in geology. But other politicians present at Bristol included Spring-Rice, the Chancellor of the Exchequer, who had attended the Cambridge meeting three years before and was on familiar terms with, among others, Greenough, Sedgwick, and Murchison. The presence of such men made the leaders of the Association acutely aware of the opportunities that the meeting offered for promoting a favorable view of science in governmental circles.[29]

On the eve of the meeting, the reporters for two rival weeklies both commented on the likely importance and popularity of the geological Section.[30] Its president was to be Buckland, and of the inner circle of active British geologists, only Lyell was missing (see fig. 2.5). Even before the meeting began, Greenough reported that the members of the Section were "looking forward with impatience" to hearing what Sedgwick and Murchison had been doing in Devonshire; after the first day's session he added that "tomorrow Sedgwick and Murchison project a fierce encounter with de la Beche."[31] Such comments leave little doubt that the leading geologists at Bristol were well aware beforehand of the likely character of what they saw as the centerpiece of their program for the week. For Murchison had been boasting beforehand about the victory that he and Sedgwick had scored. De la Beche complained later to Sedgwick that when he arrived he found it "currently reported that the North Devon [map] was little better than an abomination, and was to be blown up by yourself and Murchison." Expecting both Lansdowne and Spring-Rice to be there, De la Beche was understandably alarmed: "Murchison himself went so far as to condole with me upon the loss of fame I was to sustain,— and hang me, if altogether I knew what was coming."[32]

What was coming duly came in the second day's session. The paper that Sedgwick and Murchison had hurriedly written, on the basis of their recent fieldwork, was entitled "A classification of the old Slate Rocks of Devonshire, and on the true position of the Culm deposits of the central portion of that county."[33] There was no false modesty here: they claimed to have found the "true" solution to the Devon problem. The paper was illustrated with a "working map," which was almost certainly Murchison's copy of Greenough's map of the region, with the new interpretation superimposed in new coloring and annotations (fig. 7.3; fig. 7.4). A diagrammatic traverse section was also displayed and was published a few days later in the *Athenaeum* (fig. 7.5); it was comparable in form to the sketch De la Beche had sent Sedgwick two years before, but radically different in content (fig. 7.6). Specimens of rocks and fossils from Devonshire were also on show in the meeting room.[34]

29. Morrell and Thackray (1981, pp. 116, 340). The local background is described by Neve (1983).

30. *Ath.*, no. 460 (20 August 1836), p. 587; *LG*, no. 1023 (27 August 1836), p. 545.

31. Greenough to "Pusseykin" [Caroline Clarke], 21–22 August 1836 (UCL: GBG, 99).

32. De le Beche to Sedgwick, 20 October 1836 (ULC: AS, IC, 2).

33. *Ath.*, no. 461 (27 August 1836), pp. 611–13; *LG*, no. 1024 (3 September 1836), pp. 564–65; Sedgwick and Murchison (1837a).

34. Fig. 7.3 is reproduced, and fig. 7.4 traced, from Murchison's copy (BGS: RIM) of the southwest sheet of Greenough (1820). Murchison's MS coloring and notes were almost certainly added at or around the time of the BAAS meeting in Bristol. Fig. 7.5 is reproduced from *Ath.*, no. 460, p. 612.

On his map, Murchison painted four main bands of color across north Devon and wrote four corresponding names offshore, together with southward dip arrows that showed they were to be read as a simple ascending sequence from north to south (fig. 7.4). Three of these names—Cambrian, Lower Silurian, and Carboniferous—are unsurprising in the light of the notes that the two geologists had made in the field. But one—"Devonian"—was a complete novelty. The word was here used for the first time as an unambiguously stratigraphical term, rather than in its usual topographical sense. Murchison clearly meant it to denote a particular set of strata, rather than just any rocks found in Devonshire. By implication, these "Devonian" strata might be found outside Devonshire, just as Cambrian and Silurian strata were here identified well outside the regions of their original definition.

The reasons for this innovation are not clear. Probably both Sedgwick and Murchison felt, on reviewing their fieldwork before the meeting, that the middle portion of the sequence of older strata in north Devon could not plausibly be matched with *either* the Cambrian *or* the Silurian. Yet in Wales they had agreed two years earlier that their respective systems graded into one another without a break (§4.5), so there was no obvious way for a new major series of strata to be intercalated between them in Devonshire. Before they read their paper, they may have realized that their provisional solution was unsatisfactory, for Murchison wrote the word "Cambrian" a second time on his map, this time bracketing both the original Cambrian and the Devonian (fig. 7.3). This relegated the Devonian to equivalence with the upper part of the Cambrian, which is just how it was shown on the printed section: "Upper Cambrian or Devonian rocks" (fig. 7.5). The title to the caption, "Ascending Series of Devonian Rocks," shows that they also retained the usual topographical sense of the word.

The strata at the base of the Cambrian, in the sense of Murchison's original annotation (fig. 7.5, *a* and *b*), were shown in the section in the archlike saddle structure that De la Beche had also detected. On the map they were marked as "Plymouth Zone" (fig. 7.3; fig. 7.4), probably to express a perceived resemblance between the fossils of the Plymouth Limestone, which Sedgwick regarded as lying far down in the Cambrian sequence in south Devon, and those found in these oldest strata of north Devon (at Linton). The overlying Upper Cambrian or Devonian strata (fig. 7.5, *c* and *d*) were said to be "abounding in organic remains" (at Ilfracombe and Combe Martin). It may well have been these fossils that led the authors to use the term "Devonian" in this novel sense; they can hardly have matched them with those of the original Upper Cambrian of Wales, since Sedgwick had signally failed to find distinctive fossils there to characterize that division.

The next set of formations, overlying the Devonian conformably, was identified as Lower Silurian: in its limestone bands "it everywhere contains vast numbers of characteristic fossils" (e.g., those they had found north of Barnstaple). In the caption to the section, these were referred to confidently as "many well-known Silurian fossils, chiefly of the lower part of the system" (fig. 7.5, *e* and *f*). Since the authors had had no opportunity to have their fossils examined in detail by specialists, this judgment must have been based on little more than Murchison's perception of a general resemblance between the Devon fossils and some of those from Wales.

B R I S T O L C
Cambrian
Cambrian
Plymouth Zone
Devonian
Silurian
Carboniferous
Trough
Coal.

Cambrian
Cambrian
Plymouth Zone
Devonian
Lower Silurian
Carboniferous
Trough of Coal
Linton
Ilfracombe
EXMOOR
QUANTOCKS
Bridgwater
Barnstaple
Bideford
Taunton
Bampton
St Gennys
Boscastle
EXETER
Launceston
S. Petherwin
Chudleigh
DARTMOOR
PLYMOUTH
Dartmouth
20 mi.
CARB.
black lst.
SILURIAN
DEVONIAN
CAMBRIAN

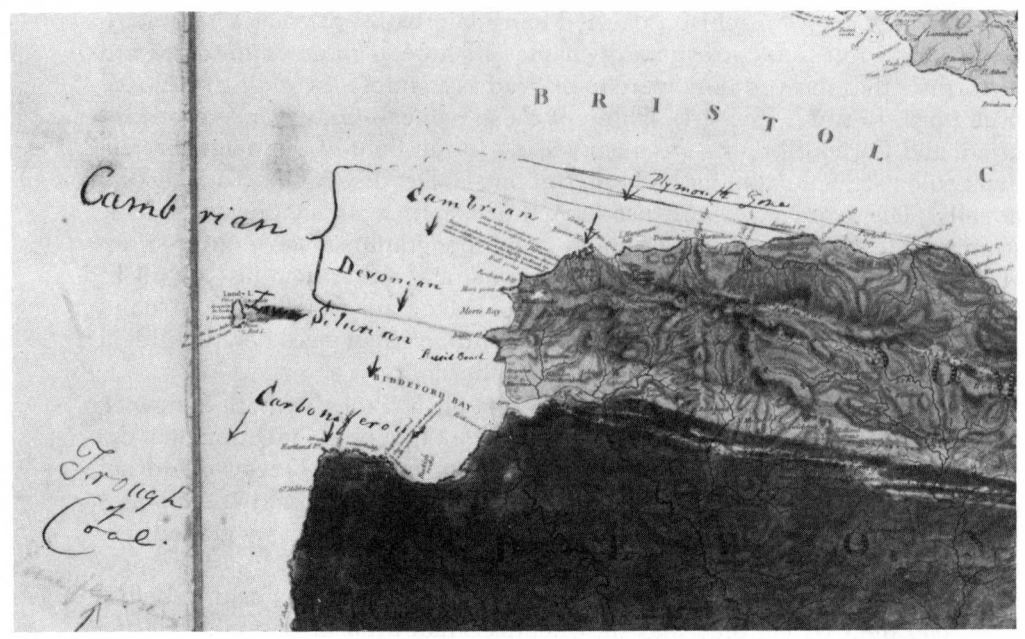

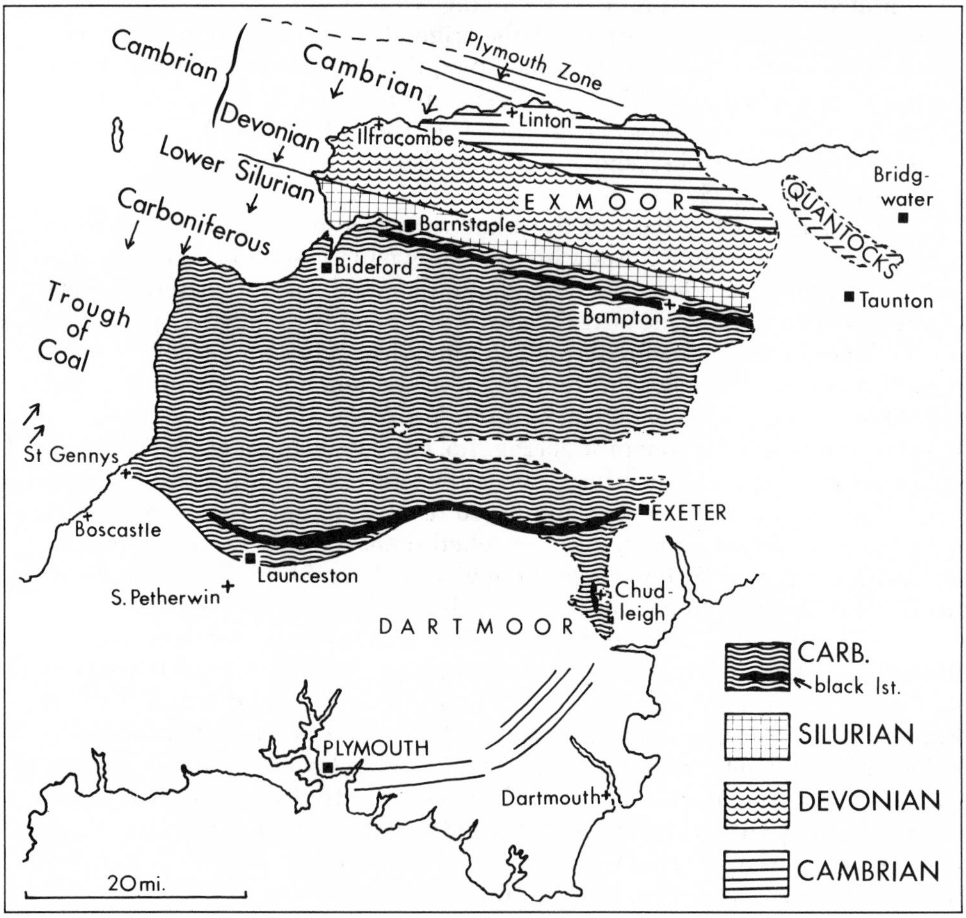

Fig. 7.3. (left) Part of Murchison's copy of Greenough's geological map, to show the annotations added for his joint paper with Sedgwick, read to the British Association in Bristol in August 1836; this map, or a sketch based on it, was displayed in the meeting room. Note the strip of "Devonian" strata sandwiched between "Cambrian" and "Lower Silurian," and the later addition of "Cambrian" in a broader sense; note also the "Trough of Coal," that is, the Culm, marked as "Carboniferous." This map records the first known use of the term "Devonian" in a stratigraphical sense.

Fig. 7.4. (left) Murchison's alterations and annotations (see fig. 7.3) on his copy of the Devonshire part of Greenough's map, redrawn for clarity. The area south of the "Trough of Coal" was left with Greenough's coloring (i.e., as Greywacke and granite: see fig. 4.10), but Murchison and Sedgwick certainly regarded the Greywacke there as Cambrian.

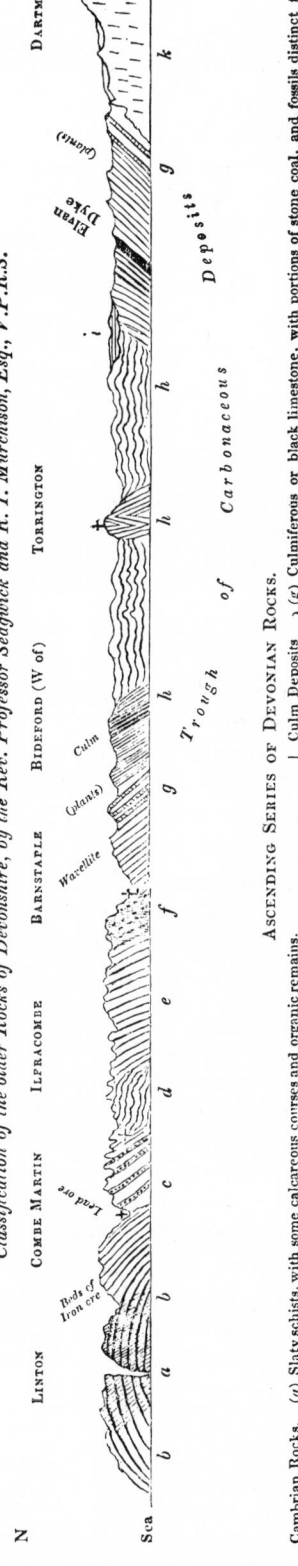

Classification of the older Rocks of Devonshire, by the Rev. Professor Sedgwick and R. I. Murchison, Esq., V.P.R.S.

ASCENDING SERIES OF DEVONIAN ROCKS.

Cambrian Rocks. (*a*) Slaty schists, with some calcareous courses and organic remains.
(*b*) Purple, red, and grey sandstones, with beds of iron ore in upper members—peculiar fossils near their junction with the overlying limestones. Veins of lead and copper.

Upper Cambrian or Devonian Rocks. } (*c*) Calcareous group of Combe Martin & Ilfracombe—fossils very abundant—slaty cleavage.
(*d*) Slates with quartzose veins and beds—incoherent schists, &c. Manganese mines.

Silurian Rocks. (*e*) Slaty sandstones and schists—cleavage passing through the beds of organic remains.
(*f*) Ditto, with concretionary limestones, and many well-known Silurian fossils, chiefly of the lower part of the system.

Culm Deposits
=Coal-field of Pembroke.
(*g*) Culmiferous or black limestone, with portions of stone coal, and fossils distinct from any found in the inferior groups. Wavellite occurs in the beds below this limestone.
(*h*) Culm beds with underlying and overlying successions of sandstone and shale often highly pyritous, with many nodules of iron ore, frequently containing coal plants, and never affected like the older rocks by slaty cleavage.
(*i*) New red sandstone resting unconformably on the carbonaceous deposits.
(*k*) Granite of Dartmoor and Elvan Dyke, both erupted through the culm deposits.

Fig. 7.5. Traverse section through north Devon, from the north coast (left) to the edge of the Dartmoor granite, based on the one used to illustrate Sedgwick and Murchison's paper, read to the British Association in Bristol in August 1836. Note that the term "Devonian" was used both in its usual topographical sense (in the title of the

caption) and in a novel *stratigraphical* sense, as a synonym for Upper Cambrian. Note also the Culm "trough"; no unconformity is shown between the Culm and the Silurian in the north (between *f* and *g*, at Barnstaple), and in the south the place of this boundary is eliminated by the intrusive granite.

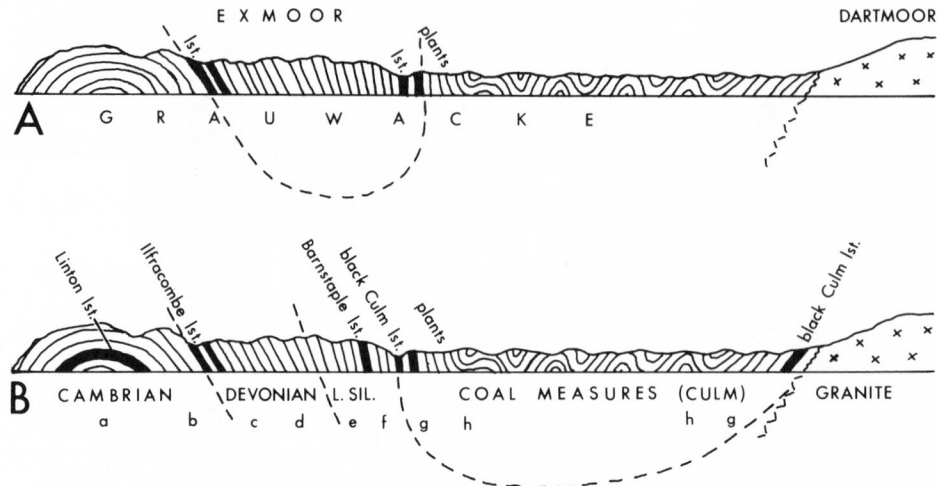

Fig. 7.6. (A) De la Beche's section through north Devon, sent to Sedgwick in November 1834, compared with (B) Sedgwick and Murchison's section, exhibited at Bristol in August 1836. The sections are redrawn from fig. 5.2 and fig. 7.5, respectively, to facilitate direct comparison between the two interpretations. Bands of limestone are marked black.

Finally, as the youngest strata of the district, came the "Culm deposits" or "carbonaceous system of Devonshire" (fig. 7.5, g and h). The Culm limestones were said to contain "fossils distinct from any found in the inferior [i.e., lower] groups," as expected on the new interpretation. The main mass of the Culm, the strata from which De la Beche's anomalous fossil plants had come, were said to be "never affected like the older rocks by slaty cleavage." Sedgwick and Murchison concluded that the Culm "may without hesitation be referred to the regular carboniferous series"; since they also stated that "these beds do not differ from the great unproductive coalfield of Pembrokeshire," it is clear that they correlated the Culm specifically with the Coal Measures.

This conclusion implied, of course, that a huge time gap separated the Culm from the underlying Lower Silurian rocks, so that some clear evidence of discontinuity in the sequence was to be expected (see fig. 7.1, A). With masterly ambiguity, however, Sedgwick and Murchison stated that the Culm "rests, partly in a conformable position upon the Silurian system, and partly upon older rocks." This formulation conceded tacitly that they had found no evidence whatever in north Devon for the major unconformity that their interpretation almost required; at the same time it implied that the appearance of conformity was illusory. Their only evidence for this was, implicitly, that they had not found any Silurian strata on the *south* side of the Culm area, and that the strata there (e.g., at South Petherwin) looked like "some of the oldest strata in the north" (§7.2).[35]

35. The appearance of unconformity in the south was accentuated on the map by the way Murchison drew strike lines for the older strata, swinging round from Plymouth toward Exeter as if to collide with a southward extension of Culm strata at Chudleigh (fig. 7.4.) But in fact their investigation of this southern area had been cursory and inconclusive.

All these strata were depicted as being in a straightforward sequence, consistently dipping southward away from the flank of the saddle on the north coast. De la Beche's inferred tight trough (through what was now termed Devonian and Silurian) had vanished without comment, so that the Culm was depicted unambiguously as overlying, and thus younger than, all the other strata (fig. 7.6, B). Disregarding its minor contortions (depicted very crudely by the engraver of the section, fig. 7.5), the Culm was said to rise again toward Dartmoor, with the Culm limestone reemerging near the edge of the granite, and having there a dip to the south. Hence the whole structure formed a "Trough of Carbonaceous Deposits" or "Trough of Coal."

Apart from their novel use of the term Devonian, the paper by Sedgwick and Murchison merely summarized in more systematic form the conclusions they had gradually reached in the course of their fieldwork together (§7.2). By drawing the section south to the Dartmoor granite, and not on the more obvious line along the coast where the exposures were much better, they carefully avoided having to show in public any specific interpretation of the embarrassingly obscure southern edge of the Culm trough in the cliffs of north Cornwall.[36] They glossed over the fact that they had completely failed to find the expected unconformity between the Culm and the other strata, and that on the south side of the trough they had even found—as Murchison had admitted privately in his notebook—a "gradual passage" between the two groups. In order to bolster their conclusion that there was nonetheless a huge time gap between the Lower Silurian and the Culm (i.e., Coal Measures), they claimed again that only the older strata were affected by slaty cleavage.

The paper was a striking presentation of a radically new interpretation. By claiming to have identified a large area of hitherto undetected Coal Measures in central Devon (fig. 7.4), Sedgwick and Murchison proposed what even laymen could see was a major revision of De la Beche's official map. They also claimed to have resolved the alleged anomaly of fossil plants of Coal Measures species being found in ancient Greywacke strata; for by reinterpreting the major structural features of the area they had also reordered the whole sequence of the strata and brought the Culm to the top (fig. 7.7, A; compare fig. 4.8, D). The argument of the paper had major weaknesses, but these were carefully concealed by a highly persuasive presentation. In fact, one reporter went into raptures at the "poetry of geology" embodied in Sedgwick's oratory, and was overwhelmed by "the originality and freshness of his images, the profundity of his scientific and general remarks, the coruscations of his fancy" and more besides.[37]

De la Beche recalled later that the "grand attack" on his map had been marked by "sections flying, geologists charging, plaudits sounding, & c. & c."[38] He was disingenuous to add that "the heavy column was on me before I had a notion of any attack having been meditated," since he knew very well from Murchison's boasts that some attack was in the offing. But it is fair to infer that he did not know in advance just what aspects of his work were to

---

36. On Murchison's map the southern boundary of the Culm had to be shown *somewhere* on the coast; it was placed without comment at St Gennys (fig. 7.4), although that provisional location had been rejected even in the field (§7.2).

37. *LG*, no. 1023, p. 552.

38. De la Beche to Greenough, 4 November 1836 (ULC: GBG, 51).

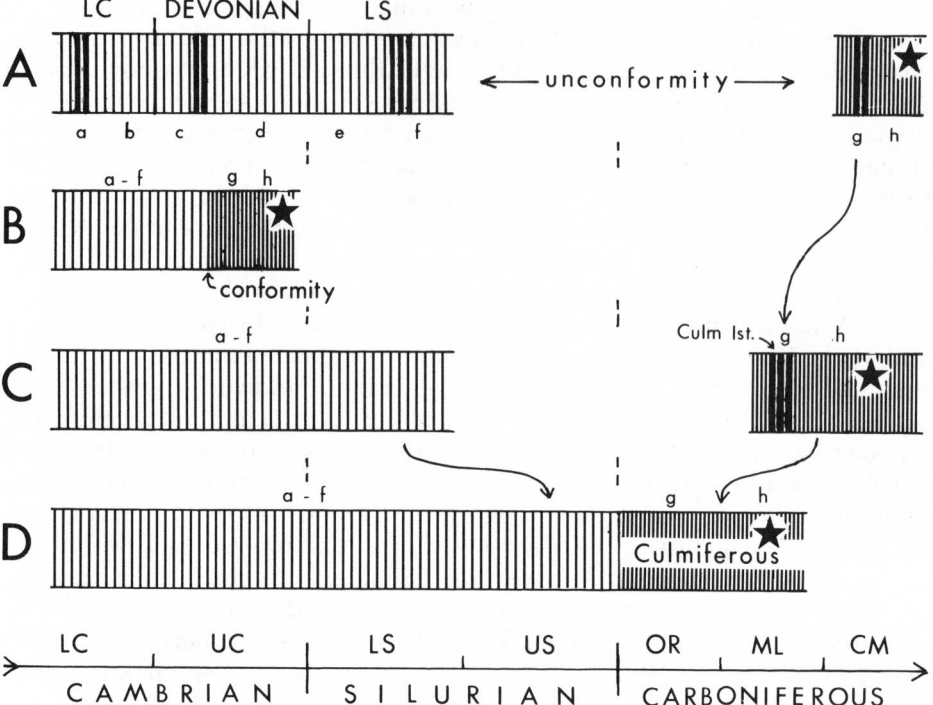

Fig. 7.7.   Diagrammatic columnar sections to illustrate interpretations of the Devon strata proposed or implied during the discussion of Sedgwick and Murchison's paper to the British Association at Bristol in August 1836. (A) Sedgwick and Murchison's interpretation (redrawn from fig. 7.1, A), as accepted also by Conybeare. (B) De la Beche's new interpretation, conceding that the Culm was uppermost (compare fig. 4.8, D). (C) Phillips's interpretation, modifying Sedgwick and Murchison's by equating the Culm limestone with the Mountain Limestone (ML) on fossil evidence. (D) Buckland's compromise or synthesis, assigning the Culm and its limestones to the "Culmiferous," here interpreted as a term for all the strata between the Silurian and the Coal Measures, i.e., as the equivalent of the Belgian *terrain anthraxifère* on its original definition (see fig. 6.3, A). The stars denote the position of the Culm plants; the letters (a) to (h) denote the formations distinguished by Sedgwick and Murchison (see fig. 7.5 and fig. 7.6, B).

be criticized. When Buckland, as chairman, opened the session for discussion, he properly gave De la Beche the first chance to reply. That De la Beche's remarks in defense of his mapping were less than impressive is hardly surprising. "I was taken most deucedly in flank," he recalled later; "my ammunition being in my magazines, and my guns dismantled, expecting nothing but peace, I made my retreat in the best manner I could."[39]

The form of his tactical retreat was to make one concession of the greatest significance: he agreed with the new structure that his opponents had claimed for north Devon, tacitly conceding that his own earlier interpretation had been mistaken in this respect. He thereby also agreed that the Culm was uppermost in the sequence; whatever its age, it followed that the fossil plants that had precipitated the controversy came from near the top of

39. Ibid.

the sequence, not from far down within it. De la Beche did not, however, concede that the Culm was therefore Carboniferous in age. Although he said that fossils were of great significance, and claimed that he "attached very little importance to mineral characters," he minimized the value of fossils by stressing again how little was known of their ranges in time. Referring to Élie de Beaumont's celebrated case in the French Alps (§5.3), he urged that geologists should not "limit too strictly" the range of the "coal" plants, adding that those from the Culm were so poorly preserved that it was not even certain that they were really the same as those from the Coal Measures. But his strongest argument, and Murchison's (and now Sedgwick's) weakest, remained what it had been from the start: he "could nowhere discover any line of separation between the carbonaceous and the older rocks," still less any unconformity to mark the vast span of time that his opponents' interpretation involved. Although he had accepted correction on the sequence of the older strata, De la Beche thus continued to regard them *all* as pre-Carboniferous, and probably even as pre-Silurian (fig. 7.7, B).

Sedgwick replied, but only with the conciliatory remark that there were no less than four parts of the sequence with limestone bands (figs. 7.5, 7.6: *a, c, f, g*), implying that the relative ages of the strata could be determined with greater confidence when the fossils had been fully studied. Conybeare insisted that "the public had exaggerated the difference of opinion then before the meeting," evidently because the audience had become excited by what one reporter saw as "a grand field day" of battle among the experts.[40] But as foreshadowed in his correspondence with Murchison almost a year before (§6.4), Conybeare was lost from De la Beche's side: he now stated in public that he was inclined to believe the new interpretation, especially in the light of the apparently similar case in Pembrokeshire.

Phillips then stated that he too was convinced by the argument for "a coal basin" in central Devon. Despite the unusual appearance of the Culm, he said, "doubts must vanish on inspecting the organic remains" laid out on the table in the meeting room. He claimed that Smith—who was in Bristol and may well have been present at this session—had not asserted that the "characteristic fossils" of one formation were *never* to be found in other adjacent formations. The respect accorded by Murchison to "the founder of English geology" was thereby preserved, while Phillips could agree with De la Beche that "organic remains alone are insufficient to point out distinctions of strata." But he then made full use of the fossil evidence after all. He repeated what he had mentioned privately to De la Beche earlier in the year, namely that the Culm strata contained fossils that he had also found in the Mountain Limestone of his native Yorkshire. But he now inverted his interpretation of this observation.

He had earlier suggested that in Greywacke times the region now represented by Devonshire might have supported species that only became widespread and "characteristic" in later, Carboniferous, times (§5.7). De la Beche had already alluded to this suggestion, which supported the dating of the Culm as pre-Carboniferous. But now Phillips used the fossil evidence to support Sedgwick and Murchison's view that the Culm strata really were

40. *LG*, no. 1023, p. 552.

Carboniferous in age. He stated that the commonest fossil shells (*Posidonia*) from the Culm limestone were identical to some found in the Mountain Limestone of Yorkshire (in the Craven area), and that they occurred there in the same kind of black limestone (fig. 6.5, A). He maintained that despite a few puzzling instances, such as that in the French Alps, "organic life must have a constant relation to the state of the actual surface" of the earth, so that fossils should generally be reliable guides to age; and he concluded that eventually "the Devon district would not offer any anomaly in geological arrangement." So he too was lost from De la Beche's side.

Phillips's speech was doubtless seen immediately as influential confirmation of Sedgwick and Murchison's position. Yet it embodied a significant modification of their interpretation, for it implied that Devon might contain, under deceptive or unfamiliar features, equivalents not only of the Coal Measures but also of the underlying Mountain Limestone (fig. 7.7, C). While in the field, Murchison had briefly toyed with the idea that the Culm limestone might be Upper Silurian, or even "a broader zone," in order to explain its complete conformity with the older, Lower Silurian strata (fig. 7.1, B). Phillips's suggestion did not lessen the difficulty posed by that conformable relation, but it did to some extent narrow the apparent gap in the sequence from the other end.

Buckland closed the discussion by making conciliatory remarks about the value of disagreement in science. He also proposed, however, that the lower portion of the Carboniferous system should be termed "Culmiferous." This was probably intended as an equivalent for d'Omalius's *terrain anthraxifère*. Although Buckland himself had reassigned many of d'Omalius's formations to the Silurian (§6.3), in a modified sense such a term might still be a useful label for all the formations between the Coal Measures and the true Transition. If this reconstruction of Buckland's meaning is correct, his suggestion was indeed a "middle course between the combatants," as one reporter put it. The whole of the Culm series—including the Culm limestone at its base—might represent *neither* the Coal Measures, as Sedgwick and Murchison claimed, *nor* the Greywacke or Transition, as De la Beche had maintained all along, but rather the formations *between* those groups (fig. 7.7, D). In much of England that intermediate position was occupied by the distinctive Mountain Limestone and Old Red Sandstone. But Phillips's recent book had shown that the Mountain Limestone was remarkably variable when traced laterally (§6.6); in Scotland the equivalent strata even included workable coal seams; and Buckland himself had implied that the Old Red Sandstone might be equally variable, since he thought it was missing altogether in Belgium (§6.3). Buckland's proposal had the eirenical effect of conceding that there were indeed some hitherto unrecognized Carboniferous strata in Devonshire, as Sedgwick and Murchison claimed, while eliminating their putative gap in the Devon sequence, against which De la Beche had protested all along. But if the "combatants" reacted to this conjectural "middle course," their comments were private and unrecorded.

So the session ended in what seemed "perfect harmony," at least to the punning reporter for the *Literary Gazette:* "the Section saw with satisfaction that, though *they hauled each other over the coals*, the geological rivals parted as good friends as when they began."[41] That the animosities of those rivals were so well concealed from the general audience was not only a tribute to their gentlemanly restraint and Buckland's astute chairmanship, but also a small triumph for the image of science that the leaders of the British Association were striving so hard to disseminate. Science was not to be divisive, as religious and political arguments so often were; disagreements in science, however heated, were to be at all costs amicable, and the community of men of science was to remain united in its search for truth.[42] The public handling of the Devon controversy at Bristol served that image building well.

The same episode also showed in action another distinctive trait of the Association, namely the concentration of effective power in the hands of the metropolitan gentlemen of science.[43] As the *Literary Gazette* commented indignantly, when reporting another moment in the Bristol meeting, those outside that elite "were most unceremoniously snubbed and set down—to listen to their superiors, no doubt wiser and better men."[44] In the handling of the Devon controversy no such crude display of power was needed. With a lively Section and no shortage of papers, the excuse of lack of time was always conveniently at hand to placate those whose offerings were refused. When some weeks later De la Beche berated Sedgwick for his part in the public criticism of his north Devon map, Sedgwick told him that the decision to give their paper at Bristol had been Murchison's: he had considered that meeting "the proper place" to make their point, because at Dublin the previous year Williams had "abused the Geological Socy. for their ignorance of the N. Devon Coalfield in the greywacke" (see §6.2). The clinching argument in favor of giving their paper in Bristol, however, had been that Williams, although not present in person, had sent a paper on north Devon, "wh. would have come on, if ours had not; & then all our views must have been brought forward in self defence."[45] Murchison and Sedgwick had thus used their power to preempt the absent Williams, whose views had been kept firmly offstage.

It is almost inconceivable that Murchison did not see and read William's paper while he was in Bristol. When writing it, Williams had probably acted on the reminder he noted for himself while still in the field, to "give localities where [fossil] plants are found" in Devon.[46] Although Murchison believed those plants all came from the Culm, he may have been uneasy about what the provincial geologist would turn up next. In any case, among the customary annual "recommendations of researches" of the geologists' committee, there was one that was clearly directed toward the resolution of the empirical problem with which the Devon controversy had begun: "the attention of observers was directed to the discovery of Plants of any kind in strata older than the coal formation."[47] Sedgwick and Murchison might claim to have found the

41. Ibid.
42. Morrell and Thackray (1981, chap. 5).
43. Ibid., chaps. 6, 8.
44. *LG*, no. 1024, p. 569.
45. Sedgwick to De la Beche, 17 October 1836 (NMW: DLB).
46. Williams, notebook for "June 1836 to Septr. 1836," notes between dated entries for 16 and 26 July 1836 (SAS: DW, box 1, notebook 9).
47. *Ath.*, no. 463 (10 September 1836), p. 656; *RBAAS* for 1836, p. xvii.

"true" solution to the enigma of the fossil plants from the Culm, but the controversy was clearly not yet over.

By comparison with the "grand field day" over Devonshire, the remaining sessions of the geological Section's proceedings were an anticlimax. But the science got a boost of publicity at the plenary session on the final day, when Buckland presented the Association with an advance copy of his long-awaited Bridgewater treatise on geology, and followed it up with a lecture that illustrated the book's argument. With characteristic animation, and probably with more than a touch of pre-Victorian scatological humor, he strutted across the lecture room, imitating the gait of the giant birds that were thought to have left their huge bipedal footprints in some of the Secondary sandstones.[48] Murchison was not amused: "the grossness of the Buffoonery," he told Lyell later, "acted on me like an emetic"; he added that Sedgwick had intended to speak in public on the final evening, "to take Science out of *the dirt* into which B. had shoved it."[49] But Buckland's lay audience was less high-minded and enjoyed his showmanship.

In any case Buckland's lecture was undeniably a vivid example of his book's consistent attempt to use fossil evidence to reconstruct the *living* habits of extinct organisms, and thereby to demonstrate the maintenance of beneficent organic design throughout the vast history of the earth. This, and not the pinched timescale of the scriptural geologists, was the vista that Buckland set before his audience at Bristol; this was the kind of science, harmoniously compatible with religious belief and practice, that the leaders of the Association sought to promote. Those who had participated in the geological Section would have been left in no doubt that this was the broader endeavor to which the resolution of the Devon controversy would contribute, by deciphering the earlier chapters of the earth's history and by displaying correctly the orderly and designful succession of its faunas and floras.

7.4   Sedgwick Explores the Cambrian in Devon (September and October 1836)

When the British Association meeting was over, Sedgwick left Bristol by coach to resume his fieldwork in the southwest, this time without Murchison for company. Although he intended to check on their new interpretation of the Culm, his primary aim was to unravel the sequence of the older strata in south Devon and to study their relation to the economically important mineral veins of the Cornish mining area. Since Murchison agreed that all the older strata south of the "Culm trough" were pre-Silurian, they appeared to lie entirely within Sedgwick's own province of the Cambrian. As already emphasized, the rich fossil faunas of the south Devon limestones held out the possibility of defining the Cambrian and its subdivisions in terms of fossils far more effectively than had been possible in Wales itself.

48. Buckland (1836, 1:259–60); *Ath.*, no. 463, p. 656; Morrell and Thackray (1981, p. 238).

49. Murchison to Lyell, 22 September 1836 (APS: CL, B.D25.L). On a celebrated occasion at Murchison's house a few years earlier, Buckland had similarly entertained a geological party by inducing a tortoise to produce a specimen of its footprints (Murray 1919, pp. 7–8), but that buffoonery had been perpetrated in gentlemanly privacy.

Working westward from Exeter, Sedgwick first had "hopes of finding the natural base of the Culm measures" to the east of the Dartmoor granite, but the boundary proved as elusive as before.[50] Turning south to Teignmouth, he met Austen for the first time on the latter's home ground. "Austen is a close [i.e., careful], good, independent workman," he told Murchison, "& wants nothing which a little practice will not give him." The tone may have been patronizing, but the assessment was realistic. The younger geologist had invaluable local knowledge, but Sedgwick saw himself as able to provide interpretative insights that Austen could not yet hope to match.

For example, in his paper read to the Geological Society earlier in the year (§7.1), Austen had described some pebbly rocks or conglomerates (at Ugbrooke Park near Chudleigh) as older than any of the Transition limestones of southeast Devon.[51] But when Sedgwick saw the place he changed all this, as he told Murchison later.

> After some good hammering we proved the limestone series to be *under* & not *over* the said conglomerate & grits; . . . and what do you think? The congloms are fine sandstone grit, & the sandy beds by the side of the lake in the Park are true coal measures with very fine plants—the very best I have seen excepting Bideford!! Austen was in raptures at the new aspect of the neighbourhood and while we were knocking out calamites on came a thunderstorm which put us in the condition of half-drowned rats.[52]

Even the cold he caught in consequence could not lessen Sedgwick's delight at having at least located the base of the Culm. But it was Austen who had shown him the spot, and it was Austen who, when he had to return home, told Sedgwick where to "look for *limestone fossils & where* for *trilobites*" near Newton Abbot. "Immediately on my arrival," Sedgwick told Murchison later, "I went to the fossil quarry & planted two workmen with good tools with direction, to bag as many fossils as they could find—I then procured another worker of slates & planted him in the trilobite quarry & I have procured such trilobites as would make your mouth water."[53] Without Austen's local knowledge, Sedgwick's fieldwork would have been much less fruitful.

Although Sedgwick regarded all the pre-Culm rocks of south Devon as Cambrian in age, he had not yet found adequate grounds for characterizing his system in terms of fossils. It is therefore hardly surprising that he recognized it in the field mainly on the grounds of rock type. Thus the "trilobite slates" just mentioned reminded him of some Cambrian slates in Cumbria (Skiddaw). After studying a thick sequence of strata further south (around Dartmouth), he concluded that it probably represented the whole Devonian or Upper Cambrian sequence in north Devon.[54] When he reached Plymouth he confirmed what he had suspected earlier in the summer, namely, that the

50. Sedgwick to Murchison, "Sunday Morg.," pmk [4] September 1836 (GSL: RIM, S11/105); Sedgwick, notebook 29, notes dated 1–3 September.

51. Austen (1836), read at GSL on 23 May.

52. Sedgwick to Murchison, 25 September 1836 (GSL: RIM, S11/106); Sedgwick, notebook 29, entry for 9–10 September. *Calamites* (fig. 5.3, C) was interpreted as the stem of a giant horsetail.

53. Ibid.

54. "The whole may represent all of N. Devon *between* the Silurians & the red beds under the Cwm Martin [*sic*] limestones": Sedgwick, notebook 29, entry for 13 September.

Plymouth limestone was at the base of a thick local sequence, far below some slates "looking as old as the Tintagel [slate]" he had seen on the north coast.[55] Sedgwick was clearly convinced that it belonged well down within his Cambrian system, just as he and Murchison had inferred earlier (figs. 7.3, fig. 7.4: "Plymouth Zone"). Its rich fossil fauna was therefore of the greatest importance for his work. Hennah later sent a collection of his Plymouth fossils, at Sedgwick's request, "to enable you," as he put it, "to decide the question hitherto in doubt," namely the true age of the limestone.[56]

Finally, after traversing still further west along the south coast of Cornwall, Sedgwick reported to Murchison that the Cornish limestones—and here he did mention fossils—seemed to him to be equivalent to a south Devon limestone (at Ashburton) that he believed to be still older than that of Plymouth; he added that he thought they might both be correlated with a thin limestone he had found well down in his Cambrian sequence in north Wales (on Arenig Fawr, near Bala).[57] All in all, Sedgwick had reason to be pleased with the way Devon and Cornwall promised to enrich the characterization of his Cambrian system beyond what had been possible in Wales itself.

Once he was in west Cornwall, however, Sedgwick's attention turned mainly to mining geology and a study of the mineral veins that traversed the ancient strata. For at least part of the time, he was joined by Austen; it may have been the first time the young geologist had ventured to do fieldwork outside his home area. Sedgwick had an amicable encounter with Boase, whose local knowledge proved unexpectedly valuable despite his wrongheaded notions of high-level interpretation (§4.6). Most important, however, was Sedgwick's meeting with De la Beche, whom, he told Murchison, was "perfectly open" with him: "he is certainly doing admirable work in Cornwall, & I believe his mining maps will be the most perfect thing in the world."[58]

Sedgwick had not, however, forgotten the still unsettled argument with De la Beche about the relation of the Culm strata to the still older rocks. He assured Murchison he would do what he could to improve their case in the limited time available before he had to return to Cambridge for the new academic year. After a brief visit to the north coast, he told Murchison he had located the southern boundary of their inferred "Culm trough" at a place (Boscastle) several miles south of the position (St Gennys) where Murchison had sketched it on the map displayed at Bristol (fig. 7.4).[59] This was no mere detail, for it implied that Sedgwick now assigned to the Culm a vast thickness of strata that he and Murchison had earlier regarded as Cambrian (§7.2). There could be no clearer indication that the distinction was far less clear in the field than they had claimed in their public presentation of their case.

Sedgwick's new location for this embarrassingly obscure boundary would be confirmed if the distinctive Culm limestone could be recognized here as it had been on the northern edge of the area of Culm rocks. "To the black limestones at their base I will give all the attention I am able," he promised Murchison; and on the eve of his return to the coast he told his Cambridge

---

55. Ibid., entry for 15 September.
56. Hennah to Sedgwick, 20 October 1836 (ULC: AS, IB, 70).
57. Sedgwick to Murchison, 25 September 1836. The Cornish limestones to which he referred were probably those that Greenough's map depicted as a band running through Fowey (fig. 4.10).
58. Sedgwick to Murchison, 29 September 1836 (GSL: RIM, S11/107).
59. Ibid.

colleague William Whewell (b. 1792) that he was about to study the point where "the *Culm measures* abut against the Cornish slate," implying that he expected the junction would be unconformable.[60] He and Austen had apparently been joined by Conybeare, who had just moved from south Wales to east Devon on becoming vicar of Axminster. When they all studied the cliffs the next day, Sedgwick noted that "it is just at this point that the black Culm measures end"; but he had to admit that he could find no trace of the Culm limestone and that he "could not descend to sea side to verify" the character of the crucial junction.[61]

On his own again, Sedgwick now tried to trace that junction eastward, though in an inland area of sparse exposures he could hardly hope to find decisive evidence. "The elements are in most dreadful disorder," he told Whewell on reaching Launceston amid violent thunderstorms; "it would vex me to the marrow of my bones to quit this country without settling the boundary of the Culm basin." However, with the help of a local fossil collector, the young Launceston solicitor Samuel Pattison (b.1809), Sedgwick did consider that he had succeeded in tracing "the Culm demarcation" from there to the edge of the Dartmoor granite, where he found the strata "well frizzled" by metamorphism. He also found "fossils without number" in what be believed to be a Cambrian limestone at the locality near Launceston (South Petherwin) that he had already visited briefly with Murchison.[62]

Finally, as he traversed round the north side of the Dartmoor granite, Sedgwick confirmed that the band of limestone marked there on Greenough's map (fig. 4.10) was indeed black Culm limestone, as he and Murchison had suspected during their earlier fieldwork (§7.2). So despite his failure to detect that limestone on the coast, Sedgwick now had further supporting evidence for their new interpretation of central Devon as a "Culm trough": the basal Culm limestone outcropped symmetrically along both the north and the south sides, just as Murchison had depicted it—somewhat prematurely—on the sketch map he displayed at Bristol (fig. 7.4). All in all, Sedgwick's six weeks' fieldwork since the Bristol meeting had been productive and reasonably successful.

7.5   Reproofs and Recriminations (September to November 1836)

"Pray don't think that our paper was an *attack on you*," wrote Sedgwick, when telling De la Beche about his fieldwork before he left Exeter to return to Cambridge.[63] But an attack was just what De la Beche felt the events at Bristol had been. In the weeks that followed the Bristol meeting he claimed repeatedly, both to Sedgwick and to Greenough, that in "the north Devon affair" he cared only for the truth, and that no "tom-noddyism" or foolish pride would prevent him from accepting his opponents' conclusions; "but I am quite cer-

60. Ibid.; Sedgwick to Whewell, 7 October 1836 (TCC: WW, Add. Ms. a. 213[22]).
61. Sedgwick, notebook for "Cornwall 1836," entries for 7, 8 October (SMC: AS, notebook 30).
62. Ibid., entries for 9–14 October; Sedgwick to Whewell, 12 October 1836 (TCC: WW, Add. Ms. a. 213[23]).
63. Sedgwick to De la Beche, 17 October 1836.

tain," he told Sedgwick, "as I consider you follow science for the sake of truth, that you would not have me agree with you if I did not see good reason for it."[64] The pointed implication, that a certain other individual was *not* so nobly motivated, would not have been lost on his correspondent. What De la Beche objected to was "the mode of attack" on his work. "Not only was it secret," as he reminded Greenough, "but the place and the time were not such as friends, if not to me, at least to the good honest cause of cooperation in science, should, in my poor judgment, have selected." He added, however, that he hoped that "all was oversight,—and the mere desire of men to signalize themselves by producing before a large audience what they thought a brilliant discovery."[65]

Embedded in this complaint were three distinct criticisms: first, that the attack had been covert and ungentlemanly, because his mapping had been criticized behind his back and the paper sprung on him without warning; second, that it was imprudent, because it could have harmed the reputation of geology and geologists in the eyes of visiting politicians; third, that it was undignified, because it made geological argument into a vulgar spectacle. Of those three criticisms, the first was dismissed by Sedgwick. "When a D----d good friend tells you anything unpleasant of another, always dock 75 p. cent from it at the least," he told De la Beche; "I am certain I never heard him [Murchison] say anything of your Devon map behind your back he would not say before your face."[66] That may have been true, but De la Beche's point was that the criticisms had been made in the presence of politicians in such a way that he had no adequate chance to defend himself. But Murchison, writing to Lyell, rejected the charge with a ruthlessness that belied Sedgwick's charitable defense.

> He [De la Beche] takes the line, commended him no doubt by some *kind, ignorant* friends, that he has been harshly treated. I have conscientiously advised him to *cancel* all the Devonshire work & start afresh. At all events you [Lyell] & all my friends can testify that *mine* is no insidious or *covered* attack but one of which he had forewarnings in abundance & *in spite of which* he finished his great map.[67]

In other words, Murchison insisted that his criticism was not just the outcome of a mere four weeks in Devon, but that the fieldwork had only confirmed the deep-seated skepticism he had expressed publicly ever since the start of the controversy.

Sedgwick had no such theory-based objection to De la Beche's interpretation of Devon, but his fieldwork had made him increasingly critical, first of routine details of the official mapping of Devon, and later of De la Beche's interpretation of the structure and hence of the sequence in the region (§7.2). That note of criticism continued when Sedgwick returned to Devon after the

64. De la Beche to Greenough, 20 November 1836 (ULC: GBG); De la Beche to Sedgwick, 20 October 1836.
65. De la Beche to Greenough, 20 November 1836.
66. Sedgwick to De la Beche, 17 October 1836.
67. Murchison to Lyell, 22 September 1836.

Bristol meeting.[68] But when they met in Cornwall, De la Beche apparently excused the deficiencies of the official map on the grounds that some of the copies in circulation—including presumably the one that Sedgwick and Murchison had carried with them—had been handcolored inaccurately and sold without authorization. Sedgwick seems to have accepted the excuse: "the *persons* who have *copied* your original," he told De la Beche, "have, I am sure, used you ill." So his praise for the "beautiful Cornish map" on which De la Beche was now working was probably unfeigned.[69] Nonetheless, he and Murchison were now putting forward an interpretation that implied a change in the map of Devon far beyond any copyist's error. It was this that was potentially so damaging to De la Beche's reputation in official quarters—the second of his three grounds for complaint.

Once he was convinced that Sedgwick at least intended him and his survey no harm, De la Beche explained openly to him why he was so apprehensive about the political repercussions of the Bristol affair. For he believed with good reason that the government's continuing patronage of science, and particularly of geology, was "very ticklish" at that moment: "they have been so often jobbed, and infernally jobbed, under the old systems, that they are always afraid of being jobbed again."[70] The party of Reform was committed, in any case, to rooting out such ancient abuses as the diversion of public funds to private advantage. But De la Beche was almost certainly alluding in particular to the notorious case of John MacCulloch (1773–1835). The Scottish geologist's vast claimed expenses for his protracted survey work in Scotland had even been the subject of parliamentary enquiry; the government had still seen no concrete return on its investment when, the previous year, MacCulloch was killed in a carriage accident while on honeymoon in Cornwall.[71] As De la Beche explained to Sedgwick, it was an embarrassing precedent.

> Hence when I arrived at Bristol, knowing that Lord Lansdowne and Mr. Spring Rice purposed to be there, and finding it currently reported that the North Devon [map] was little better than a geological abomination, and was to be blown up by yourself and Murchison, I anticipated that Government (not being versed in these matters) would not consider the matter at issue as a difference of Doctors' opinions, but rather in the light of a Geographical Survey, in which the surveyor was pleased to lay down a large area as land which in fact was covered by a great sea.

It was not unrealistic for De la Beche to fear that politicians might imagine that a geological map was as unproblematical a visual representation as a topographical map. His mistake over the Culm—if mistake it were—would then be regarded as a sign of total incompetence. He told Sedgwick that "very

---

68. For example, he found that at one point (near Chudleigh) the edge of the Dartmoor granite—as easy a boundary as any to trace in the field—was misplaced on the official map by two or three miles; he inferred that this was the result of its being sketched "from a distant view" of the topography—an inexcusable cutting of corners in fieldwork practice—rather than by hammering the rock exposures on the spot: Sedgwick, notebook 29, notes for 9 September; Sedgwick to Murchison, 25 September 1836.

69. Sedgwick to De la Beche, 17 October 1836.

70. De la Beche to Sedgwick, 20 October 1836.

71. V. A. Eyles (1937); Cumming (1985) makes a rather unpersuasive case for MacCulloch's rehabilitation.

disagreeable results" might have followed, had not his friends—doubtless Greenough was one—"explained matters in the proper quarters."[72]

Fearing, nonetheless, that rumors of his incompetence might circulate in those quarters, De la Beche took the precaution of reporting to the Board of Ordnance how his mapping work had been improperly criticized at Bristol. Colby, the director of the Survey, tried to reassure him by calling it a "little difference of opinion" that geologists would eventually regard as "trivial." Once De la Beche published his memoir to explain his mapping of Devon, Colby assured him, the public would "find enough of practical applications of your geology to assure them that your system is not without utility."[73] But this was to divorce useful "standard facts" from "systems" of theoretical interpretation in a way that De la Beche knew his opponents would find untenable; the argument was not as trivial as the director assumed. Even if De la Beche could not share Colby's utilitarian empiricism, however, the director's letter may have had the desired effect of reassuring him that the future of the geological survey, and of his own livelihood, was not in any immediate danger. Furthermore, Sedgwick promised later to write to Spring-Rice, "to dispel any bad impressions he may have carried away with him." When De la Beche heard that Sedgwick had in fact done so, it confirmed his opinion that "he is, as I have always considered him, an honest fellow."[74] Sedgwick's collaborator was, by implication, quite another sort of fellow.

The third and last source of De la Beche's sense of grievance over the affair at Bristol was that, quite apart from the chance presence of politicians, the British Association was an inappropriate arena for such an argumentative discussion among geologists. "Mooting the matter before the British Association in the way it was," he told Sedgwick, "was quite a different thing from bringing it before the Geological Society." The difference lay in the audiences (§2.4). As De la Beche insisted to Sedgwick, the "cause of truth" was better served by the confidential exchange of views among competent experts such as themselves, than by showy battles in the presence of audiences "who only attend for the purpose of witnessing the fun of a wordy fight."[75] At the very least, a difficult technical problem such as the Devon controversy was better debated and evaluated before what Greenough called the "much fairer tribunal" of the Geological Society.[76]

To his London protector, De la Beche complained with calculated rhetoric about his opponents' misuse of the more public arena.

> You, my dear Greenough, who so long participated in and must well remember the good old way we had of assisting one another's labours at the Geological Society, will, I am confident, not approve of the new system of bothering and snatching away one's labours like so many dogs at a bone, for the effect of a few momentary plaudits, or to promote the clap trap nonsense of life. What the deuce does it signify whether A. B. or C. find out a thing or not. If science be, as it ought, the pursuit of truth, for heaven's sake let it be followed in a way to promote the

72. De la Beche to Sedgwick, 20 October 1836.
73. Colby to De la Beche, 15 September 1836 (NMW: DLB).
74. De la Beche to Greenough, 4 and 20 November 1836 (ULC: GBG); Sedgwick to De la Beche, 20 November 1836 (NMW: DLB).
75. De la Beche to Sedgwick, 20 October 1836.
76. Greenough to De la Beche, 13 November 1836 (NMW: DLB).

ends of truth, and not those of interest—for the mere object of vulgar applause, or position among the mass of people who know nothing about Science, and are ever ready to gaze open mouthed on those who talk most about their grand discoveries.[77]

The "good old way" at the Geological Society may have been recollected through rosy spectacles, and De la Beche certainly made the most of the contrast between his own love of truth and his opponents' baser motives. But others would have agreed with his criticism of the show-biz character of the British Association, which Sedgwick and Murchison had undoubtedly exploited to the full.[78]

7.6 De la Beche Consolidates His Defense (October and November 1836)

When Sedgwick wrote to De la Beche, just before his return to Cambridge, his remarks on the Bristol affair were strictly subordinate to his report on his fieldwork. Throughout the following weeks, De la Beche's repeated protests about his treatment at Bristol were expressed in parallel with the consolidation of his defense on the substantive grounds of the controversy.

In the letter that initiated this further and private unfolding of the controversy, Sedgwick summarized under three points the fieldwork he had done since meeting De la Beche in west Cornwall (§7.4). First, he claimed to have found the base of the Culm on the south side of the trough and to have traced its junction with the older rocks from the north Cornish coast eastward to the Dartmoor granite. Second, he had found that all round the northern side of Dartmoor "the Culm measures are in *contact with the granite.*" Third, he considered that all the limestones east of Launceston (i.e., toward Dartmoor and around its northern edge) were Culm limestones. "Can you swallow all this?" he asked.[79]

"You certainly do call upon me to swallow many things which my gorge rises at in the present moment," De la Beche replied, "but I suppose that like Physic-taking one can get accustomed to any thing however unpalatable it may be at first."[80] But in fact it was only the third component of Sedgwick's medicine that he refused. First, since he had conceded at Bristol that the Culm strata were uppermost in north Devon (§7.3), he can hardly have been dismayed at Sedgwick's reported tracing of the southern boundary of the putative Culm trough, though he might disagree on its course. Second, he was as firmly convinced as Sedgwick about the intrusive origin of the granites in the region, and can hardly have been surprised to learn that the Dartmoor granite had intruded the Culm. Even his opponents' claim that the Culm was really Coal Measures only implied that the granite—which had "baked" the Culm—was much younger than any geologists had hitherto supposed, but not implausibly so.

77. De la Beche to Greenough, 4 November 1836.
78. Orange (1975); Morrell and Thackray (1981, pp. 157–63).
79. Sedgwick to De la Beche, 17 October 1836.
80. De la Beche to Sedgwick, 20 October 1836.

Sedgwick's third point was the only one that was directly opposed to what De la Beche had stated both in private and in public, namely, that the limestones outcropping east of Launceston and round the north side of Dartmoor were the eastward extension of the fossil-bearing rocks he had traced westward (from Tintagel to Padstow) along the north coast of Cornwall (§6.5). This was no mere local detail. He had regarded these strata, at least by implication, as some of the oldest in the Greywacke, so it is hardly surprising that he now denied Sedgwick's claim that some of them belonged to the Culm. "This, no doubt, may all be very wrong, like the greater proportion of human conclusions," he conceded with his habitual show of modesty; and he admitted that he might change his mind after further fieldwork in the relevant areas. At least for the time being, however, he remained unconvinced by Sedgwick's arguments and continued to regard *all* the older rocks of Cornwall and south Devon, including the Culm, as a single series of ancient Greywacke formations.

Feeling he could trust Sedgwick, De la Beche now "got upon the confidential system" with him and revealed his latest discovery, which indirectly supported his interpretation. In his survey of the south Cornish coast, he had just found some black Culm-like strata (at Manaccan) in what he regarded as the "continuation of the Plymouth system," or low down within the ancient Greywacke rocks. He promised to let Sedgwick know if he found fossil plants as well. This he fully expected to do, because he regarded these "sooty culmy things" as genuine Culm strata, "exactly in their place" in the local sequence, assuming that, as in north Devon, the Culm was "a portion of the grauwacke series."[81]

That assumption, however, was just what Sedgwick rejected. De la Beche sent him no confirming report of having found the distinctive fossil plants. When, a month later, Sedgwick got round to replying, he pointed out that in any case this alleged discovery of a new area of Culm strata in south Cornwall would "prove nothing, unless you could shew some greywacke *over* them." Until such field evidence was forthcoming, the Culm was still, in Sedgwick's opinion, firmly established as the youngest formation in the sequence in Cornwall and Devon, overlying what he regarded as far older Cambrian rocks. He agreed with De la Beche that the "great zone of calcareous slates" along the south coast of Cornwall (at Looe and Fowey) and along the north (at Tintagel and Padstow) belonged low down in that sequence of older rocks; much lower than the Plymouth limestone, he believed, and apparently equivalent to another limestone in south Devon (at Ashburton). But Sedgwick insisted that the black limestones around the north of Dartmoor were "totally distinct" from all these: "they are distinct in position, structure and fossils." He claimed that they were "true culm limestones" and that they were "perfectly identical" with those along the north side of the Culm trough (at Barnstaple, Bampton, and Holcombe Rogus). "In confounding these with the older limestones you have made a mistake," Sedgwick told him bluntly, emphasizing that he was as certain on this point "as I am of anything I have ever examined since I first used a hammer."[82]

81. Ibid. Two weeks later he reported the same discovery to Greenough, sketching a traverse section to show how the "culmy beds" were part of a thick conformable sequence in the area, but he did not report having found any fossil plants in them: De la Beche to Greenough, 4 November 1836.

82. Sedgwick to De la Beche, 20 November 1836.

Blunt though this was, Sedgwick was merely enlarging on what he had told De la Beche immediately after his fieldwork. His sharp distinction between the black Culm limestones—it was the first time he had alluded to their fossils—and all the older limestones of Cornwall and south Devon simply helped to confirm what he and Murchison had suspected during their joint fieldwork, and what Murchison had provisionally shown on the sketch map displayed at Bristol (fig. 7.4). The detection of the black Culm limestones north of Dartmoor strengthened the evidence for the reality of the huge trough structure they had postulated, by demonstrating that the formation outcropped symmetrically along both sides of the trough. Thereby it also strengthened the case for regarding the Culm as a whole as the uppermost and youngest formation in the region, whatever its true age might be.

De la Beche did not wait for this elucidation of Sedgwick's position before strengthening his own. Only two weeks after Sedgwick left Devonshire, De la Beche wrote to Greenough about his own work, recalling his tactical retreat in the face of the "grand attack" at Bristol. "Falling back, however, upon my resources," he added, "I found myself far stronger than I anticipated, in fact capable of taking up a good position—a position that seems to become stronger every day." His immediate reaction had been to accept his critics' interpretation; but on reflection, he told Greenough that the whole matter "appears to me little else than an ingenious endeavour to get over an apparent difficulty."[83]

"Murchison had most firmly maintained that the culm of Bideford could not be in the grauwacke, because he had not found any [culm] in the Silurian System, and Sedgwick none in the Cambrian," De la Beche recalled; "an odd reason certainly—because it went [i.e., amounted] to the assertion that plants did not exist before the period of the coal measures." As at the start of the controversy, De la Beche dismissed that assumption as "monstrous and not vastly philosophical" and as contrary to the Continental evidence. When Murchison had at last gone to Devon, and "found that the culm beds were at all events in the body of the rock of the country," he had been obliged, if he were to save his theoretical assumption, to turn the Culm into Coal Measures. "Now all this, I am confident, will not do," De la Beche insisted; "these culm beds graduate into the beds (acknowledged to be grauwacke) beneath them quite as much as the coal measures do into the carboniferous limestone, or the carboniferous series into the grauwacke series."[84]

So far, this defense did not differ from what De la Beche had offered from the start of the controversy. But in "taking up a good position" since the Bristol meeting, he had not retracted the highly important concession he had made to his critics on the spot (§7.3): he accepted the validity of their interpretation of the *structure* of Devon, and hence of their conclusion that the Culm was uppermost in the sequence, and not well down within it as he had originally claimed. At Bristol, however, he had said nothing to modify his earlier inference that the Culm belonged nonetheless to the older Greywacke

83. De la Beche to Greenough, 4 November 1836.

84. Ibid. A few weeks later he summarized these objections for Colby, who wanted to refer to the controversy in his presidential address to the Geological Society of Dublin: Colby to De la Beche, 29 November 1836; De la Beche to Colby, 23 December 1836 (NMW: TS, II, pp. 172–85).

(fig. 7.7, B). Since then, however, De la Beche had reflected on the implications of Murchison's identification of Silurian strata in north Devon, which he had no reason to contest; he now suggested a new scheme of correlation that took that discovery into account. "Altogether it seems clear to me [that] the culmiferous rocks of N. Devon are decidedly grauwacke—constituting the upper part of the series as developed in Devon," he told Greenough; "and if Murchison be right about the rocks N. of Barnstaple being his lower Silurian affairs, why then this carbonaceous deposit is equivalent to the upper Silurian" (fig. 8.1, A).[85] In other words, since there was no detectable unconformity between the two sets of strata, by the accepted canons of geological reasoning their periods of formation must also have been in immediate sequence.

This represented a significant modification in De la Beche's interpretation, but it would not have appealed to Murchison, when and if he heard about it. By attributing the Culm to the Upper Silurian instead of to the lower Greywacke or Cambrian, De la Beche did nothing at all to remove the grounds of Murchison's original objection. If anything, De la Beche's modified interpretation would vex Murchison even more than its original form, for it placed the fossil plants of the Culm squarely in Murchison's own system, where in his opinion they had no business to be.

De la Beche, however, remained critical about the way the Silurian and Cambrian systems were being promoted and skeptical about their validity outside the regions after which they were named. "I suspect that the Silurian and Cambrian systems won't exactly fit in Devon; in fact, their authors have already been obliged to invent a new name, the Devonian System, for what they find in Devon," he reminded Greenough. "Sedgwick also let out to me in conversation that Silurian fossils were found out of their places [i.e., in the "wrong" strata] in Devon," he reported, "and that he did not believe that the grauwacke series could ever be characterised or divided by aid of fossil remains as the new red sandstone, the oolitic, and cretaceous may be." Whether or not Sedgwick really was having such doubts, nothing further was heard of his and Murchison's proposed "Devonian system" in the following months. "Depend upon it, they want to employ an ingenious way of explaining matters in Devon," De la Beche commented, "so that the naming of the systems should remain in their hands, and the assertion that the culmy system be equivalent to coal measures stand as long as it can."

He added a more strictly geological reason why he found their interpretation implausible, namely, that the perfectly "normal" Carboniferous strata in the Mendip Hills were only separated by quite a short distance from the "violently changed appearance" of the nearest outcrops of Culm strata, which then remained constant in character throughout central Devon and into Cornwall (see fig. 3.5). "Really the slapdash system will not do," De la Beche concluded, "patience and work are essential." He himself had been mapping Devon for four years, his opponents for no more than a few weeks. "I really don't think Sedgwick intended me much harm," he added, "as to Murchison, I don't know what to think."[86]

85. De la Beche to Greenough, 4 November 1836.
86. Ibid.

Greenough replied promptly to his protégé, praising the way he had handled the affair at Bristol: "your retreat under the circumstances was perhaps wise, but I did not at the time, nor do I now consider the question settled."[87] Greenough emphasized that, in a region of such complex structure, only a careful study of dips and strikes would clinch the case effectively. He advised De la Beche not to be distracted by the problems of south Cornwall, but to concentrate on elucidating the structure of north Devon, where the controversy had started, and to improve his field evidence for the true sequence of the strata there. "The other argument, that founded upon fossils, is in my estimation less decisive than you seem to consider it," he added; "we know that this is only an empirical character, and that where other analogies fail, this may fail also." Far from being an unthinking rejection of fossil evidence, Greenough was simply reaffirming what Sedgwick would have wholeheartedly approved, namely, the primacy of the structural evidence of superposition.

Greenough's advice was the more necessary, because he knew the controversy would continue during the session of the Geological Society that had begun the previous week. But De la Beche had no chance of participating in the debate, at least until his next trip to London, which he expected to be for the anniversary meeting of the Society in three months' time. As he told Greenough ruefully, "those who are paid for work, must work; else their employment will be cancelled." However unproductive winter fieldwork might be, and however unpleasant the cold and wet weather, the reluctant professional had only "a hard winter's work" to look forward to (fig. 7.8).[88] Not for De la Beche his colleagues' seasonal retreat to a warm study, to reflect in comfort on maps and sections, specimens and field notes; nor could he anticipate the regular stimulus of debate at the meetings of the Geological Society, or of geological gossip at the dinner table of its Club. More than ever before, De la Beche felt acutely the intellectual isolation that his employed status now imposed, and he depended more than ever on the characteristically spiced information which Greenough's letters regularly provided.[89] "The hare which Sedgwick & Murchison have started," Greenough commented, "must now be hunted down, let her double as she will."[90] But the London season had opened with De la Beche not among the huntsmen.

87. Greenough to De la Beche, 11 November 1836 (NMW: DLB).

88. De la Beche to Greenough, 20 November 1836 (ULC: GBG). Fig 7.8 is reproduced from a letter written a few weeks later: De la Beche to his daughter Bessie, 12 January 1837 (NMW: DLB).

89. For example, after the first meeting of the new session of the Society, on 2 November, Greenough mentioned, among a miscellany of items reported in a similar vein, that "Darwin dined at the Club and is about to publish a Volume of incredibilia on the effects of Plutonism in the countries he has traversed; he is an intelligent agreeable man": Greenough to De la Beche, 11 November 1836. Darwin had returned from the *Beagle* voyage in October and was elected to the Society two meetings later, on 30 November.

90. Ibid.

Fig. 7.8. De la Beche's sketch, drawn for his daughter in January 1837, showing himself sitting disconsolately in his lodgings in Cornwall and looking out at the heavy winter rain. Note the hammer, cloak, and broad-brimmed hat on the left, the map case and compass hanging on the wall on the right, and the field notebook on the edge of the table. The mice have no geological significance.

Chapter Eight

INTERPRETATIVE MANEUVERS

8.1 Preparing a New Attack (December 1836 and January 1837)

Greenough's metaphor of hare and hunters was probably based on a rumor that Sedgwick and Murchison were preparing a revised version of their Devon research to present to the Geological Society during its new session. Almost as soon as Sedgwick got back to Cambridge after the end of his fieldwork, Murchison started nagging him to get on with his share of the writing; more than once Sedgwick had to remind him that he was not a man of leisure, and that he did have academic and other duties to attend to as well.[1] He also warned his collaborator that he would not be able to attend the Geological Society until well into the new year. Since Murchison was convinced his presence was essential, they agreed to try to complete their projected paper in time for some meeting the following spring.

Early in the new year, however, Murchison suddenly felt a greater sense of urgency, when De la Beche wrote to the Society to ask if he might present a paper of his own at the meeting scheduled for a few days after the anniversary in February. On the face of it, the request was reasonable enough. De la Beche wanted to make the most of a rare visit to London by combining attendance at the anniversary celebration with a new presentation of his case on Devon and Cornish geology. But doubtless he was also aware that an early reading of his own paper would effectively upstage what his opponents were busy

1. Sedgwick to Murchison, 1 November 1836 and 23 January 1837 (GSL: RIM, S11/109,115).

183

preparing. Certainly Murchison realized that: he consulted the president and the Society's key administrator immediately, and reported to Sedgwick that "they both agreed in the validity of my reasons for having you present, but at the same time *they both* saw worse mischief in delay." Murchison explained that "our friends Lyell and Lonsdale think that this paper by De la Beche will be an ingenious twist of his old Devonian blunders into a right view of the subject founded upon *his discoveries* in Cornwall; which discoveries they agree with me in thinking will have been elicited (inter pocula) with a certain Adam Sedgwick."[2] In other words, both Lyell and Lonsdale—the latter had not appeared previously in such a partisan role—agreed that De la Beche's paper must be upstaged in turn, because it would embody what Sedgwick had unwisely *"let out"* when he met De la Beche in Cornwall (§7.4). Murchison warned his collaborator that he was in danger of being robbed of the credit of discovery, and not for the first time; but he added pointedly that on this occasion he himself had "a *joint stock interest*," and therefore could not stand idly by. "I do feel most strongly," he told Sedgwick, "that we ought to place our *whole* view before the public, ere any of the *pirates* can rob us of our bark."[3]

In order to establish their priority, Murchison proposed that their paper should now be "rigorously restricted to a *clear general view* of the classification & succession which we intend to adhere to 'per fas et nefas'." Such a "condensed sketch" could be prepared more rapidly than the fuller account they had hitherto been working on, and it could discreetly omit those aspects about which the two authors were not yet agreed or still uncertain. Murchison proposed that the fuller supporting details or "pièces justificatives" should be presented separately and later, which would give them a further opportunity of defending their "grand positions" if they were attacked in the meantime by De la Beche. "Let therefore the forthcoming bolt be ringing, clear & sharp," Murchison ordered, "but not encumbered with the cwt of culm and sandstone I could ornament it with." Assuring Sedgwick that this plan had Lyell's and Lonsdale's approval, he added that "if this is not done T D[e la] B[eche] will, in purloining your ideas, make himself a *greater reputation than ever* & just at the very period when he ought to be for ever damned for his loose (nay dishonest) work & absence of all philosophical views."[4] Murchison mentioned that he had just declined "a pleasant invitation to kill pheasants" in order to complete *his* contribution to their joint paper; he urged Sedgwick to make a comparable sacrifice and to send him as quickly as possible a set of traverse sections to illustrate their interpretation.

Lyell, who was now in his last weeks as president, agreed that this abbreviated paper should be read at the last meeting *before* the anniversary—and thus safely before De la Beche could present his—even though Sedgwick

2. Murchison to Sedgwick, n.d. [apparently written on or near the day of a GSL meeting; probably, by internal evidence, the one on 17 January 1837]. The word "Devonian" was used here in its everyday, topographical sense. The original of this important letter has been lost since it was quoted (in part) in Sedgwick's Victorian biography (Clark and Hughes 1890, 1:474); the surviving transcript is defective (ULC: AS, IIID, 19).

3. Ibid. The earlier examples of plagiarism alleged by Murchison were "Weymouth," probably Buckland and De la Beche's joint paper (1835) on that part of Dorset; and "Yorkshire Carboniferi," probably a reference to Phillip's monograph (1836). The grounds for these allegations are not clear.

4. Ibid. The "cwt [hundredweight] of culm and sandstone" were the local details.

could not be there on that date. To avoid the charge that he was acting unfairly, Murchison took the precaution of writing De la Beche what he later called "a kind and I hope gentlemanlike letter," to give him due warning that a paper criticizing his work would be read in only two weeks' time; but he insisted bluntly that between their respective views "there was no mezzo termine."[5] Ten precious days passed before De la Beche received this letter on his return from a fieldwork trip, and only five then remained before his opponents' paper was to be read. De la Beche was unimpressed by Murchison's "gentleman-like" manner and replied with a blunt statement of what he believed to be the real motives for presenting their paper at such short notice. Murchison was indignant at what he called the "untrue and pot-house style" of De la Beche's reply; its whole tone, he told Sedgwick, was "so ungentlemanlike (I could almost say blackguard) that I do not care if I never write or speak to him again."[6] The indignation may have been heartfelt, but it must have been inflamed by the knowledge that De la Beche had not been deceived.

"The whole affair is transparent enough, and I could wish it were not so," De la Beche commented to Greenough, on receiving Murchison's letter; "the affair is brought forward before the anniversary, for the sake of the annual speech, when I expect the notion, which I firmly believe to be founded in error, will be warmed up by friend Lyell." Not only had De la Beche seen through his opponents' plan, but, more seriously, his anxiety about his own livelihood had been aroused again. He had almost finished the map of Corn-wall and had applied for funding to continue his work in the south Wales coalfield.[7]

> Now whatever ill will there may be to me personally, why the deuce attempt to bother the Geol. Survey, which at all events is good in itself. The cry has usually been that our Government would do nothing for science—no sooner however do they extend their patronage to geology than attacks are made on our survey before we can open our mouths, and the impression has got around that the Geological Society are against it—now although I don't believe this to be the case as regards the members generally, public impressions are awkward things to deal with. It does not signify how unfounded or side winded [i.e., indirect and unfair] attacks may be—an attack supported or apparently supported by a public body goes a great way with folk in power.[8]

De la Beche rejected as "perfect rubbish" the claims of some geologists to exclusive rights in certain areas—doubtless he had Murchison's "Siluria" in mind—and denied that he himself had any such proprietorial attitude toward Devonshire. But he protested that attacks on the Survey were in this instance unfair because the work was still "in an unfinished state," since the memoir to explain the map was not yet published. "Our old hearty way of cooperation," he commented, echoing an earlier plea, "was surely better than this snap snap

5. Murchison's letter of 17 January is referred to in De la Beche to Greenough, 27 January 1837 (ULC: GBG, 53) and in Murchison to Sedgwick, 30 January 1837 (defective transcript in ULC: AS, IIID, 12).

6. Murchison to Sedgwick, 30 January 1837.

7. De la Beche to Greenough, 27 January 1837. He must have heard soon afterward that his fears had again been unfounded, for on the same day the Treasury approved the estimates that allowed the geological survey to move next to south Wales: Colby to Ordnance, 13 January 1837; Ordnance to Colby, 27 January 1837 (NMW: TS, II, pp. 191–93).

8. De la Beche to Greenough, 27 January 1837.

system, which belongs more to personal vanity, than a true love of science for its own sake."

Once again De la Beche stressed to Greenough that his basic objection to his critics' case was that it depended on an unfounded assumption about the fossil plants from Bideford—"as if we know for a certainty that the vegetables of the grauwacke *must* necessarily be different from those in the coal [measures],—when we should expect that they were likely to be similar." As in the public discussion at Bristol, however, De la Beche sought to minimize the significance of the plants, on the grounds that few of the specimens were well enough preserved for reliable identification (§7.3). He claimed that his opponents' interpretation entailed a "slap dash *a priori* argument" based on a meager "tenth part" of the fossil evidence. He also repeated his earlier comment that "the attempt to foist in the term Devonian System because the Silurian and Cambrian systems don't happen to fit in Devon, though they are represented to *pass* into each other in Wales, will never do." He concluded sarcastically that "at this rate we shall have a new grauwacke system in every country—in fact every gentleman his own system maker."[9] Having thus tacitly supplied Greenough with ample rhetorical ammunition for the coming battle, De la Beche returned to the more prosaic perils of fieldwork: "I narrowly escaped losing my life today down a cliff near this place [St Austell]," he told his patron ruefully, "and was all but swamped near the Dodman Head the day before yesterday."

8.2   Problems at the Eleventh Hour (January 1837)

Meanwhile, Murchison and Sedgwick's hurried attempt to put together a "condensed sketch" of their interpretation of Devon had run into difficulties, which led the two geologists into an exchange of correspondence of unprecedented intensity in the last ten days before their paper was due to be read.

Their difficulties were precipitated in part by an unexpected report from one of the local Devon geologists. Sedgwick in particular had done much to cultivate those invaluable informants the previous autumn (§7.4) and had been well rewarded for his efforts. Hennah sent him fossils from the Plymouth Limestone; Pattison later gave him details of the boundary of the Culm around Launceston; and, although not relevant to the main issue, Major Harding sent specimens from a raised beach in north Devon. Another local collector, William Bilton (b. *c.* 1798), a clergyman whose married sister lived near Bideford, sent him plant fossils from the Culm, hoping they would convince the "sceptics" who doubted its correlation with the Coal Measures.[10] But Sedgwick and Murchison had no monopoly on local geologists such as these. De la Beche

9. Ibid.
10. Hennah to Sedgwick, 20 October 1836; Bilton to Sedgwick, 28 December 1836; Harding to Sedgwick, 12 January 1837; Pattison to Sedgwick, 9 February 1837 (ULC: AS, IB, 70, 88, 87, 90a). Harding's specimens were presumably intended to illustrate a paper by Murchison and Sedgwick (read at GSL on 14 December 1836) on a raised beach near Barnstaple. This was totally unrelated to the Devon controversy, but was another issue on which they found themselves in conflict with both Greenough and Buckland: see Sedgwick to Murchison, 3 January 1837 (GSL: RIM, S11/114).

too had met Bilton and Harding on a recent trip to north Devon and had been told it was *they* who had shown Sedgwick and Murchison much that those geologists were likely to claim as their own discoveries. In particular, Bilton and Harding had pointed out to them, near Bideford and Barnstaple, respectively, what they regarded as outliers of "Green Sand"—a Secondary (Cretaceous) formation far younger than any of the rocks under dispute in Devon and not obviously relevant to the main controversy (see fig. 3.2).[11]

Less than two weeks before the much-vaunted paper was due to be read, Harding apparently reported finding fossil plants in this alleged Green Sand near Barnstaple. Like the outliers themselves, this was not in itself implausible; but judging from Murchison's reaction, Harding must have mentioned that these plants looked to him like those from the Culm. For Murchison, this would have been an anomaly as incredible as finding similar plants in the older Greywacke—as De la Beche had originally claimed. "The Major's letter *upsets the case* if true," Murchison wrote to Sedgwick, evidently in some agitation.[12] "I don't believe a word of the Major's greensand," Sedgwick retorted; "it is no doubt part of the Caradoc."[13] In other words, he rejected the local geologist's competence to judge whether the relevant sandstone really was an outlier overlying far older rocks, let alone to pronounce on its true age. He assumed it was just a part of the local sequence of Greywacke strata, and more specifically those that Murchison had assigned to his Lower Silurian. In Sedgwick's view no "case" was at risk: "I have no objection whatever to the discovery of plants in the Caradoc system," he told Murchison. "*But I have*," commented Murchison emphatically in the margin of Sedgwick's letter.[14]

Here was the nub of the matter. Murchison had tried to expel De la Beche's plant-bearing Culm at Bideford from the category of ancient Greywacke strata, because in his view no plant fossils could be expected in strata older than the Carboniferous (§5.2). He had resolved this anomaly to his own satisfaction when he and Sedgwick corrected the structure and hence the sequence in north Devon, reclassifying the Culm as altered Coal Measures (§7.2). But now, to Murchison's evident dismay, Sedgwick was accepting Harding's report of plant fossils in what Murchison himself had identified as Lower Silurian strata (fig. 8.1, B). For Murchison, this simply erected another anomaly as monstrous as the first, whereas Sedgwick (like De la Beche!) saw no reason to doubt the possible occurrence of plants as low in the sequence as the Silurian, even if Murchison himself had not yet found them in "Siluria." Local geologists like Harding could easily be dismissed if they presumed to offer theoretical interpretations of their fieldwork; it was not so easy to ignore them, if what they reported were merely some reliably located specimens.

11. De la Beche to Greenough, 27 January 1837. De la Beche's ready acceptance of the report indicates its plausibility: no such outliers had been found hitherto so far west in Devon, but the Green Sand was well known in the east of the county. As De la Beche suspected, Murchison *was* intending to claim this as his own and Sedgwick's discovery: Murchison to Sedgwick, n.d. (see note 2).

12. Quoted in Sedgwick to Murchison, 23 January 1837 (GSL: RIM, S11/115); Murchison's original letter and Harding's letter have not been located.

13. Ibid. In an equally scornful tone, Murchison rejected Bilton's claim to have *discovered* the outlier near Bideford, saying that Bilton "really did not know whether it was green sand or granite!!"; Murchison to Sedgwick, 30 January 1837 (defective transcript in ULC: AS, IIID, 21).

14. Ibid.

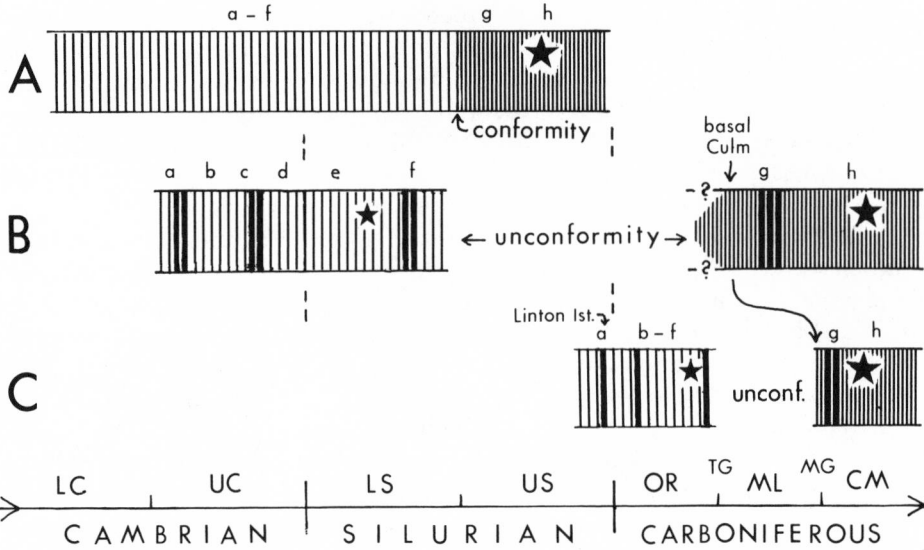

Fig. 8.1. Diagrammatic columnar sections to show the modified interpretations of Devon geology put forward privately (A) by De la Beche in the autumn of 1836, following the Bristol meeting of the British Association; (B) by Sedgwick in January 1837, in response to (C) Murchison's radically new interpretation, also suggested in January 1837, a few days before his and Sedgwick's joint paper was due to be read at the Geological Society. When the paper was eventually read, in the summer of 1837, they put forward a joint scheme similar to (B), while glossing over its problems. As in previous diagrams, the large stars indicate the position of the Culm plants that had precipitated the controversy; the small stars denote the fossil plants reported by Harding in January 1837, the discovery of which prompted Murchison's scheme (C). TG denotes Phillips's "transition group" of "lower limestone shales" at the base of the Mountain Limestone; MG, the Millstone Grit at the base of the Coal Measures in other regions.

Nor were Harding's fossil plants the end of Murchison's troubles. Only ten days before their joint paper was due, Sedgwick revealed that his interpretation of the sequence in Devon now differed significantly from his colleague's. During the public discussion at Bristol they had both stuck to the interpretation set out in their paper, namely that the Culm (including tacitly the black limestone near its base) was really an altered form of Coal Measures, resting with a deceptive appearance of conformity on Lower Silurian and Cambrian rocks (§7.3; fig. 7.7, A). But after the meeting Sedgwick had modified this interpretation in the light of his final days of fieldwork. While he was in Cornwall, he had still referred to the Culm limestone as the "base" of the whole Culm sequence (§7.4). As on his earlier fieldwork with Murchison, however, he was again confronted with an embarrassing failure of evidence for the expected unconformity. As he tried to trace the junction eastward past Launceston, he had found instead a thick series of coarse sandstones and shales, lying *conformably* below the distinctive Culm limestone. De la Beche could of course have claimed that discovery as supporting his view that the Culm was an integral part of a single conformable sequence of Greywacke formations. Indeed, according to Sedgwick, De la Beche had mistakenly equated these very "grits" with a quite different set of sandstones far lower in the sequence, near the Plymouth limestone. But Sedgwick had neatly saved his own and Murchison's interpretation from empirical collapse by annexing these thick "lower grits and shales" to the Culm, thereby tacitly displacing the anticipated but elusive unconformity between Culm and Cambrian to a still lower—and unexamined—part of the sequence. "The black limestones," he had told De la Beche afterward, "[do] not form the actual base of the culm series."[15]

As he and Murchison corresponded frantically in the final days before their paper was due, Sedgwick referred back to this modified conception of what the Culm comprised.

> Our Culm system is at the top; & some part of it contains (if we are not deceived by the Botanists) plants identical with the carboniferous fossils.—But there is an enormous development of the lower members (many 1000 feet in thickness) *under* the culm limestone,—especially on the south demarcation [of the Culm trough]—what then is the geological base of the whole culm series? At this moment I don't know— and this opinion I mean decidedly to state in what I write.[16]

"This is very loose & I do not believe it," wrote Murchison against Sedgwick's remarks.[17] Not only did his collaborator's uncertainty threaten to make their paper less than "ringing, clear and sharp." Even more disturbingly, Sedgwick's readiness to extend the Carboniferous attribution indefinitely down the sequence seemed to Murchison to jeopardize their earlier insistence on a major break between Culm and Cambrian (or, in the north, Silurian). He wrote back by return post, protesting at Sedgwick's apparent intention to turn the strata immediately below the Culm limestone from Silurian into Carboniferous.

Sedgwick was in Norwich, feeling depressed by the cold formality of the cathedral services he was required to conduct. Murchison's letter reached him just in time for him to reply with a postscript to the comments he was about to send on the sections for their paper. He assured Murchison that the strata in question were those they had found in the north *between* the Silurian and the Culm limestone (fig. 7.5, g, "wavellite" rocks). That part of their joint interpretation therefore remained intact after all. "All I contend for," he told Murchison, "is that our base line [for the Culm sequence] must not be fixed too dogmatically; my opinion is that the base is below the limit of the mountain limestone." This shows that on reflection he had adopted the suggestion Phillips had made during the public discussion at Bristol, namely, that on fossil evidence the Culm limestone might be equivalent to the Mountain Limestone (§7.3; fig. 7.7, C). For then it followed that a still lower position in the Carboniferous sequence would have to be found for the great thickness of Culm-like strata that Sedgwick had detected *below* the Culm limestone in the south. "What is their exact age? I don't know with anything like assurance," he told Murchison; "they *may* be on the parallel of the bottom carboniferous shales: but we want [i.e., lack] clear proof of it—All I want to do is to write cautiously. We do only harm by overstating our case."[18] Sedgwick thus conjectured that the Culm sequence in Devon might even encompass the equivalents of the

15. Sedgwick to De la Beche, 20 November 1836 (NMW: DLB).
16. Sedgwick to Murchison, 23 January 1837.
17. Murchison's annotation on ibid.
18. Sedgwick to Murchison, 25 January 1837 (GSL: RIM, S11/116).

"transition group" that Phillips had described (§6.6), lying between the Mountain Limestone proper and the Old Red Sandstone (fig. 8.1, B).

Against Sedgwick's previous reference to these strata at the base of the Culm, Murchison had written "Millstone Grits." This indicates that, unlike Sedgwick (and Phillips), he still regarded the Culm limestone as equivalent to the Coal Measures, not to the Mountain Limestone (see fig. 7.7, A). Now, in response to Sedgwick's untroubled acceptance of Harding's plant fossils from the misnamed "Green Sand," Murchison proposed a radical alteration of the *other* half of their earlier joint interpretation. Judging from Sedgwick's response, Murchison suggested that most of the strata below the Culm in north Devon should be regarded as the equivalents of *the Old Red Sandstone itself*. Certainly there were rocks of the appropriate kind, which Williams had even dared to term "old red sandstone" (§7.1), though Sedgwick and Murchison had ascribed them to the Cambrian (fig. 7.5, *b*). Murchison's latest conjecture would at once resolve his embarrassment at the "vegetables in the Major's greensand," because it would enable him to assign those fossil plants to the Old Red Sandstone, and they would therefore be safely Carboniferous in age (fig. 8.1, C). But since Harding's fossils came from the strata that Murchison had hitherto ascribed to the Lower Silurian, the penalty for resolving the anomaly of the plants was to lose the putative Silurian from that part of the sequence altogether! Pursuing the logic of his conjecture, however, Murchison apparently proposed a final correlation that partly restored that loss. If most of their former Silurian and Cambrian strata in north Devon (including in the latter the so-called Devonian) were really equivalent to the Old Red Sandstone, then the lowest of their "Cambrian" formations (fig. 7.5, *a*) might represent at least the top of the real Silurian. The fossils in that lowest formation, he added, did seem to him in retrospect to have a Silurian character: *"recollect there are orthoceratites at Linton,"* he told Sedgwick.[19]

If Murchison had combined this latest proposal for the older strata with *Sedgwick's* suggestion about the dating of the Culm rather than his own, their awkwardly elusive unconformity could have been eliminated altogether (fig. 8.1, compare B and C). But Murchison did not make that move. Evidently he was more concerned to avoid the anomaly posed by Harding's plants than to concede to De la Beche that the whole sequence was conformable. As far as can be inferred from the extant evidence, he left the Mountain Limestone without equivalent in Devon and thereby tacitly retained the putative unconformity (fig. 8.1, C).

Murchison's letter arrived just as Sedgwick was about to send him the watercolored sections of south Devon to illustrate their paper. "It makes me laugh," Sedgwick told him bluntly: *"I don't believe one word* (as old Buckland says) of your interpolated mass of old red"; not least, he added, because the "good red sandstone" to which Murchison alluded was *"under* the zone we considered as Silurian," not above it as its correlation with the Old Red Sandstone would require. Harding's find of fossil plants, Sedgwick insisted, was

19. This tentative reconstruction of Murchison's letter, which has not been located, is based on quotations from it in Sedgwick's reply: Sedgwick to Murchison, "Thursday Morg.," pmk 27 January 1837 (GSL: RIM, S11/117). The suggested redating of the Linton strata was not as ad hoc as it might seem, since Murchison's *first* impression on seeing them in the field had been to assign them to the Silurian (§7.2). Murchison's "Millstone Grits" annotation is on Sedgwick to Murchison, 23 January 1837.

no good reason for rejecting those strata from the Silurian: "there must always have been vegetables during the time there were animals; it is physically impossible there should not." So once again Sedgwick used against Murchison one of the arguments that had originally aligned him on De la Beche's side in the whole controversy (§5.5). "I care not one fig for the fossils," he concluded, if they went against the plain evidence of superposition.[20]

In view of the major disagreement and uncertainty that had emerged, Sedgwick now proposed a change in their plan of action as drastic in its own way as Murchison's new scheme of correlation. "We should be mad to bring our paper on [at the Geological Society] before we have made our minds up," he told Murchison; "throw the blame on me as much as you like or blame the influenza." At most, he proposed, Murchison should present no more than "a kind of *resumé*" to explain their sections, "best of all viva voce," which would not commit their views prematurely to print in the Society's *Proceedings*. Having dispatched that proposal, Sedgwick "*struck work*" on the paper and retired to his sickbed.

When his letter reached London, however, Murchison evidently could not believe Sedgwick was serious in proposing that the reading of their paper should be abandoned or drastically curtailed. He wrote back at once, appealing again to Sedgwick to send a text for him to read at the Society two days later. "I wait patiently like a lamb for the sacrifice," he told him, hoping no doubt to arouse Sedgwick's conscience, "& sacrificed I most assuredly shall be without your aid." He hinted darkly that he saw signs of a conspiracy by De la Beche to mobilize the opposition to their paper: "I will drink the best part of a bottle of sherry," he told Sedgwick, "to screw me up to face Buckland, Greenough, Yates & the [Board of] Ordnance forces which are to be brought against us."[21]

Sedgwick would tolerate none of this talk of conspiracy. Replying once more, on the eve of the meeting, he urged Murchison to refer to the absent De la Beche only with "expressions of respect and courtesy," and to conciliate their opponent's allies by emphasizing that the official map had been made "according to the information of the day," whatever its shortcomings had since turned out to be. "Pray don't *attack* T.H.D[e la Beche]," he urged; but since Murchison had already told him that their opponent's "ungentlemanlike" letter had "stirred up all my bile & roused me," it was a forlorn appeal for peace.[22]

Most of Sedgwick's final letter of advice, however, was devoted to a straightforward rehearsal of what they could confidently claim to have discovered, namely, the sequence of formations in both north and south Devon. On the Culm itself, there was little to add to what they had already said at Bristol, except to emphasize the improved evidence for its southern boundary and hence for the trough structure they had postulated. This put the Culm unambiguously at the top of the sequence and strengthened their claim that it was Carboniferous, much younger than the strata below it. Sedgwick dis-

20. Sedgwick to Murchison, 27 January 1837.

21. Murchison to Sedgwick, 30 January 1837 (extract printed in Geikie 1875, 1:253; defective transcript in ULC: AS, IIID, 12). James Yates (b. 1789), who had recently given up his Unitarian ministry, was a former Council member at the Geological Society and active in its affairs.

22. Ibid.; Sedgwick to Murchison, 31 January 1837 (GSL: RIM, S11/118).

creetly omitted all reference to his new ideas on the precise age of the different parts of the Culm series. As for the older strata, he summarized the thick sequence of formations he had unraveled in the south of the county and compared it with what they had found in the north. "In a general way," he concluded, "we therefore identify N. & S. Devon; & provisionally we also identify them with the *upper Cambrian* or great system of Plinlimmon & N. Wales." Significantly, the term "Devonian System" was dropped altogether. About the overlying formation in the north, which they had previously attributed to the Lower Silurian—"the Major's greensand inclusive!"—Sedgwick left his collaborator to "say what you like—you still call it Silurian I suppose." Sedgwick had thus completely rejected or ignored the radical alternative correlations that Murchison had suggested a few days earlier: "all your notions of superposition were in much confusion last week," he commented dismissively.[23]

With this detailed summary, Sedgwick could reasonably claim to have given his collaborator ample material for an informal *viva voce* exposition of their sections, which Murchison had got the artist Scharf to enlarge to a suitable size for display in the meeting room. But Murchison was expecting a formal text; without one, he felt indeed a defenseless lamb for the slaughter.

8.3   *Hamlet* without the Prince (February 1837)

Murchison received Sedgwick's suggestions by the overnight mail from Norwich. "But in vain I looked through the parcel for the document to be read," he complained later. Of course there was none to be found, because Sedgwick had written none. When Murchison went to Somerset House that afternoon he still hoped Sedgwick might have sent a formal paper direct to the Geological Society, but was again disappointed. He found instead "the whole room decorated for the fight" with their colored sections; Buckland and others had already arrived, and a large attendance was expected. "What was to be done?" he asked rhetorically after the meeting, for he was not prepared to extemporize, even from Sedgwick's ample notes. Referring back to his earlier hint of a conspiracy, Murchison told his collaborator that "Fitton & Lonsdale, considering what had been said & *done covertly* on the other side, & looking to *the fact* of the non arrival of the despatch, counselled me to give up the thing, which I resolved to do to *the very great annoyance* of the President & of all the others who came to hear."[24] Two other papers were read as planned, one of them by Buckland (on a quite unrelated topic). From the presidential chair, Lyell told the members that Sedgwick had not sent the expected paper owing to a misunderstanding; Murchison kept his mouth shut.

After the meeting was over, Fitton—the "impartial Censor" of the Society, as Murchison had dubbed him on an earlier occasion (§5.3)—told him that although he thought the joint authors were probably correct in their geology, there was a serious moral objection to their behavior. "It is currently

23. Sedgwick to Murchison, 31 January 1837.
24. Murchison to Sedgwick, 2 February 1837 (defective transcript in ULC: AS, IIID, 13; extracts printed in Geikie 1875, 1:254–55, and in Clark and Hughes 1890, 1:476–77).

reported," Fitton said, "that you went to Devonsh. in a private way unknown to him [De la Beche] & then endeavoured to ruin him without his being able to make any defence."[25] Murchison denied the charge vehemently, insisting that he had made no secret of his intention to go to Devon, ever since De la Beche had publicly challenged him to do so, when the controversy began more than two years earlier (§5.3). Sedgwick's original plan for their joint fieldwork in 1836 had even included calling on De la Beche before they went to north Devon; their failure to do so in the event could be excused as a regrettable but innocent effect of the practical execution of that plan (§7.1, §7.2). But the rumor that Fitton reported was probably based rather on what Murchison had told some of his correspondents in strict confidence in the autumn of 1835, namely, that at that time he *was* planning to visit north Devon surreptitiously (§6.4). Furthermore, there was the now well-known fact that at the Bristol meeting in 1836 Murchison had boasted of his forthcoming victory, even before De la Beche had arrived (§7.3).

Once Murchison had "undeceived" Fitton on the damaging rumor about his earlier conduct, they went together into the usual meeting of the Council. Murchison told Sedgwick that Greenough immediately "opened a regular battery" of criticism at them for having planned an unfair attack that afternoon. Murchison responded with equally "*sharp weapons*," condemning those who were propagating rumors about their earlier conduct as "guilty of a base and malignant falsehood." He was supported, predictably, by his friend Egerton and by Lyell. But Greenough then started what Murchison called "close pla-tooning" against the two authors, criticizing them for having "dragged [De la Beche] before the *Bristol tribunal*," with a potentially disastrous effect on the politicians present, before the Devon map was published. Murchison was indignant at "this false charge so artfully got up." He claimed that the relevant map had been published well before he went to Devon; that the geological Section at Bristol had not been an "incompetent tribunal" but "one of the largest and best Geological meetings ever known"; that after three years' work in Devon, De la Beche *ought* to have been "furnished with materials for replying (if he had them) to a plan for changing all his views from top to bottom"; that it was "essential to have this great point settled in full conclave, & well argued out"; and lastly, that it was not a personal quarrel, but "a great geological question, & that truth was not to be bottled up on account of any jobs [i.e., private financial advantages] &c &c &c."[26]

According to Murchison's own account, his friends considered that this counterblast had successfully beaten back the attack. But Greenough had been warned by De la Beche about the ulterior motive for the unseemly haste to get the paper read. He therefore pointed out to the Council that since the paper had not been read that afternoon, for whatever reason, Lyell could not properly refer to it in his forthcoming review of the Society's recent work. Lyell retorted, however, that "as a great & important discovery *had been made*" he intended to comment on it, whether or not it had been publicly

---

25. Quoted in ibid.

26. Ibid. The minutes of the Council of GSL contain no trace of this quarrel (GSL: BP, CM 1/4, pp. 246–51). Although Murchison's letter was naturally slanted in his own favor, there is no reason to doubt its essential veracity; his report of Greenough's arguments is confirmed independently by the similar points made in Greenough's earlier correspondence with De la Beche (§7.5).

read before the Society, and that he had already written the relevant part of the speech. According to Murchison, Lyell was "extremely vexed by this diplomacy by which the ends of science & the fair work of individual enterprise is to be checked." That last clause reveals clearly what was now emerging as the deeper reason for the emotional intensity of the conflict. Murchison saw it as a struggle between the "individual enterprise" of independent, gentlemanly specialists like Lyell and himself, on the one hand, and the institutional "Ordnance forces" of governmental science, on the other. De la Beche's growing personal empire—however small and insecure it might appear to the director himself—seemed to Murchison to threaten the future of gentlemanly careers in geology.

After this heated argument at the Geological Society, a supper at Lord Cole's London house gave Murchison an opportunity to get back on good terms with Greenough and Buckland: "all is as smooth as oil," he assured Sedgwick. But his anger with De la Beche was still boiling over when he reported on the meeting the following day:

> De la Beche is a dirty dog,—there is plain English & there is no mincing the matter. I knew him to be a thorough jobber & a great intriguer & *we* have proved him to be thoroughly incompetent to carry on the survey... *He writes in one style to you and in another to me....* I confess that a *very little matter* would prevent my having further intercourse with De la B. If I can trace to him the origin of those falsehoods he shall smart.[27]

Murchison was also exasperated however, if not angry, with his collaborator for having failed to send the paper he expected. "The part of Hamlet being *omitted*," he told Sedgwick, "the play was not performed, & all the scenic arrangements [i.e., their sections] which I had laboured at were thrown away though the room looked splendid." Why, he asked Sedgwick, had no paper arrived in time? Was it a misunderstanding, as Lyell had stated in public? "Did you really imagine that I was to dramatise the whole thing without a sermon before me?" he asked rhetorically, "or have you been written to by Greenough or some of the dark school?" Murchison pointed out that even if the paper was unfinished, an outline could have been read and "an *abstract* made which would have served all purposes."[28] But here he virtually admitted the real reason for his concern to have a formal paper read, if only in outline, rather than the informal viva voce presentation Sedgwick had suggested. For a formal paper, however brief, would automatically be printed in the Society's *Proceedings*, and the main points of their interpretation would have stood on public and dated record as their achievement.

On the other hand, any such staking of their claim to priority required a clear statement of the correlations they proposed, and this was just what they had failed to agree on in the days leading up to the meeting. So although Murchison told his collaborator he thought the "true causes" of the absence of a paper were "my doubts and your influenza," the scientific cause was certainly more significant than the medical. "I was afraid you were going to carry the [? Culm] down into the Silurians & [? Cambrians]," Murchison

27. Ibid.
28. Ibid.

admitted.[29] "We must have a fair stand-up fight," he insisted, at some later meeting when Sedgwick could be present; but if they were to "knock down" their opponents they would first have to find a united front on the continuing puzzles of Devonshire geology.

When Sedgwick received Murchison's lengthy report of the abortive meeting, he claimed to be surprised at his collaborator's reproach, but told him he thought it best that "our Devonshire affair stands where it is."[30] He agreed that it was "absurd" of Greenough to accuse them of unfairness. De la Beche himself had from the beginning of the controversy taunted his critics with their failure to go to Devonshire, not least by means of the now celebrated caricature of the nonexistent nose (fig. 5.4); so he could hardly complain when at length they did go to look for themselves. Sedgwick recalled how he himself had initially taken De la Beche's side in the argument; it was only the collapse of "Weaver's case in Ireland" (§5.6), followed by the persuasive force of Murchison's analogy in Pembrokeshire, that had made him switch to Murchison's side (§6.4). "I altered my first opinion," he recalled, referring implicitly to the time *before* he saw the evidence in Devon for himself, "& had very little doubt that a part of the Bideford [Culm] system would prove to be carboniferous." That opinion had of course been confirmed by their joint fieldwork in Devon.

Sedgwick now criticized De la Beche's fieldwork there far more strongly than ever before. "His views of the stratification of Devonshire were *utterly wrong*," he recalled; and this was the result of making inexcusable errors of correlation "*in the very teeth of fine natural sections*" in the coastal cliffs. Again Sedgwick mentioned the similar error he believed De la Beche had made in south Devon, confusing the "grits" at the base of the Culm with some far older strata (§8.2). "I mention this," he told Murchison, "*to prove the utter confusion of thought in De la Beche's mind* & *the utter impossibility of his classifying Devon* till his views have undergone a *purgation*."[31] Sedgwick might not be personally or emotionally hostile to De la Beche, as Murchison certainly was, but on the substantive issue he was even more determined than before to make their opponent swallow some distasteful medicine in Devonshire (§7.6).

8.4 Lyell Saves the Phenomena (February 1837)

The day after the abortive meeting, Lyell asked De la Beche for his side of the argument: "in a few words," he suggested, "you can set me right on points I require for my anniversary speech." That the matter could be dealt with so simply was implausible, however, in view of the opinion that Lyell immediately went on to express, namely, that "the culm question . . . for the last 2½ years has appeared to me one of the most important in a theoretical point

---

29. Ibid. As in other defective transcripts of lost Murchison letters (ULC: AS, IIID), the uncertain words were left blank, probably because they were technical terms unfamiliar to the transcriber; they are here inferred from the context and from Murchison's earlier letters.

30. Sedgwick to Murchison, 3 February 1837 (GSL: RIM, S11/119).

31. Ibid.

of view ever discussed at the G[eological] S[ociety]."[32] There could have been no clearer indication of the perceived importance of the continuing controversy.

Lyell now came out into the open with a clear expression of his own objections to De la Beche's interpretation, having hovered in the wings ever since the argument over the fossil plants from Bideford first erupted (§5.2). On that occasion, Lyell recalled, "I objected that the presence of coal plants made out, at least to my mind, a strong primâ facie case of true coal [measures], & I asked what evidence there was to overcome this strong presumption." The "mineral character" of the strata, he claimed, was no sufficient criterion of their age; De la Beche had not produced what Lyell regarded as adequate evidence, namely, that there were "Silurian or known transition shells or corals intermixed with the plant-bearing beds," or at least that the Culm was overlain by upper Greywacke strata. He demanded that De la Beche should show him that "there were or are proofs, or that you once supposed you had them, sufficient to outweigh the fair and natural inference from analogy & previous experience," that any fossils from the lower Greywacke (Sedgwick's Cambrian) would differ radically from those of the Coal Measures. Lyell repeated his opinion that it was "unphilosophical" to admit on any but the most rigorous evidence "such an anomaly as the identity of the vegetation of two very remote eras." Referring to Weaver's case, he stiffly requested authority to state publicly that De la Beche had likewise changed his mind, or failing that, that De la Beche should give "far more overwhelming and unequivocal proof" for his inference. Finally, Lyell denied that the matter was one of "personal controversy."[33]

Lyell's letter was, on the most charitable reading, coldly correct; De la Beche could have been left in no doubt that the president was preparing a withering public criticism of his work. But as his self-defense required, he replied promptly and, as he put it, "candidly and in the spirit of truth." He could not resist complaining first, however, of the unfairness of the Bristol episode. He accepted that the official geological map, as a reconnaissance survey, was bound to be found defective in many details, "for we do not pretend to produce that monstrosity a perfect work"; and he explained that he regretted that the map had been published without its explanatory memoir, for reasons beyond his control. He then got down to the substance of his defense. He openly conceded that he now agreed with Sedgwick and Murchison's new interpretation of the *structure* of Devon, so that "the disputed culmiferous plant-bearing beds are the highest of a series," whatever their age might turn out to be. But he repeated that, at least in north Devon, there was no sign of the unconformity between the Culm and the older strata which it was reasonable to expect if the Culm were really of Coal Measures age. There was nothing, he insisted, "to show that we have two distinct deposits formed at periods very distant from each other," whereas there was much to suggest that "the two supposed distinct masses constitute portions of one great whole."[34]

32. Lyell to De la Beche, 2 February 1837 (NMW: DLB; Lyell's copy is in ULE: CL, 1/3732–3); Lyell misdated the opening of the controversy as 1835.

33. Ibid.

34. De la Beche to Lyell, 5 February 1837 (APS: CL).

De la Beche then turned to the crucial question of the proper mode of reasoning from fossils. The fact that the few Culm plants well enough preserved to be identified at all were of Coal Measures species was, in his opinion, poor evidence on which to conclude that the whole formation must be of Coal Measures age, since so little was known about the overall history of land plants (apart from the unusually rich flora of the Coal Measures themselves). If such reasoning were applied generally, he pointed out, "it is clear that we shall never arrive at the information required." In other words, it would lead to a *logical* impasse, which would block the accumulation of real evidence of the history of land plants, or, for that matter, of any organisms whatever.

> Here perhaps some may exclaim, if this kind of reasoning is to have force, what becomes of all our facts connected with the great changes which, we learn from fossils, have taken place in organic life on the surface of our planet. This is precisely what I want to arrive at,—I want to learn what really is known, from facts, of that great assumed difference in the vegetation of these British latitudes between the time of the coal measures and those of the grauwacke.[35]

De la Beche hinted that Lyell, on his own principles, should agree with him that the history of land plants *might* have been different from—and more stable than—that of the marine animals whose fossil remains provided the "standard" of organic change; as usual he mentioned Élie de Beaumont's Alpine anomaly and also referred to the force of the other examples of Greywacke coal, such as those in Brittany. "Would it not be supposed, by the commotion which the discovery of plants at Bideford and other places in the series has caused," he asked rhetorically, "that the fact, as I believe it to be, stood alone, was opposed to all rational theory, and should be scouted [i.e., dismissed]?"

"Let us see if he be honest," De la Beche commented, sending Greenough a copy of what he had written for Lyell; "I begin to smell a rat." For if the Culm were inseparable from the alleged Lower Silurian strata, as De la Beche claimed, the natural inference would be that it too belonged to the Silurian (fig. 8.1, A), as he had suggested to Greenough several months earlier (§7.6). The only conceivable reason why that inference was unacceptable to his opponents, he suggested, was that if the Culm plants were Silurian they would upset Murchison's conception of that system: "proceeding upon fossilology he now suspects it may not be quite the thing, and hence such zeal against the North Devon coal plants."[36]

Meanwhile Sedgwick had taken steps to ensure that his and Murchison's views would also be correctly represented in Lyell's forthcoming anniversary address. "If he likes we might *give him a very condensed view of the Devonian system*," he suggested to his collaborator. The following day he duly offered Lyell a "condensed synopsis" of their general interpretation of Devon geology.[37] But first he expressed his uneasiness about the argument at the Council meeting which Murchison had just reported to him. "I fear these

---

35. Ibid.
36. De la Beche to Greenough, 5 February 1837 (ULC: GBG, 54).
37. Sedgwick to Lyell, 4 February 1837 (printed in Clark and Hughes 1890, 1:477–80). The phrase "Devonian system" clearly referred to the geology of the older rocks of Devonshire as a whole, not to the now abandoned name for the upper Cambrian strata of north Devon.

repeated sparring matches will end in ill blood," he warned the president, "and ruin the harmony of our Society." Like Murchison, he claimed that any charges of unfairness to De la Beche were unfounded, because Murchison had never made any secret of his intention to check De la Beche's interpretation, while he himself had long intended to go there before publishing his proposed summary of all the older rocks of Britain.

> Are all these labours to stop, because an official person has had the misfortune to publish a bad map? The map ought to be withdrawn without loss of time. It cannot, by any tinkering, be brought into order. It must start on new principles. . . . In our opinion Devonshire is radically wrong, as it is now published. Are we to shut our mouths, and let the error continue to be propagated?[38]

Sedgwick rejected the charge that their joint paper was a personal attack on De la Beche, because it could just as well be called an attack on all the other geologists—Greenough, Buckland, and Conybeare among them—who had interpreted the Culm as Greywacke in earlier years. He told Lyell, as he had already told Murchison, that he had no regrets that their paper had not been read at the meeting earlier in the week: "We have a good case, and want to steal a march on no one."

Lyell at once accepted the offer of help with his speech. Sedgwick explained that De la Beche had made major errors, equating quite distinct formations, "in the teeth both of sections and fossils," in both north and south Devon. "I never saw more utter confusion of thought," Sedgwick told the president, "but *of this mum*." What he did wish Lyell to state in public, indeed to emphasize more clearly, was how the correction of these mistakes had made possible a radical new interpretation of the whole structure of Devon: "we not only ascertained the nature, extent &c. of the Culm system, but we showed it to be *superior* to all the *older rocks of the county*." Their main clue had been that the Culm limestone was "dark and carbonaceous, absolutely distinct, both in character & fossils, from all the transition limestones, tho' confused with them by T.H.D." But on the age of all the Culm formations, Sedgwick discreetly suppressed his own speculations of the previous week (§8.2) and just left the Culm implicitly assigned to the Carboniferous.

On the crucial question of the relation of the Culm to the older strata, Sedgwick stood by his earlier claim that there was "no *passage* from the Culmiferous beds into the lower systems," and that in north Devon "they rest on, but do not pass into," the older strata. But he made their case appear stronger than their fieldwork warranted, for he claimed that on the south side of the trough the Culm strata "rest on the edges of the old slates of Cornwall," whereas in fact he had failed to find any clear evidence of an unconformity there (§7.4). As for the older strata themselves, Sedgwick had been unaffected by the radical doubts Murchison had expressed a week or two before (§8.2). The uppermost formation in the north, immediately below the Culm, Sedgwick still regarded as Lower Silurian ("very good Caradoc sandstone"), though "Murchison seemed staggered [i.e., wavering] about it when he last wrote." All the other older formations he now ascribed to "the *Upper* Cambrian"; the "Devonian System" of their Bristol paper had vanished without comment.

38. Ibid.

All in all, Sedgwick's summary to Lyell was essentially what he had sent Murchison on the day of the abortive meeting; it clearly represents what the Geological Society would have heard on that occasion, had the promised paper materialized.

The following week, Lyell duly gave his anniversary address. As expected, the fact that Sedgwick and Murchison's paper had not been read to the Society did not stop Lyell from commenting on what they had announced at Bristol the previous summer.[39] He claimed that their work had "settled" the true age of the disputed plant-bearing strata of Bideford, which he termed a matter of "the highest interest." Unlike Murchison, Lyell did not deny that there had been any land plants in the Transition periods; he only claimed that by analogy with the marine faunas it was to be expected that they would have been significantly different from those of later periods. De la Beche's report had therefore constituted a great "anomaly." Lyell recalled how Weaver had publicly retracted his analogous claim about Ireland (§5.6), and he defended his own and Murchison's original skepticism about De la Beche's report against Greenough's criticism that they had been too much swayed by "preconceived theory" (§5.2).

Lyell denied that his argument implied any "distrust of Mr. De la Beche's skill or experience in geological surveying," and asserted that his skepticism would have been the same if the claim had been made by Sedgwick or Murchison. If, in a hypothetical analogy, land plants were ever to be found in strata of the same age as the Chalk, he would not expect them to be the same species as those of the (Tertiary) London Clay, because the marine faunas of those two formations were markedly different, and he believed they were separated by a vast span of time. "There is a like presumption from analogy," Lyell argued, "against the conclusion that the same vegetation continued to flourish on the earth from the period of the lower greywacke to that of the coal, because we know that in the course of the intervening epochs the testacea, zoophytes, fish, and other classes of organic beings were several times changed."[40] But Lyell's clinching point against De la Beche's interpretation was again that he had failed to show that the plant-bearing Culm strata were anywhere intercalated with strata containing indisputably Transition fossils. This was the decisive evidence that he had earlier challenged De la Beche to produce (§5.4), but which was still not forthcoming.

This much of Lyell's speech had already been written before the previous meeting of the Society. After receiving De la Beche's letter clarifying his viewpoint, Lyell added a summary of its contents. He emphasized pointedly that De la Beche intended to restudy Devonshire and would not hesitate to alter his opinions if necessary. He summarized Élie de Beaumont's account of the Transition coal in Brittany, but claimed that "the testimony is quite incomplete." For the relevance of the analogy depended entirely on the age of the plant-bearing *terrain de transition supérieur*. In Lyell's opinion it might

---

39. Lyell (1837), read at GSL on 17 February. The Devon controversy is discussed on pp. 489–96.

40. Ibid., p. 492. "Testacea"" denoted shelled animals such as mollusks and brachiopods; "zoophytes" denoted corals and related organisms. On Lyell's conception of slow piecemeal "revolutions" in faunas and floras, and his inference of a vast time gap between Cretaceous and Tertiary, see Rudwick (1978).

be equivalent to the Silurian or Old Red Sandstone, or even to "a lower part of the true carboniferous system, of which the strata had been disturbed before a higher portion was superimposed." Until the age of that *terrain* was settled by independent evidence, Lyell argued, its fossil plants signified nothing. With heavy sarcasm he referred to the Alpine analogy as a mark of De la Beche's credulity and lack of judgment. "You will readily understand that a geologist, who is once persuaded that the same plants flourished in European latitudes from the period of the true coal to that of the lias, will be ready to concede without difficulty the probable existence of the same plants at an era long antecedent to the coal."[41] But such an extreme stability in the land flora, Lyell asserted, was contrary to all the conclusions of the botanists. He had studied the relevant specimens in Paris and had later seen for himself the extremely complex disturbance of the strata in the Alps (§6.2), so it is not surprising that he dismissed Élie de Beaumont's celebrated case as indecisive.

De la Beche was present to hear the speech, but complained afterward that Lyell had glossed over the crucial point about the lack of any field evidence for an unconformity in Devon, "independent of the consideration of organic remains." He claimed once more that, except in a few places, the conformity was as perfect as—taking a well-known example from much younger rocks—that between the New Red Sandstone and the Lias. De la Beche protested that Lyell should have publicly acknowledged the many well-known Continental analogies to the Bideford case, which showed that "there is not so much anomaly as might be supposed."[42] Lyell probably did alter the printed record to meet this objection; but, as already mentioned, he rejected the validity or the conclusiveness of all the analogies and presented De la Beche's claims as more anomalous than ever. On the awkward question of the lack of evidence for any unconformity in Devon, however, Lyell kept silent. His printed address—and probably the spoken version too—therefore ended with a ringing declamation of the "great reform" that Murchison and Sedgwick had effected in Devonshire. "According to their survey and sections, the coal plants of Bideford, so far from constituting any anomaly, so far from affording any objection to the doctrine that particular species of fossil plants are good tests of the relative age of rocks, do in reality, from the place which they occupy, confirm that doctrine."[43] The anomaly was thus eliminated and high theory saved. Lyell had made good use of his last appearance as president of the Geological Society.

Lyell knew perfectly well, however, that the Devon controversy was not really "settled" at all. Only three weeks earlier he had urged his successor to be present at the meeting at which Sedgwick and Murchison's paper on Devon had been due to be read; "the controversy on that point will certainly be carried on in your reign & not end here."[44] Although the expected conflict had not materialized at that meeting, Lyell knew it had merely been postponed.

41. Lyell (1837, p. 494). Lyell's addendum to his address begins on p. 493.

42. De la Beche to Lyell, 22 February 1837 (APS: CL); as usual he took the precaution of sending Greenough a copy (ULC: GBG, 66). Lyell had sent him proofs of the relevant pages to mark what he objected to: Lyell to De la Beche, 20 February 1837 (copy in ULC: GBG).

43. Lyell (1837, p. 496).

44. Lyell to Whewell, 25 January 1837 (TCC: WW, Add. Ms. a. 208[126]).

8.5   The Spread of Ugly Rumors (February to April 1837)

Lyell's first choice for his successor as president of the Geological Society had been Northampton, who had been effective as the unrehearsed president of the British Association at Bristol; Buckland was also an obvious candidate, but both Murchison and Sedgwick felt he had disqualified himself by his undignified performance at Bristol (§7.3). In the end, Lyell had asked Whewell, but Sedgwick's Cambridge colleague had needed some persuasion before he would accept the nomination. Lyell and Sedgwick had urged upon him the importance to geology of having a distinguished "Universal" and man of "exact science" as president; and Whewell's contributions to geological science, broadly defined, were far from negligible. But with a fine sense of the Society's internal politics, Whewell had hesitated until he heard that Greenough too was in favor; only then had he agreed.[45]

Just as Lyell had urged Whewell to be present when Sedgwick and Murchison's paper on Devon was to be read, so likewise, after that abortive meeting, Murchison lost no time in getting the future president's backing. "Geologists ought to know better," Whewell commented, on hearing from Murchison about the criticism being made about his conduct. Unlike "mere men of the world," men of science, in Whewell's view, "should know that there is a right and a wrong," and that "if this question is frivolous, geology altogether is frivolous." But of course this did not touch the nub of the criticism that had been leveled at Murchison, which was not that he had been overzealous in pursuing the Devon problem but that he had done so in an improper manner. However, Murchison was at least assured of the next president's support in the affair: "I do not doubt," Whewell told him, "you will rout all such adversaries horse and foot as soon as their muzzles are visible."[46]

Murchison had threatened to make De la Beche "smart" if he had been the one who had spread the accusation that had surfaced after the abortive meeting (§8.3). But he soon discovered it was Buckland who was responsible, doubtless because Buckland remembered how Murchison had disclosed to him his plans for surreptitious fieldwork in Devon in 1835 (§6.4). Sedgwick said he was astonished to hear Buckland was the culprit, but admitted that if he were in Murchison's position "I should inevitably fly out like a rocket."[47] He agreed that the refutation of the charge would have to be *"pushed to the hilt,"* but this would have the desired effect because "the canon is too cunning to stand out when he is in the wrong."

As ammunition for Murchison's refutation, Sedgwick sent him one of De la Beche's letters to show to Buckland. This made it clear that De la Beche had accepted that it would be *"fair to attack"* his Devon map, at least in the

45. Murchison to Lyell, 22 September 1836 (APS: CL); Lyell to Whewell, 5 and 22 October 1836, and Sedgwick to Whewell, 12 October 1836 (TCC: WW, Add. Ms. a. 208[124–5], 213[23]); Whewell to Lyell, 23 October 1837 (APS: CL); Lyell to De la Beche, 1 December 1836 (NMW: DLB).
46. Whewell to Murchison, 9 February 1837 (GSL: RIM, W4/16).
47. Sedgwick to Murchison, 25 February 1837 (GSL: RIM, S11/121).

checked and corrected copy held at the Ordnance headquarters in the Tower of London. This could hardly be squared with what De la Beche was now claiming, namely, that what had been criticized in public was a map that was not yet properly published.[48] In Sedgwick's opinion, this amounted to blaming the Ordnance mapseller for the "blunders" on the unchecked copy that Murchison had bought for their fieldwork (§7.5). But in Sedgwick's view such excuses were quite inadequate: he summarized for Murchison a whole series of De la Beche's earlier letters—"most fortunately preserved"—which made it clear that De la Beche "did not understand one particle of the structure of Devon when his map was finished."[49] The difference between their respective interpretations was therefore not, as De la Beche claimed, confined to the question of the age of the Culm, but threw De la Beche's competence as a geological surveyor into radical doubt.

That doubt was just what Murchison had already started dropping into various influential ears, soon after the meeting at which their criticism of the government surveyor had failed to materialize in public. To Harcourt, Phillips's patron in York, Murchison even suggested that Phillips would be a better man for the job, since he would carry out the survey "*on a great scheme* and on sound principle and would not have started in *chaos.*" Far from acting improperly toward De la Beche, Murchison protested that he had gone to Devon "*at his own request*" and with the map that De la Beche had published a year before; he emphasized that he and Sedgwick, "so far from having changed *an iota*" of their interpretation, could "now give *the whole picture.*" The "carrying out of one of the most important generalizations in Science, and the reading off of obscure pages in the fossil kingdom," Murchison insisted, was not to be "quashed and silenced" just because De la Beche had "bungled his first piece of *handy* work."[50]

That same accusation of De la Beche's gross incompetence came in due course to the ears of officialdom. At the beginning of the year, the director of the Ordnance Survey had recommended that, when the geological survey of Cornwall was completed, De la Beche and his assistants should move to south Wales to survey the great coalfield there.[51] That proposal had been approved (§8.1); the criticism of the Devon map the previous summer had apparently not done the damage De la Beche had feared (§7.5); and it seemed as if his little team might soon take another step toward becoming a permanent establishment. Early in April, however, De la Beche received an ominous letter from Colby, telling him that "extremely unpleasant but not tangible reports" of the inaccuracy of his work were circulating in official quarters, where a lack of understanding of the character of geological mapping made them difficult to rebut. The director urged De la Beche, as he had done in the aftermath of the Bristol meeting, to scotch these damaging rumors without delay by publishing the memoir to explain his Devon map, thereby "showing its utility." De la Beche's desire to give the memoir "the last polish" was a luxury that could no longer be afforded; "the Government feel they have paid for a portion of work, which is not before the public," Colby told him, "and

48. De la Beche to Lyell, 22 February 1837.
49. Sedgwick to Murchison, 28 February 1837 (GSL: RIM, S11/122).
50. Murchison to Harcourt, n.d., pmk 1837 [written less than a week after GSL meeting on 1 February] (HHO: WVH, Buckland volume). I am indebted to Mr J. B. Morrell for showing me his transcript of this important letter.
51. Colby to Ordnance, 13 January 1837 (NMW: TS, II, pp. 191–92).

when their work is impugned, they dread their responsibility." While conceding that the disagreements about the geology of Devon were a matter for geologists to settle, he warned that the issue had now become "the theme of popular conversations," and therefore a serious threat to the future of the geological survey.[52]

De la Beche ignored Colby's disingenuous claim to be unaware of the origin of the rumors, and wrote back saying the director was "no doubt aware" that they had arisen from Sedgwick and Murchison's paper at Bristol the previous summer. He repeated his usual complaint that he had been treated unfairly there, and also his more recent line that his opponents had worked from an unauthorized copy of the map, released "inadvertently" by the Ordnance mapseller. He played down the importance of the geological point at issue, referring to it as a "mere difference of opinion" about the age of some of the Devon rocks, as if this and "one or two minor omissions" were the sum total of the points on which his map "indirectly differed" from his opponent's conclusions. He did admit that a matter of "theory" was also involved, but only with his usual implication that this was an optional extra quite distinct from observation and map making. He suggested that the charges of inaccuracy could be met by an independent inquiry by "some competent and disinterested person." As so often before, however, De la Beche's dominant note was his sense of grievance at the way his "personal enemies" were continuing to "buzz their accusations about" in the quarters where they could do the most mischief, "while I toil day after day without rest, endeavouring to do my duty to the public." He told Colby he would like to resign, if it were not that such an action would look like an admission of incompetence.[53]

As usual, De la Beche kept Greenough informed of what was going on, urging him to lobby in influential quarters on his behalf. He asked his protector to approach two members of Parliament, who were also active in the Geological Society, to warn them to be ready to "help an old friend by attending and watching to see justice done," should the geological mapping be queried in the House of Commons when the funding of the Ordnance Survey came up for review. He urged Greenough to talk to the Ordnance authorities and to politicians, such as Lansdowne, Spring-Rice, and Baring, not only to counteract the damaging rumors about his work but also—with the help of Buckland—to see if "some little post of £300 or £400" could be found to enable him to continue his scientific research "out of the meddling reach of jealous men." Such a government sinecure, he warned Greenough, now seemed the only way he could avoid having to abandon a life of science and return to live on his estate in Jamaica, if his enemies were to force him to resign from his survey position.[54]

On that pitiable note, De la Beche sought to urge his allies to stir themselves in his defense. Another London friend, hearing the rumors and suspecting that Lyell and Murchison were behind them, urged De la Beche to write directly to the Board of Ordnance to explain that "there were few theories in Geology about which there were not differences of opinion"; if that did not satisfy the Board they should appoint "a committee of scientific

---

52. Colby to De la Beche, 3 April 1837 (NMW: DLB).
53. De la Beche to Colby, 8 April 1837 (NMW: TS, II, pp. 199–200; De la Beche's copy in DLB).
54. De la Beche to Greenough, 8 April 1837 (ULC: GBG, 56).

persons" to evaluate the survey impartially.[55] Greenough agreed with this proposal, urged De la Beche not to think of resigning, and said he would prime Henry Warburton (b. *c.* 1784) to defend him if necessary in the House of Commons.[56] But De la Beche's colleague Trenham Philipps, who was in charge of his embryonic Museum of Economic Geology, advised him against making an official protest, warning him that in his experience "among official people" it only did harm to draw attention to such attacks, "however atrocious might have been their character."[57] In the event, De la Beche seems to have followed that advice; he took no official action to quash the rumors about his work.

So once again, partly as a result of Greenough's effective lobbying, the threat to De la Beche's career and livelihood blew over. Indeed, his prospects even improved, since Colby proposed that the long-awaited memoir on Devon should be published by the government and not at De la Beche's personal expense.[58] Nonetheless, the "differences of opinion" on the geology of Devonshire remained unresolved; nor could they be dismissed as easily as De la Beche and Colby claimed, as "mere" theoretical arguments unconnected with the useful business of geological mapping. Their resolution would necessarily affect the whole appearance of his map and thereby reflect one way or the other on his competence as the government's chief geologist.

8.6   The Ambiguities of Critical Evidence (March to May 1837)

This latest round of hostilities, Greenough told De la Beche, had been occasioned by the "notes of dreadful preparation" that Murchison had been sounding in official ears about his own and Sedgwick's forthcoming paper on Devon geology.[59] The reading of that paper had first been brought forward to the beginning of February, in order to upstage De la Beche's intended reading of his own paper; but that maneuver had been frustrated by Sedgwick's failure to send a formal text, coupled with Murchison's unwillingness to extemporize in his absence (§8.3). Perhaps for that reason, De la Beche dropped his plan to present his riposte at the first ordinary meeting after Lyell's anniversary address, even though he was still in London on that date. But his opponents went ahead with their own plans to present their joint paper a month later, toward the end of March. As that meeting approached, however, it was Sedg-

55. Ker to De la Beche, 12 and 14 April 1837 (NMW: DLB). Charles Henry Bellenden Ker (b. 1785?), a notable lawyer and reformer, was the editor of the *How to Observe* series of elementary scientific books for which De la Beche had written the *Geology* volume (1835).

56. Greenough to De la Beche, 14 April 1837 (NMW: DLB). Although not one of the founders, Warburton had been one of the earliest members of GSL. Egerton, the other M.P. De la Beche had suggested, was too much on Murchison's side (§6.4). Greenough confirmed that one potent factor in the government's unease was their fear of being "jobbed" again as they had been by MacCulloch (§7.5).

57. Phillips to De la Beche, 19 April 1837 (NMW: DLB). The Museum (§5.8) was still no more than a cramped repository for specimens and was not open to the public; De la Beche had just been given the honorary title of Director (McCartney 1977, pp. 34–35; BGS: GSM, 1/12, 4; 1/1, 25).

58. Colby to De la Beche, 17 and 18 April 1837 (NMW: DLB).

59. Greenough to De la Beche, 14 April 1837.

wick's turn to despair at Murchison's failure to send materials for it, and he had to explain that the first meeting in April was impossible because he was due to examine for scholarships at Trinity that day. "Somehow or other," he commented, "the gods grumble at our Devonian work."[60] Beyond the gods' grumbling, however, was the continuing failure of the two collaborators to agree on several crucial points. In particular, there was the problem of how to account for the fossil plants that Harding had found in north Devon in the strata that Murchison had assigned to the Lower Silurian (§8.2). Since Murchison insisted that there were no authentic records of any land plants as old as the Silurian, Sedgwick now put his collaborator on the spot by asking him to write that part of their joint paper.[61]

Before that letter reached him, however, Murchison was further alarmed to hear from Lonsdale that Williams had found more of these awkward fossil plants—and probably some much better specimens—and that he was bringing them to London for the next meeting of the Geological Society, at which the long-awaited joint paper was due to be read. Sedgwick wrote back to ask just what "the Parson's" new specimens were and whether they were the same species as those in the Culm.[62] Williams had in fact recorded privately a fortnight before how he had had "a most successful and satisfactory day" in north Devon, finding "an abundance, & great variety of fossil wood and plants" in the sandstones underlying what he called the "Trilobite Slates." He congratulated himself on the reasoning that had led to his successful search: he thought the sandstone must have been formed from "the wreck and waste of some more ancient land," which would surely have been "clothed with vegetation." Likewise he had already found many of "the fossils of the Dudley limestone"—Murchison's Upper Silurian (Wenlock)—in the overlying strata; "I inferred that the Trilobites must be there also—I searched diligently & soon found them plentifully."[63] The provincial geologist's private record of discovery was touching in its naive candor; but the "wise master builders" in London would have to take note of his specimens, if not of his confident inferences about their dating.

By the time Sedgwick heard about Williams's latest work, however, he was down with influenza again; as if he feared his illness might be thought just an excuse, he told Murchison shortly afterward that he had had a riding accident as well.[64] So at the Geological Society meeting a few days later their joint paper again failed to appear. But Greenough took the opportunity to point out to the new president that when eventually the paper was read it would have to be retained by the Society, and he asked Whewell to remind Sedgwick that this was what the rules required.[65] Clearly he feared that otherwise the two authors would withdraw their paper for still further revision and thereby avoid having to lay it open to formal criticism.

---

60. Sedgwick to Murchison, 6, 15, 16, and 21 March 1837 (GSL: RIM, S11/123–26).
61. Sedgwick to Murchison, 9 April 1837 (GSL: RIM, S11/128).
62. Williams to Lonsdale, 6 April 1837 (GSL: BP, LR3/49); Sedgwick to Murchison, 14 April 1837 (GSL: RIM, S11/129).
63. Williams, notebook for "Octr. 1835 to June 1837," entry dated 5 April (SAS: DW, box 1, notebook 10, pp. 64–65); Williams to Lonsdale, 6 April 1837. The plants and trilobites came, respectively, from Sedgwick and Murchison's groups (e) and (f) (fig. 7.5).
64. Sedgwick to Murchison, 14 and 16 April (GSL: RIM S11/129, 130).
65. Greenough to De la Beche, 21 April 1837 (NMW: DLB).

Greenough urged De la Beche to come to London for the following meeting, early in May, when it was now hoped the paper would at last be read, and he tactfully reinforced Colby's advice to get the memoir published as quickly as possible. "Instead of condoling with you therefore in not being immediately set to work in Wales," he told De la Beche, "I congratulate you on the invitation held out to you to answer your opponents at once and at the public expense." Greenough claimed in fact to have an open mind on the controversy: their opponents' "reading of the Devonian formations" might be correct, but he himself had yet to be convinced. If the revised edition of his own geological map had to be published immediately, he told De la Beche, he would still color Devonshire "the old fashioned way," that is, as Greywacke throughout, as being "the less hazardous" of the two contested interpretations. And for the time being, he remained skeptical about claims that Murchison's Silurian system and Sedgwick's Cambrian were of more than "purely local" validity. "John Bull," he commented sarcastically, "will lord it in geology as in everything else, and make all the world submit to his dictation." Yet with characteristic internationalism, he appreciated that the controversy was of "very great geological importance" precisely because it could *not* be limited to Britain, as the Silurian and Cambrian might have to be. "It involves not only the north of Devonsh. but all the transition or supposed transition districts of Europe," he commented to De la Beche; "sure I am that it branches out into too many ramifications to be properly disposed of in the half or three quarters of an hour which are allowed us for discussion."[66]

Meanwhile De la Beche had been restudying the disputed parts of north Devon, and soon told Colby he had confirmed his earlier views. But Colby advised him to keep quiet about this and to complete his memoir, "because the object of the Survey, is not to establish or controvert theoretical opinions, but to collect and arrange facts." Pursuing that characteristic line, he recommended that the "inferences and theoretical deductions" should be rigidly separated from both general and detailed "facts" and relegated to the end of the memoir.[67]

Colby's advice to De la Beche to say nothing about his latest fieldwork arrived too late, for he had written the same day to Sedgwick. He told Sedgwick he had been in north Devon with Bilton and Harding, who had shown him the exposures they had visited the previous summer with his two rivals. "Of course," De la Beche remarked, one item on his agenda had been to look at the relation between "the so-called culmiferous series" and the "acknowledged grauwacke." Sedgwick's and Murchison's interpretation implied that there would be a sharp and unconformable boundary; De la Beche had maintained that there was an imperceptible transition or passage between what he regarded as "rocks of a consecutive series." He now told Sedgwick that on the Taw estuary near Barnstaple (at Fremington Pill) he had found just the right kind of "cross section" to settle this critical question (fig. 8.2). Here, he claimed, "the black limestone with abundant remains of Posidonia, which we all agree is in the culmiferous series," could be clearly seen in relation to the "fossiliferous slates" which—so Harding had told him—Murchison attributed

66. Ibid.
67. Colby to De la Beche, 25 April 1837 (NMW: DLB).

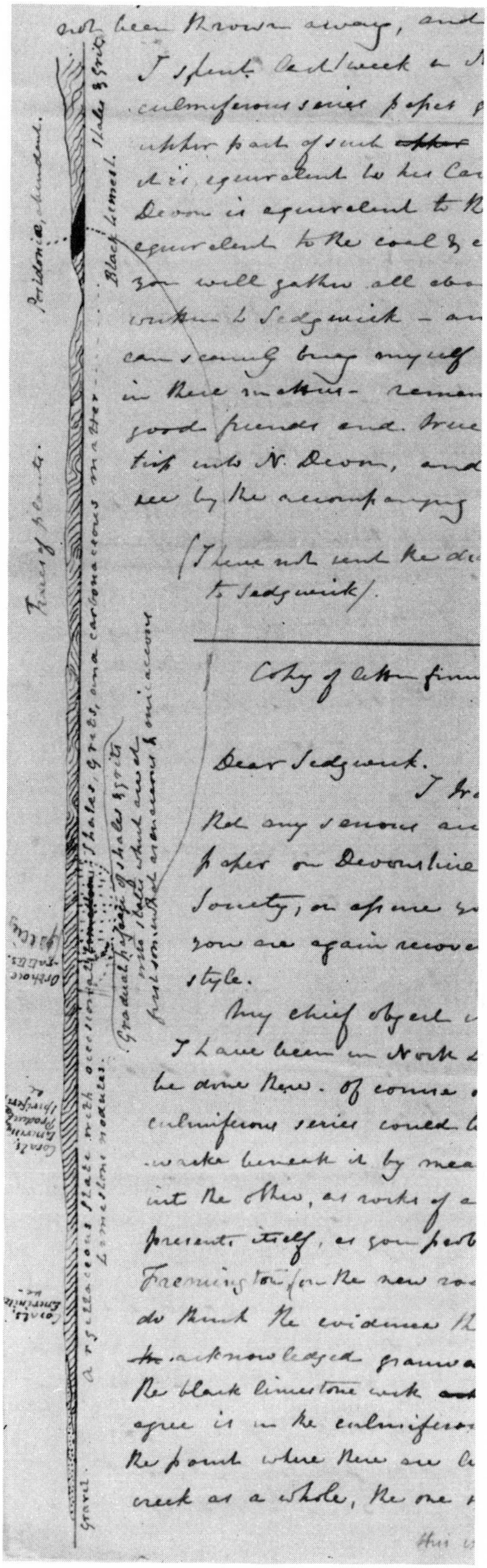

Fig. 8.2. De la Beche's sketch section of exposures on the Taw estuary in north Devon, drawn for Greenough in April 1837, showing what he claimed as crucial evidence for the complete conformity and "gradual passage" (center) between Murchison's Lower Silurian strata with corals, brachiopods, and "Orthoceratites" (left), and the Culm strata with "traces of plants" (right), including a band of Culm limestone with "Posidoniae" (black, far right).

to his Lower Silurian "Caradoc beds." This "very good" section, De la Beche asserted, proved that "the one system gradually fades with the other." He mentioned comparable transitions from other parts of the total sequence of formations, also in southwest England: between the Coal Measures and the Mountain Limestone, between the Mountain Limestone and the Old Red Sandstone, and—a case he had already used with Lyell (§8.4)—between the Lias and the New Red Sandstone. In none of these cases would any geologist (except perhaps Lyell) have postulated a huge break in the sequence of the kind that Murchison and Sedgwick *had* proposed between the Culm and the older Greywacke in Devon.[68]

With this potentially decisive evidence in his favor, De la Beche now disclosed to Sedgwick the new interpretation of the sequence that he had suggested in confidence to Greenough several months earlier (§7.6): "I should say, if the passing of one rock into another be worth any thing, that the culmiferous series represents in Devon the Upper Silurian rocks of Wales" (fig. 8.1, A). Only bad weather and a rough sea, he told Sedgwick, had prevented him from confirming his suspicion that the two sets of strata were equally conformable in the cliffs on the south side of the Culm trough (at Boscastle), but he maintained that inland exposures further east (near Launceston) did confirm this point. Furthermore, Harding had now told De la Beche what Murchison had reported to Sedgwick with such alarm some three months earlier (§8.2), namely, that there were fossil plants in the "acknowledged grauwacke" that Murchison had assigned to his Lower Silurian. De la Beche was quite unaware of the worry that Harding's discovery had caused Murchison; like Sedgwick, he regarded it as an unsurprising confirmation of a general belief that fossil plants of *some* kind might be found in these older strata. One specimen Harding showed him had "strongly reminded" him of a Coal Measures species, but this too was unsurprising, given his belief that particular plant species might have had very long ranges in time. He also mentioned that he had recently found fossil plants in Cornwall (near Fowey and St Austell), in the "grauwacke (acknowledged)" which Sedgwick the previous year had provisionally assigned to the Cambrian; so Harding's discovery was not even unique. Finally, while asking Sedgwick for a list of corrections for his map, to be incorporated in the revised version and in the memoir that would accompany it, De la Beche now conceded that the "culmiferous series" was so distinctive that it should be depicted separately on the map. But of course he did not concede that it was Carboniferous in age.[69]

As usual, De la Beche took the precaution of sending Greenough a copy of his letter to Sedgwick with a covering letter to explain it. In this he drew the conclusion—which he had added only as an afterthought to Sedgwick—that his interpretation of the Culm as Upper Silurian would also align the Culm with the similar plant-bearing Greywacke strata of Normandy and Brittany. As Élie de Beaumont had stressed, this *terrain de transition supérieur* was quite distinct from, and overlain unconformably by, the true Coal Measures strata or *terrain houiller* (§5.5; fig. 5.6). Once again, this placed the disputed Devon strata in an international context and showed that the case

68. De la Beche to Sedgwick, 25 April 1837 (ULC: AS, IC, 3).
69. Ibid.

was not as unique as Murchison (followed by Lyell) repeatedly claimed. De la Beche thanked Greenough and his other friends for their efforts on his behalf and added that he could scarcely believe that Sedgwick was personally antagonistic to him. Nevertheless, it is significant that he had deliberately withheld from Sedgwick one important piece of information that he did send to Greenough, namely, his sketch of the crucial section (fig. 8.2). He had of course described it briefly in words, but the section would have given his opponents too great an advantage to be thrown away at this stage. "All this bother about endeavouring to force a few fossil plants down our throats," he commented to Greenough, "assigning dates to their appearance and disappearance as if we knew the whole thing to a nicety," was the kind of issue in geology that was bound to have come to a head sooner or later. So the Devon controversy, "however unfair its origin," had at least spurred him into looking more closely at the critical evidence. "After all," he concluded philsophically, "good does seem to come out of evil when we least expect it."[70]

In the event, De la Beche did not need to go to London to face his adversaries, for Sedgwick again used his various duties, and his illness and accident, as excuses for postponing the paper until the middle or end of May.[71] He explained the delay to De la Beche and asserted that he and Murchison had already seen the locality De la Beche had mentioned, though in fact it is very unlikely that they had. But he denied that the field evidence was decisive at all, telling De la Beche, "you attribute more importance to a *passage* than I should do."[72] To add insult to injury, he grounded his skepticism in a reference to De la Beche's indirect predecessor as government surveyor, who was regarded on all sides as having been an incompetent rogue: "Macculloch saw 100 such *passages*, & on those accounts made the *old red* [sandstone] of Scotland *primary*, till Murchison & I put things right," he reminded De la Beche; "& then he adopted our views & swore they were his own *ten years* before!"[73]

Sedgwick was clearly countering De la Beche's latest attempt to detach him from Murchison, for he went out of his way to emphasize their joint work and joint conclusions. He admitted that they had not been wholly in agreement about the "the Culmiferous *base line*" in Devon, namely, about the place of the strata below the Culm limestone (§8.2). But "as to the main facts," he added, "we were quite agreed, & I think before long there will be no *third opinion*." He made a point of emphasizing how far De la Beche had already stepped into line with them: "you now give up the idea of making the *culm system* the *old greywacke* & the equivalent of the north Welsh series [i.e., the Cambrian]," he pointed out; "in short *you must* put the culms in a *trough* at the *top*, because such is the section." De la Beche had in fact already conceded this, but Sedgwick pressed his attack home by pointing out the implications of his opponent's new interpretation: "just see what a cleft stick these passages will put you in." De la Beche had claimed that the Culm passed conformably down into the Lower Silurian (Caradoc) on the north of the

70. De la Beche to Greenough, 25 April 1837 (ULC: GBG, 58), from which fig. 8.2 is reproduced.

71. Sedgwick to Murchison, 24 April 1837 (GSL: RIM, S11/131).

72. Sedgwick to De la Beche, 9 May 1837 (NMW: DLB). Their field notebooks contain no mention of the locality in question, let alone the detailed notes it would surely have received in view of its importance.

73. Ibid., referring to Sedgwick and Murchison (1829).

trough, *and* into the still older rocks on the south. But "things which= same = one another," Sedgwick insisted, or, in other words, "Devon & Cornwall slate = Caradoc Sands[tone]. If you admit this good and well. If *not*, what becomes of your *passage*?"[74] The point was a fair one, unless of course De la Beche could find some evidence that on the south, as on the north, the strata immediately below the Culm were Silurian and not Cambrian as Sedgwick supposed.

That, however, raised again the wider issue of the true age of all the strata below the Culm. De la Beche was using the newly discovered plants in the alleged Silurian as further evidence that those strata passed upward without a break into the similarly plant-bearing Culm. Sedgwick countered by distinguishing them sharply: "the *plants* in the *Caradoc sand[stone]* are not *coal* plants, so they say in London," he told De la Beche; by contrast, "the plants at Bideford *are* coal plants," and came from thousands of feet further up the sequence.[75] For Williams had now brought his fossils to London, where the plants had been identified by Lindley as distinct from those of the Culm, although belonging to genera that were also well known in the Coal Measures.[76] So in Sedgwick's view—though not in Murchison's—the newly discovered plants no longer presented any anomaly: they were simply pre-Carboniferous, and therefore not surprisingly of different species from those of the Coal Measures.

De la Beche was unimpressed, at least by Sedgwick's high-handed rejection of the field evidence for a conformable "passage" between the two sets of strata. He told Phillips he thought it "no wonder" that his opponents disagreed among themselves on where the base of the Culm should be drawn; "for the said beds so graduate into the acknowledged grauwacke beneath," that, as he put it sarcastically, "any gentlemen may fix upon any line for his supposed base that may happen to suit his fancy at any time."[77] Sedgwick's insistence on the ambiguity of supposed "passages" showed, however, that it would take more than field evidence to resolve the controversy.

8.7   A Damp Squib on Devonshire (May and June 1837)

In the middle of May, while Williams was still in London, Sedgwick and Murchison's paper once again failed to appear. Time was now running short for the two authors. Since it was customary to spread the reading of long and major papers over two successive meetings, it was imperative for them to have their paper ready at the latest by the end of May, or its presentation would not be completed before the Society went into its summer recess. Only four days before the penultimate meeting, the paper must still have been far from

74. Sedgwick to De la Beche, 9 May 1837.
75. Ibid.
76. Williams, undated notes between dated entries for 5 April and 27 May 1837 (SAS: DW, box I, notebook 10, p. 92). Before he left London, he asked Phillips to look at some other fossils; his note of "Goniatites—Posidonia—Unio—Anodon" suggests that what he had brought were fossils from the Culm limestone: ibid., and Williams to Phillips, 19 May 1837 (UMO: JP 1837/25).
77. De be Beche to Phillips, 31 May 1837 (UMO: JP, uncat.).

finished, for Sedgwick wrote from his seaside convalescence in Essex to ask Murchison to prepare comments on the fossil assemblages from the various formations; but when the day arrived the reading was at last "commenced."[78] Further revisions must have continued subsequently; only a week before the Society's last meeting of the season, Sedgwick asked his collaborator to make further visual aids and to get Lonsdale to identify Hennah's fossils from the Plymouth limestone.[79]

Almost at the same time, however, Murchison's final preparations were disturbed by an incident that caused him to attack De la Beche, for the second time, with great moral indignation. Bilton, who had acted for both sides as local guide to the Bideford area, passed through London en route to Norway, and Murchison entertained him to dinner. Murchison was dismayed to hear that Bilton had sent all his specimens from Devon to the British Museum, at De la Beche's request, and not to the Geological Society, as Murchison thought Bilton had promised. On inquiring at the museum, he was then furious to learn from Charles König (b. 1774), the relevant curator, that De la Beche had given special instructions that the fossils should not be shown to anyone until the official survey was complete. Murchison wrote to De la Beche in high indignation, accusing him of "stopping the march of science with a vengeance." He claimed that both Bilton and their Ilfracombe guide Harding had been *"our* friends," that is, his and Sedgwick's, before De la Beche ever met them; and he complained that Bilton had *"promised faithfully"* to send any further fossils to them, since he had already acted as "our purveyor of the said plants."[80]

This new collection of Culm plants was potentially of crucial importance in the continuing controversy, because, as De la Beche had emphasized repeatedly, so many of the earlier specimens had been too poorly preserved to provide decisive evidence. Murchison knew it was imperative for him to see them, and to have them authoritatively identified, before what he called the "great and long talked of paper on the structure of Devon" was finally read at the Geological Society the following week. He therefore threatened that at the meeting he would expose the "naked circumstances" of De la Beche's underhand conduct, unless his opponent instructed König at once to let Lindley and himself see the fossils.[81]

De la Beche denied the charge promptly and stiffly: since he had met Bilton and Harding independently, he had as much right as Murchison to their assistance. He explained that the specimens he had asked Bilton to send to the museum were additional to those Bilton had promised to give Murchison, and that the museum collection also included specimens De la Beche himself had collected, so that he had a perfect right to restrict access to it. He deplored Murchison's use of a threat; nonetheless, he did instruct König to let Lindley and Murchison see the specimens, though he challenged his

78. Sedgwick to Murchison, 27 May 1837 (GSL: RIM S11/132); *PGS* 2:556. Since the paper had not yet been read in full, there would have been no discussion of it. The other major paper at the same meeting (31 May) was Darwin's first public formulation of his theory of crustal movement as revealed by coral reefs (see Rudwick 1982b).

79. Sedgwick to Murchison, 7 June 1837 (GSL: RIM, S11/133).

80. Murchison to De la Beche, 6 June 1837 (copy by De la Beche, in De la Beche to Greenough, 9 June 1837 [ULC: GBG, 59]. On König, see Campbell-Smith (1969).

81. Murchison to De la Beche, 6 June 1837.

opponent to find a single species there that was not already in the Society's collection.[82] He took the usual precaution of sending copies of this exchange to Greenough and told his protector that he hoped Murchison would be rash enough to accuse him in public, for then Murchison himself would be exposed. Bilton had left for Norway and could not be asked to speak in his defense, but De la Beche did ask Harding to exonerate him from the "mal-practices" of which he had been accused. De la Beche told Greenough he planned to come to London in time for the meeting, but was keeping that fact secret so that he could spring a surprise on his opponents by confronting them unexpectedly in person.[83]

After all these charges and countercharges, the expected fireworks turned out to be a damp squib. The reading of Sedgwick and Murchison's long-awaited paper was duly completed at the final meeting of the Geological Society's season, more than four months after it had first been expected to appear. Its title gave no hint of its highly controversial contents: it was "On the Physical Structure of Devonshire, and on the subdivisions and geological relations of its old stratified deposits."[84] On the surface, it comprised no more than a restatement of the interpretation the two authors had put forward at Bristol nine months before (§7.3), amplified by the results of Sedgwick's later fieldwork (§7.4). But a closer scrutiny of the published summary reveals how Sedgwick and Murchison had papered over the cracks of their disagreements and their mutual uncertainties, and how they had modified their interpretation under the impact of De la Beche's criticisms, though of course without acknowledging that fact.

In north Devon they distinguished five successive groups of formations, with a combined thickness of many thousands of feet. The structure and sequence in south Devon, by contrast, had now lost the simplicity that the authors had attributed to the area when they first visited it. The reliable sequence was in fact reduced to only three great groups. The lowest, though primarily composed of slates, included the Ashburton limestone near the base and the Plymouth and Tor Bay limestones at the top; it was correlated with the lowest in the north, on the basis of their common rock types. The second group in the south, the thick, reddish sandstones above the Plymouth limestone that made such an impression on Sedgwick when he first saw them (§7.2), were now correlated with the similar reddish rocks in the second group in the north, which Williams had dubbed "old red sandstone" (§7.1). The third and uppermost group in the south was correlated with the third in the north, although it lacked the limestone bands of the latter. The fourth and fifth groups in the north were left without equivalents in the south.

Sedgwick's and Murchison's continuing uncertainties and latent disagreements about the dating of these sequences can be detected under the surface of their blandly factual summary. All the strata just mentioned were said to be younger than the rocks of north Wales and the Lake District, which were termed "the lowest part of the Cambrian system"; with the minor exception of the uppermost group in north Devon, they were referred to as "a

82. De la Beche to Murchison, 9 June 1837 (corrected copy by De la Beche, in De la Beche to Greenough 10 June 1837 [ULC: GBG, 60]).
83. De la Beche to Greenough, 9 and 10 June 1837 (ULC: GBG).
84. Sedgwick to Murchison (1837b), read at GSL on 31 May and 14 June.

magnificent development of the Upper Cambrian." This dating was based only on a general (and not very precise) similarity of rock types, together with the belief that the Devon sequence passed upward without a break into Lower Silurian strata. It *could not* be based on any direct fossil evidence, because the Cambrian strata of Wales itself provided far less evidence of the organisms of that period than the Devon strata that were being compared with them. The evidence could only be indirect, namely, that the fossils of the south Devon limestones were "indeed so very dissimilar from those of the Silurian System that they cannot have been formed in that aera."[85]

Furthermore, even the "provisional" Silurian attribution of the uppermost group in north Devon was now put forward with much more hesitation than at Bristol; the strata were merely said to be similar in *rock type* to some of the Lower Silurian strata in Pembrokeshire. The possibility of dating them by means of fossils was minimized both by an emphasis on the poor state of preservation of the fossils, and on the grounds that the distinction between Silurian and Cambrian faunas was not yet well defined anywhere. The nearest the authors would come to a positive statement was to claim that "among the shells are two or three which cannot be distinguished from Lower Silurian fossils." The only other positive piece of fossil evidence about these strata was mentioned with no discussion of its anomalous implications. As mentioned earlier. Lindley had identified their fossil plants as species distinct from those of the Culm, although the genera cited (*Lepidodendron, Sternbergia,* and *Calamites*) were well-known Carboniferous forms. To Sedgwick, the finding of such fossils in the putative Lower Silurian had posed no problem, since he believed there must have been land plants of *some* kind at that period, whereas Murchison had been thrown almost into a panic at their first discovery (§8.2). Yet no sign of Murchison's continuing objection to finding *any* land plants in the Silurian was allowed to ruffle the surface of his and Sedgwick's presentation. The inconsistency was simply left without comment.

Turning to the Culm, which was of course treated as a completely separate group of strata, the presentation again showed subtle but significant changes. The Culm strata were once again said to form "a great trough," so that "whatever may be their age" they were definitely younger than all the strata previously described; but they were now divided into two major groups. The upper contained the plant fossils that had first precipitated the controversy; as before, their close similarity to the Coal Measures of Pembrokeshire was strongly emphasized. The fossil plants were said to "differ essentially from those found in the older rocks" (i.e., those in the alleged Silurian strata), and to be identical with ordinary Coal Measures species. Hence the authors maintained, as at Bristol, that they had "no hesitation" in concluding that "these culm-bearing strata of Devon are indistinguishable from the coal measures of Pembrokeshire." With this conclusion, the original anomaly was, they claimed, finally dissolved in a simple explanation. The lower part of the Culm, now more clearly defined than at Bristol, was characterized by the distinctive black Culm limestone, though most of the strata were dark shales. This formation was said to have "no exact parallel in England," many of its marine fossils being still undescribed. But the authors noted that its *Posidonia* shells

85. Ibid., p. 560.

resembled those of certain parts of the Mountain Limestone, as Phillips had commented at Bristol (§7.3), and they also mentioned that the strata contained distinctive chambered shells (*Goniatites*) known from Carboniferous strata elsewhere but never from the Silurian (fig. 6.5, A, B). The implication, that the Culm limestone was roughly equivalent to the Mountain Limestone, was one in which Sedgwick had evidently persuaded Murchison to join him.

When they dealt with the base of the Culm and its relation to the older strata, however, their uncertainties and disagreements were once again carefully concealed. The great thickness of Culm strata *below* the Culm limestone in south Devon, which Sedgwick had so strongly emphasized earlier (§8.2), had now vanished without comment. But even with the base of the Culm redefined as not far below the base of the Culm limestone itself, it remained problematical. Sedgwick and Murchison admitted that the stratigraphical position of this base was "not yet completely ascertained," but as before they could only maintain their overall interpretation by insisting that there was a sharp break at that point. In the conclusion to their paper, they claimed to "distinctly prove that it [the Culm] never passes down into the older rocks on which it rests," but in the main text this bold conclusion was reduced to no more than italicized assertion. They claimed that since the Culm "rested on" the Lower Silurian in the north but on some of the older Cambrian in the south, in both cases it could only be an apparent conformity, not a real one. The Culm strata, they asserted, *"cannot form (whatever be the mineralogical appearance) a true passage into the different schistose masses on which they rest."*[86] De la Beche's claimed field evidence for a "gradual passage" in the north (§8.6) was tacitly dismissed. The sharp distinction between the Culm and the older strata was simply reasserted, and hence also the conclusion that there was a huge time gap between the two groups. The interpretation first put forward at Bristol was thus maintained with only minor modifications, and the awkward problems it raised were successfully concealed (fig. 8.1, B).[87]

"We had a grand battle at the Geological Society last night," Sedgwick reported to one of his Norwich colleagues after the paper was read at last; "I bore the brunt on our side, but though well banged I was not beaten."[88] No more substantive account of the discussion has survived. In the event, Murchison seems to have made no charge of misconduct against De la Beche, and his opponent did not have to defend himself in that way. As the Society went into recess for the summer, the Devon controversy remained unresolved. It was clear that only some further input of more persuasive evidence would break the deadlock.

86. Ibid., p. 561.
87. Sedgwick's (1837) new edition of the printed *Syllabus* of his lectures at Cambridge included a summary of Devon geology, essentially the same as in the paper, as part of a treatment of the older rocks greatly enlarged from the previous edition (1832).
88. Sedgwick to Wodehouse, 15 June 1837, quoted in Clark and Hughes (1890, 1:483). Charles Wodehouse was a canon of Norwich.

Chapter Nine

SHIFTING THE ARGUMENT

9.1 Putting a Provincial in His Place (July to September 1837)

Leaving aside the higher theoretical issues that the Devon controversy had raised, there were at least two distinct problems that remained unresolved. One was the relation between the Culm and the older strata; the other was the character of the fossils from those older strata, and their significance for dating that part of the sequence. Despite his rejection of De la Beche's field evidence for a conformable "passage" at the base of the Culm (§8.6), Sedgwick was well aware that his own and Murchison's case was still very weak at this point. His fieldwork the previous autumn, although hampered by vile weather, had made him suspect that his best chance of finding good evidence for an unconformity would be on the southern edge of the Culm trough, rather than on the north where De la Beche's claimed conformable exposure lay (§7.4). So it was to the south that he returned when he began his summer fieldwork about a month after the "great" paper had been read in London.[1]

On the way he called to see Austen, who showed him what he had been doing to plot the margin of the Culm in southeast Devon; Austen then accompanied him across Dartmoor to the Launceston area. Austen admitted later to De la Beche that in north Cornwall they found that between the Culm and the older strata "the exact line of junction cannot always be hit." That

1. Sedgwick to Murchison 10 July 1837 (GSL: RIM, S11/134); Sedgwick to De la Beche, 10 July 1837 (NMW: DLB).

215

supported De la Beche's opinion that the junction was elusive because it was completely conformable. But Austen also reported that a new road cutting near Launceston had revealed that the junction there *was* unconformable, "the old rock being very rich in characteristic fossils". When they reached the coast (at Boscastle), where bad weather the previous year had prevented Sedgwick (or De la Beche) from studying the cliffs from sea level, they had found that "the one formation cuts slightly unconformably on the others."[2] Thus far Sedgwick had strengthened his interpretation, though it was still open to De la Beche to insist that the conformable junction on the *north* side of the Culm trough was what mattered in settling the main issue.

The second outstanding problem was brought back into the arena by Williams with the fossils he had found in the pre-Culm formations in north Devon earlier in the year (§8.6). He and Harding had been referred to in Sedgwick and Murchison's paper as mere "zealous collectors," but Williams had no intention of being limited by that patronizing label.[3] In July he gave Murchison and Sedgwick notice—as they had notoriously failed to do to De la Beche the previous year—that at the British Association meeting he intended to criticize their interpretation by displaying a section of Devon and the best fossil specimens he had found. "The Culm field is open and debatable ground," he insisted defiantly, "and unless you consent to a compromise by including it in the upper grauwacke I will dispute every inch with you."[4] But that interpretation was only a compromise in the sense that it shifted the Culm from the lower to the upper Greywacke, or from Sedgwick's Cambrian to Murchison's Silurian. De la Beche had already made just that move, but it was a compromise that was totally unacceptable to Murchison, if not to Sedgwick (§7.6). A month later, when Williams gave Phillips formal notice that he wanted to speak at Liverpool, he commented pointedly that "the true age of the Culm of Devon & its associated flora was rather dispatched than determined in Bristol." As for the older strata, he told Phillips he would fulfill the geological Section's formal recommendation of the previous year (§7.3) by displaying the fossil plants he had now found "very low down in that group, certainly two miles in the vertical direction below the Culm field," and hence incontestably within the Greywacke.[5]

When the British Association met in September, the highly popular geological Section was assigned the large lecture theatre at the Mechanics' Institute, where the gallery was considered appropriate provision for the many ladies who were expected to attend—as spectators, of course. Lyell was away in Paris, after a trip to Denmark in pursuit of the Tertiary. Buckland had just attended the field meeting of the Société géologique in Alençon, where he had entertained the French geologists with an "improvisation brillante et animée" on the sublime grandeur and practical value of geology. He had also identified Lower Silurian strata in that region, as he had the previous year in Belgium (§6.3). But most of the leading British geologists were in Liverpool

2. Austen to De la Beche, 23 August 1837 (NMW: DLB). This letter is the only direct evidence available for Sedgwick's fieldwork, since his notebook for this period has been lost.
3. Sedgwick and Murchison (1837b, p. 561n).
4. Williams to Murchison, 12 July 1837 (GSL: RIM, W6/1).
5. Williams to Phillips, 16 August 1837 (UMO: JP, 1837/41).

for the sessions of the geological Section, this year under the presidency of Sedgwick.[6]

At the opening session, Whewell deserted his usual arena in the Section for mathematics and physics and, as president of the Geological Society, was given pride of place to deliver a paper on recent changes in sea level—a crucial topic for the parallel but separate controversy on crustal elevation. The following paper (on New Red Sandstone) brought the geologists back, however, to their central business of stratigraphy. In the subsequent discussion, Sedgwick urged the need to "sink local distinctions" in order to reach correlations that would be valid internationally, which was just what he hoped his own Cambrian and Murchison's Silurian would achieve among the older strata. Greenough continued the international note the following day, when he displayed an advance copy of the great geological map of France; since it had been inspired by his own similar map (§4.6), he doubtless took some credit for the fact that the Association had been sent one of only two copies so far released, the other having gone to the French government. As Griffith then mentioned, the revised edition of Greenough's map was likewise almost complete; what was not mentioned was that the Devon controversy would need to be resolved before he could publish it, because—as Greenough himself had recognized—it was bound to affect the whole appearance of the map in that part of the country (§8.6).[7] Later in the week, a paper by Griffith on the Carboniferous strata of Ireland prompted Sedgwick to comment on the problems of correlating individual Carboniferous formations across wide distances by fossil criteria, because, as Phillips's work had shown (§6.6), that group as a whole was evidently so variable. This too impinged implicitly on the Devon controversy, since Sedgwick's own tentative correlation of the Culm limestone with the Mountain Limestone depended on just such a recognition of the lateral variability of those formations (§8.2).[8]

So at last, with a reading of the "more important" parts of Williams's paper, the Devon controversy finally emerged in explicit form.[9] As he had told Phillips beforehand, Williams said he was presenting it in response to the committee's recommendation the previous year, that geologists should search more thoroughly for fossil plants in the Greywacke (§7.3). He displayed not only the specimens he had found, and a traverse section of Devon, but also his own geologically colored Ordnance Survey maps to match.[10] His main focus was on the north, where he showed a "series of successive emergence" or sequence of nine formations, which can be equated almost directly with Sedgwick's and Murchison's; but like De la Beche, he claimed that there was

6. On the BAAS at Liverpool (11–16 September 1837), see *Ath.*, no. 516 (16 September), pp. 670–71; on Lyell, see K. Lyell (1881, 2:13–27); on Buckland with the SGF at Alençon (3–10 September 1837—he was the only foreigner in a party of twenty-four), see *BSGF* 8:353–67.

7. *Ath.*, no. 517 (23 September 1837), pp. 696–97.

8. *Ath.*, no. 519 (7 October 1837), pp. 749–51.

9. Ibid.; Williams (1838), read at BAAS on 15 September 1837. The abridgment was probably not, or not entirely, due to covert censorship, since the Section was genuinely so "pressed for time" with unread papers that an extra unscheduled session had to be arranged for the final day.

10. Fig. 9.1 is reproduced from the section published in *Ath.*, no. 519, p. 750. William's maps (SAS: DW, R/4) are undated, and in parts (especially in the south) the coloring and labeling are certainly later than the Liverpool meeting; nonetheless, they are almost certainly the sheets he had told Phillips he wanted to display (see above). Unlike De la Beche's, his mapping depicted the outcrops of all the formations separately, and must have been his own work.

no unconformity within the sequence, nor any major change in the fossils. His section and map showed some repetition of the sequence to the north and south of the main Culm area of central Devon (fig. 9.1, 9). But the section was so crude in style and showed so little attention to features such as the dip of the strata, that the other geologists would have been highly skeptical about Williams's claim that he had discovered the "deep trough" in Devon *before* Sedgwick and Murchison, and that he had alluded to it when he presented his first paper to the Association in Dublin in 1835 (§6.2).[11]

In his usual provocative style, Williams denied that fossils bore any "authority" in the dating or correlation of this sequence. He rejected Phillips's argument that the fossils in the Culm limestone showed it to be equivalent to the Mountain Limestone, on the grounds that the fossils concerned were not known to be *confined* to Carboniferous strata, and that it was implausible to suppose that the massive Mountain Limestone should be represented in Devon only by these few thin bands of limestone. Such arguments, he concluded, "throw such accumulated weight into the scale of the opposite hypothesis"—namely, that the Culm and its fossil plants "belong at least to the upper grauwacke (below the old red sandstone)"—that it should remain there "provisionally" until much better evidence was forthcoming to the contrary.[12]

Murchison opened the discussion by flatly disagreeing. He claimed that the Culm strata formed a "regular coal basin" like that of Pembrokeshire; and he insisted that they could not be Silurian (or in Williams's terms, "upper grauwacke") because they contained Carboniferous shells. This represented a total impasse, since neither geologist would give way to the other on the basic issue of the use of fossils. On the challenge presented by Williams's new plant fossils extending down into the putative Cambrian, Murchison avoided making any comment at all, even though the paper was explicitly focused on the specimens displayed. Greenough spoke up for De la Beche, saying he agreed with him on this issue—and therefore implicitly with Williams too—and stressing once more the importance of the foreign analogues, which showed no major contrast in flora between Transition and true Coal Measures. Sedgwick insisted as before—doubtless recalling his recent fieldwork with Austen—that there was a real "hiatus" in the Devon sequence between the Culm and the older strata. As at Bristol, only the usual gentlemen of geology contributed to the public discussion, and their virtual neglect of what the provincial geologist had said would have left the rest of the large audience in no doubt about their opinion of his work. As Murchison told the editor of the *Athenaeum* later, when asked to check that newspaper's report of the session, the paper by "Parson Williams" would "occasion much merriment among geologists by the blunders therein." [13] It remained to be seen, however, whether the fashionable Londoner or the Somerset clergyman would have the last laugh.

11. The recorded summaries of William's Dublin paper do not explicitly mention a trough. The claim was bound to incense Sedgwick and Murchison, who considered themselves the true discoverers of the trough structure, in 1836 (§7.2).

12. *Ath.*, no. 519, pp. 749–51.

13. Ibid.; Murchison to C. W. Dilke, 11 October 1837 (ULE: RIM, Gen. 523/4).

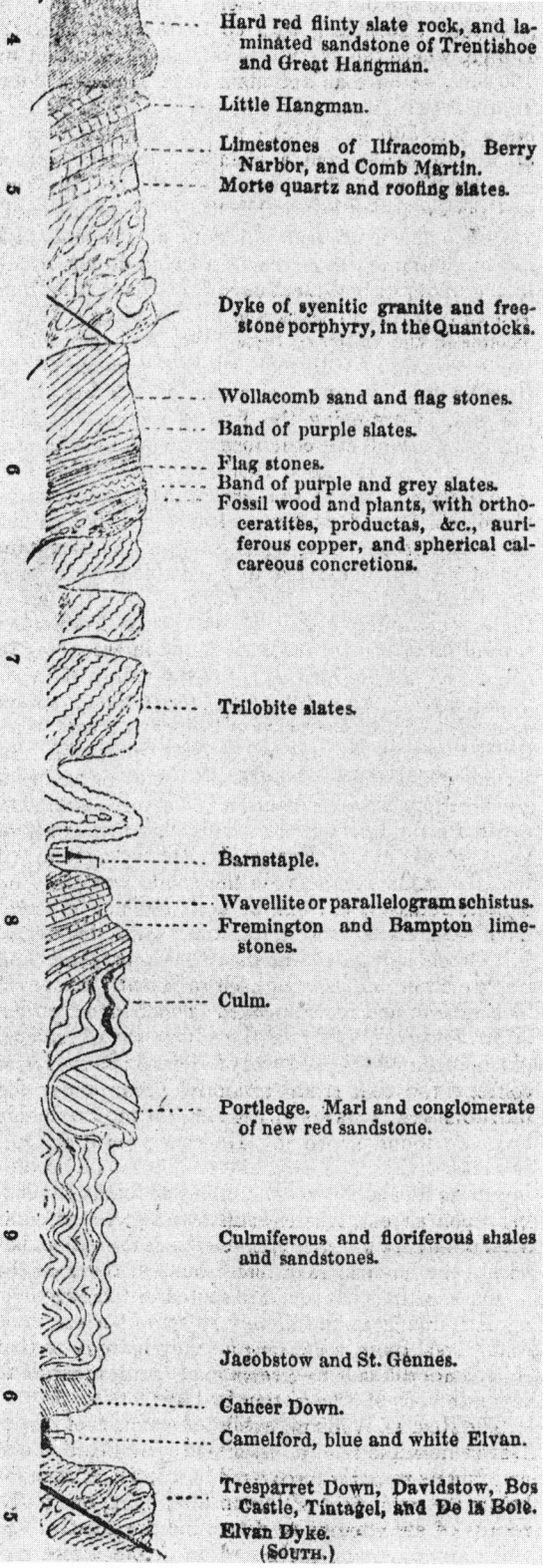

Hard red flinty slate rock, and laminated sandstone of Trentishoe and Great Hangman.

Little Hangman.

Limestones of Ilfracomb, Berry Narbor, and Comb Martin.
Morte quartz and roofing slates.

Dyke of syenitic granite and freestone porphyry, in the Quantocks.

Wollacomb sand and flag stones.
Band of purple slates.
Flag stones.
Band of purple and grey slates.
Fossil wood and plants, with orthoceratites, productas, &c., auriferous copper, and spherical calcareous concretions.

Trilobite slates.

Barnstaple.

Wavellite or parallelogram schistus.
Fremington and Bampton limestones.

Culm.

Portledge. Marl and conglomerate of new red sandstone.

Culmiferous and floriferous shales and sandstones.

Jacobstow and St. Gennes.

Cancer Down.

Camelford, blue and white Elvan.

Tresparret Down, Davidstow, Bos Castle, Tintagel, and De la Bole. Elvan Dyke.
(SOUTH.)

Fig. 9.1. Part of Williams's traverse section of north Devon, showing a sequence of thick formations, *all* of which he regarded as Greywacke, i.e., older than the Old Red Sandstone. This crude section illustrated his paper to the British Association in September 1837. (9) is the Culm proper; (8) the Culm limestone; (7) his "Trilobite Slates" (Murchison's Lower Silurian); and (6) the formations in which he had found the oldest fossil plants.

9.2   The Provincials Move Center Stage (October and November 1837)

In the months that followed the Liverpool meeting there were further attempts to break the evident deadlock by finding more convincing field or fossil evidence. Williams, undeterred by the opposition he had encountered, returned to north Devon and, as he reported to Lonsdale, "discovered a true and undeniable passage" or complete conformity between the Culm and the older strata. "I have no doubt you were convinced of the fact," he told De la Beche later, "but riddle my ree cd. you prove it!" Williams claimed, however, that the paper he was now planning to write on "the intermediate or passage beds" would indeed prove the matter, and enable him to maintain his position "against all the Geologists in the world."[14]

While in the Barnstaple area, Williams acted as guide to Weaver, who now reentered the controversy in a new role. Weaver had played no active part in the argument since he "recanted" his original conclusions about the coal in the Irish Greywacke nearly three years before (§5.6). But he had certainly followed the closely parallel debate about the Devon culm; having been persuaded of his error in Ireland, he was a natural recruit to Murchison's side of the Devon controversy. Weaver had settled in north Somerset (at Wrington), only a few miles over the Mendip Hills from Williams's parish. Having heard about the debate at Liverpool—perhaps from Williams himself—he decided to try his own hand at Devon geology.

Williams was unimpressed by the interloper: Weaver, he told De la Beche afterward, "knows nothing of the Country beyond what I showed him and told him."[15] Weaver had gone there convinced that Sedgwick and Murchison were right about the Coal Measures age of the Culm, and that there must be a major unconformity at its base, although the other geologists had failed to find it. Williams took him on a traverse across country, southward from the older strata on to the Culm, doubtless pointing out the "passage beds" that in his opinion proved the junction to be one of perfect conformity. Only after they were well *into* what Williams regarded as the Culm area did Weaver find what he took to be the unconformity he was looking for. "You are over the Bridge, my friend," Williams recalled telling him scornfully at that point; "you may now find out as many such junctions as you please, for the whole region is full of them."[16] In Williams's opinion, Weaver was so unfamiliar with Devon geology that he had identified as a major unconformity at the base of the Culm what was really no more than a common and minor feature of irregular stratification *within* the Culm itself.

Weaver, not surprisingly, had a higher opinion of his own abilities. He told Phillips afterward that Williams had been "a little disappointed when

---

14. Williams to Lonsdale, 17 November 1837 (GSL: BP, LR3/197); Williams to De la Beche, 21 November 1837 (NMW: DLB).

15. Ibid.

16. Williams, notebook for "October 1837 to April 1838" (SAS: DW, box I, notebook 12). These undated notes (later than an entry dated 19 December 1837) come after a reference to "Mr. Weaver's Paper," and may have been for a speech Williams planned to make (or did make?) when that paper was discussed at GSL on 3 January 1838 (see below).

he found I could not support him in his views," adding, in a distinctly patronizing tone, "I reaped great advantage from his knowledge of localities."[17] Sending a paper on the subject to the Geological Society, Weaver told Murchison with even greater assurance, "I conceive I have *demonstrated* the correctness of the opinion advanced by you & Professor Sedgwick as to the true aera of the Devonshire coal."[18] So with Williams's paper also in preparation, the front-line confrontation between De la Beche and the Sedgwick-Murchison team was all set to be matched during the Society's new season by a parallel confrontation between their respective Somerset supporters, Williams and Weaver.

Colby's proposal that De la Beche's memoir on Devon and Cornwall should be published at government expense (§8.5) was approved by the Treasury shortly after the Liverpool meeting; but Spring-Rice warned Sedgwick to "remember, if I get my head broken by some fragment of a Silurian rock you Gentlemen philosophers must take your share of responsibility for the recommendation you gave me."[19] The political importance of ensuring that the memoir was as free as possible from error needed no underlining. De la Beche therefore returned to the disputed region later in the autumn, primarily to check matters for himself for the last time before his *Report* went to press. After working in Austen's home area, he moved to north Devon and took Austen with him. Austen was carefully keeping clear of the personal animosities in the controversy by assisting those on both sides evenhandedly. Just as he had reported to De la Beche, after his fieldwork with Sedgwick, that they had found an unconformity on the south side of the Culm (§9.1), so likewise he reported to Murchison, after his fieldwork with De la Beche, that "it appears to me that the culms pass into the older beds along the northern line."[20] His conclusion was probably the result of being shown, by De la Beche, the best locality for that "gradual passage" (Fremington Pill: fig. 8.2). Although Murchison heard soon afterward that Weaver believed he *had* found the anticipated unconformity in the north, Austen's contrary report may have made him uneasy about his own and Sedgwick's interpretation; for Austen was clearly becoming a geologist whose judgment had to be taken seriously, whereas Weaver's credibility, after the Irish affair, was more questionable.

Certainly De la Beche was in no doubt that his opponents' interpretation would not stand the test of more thorough research: "it is a sad piece of slap dashing, and will do them little credit," he commented to Phillips; "it cannot possibly stand, even in the grauwacke proper." Even Sedgwick, according to De la Beche, was now skeptical about the existence of any real Silurian in north Devon; for the more thoroughly the fossils were studied, the more it seemed that those of "the so-called Cambrian rocks" were the same as those of the alleged Lower Silurian. "This business has been over done, depend upon it," De la Beche insisted; he told Phillips that the Silurian and Cambrian seemed to him "good as local names, bad as generals," and

17. Weaver to Phillips, 29 December 1837 (UMO: JP, 1837/67).

18. Weaver to Murchison, 25 November 1837 (ULE: RIM, Gen. 523/4, 86); cf. Weaver to Phillips, 18 December 1837 (UMO: JP, 1837/65).

19. Spring-Rice to Sedgwick, 30 September 1837 (ULC: AS, IB, 108).

20. Austen to Murchison, 5 November 1837 (ULC: AS, IB, 126), responding to Murchison's request for Austen's opinion on this point; De la Beche to Colby, 28 October 1837 (NMW: TS, II, p. 244).

that he himself would therefore "stick to Upper and Lower Grauwacke." For, as he told Phillips, more and more of the fossils he was finding in Devon were incompatible with the sharp distinction his opponents had drawn between the Culm and the still older strata. Like Austen, but in a different area, he was finding fossil plants in "the regular Cambrian rocks (according to Sedgwick)." But since the value of fossil plants for dating purposes remained debatable, what was even more significant was the character of the animal fossils. Here too the distinction seemed to be breaking down. De la Beche reported that he had found "many curious things," including particularly the distinctive chambered shell *Goniatites*, which was generally regarded as a Carboniferous genus (fig. 6.5, B), in the "unquestioned grauwacke of [South] Petherwin," the locality near Launceston that Sedgwick had firmly attributed to his Cambrian (§7.4).[21]

A full year before, De la Beche had asked Phillips if he would identify fossils for the geological survey on a regular basis; in the months that followed De la Beche had duly begun to send him boxes of fossils from Cornwall.[22] Phillips was as well qualified for this work as anyone in England; in particular, after his monumental work on the Mountain Limestone fossils of Yorkshire (§6.6), he was uniquely well equipped to evaluate the Carboniferous character of any of the strata in Devon. In the discussion at Bristol, he had commented on the similarity between the fossils of the Culm limestone and those of certain varieties of the Mountain Limestone (§7.3). Now, however, a similar affinity was being claimed for the fossils from apparently much older strata, namely, the limestones of south Devon and Cornwall.

Those fossils had long been a puzzle: as Greenough had commented to De la Beche a decade earlier, they seemed to be a mixture of Mountain Limestone and Transition types (§4.6). In the discussion at Namur in 1835 about correlations between Britain and Belgium, the similarity between the Devon fossils and some of the Belgian ones had been noted; but they did not seem to match those of either the Mountain Limestone or the Silurian (§6.3). In the spring of 1837, Austen had written to answer Sedgwick's queries about the various limestones in his home area of southeast Devon. Commenting on the fossils from what he considered the uppermost limestone (around Tor Bay), Austen had asked Sedgwick, "can it be Mountain Limestone?"[23] The suggestion was quite incompatible with the interpretation Sedgwick was busy putting into his joint paper with Murchison; perhaps for that reason—and because he was a poor correspondent at any time—Sedgwick had not even replied. After waiting more than two months, Austen had therefore written again. "Do you see any objection," he asked Sedgwick for the second time, "to the great limestone of this district being the equivalent of the Mountain Lime[stone]? So many of its fossils are the same."[24] Whether or not Sedgwick replied this time, he certainly did see an objection: in his view the south Devon limestones were clearly an integral part of a conformable sequence of immensely thick formations, all of them well below the Culm and inferentially

21. De la Beche to Phillips, 29 October 1837 (UMO: JP, uncat.).
22. De la Beche to Phillips, 6 November 1836, and 21 May, 3 June, and 28 September 1837 (UMO: JP, 1836/61.2 and uncat.).
23. Austen to Sedgwick, 8 March 1837 (ULC: AS, IB, 114).
24. Austen to Sedgwick, 22 May 1837 (ULC: AS, IB, 115).

of Cambrian age (§8.3). So Austen's suggestion had found no echo in the paper that Sedgwick and Murchison put on record at the Geological Society a few days later (§8.7).

Austen was undeterred, however, and continued with his own plans for publishing a full-scale paper on the geology and fossils of his home area. That plan was a slight embarrassment to De la Beche, who naturally wanted Phillips to be able to study as many as possible of the invaluable collections belonging to the various local geologists. He hoped to persuade Williams, Harding, and Hennah, for example, to lend their fossils for the purpose, but Austen's plans made it awkward to ask him for the loan of his now "famous" collection, although it was "better than all." Still, Austen was happy enough to show his fossils to De la Beche. After seeing them, De la Beche told Phillips that each locality seemed to have its own distinctive assemblage, and that fossil shells resembling those of the Mountain Limestone were mixed with those found in "the grauwacke (undoubted)" elsewhere in Devon and Cornwall. After De la Beche had sent him some to look at (not from Austen's collection), Phillips told him he agreed they were "*strangely* analogous to Carb. Limestone" fossils, and asked to see more of them.[25] Coming from the foremost authority on such fossils, that judgment could not be dismissed as casually as Sedgwick had ignored Austen's suggestion to the same effect. Nor would it be easy to conclude simply that the older strata of south Devon were Carboniferous in age. For, as De la Beche made a point of emphasizing to Phillips, the limestone in which "the said carboniferous gentry" had been found was quite clearly within the Greywacke sequence, and indeed "somewhat deep down in that series."[26]

The animal fossils of the south Devon limestones were thus all set to become as contentious as the plant fossils from the Culm of north Devon. The fundamental issue was not *whether,* but *how* fossils should be used to determine the relative ages of formations. De la Beche attributed all the apparent anomalies that had arisen in his region to the diversity of local environmental conditions at "the epoch of the fossil potting" there. Phillips agreed wholeheartedly. He contrasted the complexity—still too little understood—of "the correlation of the changes of life & the variation of physical conditions" with the way that some geologists were currently using fossils. "The subject is really *opening,*" he commented to De la Beche; more thorough research would yield results "a thousand-fold richer than the mere bald and trifling notion of the 'identification of Strata' by their organic Contents." Not for the first time, Phillips thus distanced himself decisively from the simplistic use of fossils that his uncle William Smith had pioneered, and that geologists like Murchison were now taking to implausible lengths—without the excuses that could be made for a pioneer. De la Beche's discovery of Culm plants was, in Phillips's opinion, a striking case in point: "it is *a fact* to be introduced into the induction, not an *anomaly* to be frightened at."[27]

25. De la Beche to Phillips, 29 October 1837; Phillips to De la Beche, 4 November 1837 (NMW: DLB). De la Beche reported later that Hennah had sent his collection to Sedgwick, but that Williams and Pattison were willing to lend theirs to Phillips: De la Beche to Phillips, 13 November, and 5 and 12 December 1837 (UMO: JP, uncat.; 1837/61.2).

26. De la Beche to Phillips, 13 November 1837 (UMO: JP, uncat.). He reported finding the well-known Wenlock Limestone species *Calymene Blumenbachii* and other undescribed trilobites.

27. Phillips to De la Beche, 4 November 1837.

The same moral applied equally to the animal fossils from the older strata of Devonshire. Since the evidence of the plant fossils continued, after three years of argument, to be ambiguous and inconclusive, the Devon controversy entered its fourth year on the public stage with the spotlight of attention turning onto the fossil shells and corals of south Devon, and onto their provincial collectors.

9.3   Austen's Enigmatic Fossils (December 1837)

Three provincial geologists had now written, or were planning to write, papers impinging directly on the Devon controversy: Williams in support of De la Beche, Weaver in support of Sedgwick and Murchison, and Austen taking an independent line. Austen's was the first to be submitted to the Geological Society. It was due to be read at the end of November, but that meeting was canceled at the last moment because the sudden death of the Society's caretaker threw the domestic arrangements into disarray. Austen stayed on in London, however, until his paper was read two weeks later. That delay not only allowed the fossil specialists unexpected extra time to examine the specimens he had brought with him, but also allowed Murchison to study the paper carefully in advance.[28]

Austen's paper appeared on the surface to be a conventional local memoir, in that it summarized *all* the rocks, not just the older strata, in his home area of southeast Devonshire.[29] Dealing with the Culm, Austen termed it the "Culmiferous or Carboniferous Series," thus maintaining his neutrality on the vexed question of its age. He referred unambiguously, however, to Sedgwick and Murchison's "rectification" of the previous "error" about its structural position, thereby showing himself to be on their side at that point. Furthermore, he described how in his area the Culm lay unconformably on what he termed the "Transition System." (His opinion that in the north of the county the relation was one of complete conformity was not mentioned, being strictly irrelevant to a local memoir on the south.) He described a sequence of five formations of "Transition" strata, which can be equated at least in part with what Sedgwick and Murchison had defined in their paper six months earlier (§8.7). Most important was what Austen termed the "Great Limestone" (at Newton Abbot and around Tor Bay) near the top, separated by a thick series of reddish strata from some much lower limestones (at Ashburton and Chudleigh).

The official summary of Austen's paper made no mention of his earlier suggestion to Sedgwick that his Great Limestone, judging by its fossils, might be equivalent to the Mountain Limestone elsewhere in Britain. The Devon limestone's "peculiar fossil shells" were mentioned, but it was placed without comment in the "Transition" sequence and not with the Culm among the

28. Lonsdale told Austen it would be read on 29 November 1837: Austen to Murchison, 5 November 1837 (ULC: AS, IB, 126). Once submitted, the paper became the property of GSL, and Murchison as a Council member could have read it without explicit permission from Austen.
29. Austen (1838a), read at GSL on 13 December 1837 (GSL: BP, OM 1/8, pp. 484–94).

"Secondary" formations. The full paper that was due to be read, however, must have contained a clear statement of Austen's earlier suggestion, for Murchison seized on this point and wrote to Sedgwick to argue in its favor.[30]

For Murchison, Austen's tentative correlation had the advantage of absorbing the upper part of the sequence of older strata in south Devon into the Carboniferous, which would explain why fossil plants were turning up in them (§9.2). Murchison's conviction that no land plants would be found in any Silurian or older strata could thus be salvaged, at the tolerable expense of abandoning his claim to have found some Silurian strata in north Devon. But he could hardly claim that *all* the older strata of Devon were really Carboniferous; for most of them, particularly in the south, were ancient-looking rocks with a fauna that was already widely regarded as characteristic of Sedgwick's Upper Cambrian. So, in effect, Murchison was obliged to argue that the upper strata in Austen's sequence were distinct from the lower. More specifically, if he were to carry Sedgwick with him, he had to convince him that Austen's Great Limestone was really quite distinct from the Plymouth limestone further west, with which Sedgwick had firmly correlated it, as well as being different from the lower limestones in Austen's area. Murchison attempted to convince Sedgwick on this point by arguments based particularly on fossil evidence. He claimed that the upper limestones were like the Mountain Limestone because they contained no trilobites but did contain brachiopod shells ("Productae") that were characteristic of Carboniferous strata (fig. 6.5, C).

Sedgwick was unconvinced: "the reasons you give are not worth a rush to prove the Torbay limestone the equivalent of the Mountain limestone." He reminded Murchison that the Mountain Limestone did contain trilobites, although fewer than the Silurian and Cambrian. Conversely, Mountain Limestone shells had already been identified from the Plymouth limestone; so, he added, "I don't care a rush for the Productae you speak of." In other words, these limestones might well contain *some* Mountain Limestone species, but that in itself was insufficient in Sedgwick's view to prove their age. "You say that *every thing* hinges on *zoological evidence*," he commented; "I deny the fact—Zoological evidence is good as far as it goes but can not upset sections." As always for Sedgwick, the field evidence had to remain paramount, and that evidence gave no support for the idea that some of the strata of south Devon were Carboniferous while others were far older. Whatever their age, Sedgwick was convinced the sequence was unbroken (apart, of course, from the Culm strata lying unconformably above, which he accepted as Carboniferous). "Are you ready then to swamp the Plymouth section on the evidence of one or two fossils?" he asked; "if so I cannot embark on the same geological bottom with you." He considered Austen's opinion on the fossils unreliable: "remember Austen has never seen, as far as I know, a cubic foot of Mountain limestone *in situ*," he pointed out. "Austen is a very good workman, but has very small experience out of his own field," he concluded; "I have no doubt his details are admirable—& I have no doubt that his conclusion is absolutely wrong &

---

30. This letter has been lost, but, as in an earlier exchange (§8.2), Sedgwick's reply contained sufficient quotations or paraphrases of Murchison's points to make a reliable reconstruction of its contents (see below).

I should be truly sorry that he should in his first great paper commit what I regard as an enormous blunder."[31]

Sedgwick was well aware of the reason for Murchison's advocacy of Austen's correlation, for it paralleled his collaborator's earlier attempt to escape the implications of Harding's discovery of fossil plants in the supposed Lower Silurian of north Devon (§8.2). Once the authenticity of that find had been established, and then confirmed by Williams's independent discovery (§8.6), Murchison had tried to minimize its significance by claiming that anyway the plants were not the same as those of the Culm. Sedgwick was now skeptical about this, and asked if Murchison had any new evidence for that alleged contrast. But Murchison had also proposed a more radical way of preserving his claim that the Silurian and still older strata were totally lacking in land plants, when he had suggested that almost the *whole* of the pre-Culm sequence in north Devon might be equivalent to the Old Red Sandstone and hence be post-Silurian (§8.2; fig. 8.1, C). "I laughed heartily," Sedgwick now recalled, "at your former letter in which the whole of N. Devon was to become the *old red* with a new face."[32] Sedgwick saw clearly that his collaborator was now proposing a very similar interpretation for *south* Devon, making at least part of the pre-Culm sequence into Mountain Limestone, likewise as it were "with a new face." Murchison's latest suggestion was as laughably implausible as his earlier one, in Sedgwick's view, because it sacrificed the clear field evidence of an unbroken sequence of formations for the sake of saving a dubious theory about the history of plants on the earth's surface.

De la Beche was also skeptical. Priming Greenough on how to defend him at the meeting, De la Beche told him he expected Austen's paper would be "a good one" in general. "His carboniferous limestone, however, near Torquay won't do," he added; "it all belongs to the Grauwacke series and I believe Austen himself now thinks so." That Austen was indeed in two minds about the matter is suggested, as already mentioned, by the fact that in his paper he grouped his Great Limestone among the "Transition" formations while claiming that its fossils looked Carboniferous. De la Beche believed Austen had rejected "the Murchisonio-Sedgwickian story" about the Culm, since being shown the "good sections" in north Devon by De la Beche himself during their recent fieldwork there (§9.2); "he is one of us I believe at present." If Austen did not mention that crucial evidence of his own accord, De la Beche suggested, Greenough should "get it out of him before the meeting," that is, in the presence of their opponents during the discussion after the paper was read.[33]

That evidence was less crucial, however, than De la Beche implied. Even if the Culm could be proved to lie conformably above the older strata, at least in north Devon, there would still be two contrasting interpretations of the age of the whole sequence. Either the strata were all Greywacke (i.e., pre-Carboniferous), as De la Beche continued to insist, so that, as he confidently expected, "the Coal Measures story of Devon will go to pot shortly." Alternatively, that story could be salvaged by assigning at least some of the

31. Sedgwick to Murchison, "Saturday Eveng.," pmk [Sunday] 10 December 1837 (GSL: RIM, S11/139).
32. Ibid.
33. De la Beche to Greenough, 11 December 1837 (ULC: GBG, 62).

lower (pre-Culm) formations to the Carboniferous too, as Murchison—following Austen's suggestion about the fossils—had now for the second time proposed to Sedgwick. But that second option was so implausible to De la Beche—as to his opponent Sedgwick—that he did not mention it even to dismiss it.

Austen's paper "On the Geology of the south-east of Devonshire" had the last meeting in 1837 to itself. Murchison was at the Geological Society to hear it read, but Sedgwick was on duty in Norwich. Greenough was present to look after the interests of De la Beche, who was in Swansea making arrangements for the geological survey's new work in south Wales. Austen's exceptionally fine collection of fossils from the Devon limestones was displayed on the table. The only strictly contemporary record of the subsequent discussion is the report that Austen himself wrote for De la Beche the following day, shortly before he returned home to Devonshire; but there is no reason to doubt its reliability. In answer to questions—probably those put to him by Greenough, at De la Beche's prompting—Austen stated publicly that in *north* Devon "the old transition rocks pass upwards into the carbonaceous beds," that is, into the Culm, with complete conformity; and that in this respect "the section admits of no equivoque whatever." In his own area, by contrast, the Culm strata were "found upon the edges" of all the older formations; but he regarded that unconformity merely as a result of "local disturbance," and considered it "only as the exception, not the rule." Austen stated publicly that in his opinion it was therefore "incorrect to separate the Carbonif. strata from the Transition," and that the Culm was "only the upper portion of a thick series" or unified sequence of formations. Austen thus maintained his sturdy independence: in insisting that the sequence was unbroken and conformable, at least in north Devon, he was agreeing with De la Beche; but in attributing the Culm to the Carboniferous, he was agreeing with Sedgwick and Murchison. "You see," he commented to De la Beche, "that I have both yourself and S. & M. against me in certain points, and perhaps between 'battlement and battering ram' my notions will be smashed."[34]

The most important of his vulnerable notions, however, was his claim that his *"great limestone"* of southeast Devon was none other than the Mountain Limestone in thin disguise. Since the Great Limestone was one of the formations truncated by the local unconformity, its age would clearly help "determine the real age of the so-called culm beds" above it. But if Austen were correct about the limestone, it would follow almost inevitably that Sedgwick and Murchison were correct in assigning the Culm to a higher part of the Carboniferous, namely the Coal Measures. "You will kick at this," Austen told De la Beche afterward, but asked him to "think the thing well over" before rejecting it. For he claimed that there was nothing in the structure or sequence in southeast Devon to preclude assigning the limestone to the Carboniferous, whereas its fossils gave strong positive evidence in favor of a Mountain Limestone age: he displayed on the table no fewer than thirty-six species of "shells" characteristic of the Mountain Limestone, which he had identified from publications.[35]

34. Austen to De la Beche, "Thursday," pmk [Friday] 15 December 1837 (NMW: DLB).
35. Ibid. The publications he mentioned ("Phillips and Sowerby") were presumably Phillip's Yorkshire volume (1836), and the serial *Mineral Conchology* (Sowerby 1812–46), which the younger Sowerby (see below) had been continuing ever since his father's death in 1822.

This evaluation of the fauna was now open for comment by two of the leading specialists on the relevant fossils. Lonsdale was there as usual in his official capacity, but so was James de Carle Sowerby (b. 1787), a professional artist and engraver of natural history illustrations, who was an expert in his own right on living and fossil mollusks.[36] Just as Lindley's expert and impartial identification of De la Beche's fossil plants had given a sharp edge to the start of the controversy three years before (§5.1, §5.2), so likewise Lonsdale's and Sowerby's expert opinions on Austen's animal fossils could now be expected to add further fuel to the continuing argument. They had both been helping Murchison to identify and describe Silurian fossils for his forthcoming book, so they were particularly well placed to evaluate the relation between the Silurian fauna and Austen's from Devon. Although Phillips was not present in person, his monograph on the Carboniferous fossils of Yorkshire, published the previous year (§6.6), provided them likewise with a reliable basis for comparison.

Austen reported to De la Beche that "Lonsdale says that the rocks from which my fossils come must be younger than any he has examined for Murchison." That bare summary was sufficient for Austen's immediate purposes: Lonsdale considered that the south Devon fossils were not only distinct from those of the Silurian limestones of the Welsh Borderland, but also *younger*, and therefore by implication Carboniferous. This supported Austen's own, less-expert conclusion, and he was impressed. "I am not, as you know, one of those who would settle any such question by reference to organic remains," he told De la Beche, "but when the rocks contain not one single species in common, I think we should attend to it."[37] Logically speaking, on finding such an unexpectedly *total* contrast to the Silurian fauna, at least on the specific level, Lonsdale could have concluded that Austen's fossils were either younger or older than the Silurian. Since all the south Devon limestones had hitherto been attributed by the field geologists to the Cambrian, he might have been expected to conclude that here at last was the distinctive Cambrian fauna that Sedgwick had sought in vain in Wales itself. Since in fact Lonsdale chose the other, less obvious alternative, he must have seen at least *some* affinity between the Devon fossils and those of the Mountain Limestone elsewhere. Certainly he was supported in this judgment by Sowerby, as Austen recalled later: "Lonsdale says that the district from which these fossils come must be younger than the Silurian," he told Sedgwick; "Sowerby says it must be Carboniferous [Limestone]."[38]

That wording did, however, hint at some disagreement between the two specialists about just where in the Carboniferous *group* the Devon fossils belonged. Still longer after the event, Austen amplified his recollection of their discussion and made the basis of the disagreement explicit. "Sowerby examined 43 species of shells," he told Sedgwick, "all of which he pronounced to be well known Carb. L[imestone] shells." But Austen added that "there

36. On the Sowerby family, see Cleevely (1974). Lonsdale's duties at the Geological Society had just been curtailed to those of librarian, to try to save him from ill health caused by overwork; although no longer curator as well, he was continuing his research in the Society's museum.

37. Austen to De la Beche, 15 December 1837.

38. Austen to Sedgwick, 18 January 1838 (ULC: AS, IB, 133). The inference that "Carboniferous" was here an abbreviation for Carboniferous or Mountain *Limestone* is supported by Austen's further reports of Sowerby's opinion (see below).

were also a great many new [species]" as well, which, by implication, were not recognized by the two specialists as having been found *either* in the Mountain Limestone *or* in the Silurian.[39] So the south Devon fauna was not after all a perfect match for the Mountain Limestone, as Austen had anticipated, but only an approximation. Austen now recalled, however, that "Lonsdale did not agree with Sowerby, as he said the corals were not Carboniferous [Limestone].[40] Like the new or undescribed species of "shells," this could have been regarded as a natural result of examining fossils from a well-known formation, collected from a region hitherto not fully studied. But Lonsdale interpreted the discrepancy differently, as Austen recalled later: "looking at the shells," Lonsdale said that in his opinion "they could not be of a much greater age than the Carb. Limestone."[41]

Lonsdale could not have meant by that phrase that the Devon fauna was in any way *intermediate* between those of the Mountain Limestone and the Silurian. For as Austen recorded, the two fossil specialists, like Murchison, "cd. not discover a single Silurian fossil" in his collection. It is far more likely that Lonsdale meant simply that Austen's fauna must belong *somewhere* within the Carboniferous group, but that its difference from the Mountain Limestone fauna was sufficient to conclude that it was not the *exact* equivalent of that formation. It could hardly be still younger, since Austen had reported that the south Devon limestones were overlain unconformably by the Culm with its Coal Measures flora; so the only alternative was that it was slightly older. The only major formation known elsewhere below the Mountain Limestone but above the Silurian was, of course, the Old Red Sandstone. But it is extremely unlikely that that formation was in Lonsdale's mind. The few and peculiar fossils known from the Old Red, and its distinctive rock type, made it a highly implausible candidate for any such correlation, as Murchison had discovered when he tried out a similar conjecture on Sedgwick (§8.2). If Lonsdale had mentioned the Old Red Sandstone at all in the discussion, the suggestion would have been so striking that Austen or others would surely have reported it. It is far more likely that what was in Lonsdale's mind, and what his hearers took him to mean, was that Austen's Devon limestone might be equivalent to the "transition group" that Phillips had described as forming locally a transition between the Mountain Limestone itself and the Old Red Sandstone beneath (§6.6). That intermediate formation, and it alone, would have seemed a plausible candidate for correlation with Austen's Great Limestone of southeast Devon, if Lonsdale were right in considering its fauna "not of a much greater age" than the Mountain Limestone.

In any case, Austen seems to have accepted some such slight modification of his own original conclusion: as he told Sedgwick after the event, "I thought this Devon Limestone might be about the age of the Carb. Limestone"

---

39. Austen to Sedgwick, 6 March 1838 (ULC: AS, IB, 145). The increased number of species mentioned may reflect further identifications by Sowerby *after* the meeting.

40. Ibid. By "Carboniferous" Austen must again have meant the Carboniferous or Mountain *Limestone:* the abbreviation makes good sense in context, whereas to say that the corals did not belong to *any* part of the Carboniferous *group* would have been nonsense in the light of Lonsdale's other comments. Austen's specimens were not the first Devon corals Lonsdale had seen: Lonsdale to De la Beche, 22 December 1835 (NMW: DLB).

41. Austen to Sedgwick, 6 March 1838.

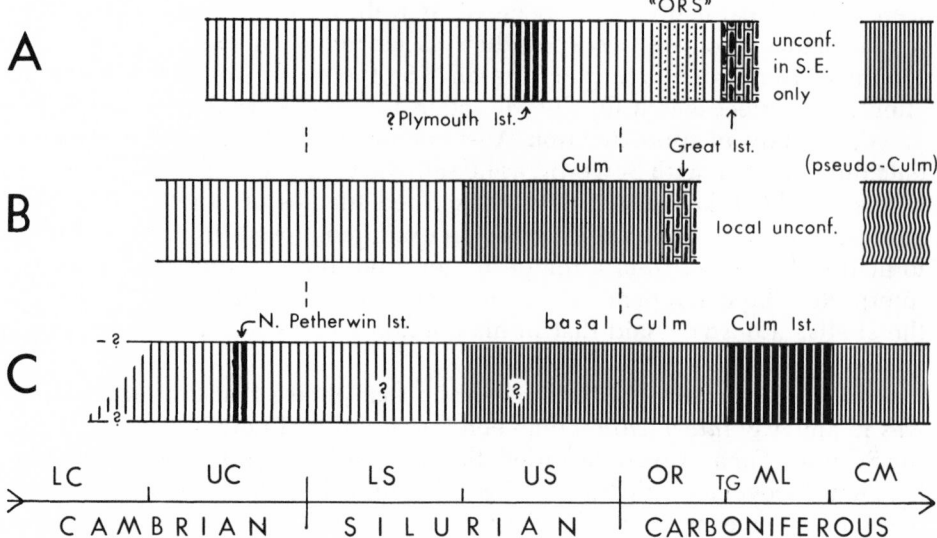

Fig. 9.2.   Diagrammatic columnar sections to illustrate three correlations of the older strata of Devon and Cornwall in early 1838. (A) Austen's correlation for southeast Devon, equating his Great Limestone with the Mountain Limestone (ML) or at least with the "transition group" (TG) at its base, the underlying reddish strata with the Old Red Sandstone (OR), and the still older formations with—probably—the Silurian and Cambrian. (B) De la Beche's correlation for southeast Devon, devised in response to the expert examination of the fossils from Austen's Great Limestone, equating it tentatively with the Old Red but putting the "true" Culm *below* it, in the Silurian, on structural grounds; this interpretation was incorporated obscurely in his official *Report* on southwest England (1839). (C) Sedgwick's correlation for Devon and Cornwall generally, as embodied obscurely in a paper read in May 1838; this modified his earlier scheme (fig. 8.1, B) by extending the lowest Culm strata down into the Upper Silurian (US), thereby eliminating any unconformity; there was no mention of the south Devon limestones, but Sedgwick retained one Cornish limestone for his Upper Cambrian (UC).

(fig. 9.2, A).[42] Even that modified conclusion, backed as it now was by the expert opinions of both Lonsdale and Sowerby, was quite enough of a palaeontological cat for Austen to have set among the field geologists' stratigraphical pigeons.

9.4   Weaver Reenters the Lists (January 1838)

Before the field geologists had time to assimilate the implications of Austen's fossils, the earlier issue of the Culm's age returned to the arena of the Geological Society with the second of the three papers prepared by provincial geologists. At the first meeting of 1838, only two weeks after the discussion of Austen's paper, Weaver reentered "the lists," as Williams had put it, "hoping to retrieve his disaster" over Ireland with his new paper on north Devon.[43] His earlier work on Ireland had just appeared at last in the new issue of the

42. Ibid.
43. Williams to De la Beche, 21 November 1837.

Society's *Transactions,* after an unprecedented interval of more than six years since its original reading, and with all reference to his original views on the Greywacke coal carefully expunged.[44] Having "recanted" on the Irish case, Weaver had now become a zealous critic of De la Beche's analogous "heresy" in Devonshire, and of his provincial supporter Williams (§9.2).

Weaver urged Phillips to come to the meeting and help "put the coal pie in its place." Endeavoring to win Phillips onto his side, he assured him that Williams would have no cogent counterargument to put forward, since Weaver himself had proved that the "supposed graduation" between the Culm and the older strata in north Devon was "quite fallacious." He added that Austen's "right clever paper" read at the previous meeting had also supported that opinion, not knowing that in the subsequent discussion Austen had maintained that there was *no* unconformity at all in north Devon, although there was in the south. Weaver's claim that Sedgwick and Murchison were now "doubly supported on this question," by himself and Austen, was therefore itself fallacious. But Weaver cannot really have believed that the matter could be "considered as at rest," for he anticipated a "melée" at the forthcoming meeting.[45]

Weaver's paper duly reported that, in his opinion, Sedgwick and Murchison had been essentially correct in their interpretation of north Devon. He defined eight successive formations, similar to theirs (and incidentally to Williams's). The lower group of six formed a continuous sequence; the upper group of two formations—the Culm strata—"though apparently in some places in a parallel (conformable) position" on the lower, were found "when thoroughly examined" to lie unconformably on them. This formulation conceded the appearance of conformity, as stressed by Williams (and De la Beche), while reducing it to the status of a deception based on insufficiently careful fieldwork. Weaver claimed that "this unconformity denotes two different aeras of deposition," which were reflected in the difference in the fossils. He denied that the plants found (by Harding and Williams) in the lower strata undermined this inference. He concluded that the Culm represented "the true coal measures of England," but he evaded the difficulties of correlating the older strata with any formations elsewhere by saying that they were a *"peculiar transition group"*—peculiar, that is, to southwest England.[46]

The only record of the subsequent discussion is far from impartial. Sedgwick, who as usual could not be present, had urged Murchison beforehand to "fight the good fight" for their joint interpretation of the Culm. The day after the meeting, Murchison wrote to tell Phillips—who likewise had not been there—how it had gone.

> The paper by old Weaver was read last night, and the fight is over. He has sided completely with S[edgwick] and [my]self. Austen, a remark-

44. Weaver (1838a); publication of this issue of *TGS* was announced by the Council of GSL on 1 January. In a lengthy footnote dated May 1837 (p. 75), Weaver asserted that *all* the American coalfields were Carboniferous in age, thereby withdrawing his support from a major group of foreign analogues that De la Beche and Greenough had relied on hitherto.

45. Weaver to Phillips, 29 December 1837 (UMO: JP, 1837/67).

46. Weaver (1838b), read at GSL on 3 January (GSL: BP, OM 1/8, pp. 489–502). An anonymous referee (probably Fitton) later recommended the paper for full publication in *TGS*, but urged Weaver to include comparisons with Dumont's work in Belgium: anon. to GSL, 10 March 1838 (GSL: BP, COM P4/2, p. 246).

ably clever young geologist, is also with us; Major Harding from the first with us. The case therefore stands thus: for the old constitution— Greenough, De la Beche and Parson Williams. On our side are the two geologists of Great Britain who have given the longest attention to the old fossiliferous strata, and their opinions are supported by every man who has gone into the tract to judge for himself.[47]

Murchison's political imagery was clearly intended to press Phillips into public adherence to what he portrayed as the victorious party of Reform. But he ignored three crucial facts: that De la Beche and Williams had spent much longer in "the tract" than Murchison and Sedgwick themselves; that he himself had been the one who had originally pronounced an opinion on the matter without going to Devon "to judge for himself"; and that he had been bitterly criticized by De la Beche on just those grounds (§5.3). Undeterred by such niceties, however, Murchison also sought to undermine the plausibility of "the old constitution" by claiming that the French analogues to the Culm case, which De la Beche had so often stressed, were now also suspect. "All the support expected from France has gone against the ancients," he told Phillips; "for Buckland (himself as unwilling a witness as Weaver) comes back from France persuaded that Élie de Beaumont's 'Greywacke coal-fields' are nothing but ordinary Carboniferous deposits reposing on Silurian rocks."[48]

Murchison continued this consolidation of his and Sedgwick's position by telling Lyell that he regarded the Coal Measures age of the Culm as "*COMPLETELY* settled," now that Weaver as well as Austen had agreed on that point. But by referring to them as "*the only 2 geologists* who have been in Devon since our hit (and a great hit it was)," he tacitly spurned the value of the fieldwork that De la Beche and Williams had done since the Bristol meeting; and by concluding that "I think no reasonable man need now contest the point," he strikingly excluded *all* their opponents from the company of reasonable men![49] But the rhetoric was, as usual, calculated. Lyell was busy preparing a book on the *Elements of Geology*, which was to present some of the material from his massive three-volume *Principles* in a single small volume. Murchison wanted to ensure that his own views would be forcefully upheld by Lyell in what could well prove a popular and influential book.

The main point he wanted to emphasize, in response to a query of Lyell's, was his own and Sedgwick's earlier discovery that the Culm was intruded by the Dartmoor granite, which was therefore still younger than what they believed to be "*true coal measures*" (§7.2, §7.4). But he also stated, almost casually and in passing, that "there is little or no true Silurian in Devon," adding in explanation that "the coal reposes at once on Cambrian."[50] The statement signaled the final demise of the confident claim he had made

47. Murchison to Phillips, 4 January 1838 (extract in Geikie 1875, 1:252); Sedgwick to Murchison, 28 December 1837 (GSL: RIM, S11/140). It is not clear how many of those mentioned had actually been present; Austen and De la Beche were certainly not.

48. Murchison to Phillips, 4 January 1838. Buckland may have mentioned this in the discussion, on the basis of his fieldwork in Normandy and Brittany the previous autumn around the time of the SGF field meeting at Alençon (§9.1).

49. Murchison to Lyell, n.d., no pmk [? late January 1838] (APS: CL); marked by Lyell "1838" and almost certainly a response to Lyell's note dated 17 January 1838, reminding himself to ask Murchison about the age of the Dartmoor granite (KHK: CL, notebook 70, pp. 1–7).

50. Ibid.

at Bristol in 1836, that he had found Lower Silurian strata and fossils in north Devon (§7.3). That claim had been greatly muted in his joint paper with Sedgwick in 1837, partly because of the unnerving effect of Harding's discovery of plants in those strata, partly because Sedgwick could with equal justice claim the animal fossils for his Upper Cambrian (§8.7). Now Murchison had virtually abandoned the Silurian in Devonshire. He might try by confident assertion to persuade Phillips and Lyell—and doubtless others too—that "the fight is over," but a growing consensus on the Coal Measures age of the Culm could not conceal a growing unease about the age of the still older strata.

9.5   The Judgment of a Universal Scientist (February 1838)

At the anniversary meeting of the Geological Society a few weeks later, Whewell gave his first presidential address in the presence of what Lyell recorded as "a grand display of talent."[51] Twelve months earlier, before Whewell took office, it had been clear that he would tend to side with Murchison (§8.5). But in the event he gave a fair and perceptive review of the Devon controversy. He applied to his address the analysis of geology that he had put forward in his recently published *History of the Inductive Sciences*. The kind of research championed by Lyell, and followed by his young protégé Darwin, was distinguished by the term "geological dynamics," while the analysis of the sequence of the strata was termed "descriptive geology."[52] Acknowledging that the latter constituted the mainstream business of the science, Whewell dealt with it first in his address and divided it in the customary manner into *"Home"* and *"Foreign"* geology. But he enlarged the first category to include the whole of northwest Europe, where the strata were well enough known for relatively confident comparisons with Britain; so defined, it included most of the research presented at the Geological Society and the Société géologique in the year under review. Yet of the space devoted to it, Whewell used no less than half to discuss the geology of Devonshire. Such was the theoretical importance that the Society's "Universal" man of science, as Sedgwick had dubbed him, attributed to the continuing controversy.

Like Lyell before him (§8.4), Whewell had evidently asked other members of the inner circle of the Society for their suggestions. When he mentioned that Austen and Weaver now agreed with Sedgwick and Murchison about the Coal Measures age of the Culm, Whewell's words echoed unmistakably those of Murchison himself. "Resting on the concurrence of so many able observers," he commented, "I should conceive that we may look upon this view as *established*, so far as the time which has elapsed allows us to use the term." Whewell maintained that the new view was "entirely different" from its pre-

---

51. Whewell (1838), read at GSL on 16 February; Lyell to Horner, 24 February 1838 (K. Lyell 1881, 2:37–39). Although De la Beche was present, Lyell did not include his name in the list of "talent"!

52. Whewell (1837, 3: book 18). Darwin had presented a striking series of papers (derived from his work during the *Beagle* voyage) at GSL meetings in 1837, and his most openly theoretical paper was due to be read three weeks after Whewell's address: see Rudwick (1982b). At the close of this anniversary meeting Darwin took up office as one of the secretaries, under Whewell's presidency, filling a position indirectly occasioned by Turner's untimely death the previous year.

decessor, and that as a result "one-third of our geological map of England will require to be touched with a fresh pencil." Certainly an exaggeration, this was also no abstract or innocent remark. Whewell was tacitly using the authority of the occasion to put pressure on Greenough and De la Beche to incorporate Sedgwick and Murchison's interpretation of the Culm into the revision of their respective maps. By praising "the vast knowledge and great sagacity" of the (unnamed!) author of "the geological map of England," Whewell hoped to make the change palatable to Greenough without loss of face; "such modifications we must ever expect to have to make of a first approximation." But he warned that to refuse to incorporate the new interpretation would be "to elude this necessity" of modifying older views; it would involve "giving up the key of all our geological knowledge of our country." Here again Whewell showed himself to have been persuaded, or at least coached, by Murchison; for he gave as the "key" nothing less than "the doctrine that there is a fixed order of strata, characterised mainly by their organic fossils." More specifically, he argued, "when certain strata of Devon have thus been identified with the coal measures of other regions, can we still term them grauwacke?"[53] Whewell's listeners would have been in no doubt about the target of that remark.

Yet Whewell did not let Murchison have it all his own way. He admitted that the fossil plants found by Harding and Williams in the older strata of north Devon posed an "apparent anomaly." Unlike Murchison, Whewell did not regard the existence of plants in supposedly Cambrian and Lower Silurian strata as an anomaly in itself. What did seem anomalous to him, and a problem for the simple identification of the Devon Culm with the Coal Measures elsewhere, was that "the same fossil plants which occur in the culm measures are said to have been detected in the subjacent strata." If they were indeed the *same* species, it would imply either that the older strata were *also* equivalent to the Coal Measures, or that these plant species were not really characteristic of the Coal Measures after all. But this anomaly and its significance were left as "questions hereafter to be decided"; the matter was simply shelved. Whewell's closing reference to "the manly vigour of discussion" at the Society, "tempered always by mutual respect and by good manners," must have been as much a prescriptive hope for the coming year as a description of what he had actually observed in his first year of office![54]

The following day Greenough wrote to answer Whewell's criticism, which he was quite certain had been directed at himself: "you gave me I am sure more praise and I think also more censure than I deserve." He warned Whewell to "look a little closer into the evidence on both sides before going to press." He conceded that Sedgwick and Murchison's view might eventually prove correct but warned that it should not be "hastily adopted." Greenough gave two reasons for caution. First, the new view tended "to change the whole system of continental geology": internationally minded as ever, Greenough still regarded the many parallel cases in France as powerful and authoritative support for his own and De la Beche's belief in the Greywacke age of all the strata in Devonshire, including the Culm. Second, he asserted that the new interpretation was having an undesirable practical effect: it was "at this mo-

53. Whewell (1838, p. 635–36).
54. Ibid., pp. 634–35, 648.

ment leading landed proprietors to [make] expensive borings for Coal in places where there is no reasonable prospect of its existence." Of course the Culm did contain a few poor seams of anthracite; but landowners, hearing that geologists regarded the strata as Coal Measures, could derive the quite mistaken impression that there might be many more and better seams still undiscovered beneath their fields. "In all probability," Greenough declared, "the new edition of my map will in respect of the disputed ground differ very little from the old"; he had no intention of coloring the Culm as Coal Measures rather than as Greywacke. He also gave it as his belief that "neither Murchison nor Sedgwick consider the matter by any means so clear now as they did in the first instance and as you represent it."[55] But in fact, although they were indeed highly uncertain about the dating of the *older* strata of Devon, they were as firm as ever about the Culm.

Greenough told Whewell he had wrongly listed Williams among those who had adopted the new interpretation; Williams himself inferred that Whewell had confused him with Weaver! Williams assured De la Beche after the meeting that "the more I see and learn the more convinced I am of the truth of your original views as to the relative age of the Devon culm." He pointed out that Weaver in his recent paper had "very warily" omitted to mention the "important passage beds" in north Devon, which Williams (and De la Beche) regarded as decisive evidence for a conformable transition from the older strata into the Culm. Not content with reaffirming the reality of that crucial conformity in the north, however, Williams also dismissed Austen's alleged unconformities in the south as "all moonshine," and maintained that the north constituted the "alphabet" that had to be mastered by Austen or anyone else before they would be fit to tackle the south of the county. On this crucial and controversial issue, Williams said he had found Murchison "as slippery as a Gladiator." The provincial geologist remained convinced, however, that he was on the winning side: he told De la Beche that, rather than Murchison being the ultimate victor, "I confidently expect [that] himself & Professor Sedgwick and Mr. Lyell & Mr. Whewell and everyone else worth having will shortly with you maintain the Grauwacke age of the Culm of Devon."[56] The "ancients," as Murchison had scornfully called them, were a party with life in them yet.

9.6 Expanding the Carboniferous in Devon (January to May 1838)

One controversial topic that Whewell omitted altogether from his address was the age of the south Devon limestones and their fossils. Although that issue might affect the coloring and whole appearance of geological maps just as much as the dating of the Culm, the omission was prudent, because even

55. Greenough to Whewell, 17 February 1838 (TCC: WW, Add. Ms. a. 203[41]). Whewell did later ask Murchison if it were true that their "Devonshire doctrines" were now less confidently held than he had stated in the address; the printed version may incorporate revisions made in the light of this correspondence: Whewell to Murchison, 4 March 1838 (GSL: RIM, W4/20).

56. Williams to De la Beche, 20 February 1838 (NMW: DLB).

Sedgwick and Murchison could not present a united front on the matter, let alone other members of the Society.

In view of the importance of the issue, Murchison had urged Austen, when his paper was read, to lose no time in getting Phillips's expert opinion as well as that of Lonsdale and Sowerby. Austen took that advice, setting aside whatever proprietorial feelings he may have had about his collection; he arranged for Phillips to be sent on loan the specimens that had been displayed in London and suggested that they should also exchange other specimens so as to enrich both their collections. So Phillips was able after all to study the best amateur collection of south Devon fossils. Austen was skeptical about the contrast that Lonsdale had detected between his Devon "shells" and those from the Mountain Limestone: he told Phillips he thought the differences were well within the range of variation of living molluscan species.[57] To Sedgwick he added that "if among living shells, species were to be established in this way, the same species might be divided ad infinitum."[58] By criticizing Lonsdale as too great a splitter in fossil taxonomy, Austen thus brought into the open the practical issue that underlay any use of "statistical" estimates of the affinities of fossil faunas. For as Lyell had already discovered in his analysis of Tertiary faunas, such estimates were of little use unless realistic allowances were made for the range of variation within any biological species.[59]

Nonetheless, it is clear that Sowerby, whose proclivities as a taxonomic lumper Austen regarded as more realistic, was in agreement with Lonsdale about the striking contrast between the Silurian fauna and that of the overlying Carboniferous formations; after all, those two fossil specialists were jointly responsible (with Phillips) for the identifications that had established the contrast in the first place. "In Phillips's Monograph and from his Documents we know of about *500 species Carboniferous*," Murchison told Lyell, not long after the public discussion of Austen's fossils; "from Murchison's Silurian upwards of *300*, the 19th/20 of which are *new*. Now, in these 800 species not 3—certainly not 6 are *common*!!!"[60] This remarkable contrast underlined the distinctive faunal character of the Silurian strata, and hence their potential validity as a widely applicable system. There was clearly no question of any *transition* between the Silurian fauna and that of the Carboniferous group directly overlying it.

That contrast, however, only highlighted the enigma that Austen's Devon fossils now posed. It was not that the Devon fauna seemed in any way transitional either; Murchison agreed with the fossil specialists that it was at least *similar* to the Mountain Limestone fauna, although not identical. But if some such age were assigned on fossil grounds to Austen's Great Limestone of southeast Devon, what was to become of the rest of the pre-Culm strata?

57. Austen to Phillips, 18 and 24 January and 25 March 1838 (UMO: JP, 1838/1, 2, 9). His proposal for an exchange of specimens revealed incidentally that his own collection contained as yet *no* Mountain Limestone fossils, indicating that he had indeed relied on published descriptions of them.

58. Austen to Sedgwick, 6 March 1838 (ULC: AS, IB, 145).

59. Rudwick (1978). At just this same period (from July 1837 into 1839), Darwin, in his *Transmutation Notebooks*, was privately wrestling with the radical implications of this problem of taxonomic practice: Kohn (1980). Darwin's note that "Lonsdale is ready to admit permanent small alterations in wild animals" is compatible with Austen's opinion about Lonsdale's taxonomic practice (Darwin, notebook C, p. 176, printed in De Beer 1960–61, p. 103).

60. Murchison to Lyell, n.d. [? January 1838: see n. 49].

Austen had "provisionally" assigned his fossils to "about" the age of the Mountain Limestone (§9.3). Tacitly, he had allowed that the lower parts of his sequence (including the lower limestones of Ashburton and Chudleigh), and most of the Greywacke rocks in the rest of south Devon and Cornwall (including the Plymouth limestone), might be far older. But this distinction, as Sedgwick pointed out to Murchison, ran directly counter to Sedgwick's conclusion that the Great Limestone of southeast Devon was *equivalent* to the Plymouth limestone further west. Had Sedgwick been at the meeting, Murchison commented to Austen, the Devon geologist would have had "the satisfaction of seeing the fair fabric of your order of superposition, battered to ruins."[61]

No mere local detail was at stake here, but rather the interpretation of all the older strata of Devon and Cornwall, and indeed of the Greywacke as a whole. From the start of his fieldwork in Devon, Sedgwick (like Murchison) had regarded the Plymouth limestone as the best fossil-bearing representative of those older strata, lying well down within the sequence of formations older than the Culm (§7.2). They had equated it confidently with Austen's similar Great Limestone further east. The fossils from all those limestones had seemed by far the best fauna by which to characterize what they had later termed "a magnificent development of the Upper Cambrian" (§8.7). Now Austen had proposed removing some of the richest sources of this fauna, with one sweeping stroke, from the Upper Cambrian to somewhere near the Mountain Limestone in the heart of the Carboniferous. That proposal had been so deeply unacceptable to Sedgwick that he had at first ignored it (§9.2). But now it was being advocated not merely by a local amateur but by at least two of the three most competent fossil specialists, and even by his own collaborator, Murchison. It could no longer be ignored or dismissed.

When the even tenor of life at his country house was restored after the Christmas and New Year festivities, Austen wrote to Sedgwick to defend his conclusions, and in particular his claim that the Great Limestone in his home area was much higher in the sequence—and hence much younger—than the Plymouth limestone further west. He discounted the resemblance in their faunas on the grounds that it was mainly confined to the fossil corals, which he regarded as of little value for correlation. "They will be found to have a great range," he claimed, alluding by way of analogy to the species of corals that straddled the otherwise abrupt faunal boundary between Secondary and Tertiary. Instead, he emphasized to Sedgwick, as he had at the meeting, the "large proportion of Mountain Limestone fossils which the Newton [Abbot] quarries contain," and in particular the "abundance of Cephalopoda, a class of animals wanting at Plymouth."[62] To this fossil evidence Austen added a different argument, more likely to appeal to Sedgwick because it was based on field evidence. For direct correlation across country, he pointed out, "we have no guide but mineralogical character"; but that criterion, "if such is worth anything," strongly supported his interpretation of the sequence, because a distinctive formation of reddish strata lay *above* the Plymouth limestone but *below* the Great Limestone in his home area. The implication was clear: Sedgwick should allow that the Great Limestone was about the same

61. Quoted in Austen to Sedgwick, 18 January 1838.
62. Ibid.

age as the Mountain Limestone, while he could keep the Plymouth limestone (and Austen's lower limestones) as Upper Cambrian if he so wished.

Austen failed to point out, however, that his correlation of even the uppermost formation with the Mountain Limestone put at risk Sedgwick's correlation of all the rest. For in the apparently unbroken sequence of formations in south Devon, if the upper part were Carboniferous it was difficult to see how the lower could be Cambrian, unless somehow the whole of the Silurian were packed in, as it were, between. Undeterred by that problem, Austen extended his provisional correlation with growing confidence during the spring of 1838. "My notion," he told Sedgwick in April, "is that our limestones, and the red shales and sandstones which dip under them, are about the age of the mountain limestone and old red sandstone."[63] He thus expanded his scheme to include the reddish strata that, in his view, separated his Great Limestone from the older limestones of south Devon (notably the Plymouth limestone). He extended downward what he regarded as the Devon representatives of the Carboniferous group, including not only the Culm as Coal Measures and the Great Limestone as (approximately) Mountain Limestone, but also the "red shales and sandstones" as *the Old Red Sandstone itself* (fig. 9.2, A).

That this was no casual or passing fancy is clear from a letter Austen wrote a little later, in which he said he hoped De la Beche too would "admit" that expanded correlation: "I find my Old Red Sandstone containing every where, abundance of fossils," he added.[64] Undeterred by the knowledge that that formation elsewhere contained few fossils, and most of those peculiar, Austen probably extended his provisional correlation still further down the sequence, to equate the Plymouth limestone or some such strata with the Silurian, and still lower strata with the Cambrian (fig. 9.2, A). For a few weeks later, referring back to just this time, he told De la Beche he "still" believed that in Devon and Cornwall "the whole series of rocks" was present "up to the Mountain Limestone period inclusive (or nearly so)."[65] However unpalatable and implausible it might seem to both Sedgwick and De la Beche, Austen had at least developed an interpretation that did justice both to the unbroken sequence in south Devon and to the experts' opinion of the fossils he had found there.

9.7   Fighting Back for the Greywacke (March and April 1838)

Austen's claim that some of the south Devon limestones were at least approximately of Mountain Limestone age had dismayed all the main protagonists in the controversy, although for different reasons. For it was no longer merely the questionable guess of a local amateur. Austen's identifications had been broadly confirmed by Sowerby and Lonsdale, however much those specialists disagreed in detail; and rumors must soon have spread that Phillips, as he received more and more collections from Devon, was also broadly in

63. Austen to Sedgwick, 21 April 1838 (ULC: AS, IB, 150).
64. Austen to De la Beche, n.d., pmk [?2] May 1838 (NMW: DLB).
65. Austen to De la Beche, 29 July 1838 (NMW: DLB).

agreement about the fossils.[66] Murchison was the least alarmed by this development, since Austen's correlation strengthened his case for equating the Culm with the Coal Measures, dissolved the anomaly of the plants that were turning up in the older strata, and did not seriously threaten the distinctive character of his Silurian fauna. Having virtually abandoned any claim to Silurian strata in Devonshire, Murchison could concentrate on getting his great book published as soon as possible, so that his central claim to have delineated a major epoch in the history of the earth and of life could be fully on public record.[67]

Sedgwick was already due to present the Geological Society with his long-awaited summary and synthesis of all the older strata of England and Wales, the first fruits of his intended collaboration with Conybeare (§4.2). This would require him to make some kind of firm decision about the strata in Devonshire. Like his and Murchison's joint paper of the previous year, Sedgwick's new paper was long and important enough to be split into two parts. In the first part, presented toward the end of March, Sedgwick must have included at least his introduction in which he set out his own "principles" for the correlation of strata. These were, not surprisingly, that neither "mineral structure" nor fossils should ever "upset conclusions drawn from the clear and unambiguous evidence of sections." Sedgwick cited the controversy about the Culm as an example of how those principles had recently been "violated." He criticized the attribution of the Culm to the Greywacke on the grounds of its rock types, but also its attribution to the Carboniferous on the grounds of its fossils; decoding the allusions, he distanced himself as much from Murchison as from De la Beche. He claimed that "true geological reasoning," by contrast, required that "anterior to either of the preceding conclusions, the true position of the culm should be determined by actual sections."[68] This was unexceptionable; but of course the controversy had arisen because the relevant sections were *not* clear and unambiguous, and because, even if they had been, they could not have settled the question of the *age* of the strata. The second part of Sedgwick's paper, which would have contained his interpretation of Devonshire, failed to appear at the next meeting, or indeed at the two that followed. The reasons for this unprecedented delay are obscure, but developments behind the scenes seem to have made Sedgwick more and more uncertain of what to say about Devon geology.

Only about a week after the reading of the first part of the paper, De la Beche, who had stayed on in London, told Sedgwick that Lindley had changed his mind about the fossil plants that had precipitated the whole controversy more than three years earlier (§5.1). Having had more extensive collections to study—including doubtless the specimens De la Beche had hoped to keep safe in the British Museum, away from Murchison's prying

66. Austen to Sedgwick, 21 April 1838.

67. The book was now virtually complete and set up in print, apart from the fossil specialists' descriptions of the fauna, and the engraving of the illustrations: Thackray (1978b).

68. Sedgwick (1838), read at GSL on 21 March and 23 May. As usual, the official record stated only that the paper "was commenced" at the first meeting (*PGS* 2:662); since it shared that meeting with a substantial paper by Richard Owen, it is unlikely that much was read. The other examples of stratigraphical practice that Sedgwick criticized were the proposed dating of the (Oolitic) Stonesfield Slate as Tertiary because it contained a few mammalian fossils, and Deshayes's (and Lyell's!) method of dating Tertiary formations exclusively on the basis of "statistical" or percentage analyses of the fossils (Rudwick 1978).

eyes (§8.6)—Lindley now considered that "a few species only" from the Culm were the same as those in the ordinary Coal Measures, while "the bulk" of the specimens belonged to species he could not match from elsewhere. The Culm fossils, as De la Beche pointed out, had thus been "somewhat prematurely dubbed coal measure plants."[69] The implication, which Sedgwick would have seen at once, was that it had been equally premature to regard the Culm as Carboniferous at all. In other words, De la Beche was claiming new and expert support for the contention he had fought for all along, namely, that the Culm was an integral part of the Greywacke sequence and certainly pre-Carboniferous in age.

Having launched that counterattack, De la Beche took the mail coach to Devon for a few days' fieldwork, which he hoped would consolidate it. Austen, who joined him near Ashburton, told Sedgwick later that De la Beche went there "to verify an impression that the Culm strata were older than all S. Devon, and younger than all the fossiliferous rocks of N. Devon." That startling new notion, literally overturning all previous interpretations of the field evidence, was clearly intended by De la Beche to reinforce his assignment of the Culm to the Greywacke, while accommodating the new fossil evidence that some of the south Devon limestones might be roughly equivalent to the Mountain Limestone. He had indeed first put this conjecture forward to Phillips soon after Austen's fossils were first discussed, explicitly as a way of explaining them. Austen, with his unrivaled local knowledge, considered that the course of the junction between the Culm and the other rocks was quite distinct in his area, "but the question of superposition not equally so." He was critical of De la Beche for relying on the evidence of dip and strike in isolated exposures of the respective strata; by contrast, Austen considered that "one good instance of super position is sufficient to settle the question, and there are several." Having shown De la Beche these decisive exposures, he thought the older geologist had left the area "much inclined" to abandon the notion with which he had arrived.[70] De la Beche, he believed, had conceded that the Culm, whatever its true age, was in any case clearly overlying—and indeed unconformable to—all the older strata in the area, including the limestone whose fossils had generated these maneuvers.

In fact, however, De la Beche had not abandoned his new and revolutionary notion at all. As soon as he got back to his survey headquarters in Swansea, he told Phillips that he considered that the "so called culmiferous rocks" *underlay* Austen's limestones, and a few days later he gave Sedgwick the same news. He admitted he might have made "one vast mistake" about this highly important point; but he thought the evidence sufficiently compelling—"supposing . . . that my eyes have not been playing humbug with me"—to urge Sedgwick to go and look at it for himself before having the summary of his paper printed. "So ho! my fine fellow," he imagined Sedgwick saying in response: what now was to become of the alleged Culm which—as Austen had shown him—manifestly *overlay* the limestones? De la Beche slipped out of what he admitted was "an awkwardness," by claiming that these rocks were quite distinct from the genuine Culm: "two things have been

69. De la Beche to Sedgwick, 26 March 1838 (ULC: AS, IC, 5).
70. De la Beche to Phillips, 3 January 1838 (UMO: JP, uncat.); Austen to Sedgwick, 21 April 1838.

confounded with each other." He conceded that these patches of younger rock, unconformably overlying the limestones that contained Austen's apparently Mountain Limestone fossils, might in fact be true Coal Measures; but he distinguished them sharply from the main body of the Culm, which he insisted was much older than the limestone (fig. 9.2, B).[71]

"I have some nice bones for the fossil-mania men to pick," De la Beche told Sedgwick as an afterthought; "something very like physical structure *versus* sweeping fossil conclusions." Whether calculated or fortuitous, the remark certainly echoed the "principles" Sedgwick had laid down in the introduction to his paper three weeks earlier; it certainly allowed De la Beche to align himself with Sedgwick, and may have been intended once again to prize that powerful figure away from his collaborator, the fossil-mad Murchison. Likewise, he assured Phillips that in his forthcoming official memoir he would avoid the controversial question of the *age* of the Culm strata: "I shall tell the story without bias, and if folk should afterwards choose to call them cream cheese, they are heartily welcome as far as I am concerned."[72]

Neither Sedgwick nor Phillips was impressed by what must have seemed suspiciously like an ad hoc elaboration in the interpretation of south Devon. Phillips thought De la Beche might have to *"bore the earth"* to get conclusive evidence for his startling new notion about the Culm. But he was more concerned to defend his own expertise in the matter: "I will wager the fossils against all the sections, provided *enough of them* can be found," he told De la Beche, while agreeing that it would be "folly" to trust only a handful of "these silent witnesses." Sedgwick, he reported, wanted to avoid having to state openly how radically he differed from De la Beche, but Phillips told his collaborator that he himself was now more firmly than ever on Sedgwick's side: from all the fossils he had so far seen, "the culmiferous beds would appear to belong to the Carboniferous *System*."[73] That final emphasis left as an open question the *part* of the Carboniferous to which the Culm was to be assigned, but on the main issue Phillips was clearly still on the other side from De la Beche, as he had been ever since the Bristol meeting. (§7.3).

Sedgwick, characteristically, was more forthright. "If you can prove the Culms to be below the Plymouth limestone," he told De la Beche, "I will for the rest of my life shut up shop." Surrounded by Trinity undergraduates taking a Latin examination, he was in no position to follow De la Beche's long-winded arguments in detail, but he assured him he would go back to Devon in the summer to look for himself; as for the vexed question of Austen's limestones, he added, "I fancy already that I see my way through the mist." De la Beche's report of Lindley's change of mind had not, he claimed, caused him to doubt his earlier opinion about the age of the Culm: he dismissed the relevant specimens as "far too imperfect to settle any great question on." But he admitted that he and Murchison were "at issue" on one crucial point: even though "all *the direct evidence*" had made them identify the upper part of the Culm as Coal Measures, he told De la Beche that he himself had "*never thought* the *base* of the *culm measures* on a level with the base of the Car-

---

71. De la Beche to Phillips, 31 March 1838 (UMO: JP, uncat.); De la Beche to Sedgwick, 11 April 1838 (ULC: AS, IC, 6); Austen to Lonsdale, n.d. [*c.* April 1838] (GSL: BP, LR 3/371).
72. De la Beche to Phillips, 31 March 1838.
73. Phillips to De la Beche, 4 April 1838 (NMW: DLB).

boniferous series."[74] Whereas De la Beche assigned the whole of the Culm to the Greywacke, Sedgwick, unlike Murchison, was prepared to concede that at least the lowest part—alluding probably to the strata below the Culm limestone—did indeed belong there (§8.2).

It remained to be seen how that concession could be reconciled with the claims that Austen and the fossil specialists were making about the Carboniferous age of some of the south Devon limestones, which—unless De la Beche's implausible new notion were adopted—certainly underlay the Culm. Despite the casual way he had written to De la Beche about the fossil plants, Sedgwick was sufficiently concerned to ask Murchison to check whether Lindley really had changed his mind. But he was quite unconvinced by De la Beche's latest notions: "it is plain he is in a dead puzzle," Sedgwick told his collaborator. He was confident that the fieldwork he was planning to do with Austen in the summer would successfully "untie all knots in that district, that disturb the great questions of arrangement."[75] But before that, he had to make a major public statement of his overall views on Devon geology. He used his examining duties as an excuse for postponing yet again the final reading of his paper at the Geological Society; among other papers read instead at that last meeting in April was one by Austen, interpreting his Devon limestones as ancient coral reefs.[76] After that, however, only three meetings were left before the summer recess. As in the previous year, Sedgwick was being pushed by the Society's calendar into a most reluctant performance.

9.8   The Disappointment of the Cambrian System (May to July 1838)

At the penultimate meeting of the season, after an unprecedented gap of two months and three meetings, the second part of Sedgwick's paper was read at last at the Geological Society. Sedgwick summarized the regional sequences of rocks older than the Old Red Sandstone, starting in Scotland and proceeding by way of northern England (Cumbria) and Wales ("Cambria") to southwest England. It is hardly surprising that he devoted as much space to Devon and Cornwall as to all the other regions put together. At the end of the paper he synthesized these various "Palaeozoic" sequences into three major "systems," in ascending order the "Lower Cambrian," "Upper Cambrian," and "Silurian."[77] Since formations from all the regions were assigned to these systems,

74. Sedgwick to De la Beche, 19 April 1838 (NMW: DLB).

75. Sedgwick to Murchison, 21 April 1838 (GSL: RIM, S11/144).

76. Austen (1838b), read at GSL on 25 April. He reported that De la Beche was convinced by the field evidence for this interpretation (Austen to Lonsdale [c. April 1838]: see n. 71). But Darwin, now in effect the GSL's expert on coral reefs, judged the paper too inconclusive to warrant full publication in TGS: Darwin to GSL, 7 September 1838 (GSL: BP, COM 4/2, p. 4).

77. Sedgwick (1838), read at GSL on 21 March and 23 May. Even the title changed between the two meetings. At the first, Sedgwick used the term "Primary" in a new and idiosyncratic sense to denote all the formations older than the Old Red Sandstone (PGS 2:662), but Murchison probably talked him out of this afterward. However, instead of adopting his collaborator's suggested "Protozoic" to denote the Cambrian and Silurian jointly, Sedgwick chose "Palaeozoic," returning "Primary" to its normal use for the oldest, nonfossiliferous rocks, and suggesting that they should become "Protozoic" if any fossils were subsequently found in them: Sedgwick (1838, pp. 684–85); Murchison to De la Beche, 6 April 1838 (NMW: DLB).

at least provisionally, it is clear that Sedgwick regarded them as the "successive natural groups" to which his title referred, which could in principle be recognized widely, at least throughout Britain. In his conclusion Sedgwick emphasized that his regional sequences were "founded on the details of actual sections," but that there was "not much that remains to be done" in that way.

For correlations *between* regions, therefore, it was "now necessary to appeal" to fossil evidence. On the Silurian, Sedgwick referred tactfully to Murchison's forthcoming book. For the two Cambrian systems, he left it an open question whether fossils would ever allow an equally detailed subdivision; but he stressed the importance of accurate identifications of the fossils from Devon, Cornwall, and Wales, as the "immediate *desiderata*" most likely to yield results. Even for Wales, after which the Cambrian was named, Sedgwick's results were disappointing to those who had hoped to hear something comparable to Murchison's Silurian conclusions. For Sedgwick could only state that the lower Welsh strata contained very few fossils, and that even in the younger formations many of the species were also known from the Lower Silurian. The "true distinctive zoological characters" of the Cambrian had therefore not yet been "well ascertained" in the Cambrian region itself.[78] Nor, when Sedgwick went on to deal with southwest England, was it clear that this great *desideratum* had been filled in any convincing way outside Wales.

Most of what Sedgwick had to say about Devon and Cornwall contained no surprises and simply reiterated his earlier public statements. He stressed the magnitude of the change that he and Murchison had introduced when they claimed that the Culm—whatever its age—was the youngest formation in Devon (§7.3); before that time, he emphasized, "the position of [the] whole series among the Devonian [i.e., Devonshire] groups had been misapprehended." On the age of the main part of the Culm, Sedgwick ignored whatever doubts Lindley had recently had about the fossil plants and insisted that its correlation with the Coal Measures should continue to stand "unless some conflicting evidence be discovered." He claimed that the age of the *base* of the whole Culm series had been "intentionally left in an ambiguous position" in his earlier paper, omitting, however, to mention that this had been in order to conceal a long-standing disagreement with his fellow-author Murchison (§8.2, §8.7). He now stated publicly that in *his* opinion the base of the Culm was "lower than the base of the ordinary English carboniferous series." This cryptic comment was at last clarified in his summary, for "the lowest part of the culmiferous series" of Devon was there listed, albeit with a question mark, at the end of his list of examples of the *Silurian* system. No fossil evidence was cited for this startling inference, but it can only imply that in Sedgwick's view the Culm extended through the *whole* of the Carboniferous group, from Coal Measures at the top, right through Mountain Limestone and Old Red Sandstone, and down into at least part of the Silurian (fig. 9.2, C).[79]

This clarification of what had evidently been Sedgwick's private opinion for about a year was tucked away so discreetly that those who heard the paper at the meeting, or read the summary afterward, may not even have noticed it. But it revealed a further reason why he had been so adamantly

78. Sedgwick (1838, pp. 679, 683–84); Secord (1986).
79. Sedgwick (1838, pp. 681, 685).

opposed to Murchison's and Austen's suggestions that the *older* (pre-Culm) strata included representatives of the Old Red Sandstone or even of the Mountain Limestone. For it was now clear that Sedgwick had his own ideas about where those formations might be found "with a new face" in Devonshire. By conjecturing that the Culm represented the *whole* of the Carboniferous group, and even some Silurian too, Sedgwick could hope to preserve almost *all* the older strata for his Cambrian against the encroachments that Austen's recent suggestions entailed.

That defensive maneuver had become the more necessary since Sedgwick had evidently concluded that the major unconformity he and Murchison had claimed between the Culm and the older strata would have to be abandoned. For if there was, at least locally (in north Devon), an unbroken sequence downward from the Coal Measures, the older strata could only be retained as Cambrian by extending the Culm downward at least as far as the Upper Silurian (fig. 9.2, C). In his paper Sedgwick noted than in north Devon the Culm strata "seem to graduate insensibly" into those below, without admitting that this represented a major volte-face, that he was making a significant concession to De la Beche (and indeed to Williams), or that he was rejecting the firm opinion of his collaborator Murchison (and indeed of Weaver). He simply pointed out that the conformity in the north, coupled with the unconformity he and Austen had found in the south (§9.1), proved that some of the north Devon formations must be without equivalents in the south. These were the strata that Murchison had originally claimed for his Lower Silurian (§7.2), but which he had subsequently abandoned on account of the fossil plants that Harding and Williams had found in them (§8.7). Sedgwick mentioned the "supposed discovery of some true carboniferous plants" in those strata, but left the anomaly unresolved and the age of the strata unclear.[80]

For the bulk of the older strata of both north and south Devon, however, Sedgwick maintained the view he had expressed in his joint paper with Murchison the previous year (§8.7); he attributed them to the Upper Cambrian, with perhaps a little Lower Cambrian too. He rejected the startling inversion of the sequence that De la Beche had suggested to him, in the interval since the reading of his paper began (§9.7), with the terse comment that to adopt it "would be to violate all the analogies of structure derived from other parts of England" and to run counter to the fossil evidence as well.[81] Yet even with that implausible conjecture dismissed, Sedgwick's assignment of most of the older strata of Devon to the Upper Cambrian still left him with serious difficulties, which are not hard to detect between the lines of his paper.

There was a striking omission in Sedgwick's summary of southwest England. He made no mention of the rich fossil faunas of the south Devon limestones—either Austen's Great Limestone or the Plymouth limestone further west—which were now being systematically studied by Phillips on behalf of the geological survey. Had Sedgwick been confident about the matter, he would surely have mentioned that Phillips's work could be expected to lead to a satisfactory characterization of the Cambrian fauna. Instead, he merely repeated the summary of the south Devon sequence that he and Murchison had outlined the previous year, adding only that certain strata contained "some

80. Ibid., p. 681.
81. Ibid., p. 682. He saved De la Beche's face by not mentioning him by name.

corals that do not appear in the mountain limestone, but are found both in the Cambrian and Silurian systems."[82] The contrast between that comment and Lonsdale's expert opinion that Austen's fossil "shells" were *not* Silurian at all, but similar to those of the Mountain Limestone (§9.3), was left unresolved.

There is only one plausible explanation for this. Sedgwick now feared that the fossil evidence might soon become so overwhelming that some or even all of the south Devon limestones would have to be surrendered from the Cambrian and transferred to the Carboniferous. For in his paper almost the only Cambrian fauna that Sedgwick mentioned unambiguously from southwest England came from the limestone near Launceston (at South Peth-erwin), which he had first seen nearly two years earlier and assigned at once to the Cambrian (§7.2, §7.4), and which he had revisited more recently with Austen (§9.1). Only at this one point in his whole paper did Sedgwick mention *any* fossil fauna in more than a phrase. Having tacitly abandoned almost all the rich south Devon faunas, this Cornish fauna was the only one left to characterize his Upper Cambrian. He mentioned its corals and crinoids; listed several genera of shells (brachiopods, and cephalopod mollusks such as *Goniatites*); and emphasized one particularly distinctive chambered shell (*Clymenia*) which his young colleague David Ansted (b. 1814) had described to the Cambridge Philosophical Society earlier in the year. In the aftermath of the discussion of Austen's fossils, Sedgwick had assured Ansted that these Cornish limestone strata were "far, very far removed in the order of their deposit from the mountain limestone."[83] Now, likewise, he concluded in his own paper that "as they occupy a position so much lower, so, as a group, these fossils are distinct from those of the Silurian system." Here at least Sedgwick evidently hoped he had established a firm bulwark for the distinctive Cambrian fauna he had failed to find in Wales itself, secure against any encroachment from Murchison's Silurian or—even more disturbingly—from Austen's alleged Carboniferous.

It is therefore not surprising that Lyell, urgently needing an example of a distinctively Cambrian fossil to mention in his elementary book before it went to press, chose one from Sedgwick's Cornish fauna. Lyell expressed confidence that further research would disclose a more "distinct assemblage" to characterize the Cambrian, but this one had to suffice for the time being. The chambered shell *Clymenia* (under Ansted's name *Endosiphonites*) duly appeared as the only Cambrian fossil to be illustrated in Lyell's *Elements of Geology* (fig. 9.3).[84]

Apart from that, Sedgwick's paper left almost no ripples of discussion. Conybeare later refereed it (along with Austen's and Weaver's earlier papers) and approved it, predictably enough, for full publication in the *Transactions of the Geological Society*. But even Conybeare referred less to the paper he

82. Ibid.
83. Ibid., p. 683; Ansted (1838), read at CPS on 26 February, published text dated 18 May (Sedgwick's comment is on p. 422). Ansted called the fossil *Endosiphonites*, but Phillips told him afterward that it had already been named *Clymenia* (Münster 1834), and the latter name therefore took precedence.
84. Lyell (1838, pp. 465–66); fig. 9.3 is reproduced from his fig. 283. Lyell's note to ask Sedgwick about "the Endosiphonite" is between notes dated 25 May (two days after the paper was read) and 15 June; the *Elements* finally went to press on 16 July (KHK: CL, notebook 71, pp. 5–12). He had already asked for Sedgwick's help in more general terms: Lyell to Sedgwick, 20 January 1838 (K. Lyell 1881, 2:35–37).

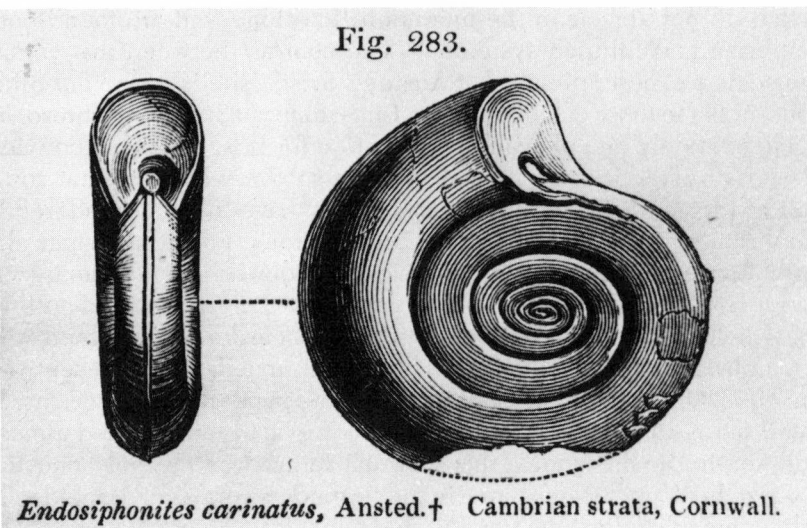

Fig. 283.

Endosiphonites carinatus, Ansted.† Cambrian strata, Cornwall.

Fig. 9.3.   The cephalopod mollusk *Clymenia* (or *Endosiphonites*) from north Cornwall, the only fossil used in Lyell's *Elements of Geology* (1838) to illustrate the fauna of Sedgwick's Cambrian system.

was ostensibly evaluating than to the others he hoped Sedgwick would produce in the future, "promising to throw a compleat & satisfactory light on the older formations of our island." The almost total lack of substantial discussion of Sedgwick's paper, either at the Geological Society when it was read, or in correspondence subsequently, reflects a profound sense of disappointment among other geologists that Sedgwick had failed to characterize his Cambrian as clearly as Murchison's Silurian. Conybeare concluded that "the Devonshire or upper portion of the great System [i.e., the Cambrian] included under the vague name of Greywacke may now be considered as fully elucidated."[85] But Sedgwick must have known that that judgment was premature. As long as the enigma of the apparently Carboniferous fossils from the south Devon limestones was unresolved, Sedgwick was bound to remain uneasy about the foundations of his faunal characterization of the Cambrian. The correlation of the Devon Culm with the Coal Measures might seem increasingly secure, but the partial resolution of that issue had now spawned another problem that seemed likely to keep the Devon controversy bubbling vigorously for some time to come.

9.9   Pressure on the Cambrian of Devon (July 1838)

As soon as his various duties allowed, Sedgwick set out on his planned trip to Devonshire, hoping to resolve the problems that still beset the interpretation of the region before he had to commit himself in print on the Cambrian

85. Conybeare to Lonsdale, n.d. (GSL: BP, COM P4/2, p. 3); dated August 1838 by Conybeare when adding a postscript on 13 May 1839, but certainly written earlier, since his report was formally received by the Council of GSL on 25 June 1838 (GSL: BP, CM 1/5, pp. 55–56).

rocks of Britain. His first task was to check De la Beche's surprising and implausible new claim that the genuine Culm in southeast Devon lay *under* the other rocks and not over them as all observers had agreed hitherto (§9.7). Austen showed him the exposures that in his opinion proved the official surveyor was mistaken, and Sedgwick was convinced. "There is not the shadow of a proof that they [i.e., the Culm strata] dip under the older system," he concluded, inferring that De la Beche's "mistake" had been to confuse similar rock types in the two formations.[86]

After Sedgwick had left, Austen told De la Beche that he too was now " decidedly of opinion, nay more, convinced" that De la Beche was mistaken on this point. "I should be very sorry if you had printed it so," he added, sending De la Beche some sketch sections that showed the Culm clearly lying unconformably *above* the other rocks, an inference recently confirmed in one case by the sinking of a well (fig. 9.4).[87] But as already noted, this "crucial" evidence was unlikely to convince De la Beche, because he had adopted an elaboration of his earlier interpretation, by which he could allow that *some* so-called Culm was younger, and perhaps even genuine Coal Measures, whereas the main body of the Culm was to be kept *under* Austen's limestones

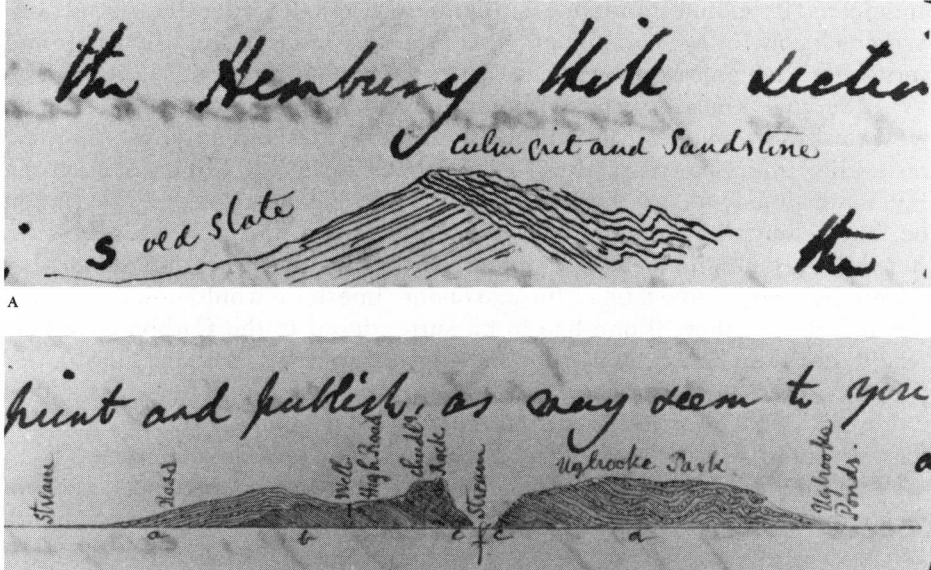

Fig. 9.4. Austen's sections, drawn for De la Beche in July 1838, to try to persuade him that the Culm lay clearly *above* the other strata in southeast Devon, with a local unconformity. The upper section depicts an unambiguous exposure of the unconformable relation. The lower section depicts a more problematic locality: Austen claimed that the Culm (*a*) overlay one of the limestones in the lower strata (right), and that the sinking of a well had clinched the case for an unconformity (left).

86. Sedgwick, notes for 11, 13, and 16 July 1838 (SMC: AS, notebook "XXXII 1838," pp. 1–3).

87. Austen to De la Beche, 22 and 29 July 1838 (NMW: DLB), from which fig. 9.4 is reproduced. The upper section is at Hembury Hill near Ashburton; the lower, at Ugbrooke Park near Chudleigh.

and hence safely in the Greywacke (§9.7; fig. 9.2, B). Not surprisingly, therefore, when De la Beche returned to Austen's area later to check the point, he remained unconvinced.[88]

Even if De la Beche were wrong on this point, however, Sedgwick still had to face the enigma of the fossils from Austen's limestones. He used a "vile wet day" to write and tell Phillips what he thought about them. He saw no reason to alter his interpretation of the structure of southeast Devon as he and Murchison had represented it the previous year, or the sequence that followed from it (§8.7). "There is however a difficulty about the Tor Bay limestone," he admitted; "if it should turn out different from the Plymouth limestone, its age must be decided by its fossils." But he found those fossils as puzzling as ever: "its spirifers are *very like* those of the *mountain limestone;* but its corals are *very different* & it has very few productae. Its orthoceratites are very numerous."[89] When that wet weekend was over, however, and he resumed his fieldwork with Austen, Sedgwick became more and more bewildered by the anomaly that the fauna posed. Although he dismissed Austen's suggestion that some of the older strata were equivalent to the Old Red Sandstone (§9.6) with the crisp comment that they were "assuredly not so," the two geologists were agreed that the limestone was not sharply separated from the strata below, still less lying unconformably on them, as might have been anticipated if the limestone were Carboniferous and the older strata Cambrian. Sedgwick concluded, nonetheless, that *"the true place* of the L[ime] S[tone] series [is] such as was determined on my former visit."[90]

Sedgwick made that note before he left Austen's area and moved on westward to Plymouth; but it probably implies that he had now confirmed his earlier inference that Austen's limestones were *not* "different from the Plymouth limestone," as he had thought a few days earlier that they might be, but belonged at roughly the same point in the south Devon sequence. But that only heightened his problem. Although there were differences between their respective faunas, the age of one limestone would now determine the age of the other: if one had to be surrendered to the Carboniferous, so would the other. Reaching Plymouth, Sedgwick took his customary day of rest and attended Sunday services at Hennah's garrison chapel; but a day's local fieldwork with Moore as his companion, and a day studying the fossils in the local museum, did nothing to ease the problem he was now facing.[91]

A mail coach took him next to Launceston, where again he had Pattison for company. But the fossils they collected from the nearby limestone (at South Petherwin), and those the local geologist gave him, offered him no comfort. In his recent paper, he had cited that limestone as the only one in southwest England that he could confidently term Cambrian (§9.8). But he now found it contained not only the distinctive *Clymenia* (or *Endosiphonites*) that Lyell had chosen to exemplify the Cambrian fauna (fig. 9.3), but also a "ribbed orthoceratite like the Newton [Abbot] species" in Austen's home area, and a

88. De la Beche to Phillips, 5 August 1838 (UMO: JP, uncat.).

89. Sedgwick to Phillips, 14 July 1838 (UMO: JP, S14). "Spirifers" and "productae" were brachiopod shells (fig. 6.5, D, C); "orthoceratites" were straight cephalopod mollusk shells (fig. 10.3, B).

90. Sedgwick, notes for 17 and 18 July 1838 (notebook XXXII, pp. 3–5), referring to the area between Totnes and Berry Head.

91. Ibid., notes for 23 and 24 July 1838, pp. 9–11.

"turbinolia" shell also found in the nearby Culm. This Cornish limestone fauna was thus beginning to look less distinct from its south Devon counterparts than Sedgwick had hoped: if they would have to be surrendered to the Carboniferous, this limestone now looked equally vulnerable. In addition, he had problems even with his preferred evidence of structure and sequence: he admitted having "some difficulty in separating the Petherwin [strata] from the *bottom Culm series.*"[92] He can hardly have suppressed some doubt about the reality of the unconformity he had earlier been so keen to discover on the south side of the Culm trough (§9.1).

Continuing this rapid review of Devonshire geology, Sedgwick took a coach to Barnstaple and the north side of the Culm trough. Harding showed him where he had found the now notorious fossil plants (at Marwood) in what Murchison had originally claimed for his Lower Silurian. Sedgwick later traversed right round the northern coastline, doubtless to check his original fieldwork of 1836 in the light of his more recent problems of interpretation. But before doing so he went to look at the locality near Barnstaple (at Fremington Pill) that De la Beche had told him was the clearest exposure of the perfectly conformable passage between the older strata (Murchison's now abandoned Lower Silurian) and the Culm (§8.6; fig. 8.2). There on the shore of the Taw estuary Sedgwick carefully measured the dips and strikes, drew a rough sketch showing an unbroken sequence, and concluded that "there is a conformable junction but no true passage."[93]

Although that final phrase weakened its force a little, the concession was still a major one for Sedgwick to make. De la Beche had claimed, ever since his first fieldwork in north Devon in 1834, that the Culm was an integral part of the whole sequence of older formations (§5.1). Murchison had at once insisted, on the contrary, that there must be an unconformity (§5.2); Sedgwick, once convinced that the Culm was largely Carboniferous, had made the same inference, although their joint fieldwork in 1836 had failed to reveal any evidence of one (§7.2). Sedgwick had tacitly anticipated having to retreat on this crucial point when he summarized the whole affair in his most recent paper to the Geological Society (§9.8). Now he had made the concession on the spot, although only in the privacy of his notebook. But the concession had only become tolerable because he had simultaneously devised an explanation for what he had hitherto treated as an unacceptable anomaly. Only with his discreet and almost evasive redating of the Culm, to include even some Silurian as well as *all* the Carboniferous, did it become tolerable to him to accept that that formation lay *conformably* on what he was determined to retain as his Upper Cambrian system (§9.8; fig. 9.2, C). Nonetheless, as Sedgwick completed his three weeks of fieldwork and traveled north to his family home in Yorkshire, he can hardly have avoided some sense of unease about his interpretation of Devon. It was no minor worry, for the increasing vulnerability of his interpretation threatened his whole project for delineating the Cambrian as one of the earliest chapters in the history of the earth and the very earliest in the history of life.

92. Ibid., notes for 25 July 1838, pp. 11–12.
93. Ibid., notes for 27–31 July 1838 pp. 12–20; sketchbook "D 1838" (SMC: AS), undated "Coast section exactly opposite Fremington Pill." Before going there, Sedgwick had evidently discussed with Harding the crucial significance of this locality: Harding to Sedgwick, 20 August 1838 (ULC: AS, IB, 194).

Chapter Ten

BIG BOOKS IN COLLISION

10.1 A Surfeit of Silurianism (August 1838)

As the leading geologists assembled for the 1838 meeting of the British Association, they were well aware that the Devon controversy was approaching a new climax; not this time at the meeting itself, but as soon as the chief protagonists got their *magna opera* into the public realm. For both Murchison's big book on the region of "Siluria" and De la Beche's massive memoir on southwest England were now in the final stages of preparation and printing. Both works would enable their authors to state more or less directly where they stood in the continuing controversy about the ancient strata of the earth's crust, and to do so in a more public and permanent way than any mere summaries of papers read to the Geological Society or the British Association.

The meeting of the British Association brought it this year to Newcastle, and the attendance—some 2400—was larger than ever before. Since the surrounding coalfield was the most productive in the world, and literally proverbial, public interest in the geological and engineering Sections was expected to be intense, and each was allotted a 1000-seat space in the town's music hall. The geological Section met under the presidency of the quiet-spoken Lyell, who had difficulty keeping the large crowd in order and probably did not enjoy the task. It was only the second time he had attended a meeting, but he had been persuaded that this kind of show of strength by geologists was essential for the public standing of the science.[1] Most of the other leading geologists were also present and dominated the proceedings in the usual way.

Austen may not have been the only geologist to suspect, as he put it to De la Beche beforehand, that the Section would be "treated usque ad nauseam to Silurianism." On the second full day of the meeting, Murchison did indeed report that his book was about to be published, and he whetted the appetites of potential purchasers by displaying its impressive colored geological map and sections. He stressed publicly, as he had already in private (§9.6), that the characteristic Silurian fossils were totally distinct from those of the overlying Carboniferous system, thus reinforcing his claim that the Silurian system was a major natural group. He announced that, having completed his book, he intended next to look for Silurian strata and fossils on the Continent, adding that he expected them to be found extensively throughout Europe and beyond. Murchison's confidence in the wider validity of his Silurian system could hardly have been illustrated more strikingly: he mentioned that the distinguished astronomer John Herschel (b. 1792)—no mean amateur geologist in his spare time—had brought fossils back from Cape Colony, on the completion of his great project of mapping the stars of the southern hemisphere, and that they had been identified as Silurian species.[2] In the subsequent discussion, Phillips acclaimed Murchison's work as providing a firm basis for future research on the Transition rocks; he mentioned that August Goldfuss (b. 1782), the professor of zoology and mineralogy at Bonn and author of the great *Petrefacta Germaniae*, had recently written to suggest that much of the Transition sequence in Germany could be equated with the Silurian on the basis of its fossils.[3] Sedgwick added that what had formerly seemed a major break between the Carboniferous and the older Transition had now been filled satisfactorily by the Silurian.

The following paper was by Griffith, who reported on the progress of his geological map of Ireland; "Silurianism" surfaced again in the discussion. Griffith reiterated in public his earlier repudiation of Weaver's claims about the south of Ireland (§5.6), but he had brought some fossils that he thought might be Silurian (from county Cork). His colleague Joseph Portlock (b. 1794) then reported finding Silurian strata in the north of Ireland (county Tyrone), clearly below the Old Red Sandstone and the rest of the Carboniferous. Buckland alluded to the parallels between Griffith's results and recent work in the Rhineland, mentioning that there too the Silurian was "fully developed." All in all, the session showed clearly how it was not only Murchison who was now treating the Silurian routinely as a major category valid throughout Europe, even without waiting for the full description of its "type" area. As Austen

---

1. Morrell and Thackray (1981, pp. 133, 196, 457); Lyell to Darwin, 6 September 1838 (K. Lyell 1881, 2:46). The Northumberland and Durham coalfields supplied the whole of eastern and southern England, including London, as far west as Cornwall—a far larger area than any other (see Arrowsmith 1830): hence the proverbial absurdity of "taking coals to Newcastle." On the local scientific background generally, see Orange (1983); on Newcastle's strongly geological Natural History Society, see Morrell (1983, pp. 234–36).

2. BAAS, 22 August 1838: *Ath.*, no. 566 (1 September), p. 631; Austen to De la Beche, 8 August 1838 (NMW: DLB).

3. Goldfuss (1826–44).

had commented tartly beforehand, "I can foresee the fate of geology for the next eight years—half the globe will become Silurian."[4]

The usual dominance of the leading geologists was amply demonstrated in the following days, both in the way they imposed their own definition of the science and in their treatment of merely provincial geologists. When one amateur, the Presbyterian minister George Young (b. 1777) of Whitby, tried to link some competent observations on his local Secondary strata and fossils to the restricted timescale of scriptural geology, Sedgwick put him firmly in his place. Conversely, the alliance between science and religion that the leading gentlemen of science were keen to promote was vividly displayed the following day. Together with a local geologist, Sedgwick led an excursion to see the Coal Measures exposed on the shore at Tynemouth; he gave an impromptu popular lecture on the shore to a vast crowd that included a large proportion of "colliers and rabble," and led them eloquently from the geology of their coalfield upward to their social, moral, and religious duties.[5]

After the excursion to Tynemouth, the geological Section had many papers still unread; and the following day there was only time for them to be summarized orally. Among papers relegated to this distinctly second-class treatment was Austen's on "geological evidences and inferences," which he had asked De la Beche to read in his absence.[6] Earlier in the year he had told Sedgwick he thought geologists were "hunting on the wrong scent" in certain respects, and that the science might be "toppled into the dust" if it did not examine its own foundations critically.[7] His paper at Newcastle revealed that what he regarded as particularly vulnerable was the use of fossils for global correlations of the kind that Murchison was now promoting. Austen was concerned that in the comparison of fossil assemblages too little allowance was being made for the likely effects not only of local environmental conditions in the past, but also of differences of latitude. Austen had derided Darwin's recent report of Silurian rocks in the Falkland Islands: "if Darwin was a joker of jokes, which he is not," he told De la Beche, "I should esteem this one of his best."[8] His caveat about such global correlations was telling enough, but the reception of his paper was mixed: "some spoke for, and others against," Sedgwick told him afterward, implying perhaps that he himself had been critical of Austen's conclusion.[9]

Austen's bid for recognition as a competent theorist in geology had failed. With his paper squeezed into a few remarks at a hurried final session, the provincial geologist, however competent in his home area, had no chance

4. *Ath.*, no. 566, pp. 631–32 (the unidentified work on Nassau was probably Beyrich 1837); Austen to De la Beche, 29 July 1838 (NMW: DLB); Griffith (1838, 1839); Portlock (1839).

5. BAAS, 23 August 1838; *Ath.*, no. 566, p. 652; Morrell and Thackray (1981, pp. 163, 240). Herschel's second-hand account of the seaside lecture is in Clark and Hughes (1890, 1:515–16). Young had been the author (in collaboration with a local artist) of a competent description of the strata and fossils on the Yorkshire coast (Young and Bird 1822).

6. Austen (1839a), read at BAAS on 25 August 1838; *Ath.*, no. 568 (15 September), p. 678, which states incorrectly that Austen's paper was delivered in person.

7. Austen to Sedgwick, 18 January and 6 March 1838 (ULC: AS, IB, 133, 145).

8. Austen to De la Beche, 8 August 1838, referring to Darwin (1839–42, 2: p. iii); see also Austen to Sedgwick, 11 September 1838 (ULC: AS, IB, 159). He would doubtless have been equally incredulous about Herschel's fossils had he been present to hear Murchison make the claim.

9. Reported in Austen to De la Beche, 10 November 1838 (NMW: DLB); see also Austen to Sedgwick, 11 September 1838.

in absentia to dent the massive authority of the leading gentlemen of science. Austen had in fact also prepared a paper on Devon, which put forward a conjectural reconstruction of ridges and basins in the ancient sea floor. He thought this would explain how "contemporaneous deposits in close juxta-position" might have resulted in widely contrasting rock types. Specifically, it would explain how the Culm of central Devon might be of the *same* age as the south Devon limestone sequence, thus reinforcing his argument that both were Carboniferous. "Now this I call a tarnation handsome bird, though I say it as shouldn't," he told De la Beche jokingly beforehand; "but do you think he'll fight?" Perhaps because the older geologist found it implausible and was skeptical about its chances, Austen's "Devonian" fighting cock was not in fact put into the ring.[10]

So the Newcastle meeting of the British Association, unlike the previous three, ended with no explicit airing of the Devon controversy. There was perhaps a tacit agreement among the leading geologists to let it lie until the chief protagonists' long-awaited books were published. But with Murchison's global ambitions for the Silurian system now fully apparent, the problems of Devon remained as urgent as ever if the early periods of the earth's history were to be reliably understood.

10.2   Continental Silurianism (September to November 1838)

Murchison's confidence that his distinctive Silurian fauna would soon be recognized more widely was not unfounded. Reports of Silurian fossils from very distant localities, such as Cape Colony and the Falkland Islands, might meet with skepticism or even derision from some geologists. But reports from the Continent were far more acceptable, since it was plausible to suppose that the area represented by present-day Europe might have constituted a single faunal province in Silurian times. Back in 1835, within a few weeks of Murchison's public naming of the Silurian, Buckland had claimed to recognize its equivalents within Dumont's Belgian sequence, and soon afterward von Buch had suggested that several formations even further afield might also be of the same age, judging by their fossils (§ 6.3). More recently, in 1837, Buck-land had made similar claims for some of the older rocks of northwest France (§9.1); while Goldfuss and other German geologists, as the English had just heard at the Newcastle meeting, were now coming to the same conclusion in their own regions (§10.1). With this spread of "Silurianism," John Bull did indeed seem set to lord it over the rest of Europe in geology, as Greenough had sarcastically predicted (§8.6).

Of all the Continental sequences, however, the Belgian remained by far the most important, thanks above all to Dumont's thorough mapping work. Dumont, still under thirty, was now a professor at Liège, and had been com-missioned by the Belgian government to extend his prizewinning survey of his native province to cover the rest of the country. Ignoring political frontiers as irrelevant in geology, he had in fact extended his survey into both France

---

10. Austen to De la Beche, 8 August 1838. Austen may have combined the two papers into one, before the meeting.

and Prussia. In a summary of his progress, read to the Société géologique in Paris late in 1836, he had reported finding a vast new area of *terrain anthraxifère* in the Eifel region to the east of the Ardennes. He had mapped the Eifel limestone, long famous for its fine Greywacke fossils, as lying in a series of troughs at the top of the sequence (fig. 10.1), and had equated it with his lower limestone group in Belgium.[11] This not only extended the Belgian sequence in a way that might later link it to regions still further to the east. It also implied that the Eifel fossils were younger than the Greywacke proper (or Dumont's *terrain ardoisier*) and that, following Buckland's suggestion, they were perhaps equivalent to the Upper Silurian (Wenlock Limestone) of England.

Dumont himself, however, was above all a field geologist and had no great confidence in the value of fossils. But a French geologist had begun to interest himself in them, taking in relation to Dumont somewhat the same role that Phillips had adopted in relation to De la Beche. Édouard de Verneuil (b. 1805), a Parisian lawyer and keen fossil collector, had been to England in 1835, with introductions from Élie de Beaumont, to collect Carboniferous and Transition fossils and to study those in the main English collections. He had also attended the British Association meeting in Dublin and had met most of the leading English geologists.[12] Later he had collected extensively in Dumont's Belgian strata and in the Eifel. Early in 1838, when the Société géologique held a general discussion on the nature of the boundaries between successive major *terrains* or systems, de Verneuil had strongly supported Élie de Beaumont's view that such boundaries were usually marked not only by unconformities but also by sharp changes in the fossil contents of the formations. As an example of such a "séparation bien tranchée," de Verneuil had emphasized the sharp contrast between the Carboniferous and Silurian faunas, even where—as in Murchison's Welsh Borderland and Dumont's Belgian sequence—there was *no* unconformity between those successive *terrains*.[13] In this opinion, at least in so far as it concerned the British sequence, he was merely echoing what Murchison had told Lyell at that time and had later reiterated in public at the Newcastle meeting (§9.6, §10.1).

Only ten days after that meeting ended, the French society assembled for its annual field meeting, this time at Porrentruy in the Jura, just beyond the French frontier into Switzerland (see fig. 3.6). Among the thirty members present, Buckland had come direct from Newcastle and represented the English as he had the previous year; Dumont's patron d'Omalius was there from Belgium; Agassiz and many others represented the Swiss, while de Verneuil was among the Parisians. Somewhat unusually, the field meeting this year was not near any of the areas of old rocks, and the excursions focused instead

11. Dumont (1836), read at SGF on 21 November. Fig. 10.1 is traced from part of the accompanying "Coupe des terrains primordiaux" (p. 77). The work was somewhat less original than Dumont claimed, since he must have known of an earlier map (Steininger 1822) which, though crude, clearly marked a series of troughs or basins of "jüngere Übergangsgebirge" within the vast Greywacke area stretching from the Ardennes through the Eifel to the Rhine and beyond.

12. Élie de Beaumont to Murchison, 2 May 1835 (GSL: RIM, E4/6); Hardy (1835).

13. Report of discussion at SGF, 5 February 1838 (*BSGF* 9:155–56), initiated by remarks by Deshayes on the Tertiary faunas.

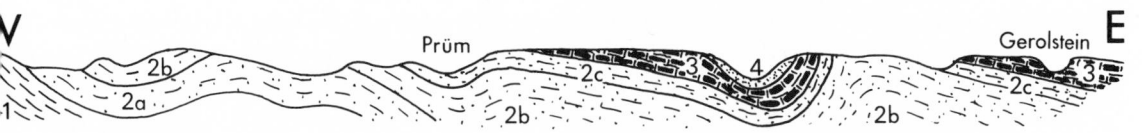

Fig. 10.1. Part of Dumont's traverse section from Belgium into the Eifel, redrawn from his 1836 progress report, showing the western side of his "Bassin Anthraxifère de l'Eifel." Dumont equated the Eifel limestone (3) near the top of the sequence with his lower limestone group in Belgium; the underlying strata (2a, 2b, 2c), with his lower sandstone group; and the basal slates (1), with the *terrain ardoisier*. (Compare with fig. 6.4, on which the formations are similarly numbered and shaded; for the location of the section, see fig. 12.2.)

on the spectacularly folded Oolitic strata of the *système jurassique*.[14] But one evening de Verneuil, in a paper unrelated to the local fieldwork, extended the Belgian sequence of older strata westward almost to the English Channel. In the Bas Boulonnais, inland from the port of Boulogne, a tiny inlier of older rocks had long been known, surrounded and overlain by Secondary strata. Its importance was out of all proportion to its size, because the strata included coal seams that supported some small mines, and its position made it plausible to suppose that it might betray a major concealed extension of the great Belgian coalfield to the east (see fig. 3.6).

Earlier in the summer, a strong team consisting of d'Omalius, Dumont, and de Verneuil had gone to England to study Murchison's and Sedgwick's sequences in Wales and the Borderland in order to compare them firsthand with the Belgian rocks. Before crossing the Channel, they had taken the opportunity to look at the Boulonnais inlier; de Verneuil had already been shown some of its fossils, which he suspected were Silurian, so the inlier could be expected to provide another example of the relation between the Silurian and Carboniferous strata. As he reported to the Société géologique at Porrentruy, this was just what they had found. They had identified what they considered to be the equivalents of all four of Dumont's *terrain anthraxifère* groups. Unlike the Belgian sequence, however, in the Boulonnais Dumont's upper limestone lay almost directly above the lower, with only a thin sandstone formation between. This was no mere local detail, because the upper limestone group included the workable coal seams and its fossils were clearly Carboniferous, whereas the lower limestone group yielded fossils similar to those of its equivalent in Belgium and the Eifel, which de Verneuil (following Buckland) considered similar to those of the Upper Silurian limestones of England (at Wenlock and Dudley).

The implications, de Verneuil claimed, were important both economically and scientifically. The Boulonnais showed that a careful attention to fossils could distinguish between superficially similar limestones, the one closely associated with coal seams, the other predating any such deposits. "Paléontologie"—the newly named subject that he himself was helping to establish in France—was thus an indispensable aid in the nation's search for

---

14. *Réunion extraordinaire* of SGF, 5–12 September 1838 (*BSGF* 9:388–96). Twenty-nine nonmembers, mostly Swiss but also including Ansted from Cambridge, brought the total attendance to fifty-nine.

new reserves of coal. Furthermore, de Verneuil pointed out that the Boulonnais also proved the reality of the sharp faunal distinction between the Silurian and the Carboniferous, even more clearly than on Murchison's own home ground; for the sequence was just as conformable and unbroken as in the Welsh Borderland, yet there was no thick Old Red Sandstone to separate one rich fauna from the other.[15] Murchison could hardly have won a more zealous supporter than de Verneuil in his campaign to establish the distinctive faunal character of the Silurian and its economic value as a limit in the search for coal.

Two months later Dumont himself continued this exercise in international correlation when he gave the Académie des Sciences in Brussels another progress report on his survey of Belgium. In the light of his summer fieldwork in Britain with de Verneuil and d'Omalius, he reported that there was "un parallelisme complet" between the British and Belgian sequences. He claimed that the "systèmes cambrien et silurien" corresponded exactly to the *terrain ardoisier* and *terrain anthraxifère* that d'Omalius had named a quarter-century earlier. But this was more patriotic than accurate; for, as his own table showed, Dumont agreed that his upper limestone group was equivalent to the Mountain Limestone of England, so that the *terrain anthraxifère* was *not* exclusively Silurian.[16] In fact, Dumont's table was virtually copied—without acknowledgment—from the one that Buckland had presented at the end of the Mézières meeting in 1835 (see fig. 6.3, B). His only modification of Buckland's scheme was to concede that the Old Red Sandstone of England might just be represented in Belgium after all; but even that important suggestion was not original. It derived almost certainly from a discussion in Paris about a year before, when both Prévost and Rozet had insisted that the thick upper sandstone group in Belgium, sandwiched between the upper and lower limestones that were agreed to be Mountain Limestone and Silurian, respectively, must surely represent the Old Red Sandstone of England.[17]

That modification simply improved the match between the British and Belgian sequences of older strata, the latter now extended geographically to include the Boulonnais to the west and the Eifel to the east. Neither Dumont's work on the structures and sequences, nor de Verneuil's on the fossils, did anything directly to solve or alleviate the continuing problems of Devonshire. Dumont was only concerned with correlating the Belgian sequence with that in Wales and the Borderland; de Verneuil did not mention the reservations that Greenough had expressed at the Mézières meeting about the simple equation of the lower Belgian limestones with the Silurian (§6.3). But after the Porrentruy meeting, when de Verneuil was back in Paris, he was visited by Austen. Continuing the expansion of his horizons beyond those of a merely local geologist, Austen had gone to France primarily to look at the geology

15. De Verneuil (1838), read at SGF on 6 September. The rather crude section (fig. 14) accompanying the paper did not show clearly how the Secondary strata *overlay* the older rocks; it reveals how much de Verneuil's knowledge of fossils needed to be complemented by Dumont's understanding of geological structures. The distinctive lower limestone had first been detected by Rozet (1828); Fitton (1827) had noted only the upper.

16. Dumont (1838, p. 638, table), read at ASB on 3 November.

17. Discussion at SGF on 18 December 1837 (BSGF 9:81–84), initiated by remarks by Collegno, claiming an unconformity between Silurian and Carboniferous in Belgium; Rozet (1830). In addition to reading the summary in *BSGF*, Dumont could have heard about the discussion from de Verneuil, who had also taken part.

of the Cherbourg peninsula immediately opposite Devonshire and not dis-similar in character (see fig. 3.6). But his meeting with de Verneuil was even more rewarding than his fieldwork. For the French geologist showed him his collection of the distinctive fossils from the Eifel, and Austen was immediately struck by their close resemblance to those from the Devon limestones near his home. De Verneuil gave him some specimens for his collection, and when he got home Austen sent some of the Devon fossils to Paris in exchange.[18] His perception of a faunal similarity between Devon and the Eifel did not, however, lessen the puzzles of Devonshire; for Austen was convinced that the Devon limestone fossils were of about the age of the Mountain Limestone, whereas de Verneuil regarded the Eifel limestone as unquestionably Silurian.

10.3   Encapsulating a Consensus (December 1838)

At the last meeting before the end of 1838, the Geological Society received a copy of Phillips's little *Index Geological Map* of Britain, It had in fact been published back in the summer, but it was not well advertised and did not come widely to the attention of geologists until the end of the year.[19] Although small in scale and unaccompanied by any text, it encapsulated a significant summary of what had now become a somewhat uneasy public consensus on the problems of Devonshire geology, on the eve of the publication of both Murchison's and De la Beche's big books.

Quick to see an opportunity to supplement his modest regular income, Phillips had realized the potential market for an inexpensive geological map of Britain. Without waiting for Greenough to publish the long-delayed new edition of his relatively large-scale map, which was bound to be costly, Phillips had drafted a small and simplified version based on the original edition. He had sent this to various geological friends, asking them to amend and improve the parts with which they were familiar. De la Beche, checking southwest England for him, copied the geological boundaries from the forthcoming index map to his revised version of the geological survey sheets. As he had told Sedgwick, he was now sufficiently convinced by his opponents' work to agree that the Culm of central Devon needed a separate color on the map to dis-tinguish it from the still older strata (§8.6). So he had no hesitation about marking the boundaries of the Culm trough for Phillips, *as if* he believed the

18. Austen to Sedgwick, 11 September and 16 October 1838 (ULC: AS, IB, 159, 160); Austen to Phillips, 28 September 1838 (UMO: JP, 1838/47); Austen to De la Beche, 10 November 1838 (NMW: DLB).

19. Douglas and Edmonds (1950) described and compared the successive versions of this important map. Their so-called first edition (n.d., wmk 1837) is known only from a unique copy preserved among Phillips's papers at Oxford. In my opinion this is probably a proof copy of an unpublished draft; certainly when Weaver asked Phillips at the end of 1837 about his progress with the map, he believed it was still unpublished: Weaver to Phillips, 18 December 1837 (UMO: JP, 1837/65). The so-called second edition (n.d., wmk 1838), of which several copies survive, is probably the version of which Austen had a copy by July 1838, and on the poor advertisement of which Bilton commented in October: Austen to De la Beche, 22 July 1838 (NMW: DLB); Bilton to Phillips, 10 October 1838 (UMO: JP, 1838/52). The so-called third edition differs only in minor ways; one of the known copies was presented to GSL on 20 December 1838. The following analysis is based on these two ("second" and "third") versions, which are listed in the bibliography under Phillips (1838a); the still later "editions" differ substantially and are listed separately.

Culm to be "good honest coal measures": its area of outcrop was of course unaffected by the dispute over its age.[20] Phillips's "tiny map" (fig. 10.2) duly showed the culm area not only as distinct from the strata to the north and south, but as Carboniferous; for on this point Phillips had been on the side of De la Beche's opponents ever since the Bristol meeting (§7.3). Indeed there was now a virtual consensus on the Carboniferous age of the Culm, with only De la Beche himself and his supporter Williams still holding out for a much older dating, and with Greenough remaining skeptical.

Phillips's color scheme for the whole country obliged him, however, to choose for the Devonshire Culm a color that would imply its correlation with a specific *part* of the Carboniferous elsewhere. On at least some copies he adopted Murchison's interpretation and got the mapmaker to color the Culm as Coal Measures.[21] But this could give the general public the highly misleading impression that a huge new coalfield—larger even than that in south Wales—had just been discovered in Devonshire. Perhaps for that reason, the Culm area was colored on some later copies as if it were equivalent to the Millstone Grit of northern England.[22] In the south of the country that formation was too thin and unimportant to be marked on such a small-scale map; but in the north it was thick and distinctive enough, and outcropped over wide enough areas, for Phillips to give it a color of its own. The Millstone Grit included some rock types similar to the Culm, lacked productive coal seams, and lay in the sequence between the Mountain Limestone and the Coal Measures. To color the Devonshire Culm as Millstone Grit was therefore an ingenious and appropriate way of depicting it as Carboniferous and alluding to its probable approximate place within that group, while avoiding the misleading implications of coloring it like a coalfield.

Phillips also got Murchison to mark the outcrop of the Silurian strata, and his map duly showed a strip of them in Wales, extending through the Welsh Borderland and westward to Pembrokeshire. They were flanked on one side by the lowest division of the Carboniferous, the Old Red Sandstone, and on the other by Sedgwick's Cambrian rocks, which Phillips termed "Clay Slate and Greywacke Slate." Phillips also used the latter color for *all* the older (pre-Culm) strata of Devon and Cornwall, thereby endorsing Sedgwick's claim that they were Cambrian in age. The strip of Silurian strata in north Devon, which Murchison had displayed on his sketch map at Bristol (fig. 7.4), was not copied onto Phillips's map, presumably because Murchison told him he no longer believed in it (§9.4).[23]

The unease caused by Austen's fossils from south Devon, on the other hand, found no echo on Phillips's map: the older (pre-Culm) strata in Austen's area, like those elsewhere, were simply colored as Cambrian, as Sedgwick had earlier been confident they should be (fig. 10.2). No alternative could be well grounded until Phillips himself had completed his thorough study of the

20. De la Beche to Phillips, 7 and 16 April, and 5 May 1838 (UMO: JP, uncat.).

21. Austen described his copy as a map on which Phillips had made "culms = coal measures and I think right[ly]": Austen to De la Beche, 22 July 1838.

22. Fig. 10.2 is traced from the copy in GSL, received there on 20 December 1838. For simplicity, the areas distinguished on the original as Coal Measures, Millstone Grit, and Carboniferous Limestone are shown here by a single pattern.

23. The Silurian in north Devon remained on Murchison's index map, which had been engraved late in 1837 (Thackray 1978b), when his book was finally published in January 1839.

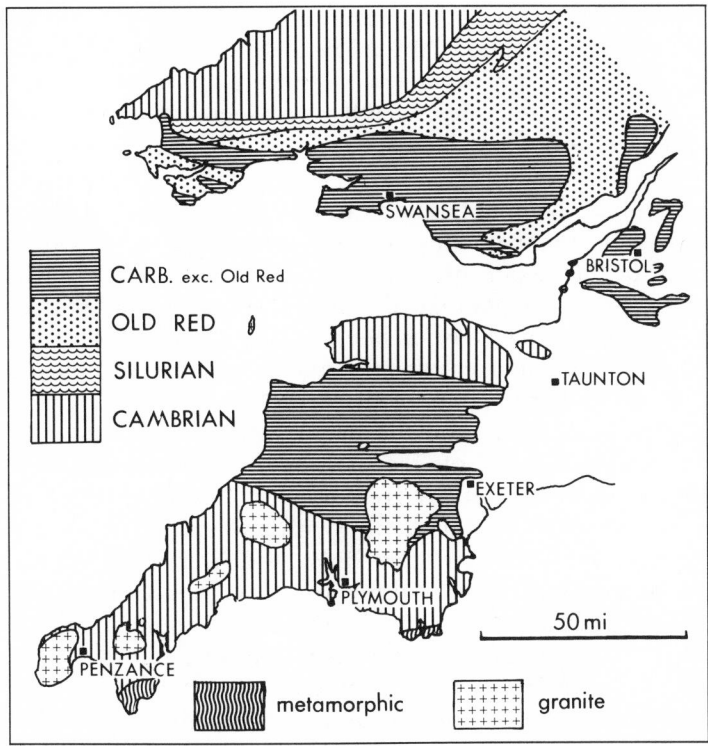

Fig. 10.2.   Phillip's interpretation in 1838 of the geology of southwest England and south Wales. For the first time on a *published* map, the great Culm trough of central Devon (see fig. 7.4) was shown as an area of Carboniferous rocks. Most of the rest of Devon and Cornwall was given the same color as Sedgwick's Cambrian rocks in Wales.

Devon fossils, which De la Beche had now persuaded the Ordnance authorities to subsidize.[24] Until his identifications were complete, Phillips avoided becoming embroiled in the arguments that the display of Austen's fossils in London had generated (§9.3). His little map therefore encapsulated and endorsed what had become a general consensus about that region; a consensus that had been disturbed but not disrupted by the discovery of fossil plants in the pre-Culm strata of north Devon and by the apparently Carboniferous character of many of Austen's fossils in the south. The Culm was attributed to the Carboniferous, and the older strata to the Cambrian; but the continuing problems of relating the two were glossed over and still unresolved.

10.4   Murchison's Magnum Opus (January 1839)

In the first days of 1839, Murchison's massive volume on *The Silurian System*, with its accompanying maps and sections, was published at last and put on sale at the substantial price of eight guineas. A copy was "laid on the table" at the first meeting of the Geological Society in the new year. Among almost

24. De la Beche to Phillips, 5 August 1838 (UMO: JP, uncat.).

400 subscribers, who received their copies at the reduced price of five guineas, those whom Murchison called "the native Silures"—the nobility and gentry of "Siluria"—were strongly represented. But the list also included almost all the Geological Society notables; wealthy provincials such as Austen, Bilton, and Harding; and foreign geologists such as Élie de Beaumont, Dumont, d'Omalius, and de Verneuil.[25]

As could have been anticipated from its history, the book was in effect a regional geological memoir: it purported to describe *all* the rocks, of whatever age, within the region of "Siluria" and southwestward all the way to Pembrokeshire; the main geological map had even retained the caption, "Silurian Region," which had earlier been planned as the title for the whole work.[26] Within the format of a regional memoir, however, Murchison naturally gave most attention to the strata of the Silurian "system" itself. Here there were many new local details but no major surprises: Murchison's sedulous publicity for the progress of his research over the past seven or eight years had robbed his final results of any great novelty. His descriptions of the strata above the Silurian were unremarkable (with one exception to be mentioned presently), and on the Cambrian strata below the Silurian he deferred tactfully to Sedgwick and said virtually nothing. What was most important about the fully published work was what no preliminary announcements could have given, namely, the descriptions and illustrations of the fossils that substantiated Murchison's claim to have delineated a distinctive new fauna (fig. 10.3). Sowerby had described the mollusks and brachiopods, and Lonsdale the corals; Charles Stokes (b. 1783), a London stockbroker and keen fossil collector, had assisted Murchison with the trilobites; and Agassiz had worked on the fish, though they were mainly from the Old Red Sandstone.[27]

Two major issues of theoretical importance were discussed in the work: the geographical distribution of the Silurian system and the nature of its temporal boundaries. Even within the region he described in detail, Murchison emphasized the priority of fossil evidence in the detection of Silurian strata. By tracing those strata laterally along the main outcrop from Shropshire into south Wales and all the way to Pembrokeshire, he claimed to have demonstrated that they retained their distinctive fauna even when the rock types changed. Alluding to the much younger (Oolitic) examples of lateral variation that had so impressed him early in his geological career (§4.1), he drew the moral firmly: "the lesson which has been already taught us by an examination of the younger deposits is repeated—that the *zoological* contents of rocks, when coupled with their *order of superposition*, are the only criteria of their age."[28] Since the Silurian could be recognized as far as Pembrokeshire, Murchison believed it could also be detected in the same way still further afield. In the introduction, which must have been revised extensively in proof in the months before publication, he summarized the way this anticipated recognition of the Silurian fauna had already begun. Élie de Beaumont and Duf-

25. Thackray (1978b, pp. 69–70); Murchison (1839, list of subscribers); GSL meeting, 9 January 1839 (GSL: BP, OM 1/9). Murchison probably continued to revise the work until shortly before publication; the preface is dated 29 November 1838.

26. *RBAAS* 1837 (trans.): 91.

27. Thackray (1978b) Fig. 10.3 is reproduced from Murchison (1839, pl. 7, fig. 6; pl. 9, fig. 2,5; pl. 10, fig. 1; pl. 12, fig. 2; pl. 15 bis, fig. 3,4).

28. Murchison (1839, p. 9).

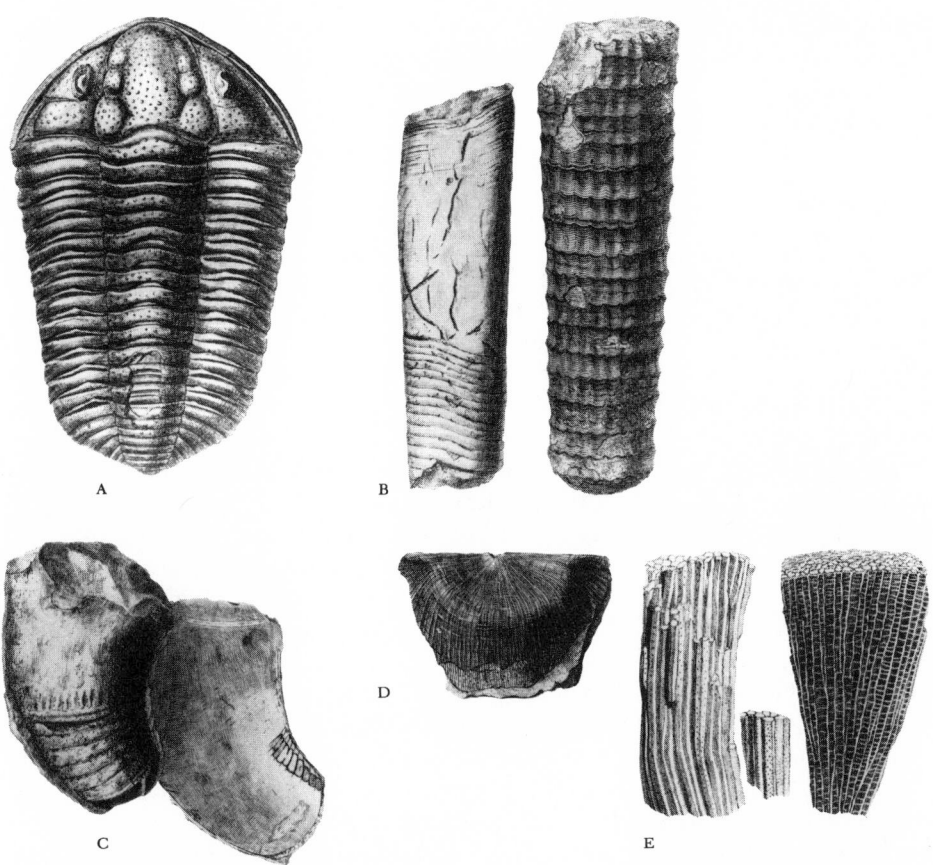

Fig. 10.3.   Some characteristic Upper Silurian fossils from the Welsh Borderland, as illustrated in Murchison's *Silurian System* (1839) and cited in discussions about the age of the strata in southwest England and on the Continent: (A), the trilobite *Calymene*; (B, C), the cephalopod mollusks *Orthoceratites* and *Phragmoceras*; (D), the brachiopod *Leptaena*; and (E) the coral *Favosites*.

rénoy had found it in France, d'Omalius and Dumont in Belgium, and other geologists in the Balkans and Scandinavia. Outside Europe, Silurian fossils had already been found in North America, South Africa, and the Falkland Islands.

None of this was news; but in the introduction to the part of the volume dealing with the fossils, Murchison at last made public the theoretical justification for his confident assertions that the Silurian would be found to be of worldwide validity. He explicitly linked it with the concept of "central heat," the high-level theory accepted by almost all the leading geologists at this time, other than Lyell. Murchison argued that the wide geographical distribution of the Silurian fauna implied a globally more equable climate than in modern times; he inferred that this could be attributed to the greater effect, in that distant epoch, of the earth's residual internal heat. This kind of inference was in no sense original, and Murchison may not even have felt strongly committed to it; but it did serve to legitimate in acceptable theoretical terms his intuitive confidence that he had delineated a new major epoch in the history of the earth and of life. It also distanced him decisively from his earlier collaborator, since Murchison emphasized that the apparent global uniformity of the Silurian fauna contrasted strikingly with the heterogeneity of modern faunas, and he argued that this indicated the limitations of Lyell's method of inferring the past from the present.[29]

On the other major theoretical issue raised by his research, namely, the nature of the temporal boundaries of his Silurian system, Murchison was revealingly inconsistent. In the introductory chapter, he recalled how he had gradually been led to believe that his Silurian strata deserved to be termed a system, and he defined the object of his book as being to draw attention to a thick sequence of formations, distinct from the older Greywacke below, but also *"essentially different both in structure and organic remains from the carboniferous strata"* above. In the introduction to the descriptions of the fossils, he emphasized likewise, as he had done repeatedly on earlier occasions (§9.6, §10.1), that the Silurian fauna was almost completely distinct from the Carboniferous, with only a handful of fossil shells in common. As with the wide geographical distribution of the fauna, so too he gave this faunal disjunction a theoretical gloss. It supported the view "that each great period of change, during which the surface of the planet was essentially modified, was also marked by the successive production and obliteration of certain races." Only a very few species, "whether endowed with powers to resist vicissitude or living in those parts where few agents of destruction were at work," managed to survive such phases of major change, and therefore crossed the boundaries between successive systems of strata. But these would be minor exceptions; in general, the distinctive character of the Silurian fauna bore witness, Murchison argued, to some decisive break in environmental conditions that had separated the Silurian from the subsequent period.[30]

This view of quasi-catastrophic faunal change, probably derived from Élie de Beaumont but in any case not original, was not consistent with Murchison's emphasis on the continuity of the sequence of strata in which the fossils were found. For he claimed that in "Siluria" there was a uniquely unbroken transition from the Mountain Limestone downward through the Old Red Sandstone into what had earlier been termed Transition or "fossiliferous greywacke."[31] At a late stage in the lengthy process of writing and revising the text, and presumably after the Newcastle meeting in August 1838 (§10.1), Murchison must have altered his view of faunal change to a more "gradualist" interpretation, which would match more closely this continuity in the strata themselves. In what reads suspiciously like a passage tacked on during the printing of the book, he suddenly turned away from "general views" about the cooling earth to more "immediate objects," namely, the relation between the Silurian and Carboniferous strata in "Siluria" itself. He claimed that this region contained not only a sequence of strata of unparalleled completeness, from Sedgwick's Cambrian right up into the Secondary strata, but also a corresponding "true zoological transition" between the Upper Cambrian and

29. Ibid., pp. 584–85; Rudwick (1971); Bartholomew (1979).
30. Murchison (1839, pp. 6–7, 582–83).
31. Ibid., p. 10. He did not mention Dumont's Belgian sequence, although he must have known about it; to do so would of course have detracted from the unique status of "Siluria."

Lower Silurian, between Lower and Upper Silurian, between Upper Silurian and the lowest part of the Old Red Sandstone, and between the upper part of the Carboniferous and the New Red Sandstone.[32]

This gradualist interpretation of faunal change was quite incompatible with what Murchison had written earlier; yet now he worked hard to make it apply plausibly to his own region by seeking to explain the one glaring exception that marred its generality. For as he had emphasized earlier, there was *no* faunal graduation between the lowest part of the Old Red Sandstone (which contained a few residual fossils like those of the Upper Silurian) and the Mountain Limestone much higher in the sequence. Between those formations lay the middle part of the Old Red, with a peculiar fauna consisting mainly of the bizarre fish that Agassiz had described for him (see fig. 13.1), and the upper part with virtually no fossils at all. Murchison now explained the lack of a faunal gradation at this point by linking it to a significantly revised conception of the Old Red Sandstone itself.

When he began writing his text, Murchison, in common with all other geologists, had regarded the Old Red Sandstone simply as the basal formation of the Carboniferous group or system. But some time in the course of his work his conception of that formation changed significantly: he elevated it to the rank of a system in its own right, "in order to convey a just conception of its importance in the natural succession of rocks"; and he claimed that it was quite distinct from the true Carboniferous. The earlier failure to appreciate the importance of the Old Red Sandstone had been due, he argued, to its absence or minor development on the Continent; by contrast, its "enormous thickness" of some 10,000 feet in "Siluria," combined with the discovery of its "very peculiar" fossil fish, had elevated this quintessentially British formation to a quite new position of international importance.[33] In this apparent postscript to his general conclusions, Murchison's new conception of the Old Red Sandstone led him to an interpretation that virtually contradicted what he had written earlier.

> When therefore it is asserted [i.e., by Murchison himself!], that the fossils of the Carboniferous aera are dissimilar to those of the Silurian, the reader must bear in mind, that the strata so broadly distinguished, are separated by accumulations of enormous thickness, and that the vast time occupied in their deposit, accounts satisfactorily for an almost entire change in the forms of animal life.[34]

Although "zoological links" had not yet been found to bridge this one anomalous lacuna, Murchison predicted confidently that "such proofs will hereafter be discovered," and would then provide "as perfect evidence of a transition between the Old Red and Carboniferous rocks, as we now trace from the Cambrian, through the Silurian, into the Old Red System."

This gradualist view of faunal change had not been put forward previously by Murchison, either in print or even in correspondence, though it would have delighted Lyell and a few other subscribers to *The Silurian Sys-*

---

32. Ibid., pp. 585–86.
33. Ibid., p. 169, cf. p. 9. This revised view was doubtless influenced by the impact of the fossil fish collected by the Scottish amateur geologist Hugh Miller in Cromarty: Murchison to Agassiz, 27 June 1838 (HLH: LA, 1419, 500). On these discoveries, see Andrews (1982).
34. Murchison (1839, p. 585).

tem. Applied to the specific task of explaining the faunal contrast between the Silurian and the Carboniferous (in Murchison's novel and narrowed sense of the latter term), it had major significance. Once the Old Red Sandstone was regarded as a system in its own right, on account of its great thickness and its putatively long period of deposition, it was possible to conceive how faunal change across that long time-span might have been as gradual as in other periods. This then implied that the poor fossil record in the Old Red Sandstone was not typical, and that the ordinary fauna that had characterized that period might be found in equivalent but contrasting strata in other regions. It therefore became a worthwhile strategy to look elsewhere for a transitional fauna that would blur the sharp contrast between the Silurian and the Carboniferous. Murchison had hitherto stressed the reality of that contrast at every opportunity, partly to accentuate the distinctive character of the system he had defined and on which he was staking his scientific career. His reasons for now effecting this remarkable volte-face were not made clear in the work itself.

Southwest England of course lay outside the region of "Siluria" and was scarcely mentioned in the text of *The Silurian System*. But Murchison could not refrain from recalling how his work in Pembrokeshire had first convinced him by analogy that De la Beche had made a major error in Devonshire (§6.2, §6.4). He described how the "Coal or Culm Measures" of Pembrokeshire locally overlapped the Mountain Limestone and Old Red Sandstone and lay directly on the Silurian; by coincidence they might then appear deceptively conformable to the older strata, but the fossils always served to distinguish them (§6.5). Were it not for the fortunate chance of some good cliff exposures, Murchison exclaimed rhetorically, "how much might have been written upon *conformability* and *passage!* and what erroneous inductions might have been drawn from these fallacious appearances!" The implicit criticism of De la Beche's emphasis on the conformable base of the Culm in Devonshire would have been clear to all his more knowledgeable readers. Likewise he mentioned in another substantial footnote that several of the fossil plants in the undoubted Coal Measures of Pembrokeshire were "*the most abundant* species in the *culm measures of Devon,*" thereby tacitly criticizing De la Beche for denying that correlation.[35] This served to support Murchison's interpretation of the Devon Culm, which, although it was outside "Siluria," was carefully if marginally recorded in *The Silurian System*. In the small index map that accompanied Murchison's main geological map of "Siluria," the Culm area of central Devon was marked unambiguously as Carboniferous. Almost all the older strata of southwest England were marked, as on Phillips's little map, as Cambrian; Murchison's map showed no hint of his repeated speculations that some of the older Devon strata might be of Old Red Sandstone age and, as Austen claimed, of Mountain Limestone age too.[36] But of course his readers would not have expected the book to contain significant revelations about Devonshire geology. Its importance lay rather in its establishment of the distinctive character of the Silurian system, at least in its "type" region.

35. Ibid., pp. 374n., 380n.
36. As already mentioned (n. 23), Murchison did retain on his index map the thin strip of Silurian strata in north Devon, the reality of which he had abandoned in private after the map was engraved.

10.5   De la Beche's *Report* (February 1839)

About a month after the publication of Murchison's *Silurian System,* and before there had been time for any substantive reaction to it, De la Beche's equally massive *Report on the Geology of Cornwall, Devon and West Somerset* emerged from its equally long gestation. Both authors had hoped to come first in the race, and the close finish was no coincidence.

In many ways, however, the two books were in sharp contrast. The copy of the *Report* sent to the Geological Society was presented not by De la Beche himself but by the Treasury, and it went on sale at the modest and subsidized price of half a guinea.[37] The contrast did not end there. Whereas Murchison's elegant volume was designed to grace the libraries of the nobility and gentry of "Siluria" and to charm its subscribers with geologically oriented vignettes of their familiar local scenery (fig. 4.7), De la Beche's book was altogether more forbidding. Colby's plea that the economic importance of the survey be seen as paramount (§8.5) had hardly been met. "Economic Geology" was relegated to the last chapter, although it was a substantial one. Excluding that tailpiece and the corresponding illustrations of mine workings, the *Report* was as much a regional geological memoir as Murchison's book, and intended equally as a contribution to fundamental science. But De la Beche's uninspiring prose style, his long-winded descriptions, and the unrelieved pages of dense print made it a far less approachable work than Murchison's. Furthermore, the crucial theoretical points that its more knowledgeable readers would have looked for were tucked away inconspicuously among the relentless catalog of local details, and were expressed with remarkable ambiguity.

As in Murchison's book, there was little high-level theorizing in De la Beche's memoir. Although the theory of a cooling earth had been the central theme in his earlier volume of *Researches,* he used it in his *Report* only as an explanation of the ancient metamorphic rocks on the south coast (Start Point and The Lizard).[38] The arrangement of the following chapters reflects a concession to Colby's plea that useful "facts" should come first and be clearly separated from theoretical inferences. The older strata and the Culm were described separately and in succession, although just "for convenience"; only after that did De la Beche tackle the contentious issues of their dating and relation to each other.

The first substantial chapter of the *Report* therefore gave a lengthy description of what De la Beche defiantly termed the "Grauwacke Group." He defended his use of the traditional term, against Murchison's criticisms (§6.5), by claiming that it was, like the term Chalk, useful and well understood, and that it carried no unwarranted theoretical overtones. Turning the rhetorical tables on Murchison, who at the start of the controversy had attacked him for reviving the outmoded doctrines of Werner (§5.2), De la Beche criticized the proposed new nomenclature of Silurian and Cambrian as containing "the relics of the old doctrine of universal formations." For as soon as these terms

37. De la Beche (1839), presented to GSL on 6 February (GSL: BP, OM 1/9).
38. Ibid., chap. 2, see p. 33; De la Beche 1834a.

were extended outside their eponymous regions, they implied a belief "that detrital matter has been strewed in exactly the same manner, enveloping exactly the same organic remains, over all parts of the world, where deposits were taking place at the same time." Nothing could be more improbable, De la Beche implied, in view of the heterogeneous spread of deposits and organisms at the present day. He acknowledged the value of Sedgwick's and Murchison's work—they were here named for the first time—but referred to their sequence in Wales and the Borderland merely as "a well-ascertained local series." De la Beche mentioned briefly the fossil contents of the Devonshire strata, summarizing the identifications that Phillips had already made in his study of collections such as those of Austen, Harding, and Pattison. He also mentioned that some of the fossil shells from north Devon (near Barnstaple) were like species from Murchison's Silurian, that Williams's fossil plants from a still lower level in the sequence were in Lindley's opinion like "species detected in the coal measures," and that Austen's fossils were remarkably similar to those from the Mountain Limestone.[39] But the discussion of the significance of these similarities was deferred.

In the following chapter De la Beche dealt at length with the Culm, which he termed, with equally noncommittal effect, "Carbonaceous Deposits." Predictably, he rejected the contrast that Sedgwick and Murchison had drawn between the Culm and the older strata and described the gradual and conformable transition between the two "systems" with a wealth of local detail. So gradual was it, he claimed, that in mapping the junction in order to mark them by distinct colors he had had to "trace an imaginary line for the sake of convenience—one nevertheless which nature has not traced." In this respect De la Beche now adhered publicly, and with a mass of supporting evidence, to the position he had taken privately ever since the Devon controversy began: whatever the ages of the two sets of strata, there was no unconformity between them. In this opinion, at least in so far as it applied to north Devon, he had been joined successively by Williams, Austen, and even Sedgwick; only Murchison and Weaver were left to deny the claim. De la Beche went further, however, than any of his supporters, except Williams, in that he maintained that the same conformable relation was to be found on the south side of the Culm area as well as on the north. His longest traverse section (from Combe Martin to Bolt Head) showed the Culm clearly overlying the other rocks, dipping inward on both flanks to form a huge trough in central Devon.[40] But De la Beche made no reference to the fact that this interpretation of the sequence and structure reversed his own original opinion in 1834 (§5.1), and that the change had been due to his acceptance—in *this* respect—of what Murchison and Sedgwick had first expounded at Bristol in 1836 (§7.3).

As in the previous chapter, De la Beche only alluded in passing to the question of the age of the strata. Turning the tables on Murchison again, he argued that a proper cause for surprise would have been if the fossil plants of the Culm had *not* been roughly like those of the Coal Measures. He pointed out that there were similar plants at a much lower level in the sequence, namely those Williams had discovered; and he referred to the Culm as older than the Coal Measures. In this deposit of "an intermediate period" it was therefore to be expected that the plants should have a similarly intermediate

39. De la Beche (1839, chap. 3, see pp. 39, 50, 51).
40. Ibid., chap. 4, see p. 102; pl. 2, fig. 1.

character. "Somewhat prematurely" they had been considered identical to those of the Coal Measures; in fact, De la Beche claimed, only "a minor portion" had been so identified by Lindley, whereas "many species, and even one genus, were entirely new to him" and differed from any Coal Measures forms (§9.7).[41] What De la Beche meant by an intermediate age remained obscure at this point, though he did mention that Phillips had found species of the mollusk *Goniatites,* identical to some from the Mountain Limestone (fig. 6.5, B), in the black Culm limestone.

Having described both the "Grauwacke" and "Carbonaceous" strata in detail, De la Beche devoted the following chapter to a fuller discussion of their dating and of the relations between them. Emphasizing once more the complete conformity between the two groups, and the lack of evidence to suggest any gap in time between their periods of deposition, he argued that the Culm would have continued to be regarded as an integral part of the Greywacke had it not been for the fossil plants that he had sent to the Geological Society in 1834 (§5.1). "Under the supposition that the species of plants detected in the coal measures must be limited to them," the Devon fossils had been regarded as anomalous. But in De la Beche's opinion, Williams's discovery of similar plants in the much older "upper grauwacke" strata of north Devon had since destroyed the validity of that argument. Parading once more the wide range of Continental analogies that he had used in private ever since the controversy began, De la Beche asserted that the Devon plants had constituted no anomaly at all. In an argument that could have been borrowed straight from Lyell's work, he maintained that coal could be formed at *any* epoch, given the right conditions; any similarities between the floras of the Greywacke and the Coal Measures could likewise be due simply to similarities of the kinds of vegetation and of the conditions under which they flourished, and not to similarity of age. Phillips's *Goniatites* were likewise explained away, in a now familiar argument, as the earliest localized occurrences of species that had been more representative of a later (Carboniferous) epoch; in effect, their value as evidence for the age of the Culm was dismissed.[42] De la Beche therefore concluded that the Culm was "upper grauwacke," thus restating in public the view he had formulated in private in the aftermath of the Bristol meeting (§7.6; fig. 8.1, A). This implied that the Culm was roughly equivalent in age to Murchison's Silurian strata; the contrast between them, in both rock types and fossils, simply showed that they were *both* merely local deposits.

Later in this involved discussion, De la Beche amplified this view in a way that makes sense of an earlier cryptic remark that the Culm might "correspond in age with the upper grauwacke of other places, and the whole of the old red sandstone series."[43] He had not been persuaded by Austen during his fieldwork the previous summer, and he still maintained that the main body of the Culm *underlay* the limestones of southeast Devon (§9.7). De la Beche stated that these limestones, with their fauna that Sowerby and Lonsdale considered similar to that of the Mountain Limestone, "might even be equivalent to the beds known as the old red sandstone." The striking

41. Ibid., pp. 113, 126.
42. Ibid., chap. 5, see pp. 131–32, 136–38.
43. Ibid., p. 115. In context the remark referred only to the "passage" strata at the base of the Culm, but that is inconsistent with De la Beche's other comments and was probably just carelessly expressed.

contrast in rock type and fossils between the Devon limestones and the normal Old Red Sandstone was for him no problem, because he regarded the Old Red simply as a *local* variety of the strata formed at that period: "we are not to suppose that over even the western European area the sandstones so named are always red, or that equivalents of them may not be argillaceous slates and large masses of compact limestones."[44] The reason for De la Beche's insistence that the Culm underlay Austen's limestones now became apparent. For if those limestones were regarded as roughly equivalent to the Old Red Sandstone, then, by their structural relation, the main body of the Culm was, as it were, safely tucked in underneath, with a dating even more firmly down in the "upper grauwacke." This only left as an anomaly a few patches of Culm-like strata in southeast Devon, which, as Austen had shown him, undeniably *overlay* the limestones with a clear unconformity. The age of those pseudo-Culm strata, now excluded from the genuine Culm, remained obscure; but De la Beche's reference to the Culm as "intermediate" is consistent with the opinion he had earlier expressed in private, that these anomalous strata might be small outliers of genuine Coal Measures (§9.7; fig. 9.2, B).

De la Beche's conclusions about the ages of the strata he described were so vague and elliptical, and so often couched in the conditional mood ("if we suppose . . . ," "if we imagine . . . ," "if we consider . . ."), that his own opinions are difficult to disentangle from those he imputed to others, real or imagined. But it is clear that he had stuck firmly to his original belief that *all* the older strata of southwest England (with the minor exceptions of the metamorphic rocks and the anomalous pseudo-Culm strata) formed an unbroken sequence of truly Greywacke age. He was now prepared to concede, however, that the sequence might extend upward as far as to include the equivalents of the Old Red Sandstone, under the unfamiliar guise of the limestones and associated strata of southeast Devon. For the rest, he apparently regarded the main body of the Culm as roughly equivalent to the equally *local* variety of Upper Greywacke that Murchison termed Silurian, and the older strata as roughly equivalent to those Sedgwick termed Cambrian (fig. 9.2, B).

De la Beche's reluctance to make such correlations, however, did not represent an outmoded lack of confidence in the value of fossils. On the contrary, it was due at least in part to his appreciation of the heterogeneity of the environments that must have characterized the earth's surface at every period in the past, as at the present, and to which the animals and plants of every period must have been adapted. This made untenable any simple assumption that each period had been characterized in all regions by exactly the same species. On this crucial theoretical underpinning for the practical business of correlation, De la Beche was thus close to his collaborator Phillips—and indeed to Lyell—and poles apart from Murchison.

10.6   A Priority Dispute Flares Up (February and March 1839)

A presentation copy of De la Beche's *Report* was "laid on the table" at the Geological Society at the second meeting after Murchison's book, but it was not available for purchase until the following week. Murchison could therefore

44. Ibid., pp. 145–50.

only get a first impression of it by looking at the index map and sections and had to rely on hearsay about the text. That, however, was quite enough to incite him to a characteristic reaction: he wrote the following day to both Whewell and Sedgwick, full of fury at De la Beche. *"The fact is indisputable,"* he reminded Whewell, that before his own fieldwork with Sedgwick in 1836 the sheets of De la Beche's map had been issued with the Culm area the same color as the rest of the Greywacke. Now, however, the Culm was colored separately and shown lying in a trough above the older strata. According to Murchison's unnamed informant—probably Lonsdale, who as the Society's librarian would have formally received the *Report*—the text made no mention of the fact that this change was entirely due to his and Sedgwick's work, and that they alone had saved De la Beche from "crass error" in this respect. By contrast with this conspicuous lack of acknowledgment, De la Beche had carefully cited all their "assistants and followers" and given them full credit, although those local geologists had been "actually instructed & put on the right scent" by their betters, before they ever met De la Beche. "This is 'right honourable' conduct!" he exclaimed indignantly to Whewell.[45]

To Sedgwick he wrote more bluntly, "I really think there is nothing in the history of Science so bare-faced & unprincipled & ungentlemanlike as this Spoliation." He claimed to be particularly angered that this "plagiarism" had been perpetrated "with John Bull's cash," but he also insisted that the "great guns" of government and officialdom must not "be allowed to silence the voice of geologists who have a reputation to sustain." His complaint was against what he called "the Ordnance Jockeyship of riding home with false weights"; for De la Beche had corrected his major mistake without acknowledgment, and "all [was] coolly laid before the public & at the Government expence, as an original work the result of T.H.D's labours." Behind all his indignation on behalf of the taxpayer, however, lay Murchison's anxiety that De la Beche's state-supported geology posed a long-term threat to his (and Sedgwick's) style of independent and gentlemanly geology. "It was always to be feared that the employment of public means & authorities would swamp our Society & individual efforts," he complained, "& here we have a crashing proof of it."

As a remedy for all this, Murchison urged both Whewell and Sedgwick to take firm action. He warned Whewell to read the *Report* carefully before preparing any reference to it for his forthcoming anniversary address; if he did not refer to De la Beche's plagiarism, Murchison threatened to tell him in public that he had failed in his duty. At the same time Murchison urged Sedgwick to publish a pamphlet addressed to the Master-General of the Ordnance, "to smash for ever such arrogant pretention & imposture" as De la Beche had perpetrated; he insisted that "the traitors and pirates must be publicly exposed" (the plural probably indicates the inclusion of De la Beche's allies, such as Greenough and Buckland, rather than his Survey assistants).[46]

Murchison had also thought of another way of attacking De la Beche's *Report* while simultaneously promoting his own *Silurian System* in the public realm. He urged Sedgwick to write a "powerful general review of the present

---

45. Murchison to Whewell, 7 February 1839 (TCC: WW, Add. Ms. a. 209[101]). The last phrase alluded to the parliamentary standards of behavior that GSL members emulated.
46. Ibid.; Murchison to Sedgwick, 7 and 9 February 1839 (transcripts in ULC: AS, IIID, 21).

state of our knowledge of the older rocks" for the *Quarterly Review*, using both publications as the customary peg for the essay. "It is not, my dear friend, the love of fame which impels me to ask you to respond to my call," he assured Sedgwick rather disingenuously; he warned him that this might be a last opportunity "for setting very many matters to rights—all of them deeply connected with your own well earned Laurels (of which they are stripping you)." Murchison was concerned above all with recognition in the wider world: memoirs published in the Geological Society's *Transactions* were in his opinion useless, because they were read by so few, and meanwhile "the public are deceived by others & we are cheated out of all that we labour for & laughed at into the bargain by the winners." He feared that De la Beche had a head start in the race, because the Board of Ordnance would be sending free copies of the *Report* to "every functionary & M.P." Only an essay in the *Quarterly*, with its wide circulation among the most influential in the land, would, in Murchison's opinion, suffice to counter this advantage and "disabuse the John Bulls of the jesuitry and errors of De la Beche." The editor of the *Quarterly* had first asked the political economist George Poulett Scrope (b. 1797) to write the review; but Scrope, the author of important geological works in the previous decade, had in Murchison's opinion "let the newest of the Science steal away from him" and felt inadequate to do it, while Lyell was "too much immersed in his own concerns to think of *our old Rocks*." So with a judicious blend of appeals to Sedgwick's self-interest, his assumed indignation at De la Beche, and his sense of loyalty to his science, Murchison pressed him to use his annual confinement in Norwich to stir himself to authorship and become "Adam redivivus."[47]

Of the three weapons against De la Beche that Murchison had pressed into the hands of his Cambridge colleagues, the first failed to fire only a week later. Whewell did not use his last appearance as president of the Geological Society to blast De la Beche from the gentlemanly scene. At the anniversary meeting he adopted a more subtle strategy: he heaped praise on the work of both Murchison and Sedgwick, and conspicuously omitted any reference whatever to the *Report* he had announced only the previous week. Apart from a skeptical reference to Austen's suggestion that some of the south Devon limestones might be ancient coral reefs (§9.7), the problems of Devonshire received no explicit comment at all in this, the fifth anniversary address since the controversy began. "The great works of Messrs. Murchison and Sedgwick" were praised at length, but with no reference to the tangled problems that had arisen in extending their Silurian and Cambrian systems to southwest England.

"The indefatigable exertions and wide views of Professor Sedgwick," added to "the rich results of the labours of Mr. Murchison," had in Whewell's view amplified still further the extent to which Nature had given Britain "an *Index Series* of European formations in full detail." The two collaborators had denoted the joint mass of their respective systems by terms—"Palaeozoic" (Sedgwick) and "Protozoic" (Murchison)—which neatly referred to the sig-

---

47. Murchison to Sedgwick, 9 February 1839. On Scrope as a geologist and an economist, see Rudwick (1974) and Opie (1929), respectively. Sedgwick had already declined Murchison's request that he should review *The Silurian System* in *The Times*: Sedgwick to Murchison, 10 January 1839 (GSL: RIM, S11/149).

nificance of those strata in the history of life. Silurian fossils had already been recognized widely abroad—Whewell here copied the localities almost verbatim from *The Silurian System*—and he suggested that an equally successful international correlation of the Cambrian rocks was only more doubtful "for the present." In the midst of this lavish praise, only one passing comment is noteworthy. Whewell referred quite casually to "the Carboniferous, Old Red, Silurian, and Cambrian systems," thus following Murchison's lead in elevating the Old Red Sandstone into a system on a par with the others. If this change in the perception of the "Index Series" of the strata were to become general, it was bound to affect any future reassessment of the troublesome correlation of the strata of Devonshire.[48]

At the close of his speech, Whewell delivered the presidency into what he termed the "abler hands" of his successor. Two months earlier he had written almost in desperation to Murchison and Lyell to ask for advice on who that successor should be. He had approached Herschel, Egerton, Lyell's father-in-law Leonard Horner, and even the chemist Michael Faraday, but all in vain; and Murchison would have been unenthusiastic about Scrope. In the end, Whewell had been obliged to fall back on "the old cycle of presidents": the abler hands were those of Buckland, and power in the Society thus reverted to those whom Murchison and Lyell regarded as the old guard.[49]

Sedgwick had been prevented by his duties in Norwich from attending the anniversary meeting. Ten days later, just back in Cambridge and as busy as ever with other affairs, he replied calmly to the letter Murchison had sent him, full of "bile" against De la Beche: "I am sorry for any entanglement, because it may take time to unloose, but we have nothing to fear as to the events." He ignored Murchison's plea that he should publish a pamphlet against De la Beche's alleged plagiarism, and he declined to write an essay for the *Quarterly*. Yet he did not gloss over the geological mistakes he believed De la Beche had made. He mentioned with scorn how at their last meeting De la Beche had equated the Culm limestone (at Holcombe Rogus) with one of the southeast Devon limestones (at Chudleigh), which in Sedgwick's view was quite distinct and far older. He referred in equally scornful terms to De la Beche's claim that the Culm strata lay *under* the latter limestones: "last summer [he] sent me on a complete fool's errand, being wrong on all the points on which he urged me to re-examine" (§9.7, §9.9). In striking contrast to Murchison, however, Sedgwick clearly regarded the specialist publication of their scientific results as more important than attacking De la Beche personally in the wider world of government and politics. He was most concerned that he and Murchison should publish in the Geological Society's *Transactions* the paper on Devonshire that they had written jointly almost two years before, but which was still unrevised and incomplete (§8.7); and he therefore offered to help prepare the sections that would effectively establish their interpretation in visual form.[50]

Murchison accepted Sedgwick's tacit reproof and agreed that they should indeed publish their revised paper with all possible speed. Since he consid-

48. Whewell (1839), read at GSL on 15 February.

49. Whewell to Lyell, 13 December 1838 (ULE: CL, 1); Whewell to Murchison, 21 December [1838] (transcript, dated in error 1839, in TCC: WW, 0.15.47[284]).

50. Sedgwick to Murchison, 25 February 1839 (GSL: RIM, S11/151).

ered De la Beche's sections were "to a great extent the same as ours," he proposed that they should publish a section showing the same structure—which they had first put forward at Bristol in 1836—but with the different names that would establish their priority in the matter: "we *prove* thereby who did the trick, though he has so shuffled the cards that I presume he will gain 'pro tempore' an advantage over us." Murchison suggested that the best "mode of our getting out of the affair" was for them to confine themselves to establishing their priority in the "geometry" or structural interpretation of the region and to leave the fossil evidence to Phillips. Murchison reported that De la Beche had secured government money for Phillips to travel in Devonshire, ostensibly to enable him to see in situ the fossils he was describing. "This is fudge!" Murchison exclaimed; the real reason, he believed, was to give Phillips a chance to "put to rights the geometry of the subject in which D[e la Beche] is defective." But Murchison saw that it might be a dangerous concession to leave the fossils wholly to Phillips: "we must consider if in doing so we may not part with our weapons."[51] For if his own Silurian researches were any precedent, it might still be the fossil evidence that would somehow win the day for them in Devonshire.

To do so, however, Murchison would first have to convince others that his Silurian system was no merely local series of deposits, with merely local fossil faunas to characterize them, but that it could genuinely be recognized further afield. Until then, many geologists would continue to share De la Beche's skepticism about Murchison's more sweeping claims. "As to a 'System'," Austen commented to De la Beche, "it's 'all my eye,' in which elegant predicament I also place the 'Cambrian'." Having at last got a copy of the *Report* through his country bookseller, Austen contrasted it favorably with the "mortal mass of heavy writing—and dry profitless mineralogical detail" in *The Silurian System*. Other readers might find the *Report* equally dry; but Austen, doubtless because he could follow the elaborate local detail more easily than most, found it so absorbing that it kept him up till late at night, dipping into it "as a boy picks out the more conspicuous sweetmeats of his cake." Undeterred by De la Beche's elliptical style, Austen found in it "plenty of the rationale and theory, and conjecture, to make the study of the bones of mother Earth, pleasant and encouraging." He claimed not to be "in the habit of buttering my friends," and his praise for the *Report* may have been as unforced as it was untypical of De la Beche's readers. But in any case it would have been clear, and not only to De la Beche, where Austen's sympathies and increasingly weighty support would go in the forthcoming conflict between the authors of the *Report* and *The Silurian System*.[52]

10.7   Continental Itineraries (February to April 1839)

In *The Silurian System* Murchison had staked his claim to have delineated a "Silurian" epoch in earth history, marked by a fauna that was quite distinct from that of the Carboniferous (§10.4); and Whewell had promoted that claim

---

51. Murchison to Sedgwick, 28 February 1839 (transcript in ULC: AS, IIID, 9). Phillips was planning to go to Devon in the late summer, after the BAAS meeting: Phillips to De la Beche, 26 January 1839 (NMW: DLB).
52. Austen to De la Beche, 19 March 1839 (NMW: DLB).

in his anniversary address (§10.6). At the Newcastle meeting the previous summer, Murchison had announced that he himself intended to take part in the expansion of the Silurian by pursuing it on the Continent (§10.1); doubtless Buckland had reported to him after the Porrentruy meeting how Continental geologists such as de Verneuil were already "Silurianising" the Continent for him (§10.2). Murchison had successfully seen his *magnum opus* through the press; and now, barely a year after the death of his demanding mother-in-law, he could afford—in both senses of the word—to turn to this new project.[53] He was annoyed that his mother-in-law had pointedly left her substantial fortune in trust to her daughter, but Murchison did not let that legal nicety cramp his style, and he treated his wife's greatly increased wealth as a major enlargement of his personal research fund.

As soon as *The Silurian System* was safely published, Murchison lost no time in proposing to Sedgwick that they should spend the coming field season working together, traveling more widely than had hitherto been possible. He suggested that they should go either to the Baltic or to France and Ireland. The attractions of either option were clear. The Baltic region offered them the lure of reportedly Silurian and other Transition strata, lying in an unfolded and unaltered condition, with abundant and well-preserved fossils. This would give them an opportunity to sort out the true sequence of faunas without the complications and obscurities of the more folded and altered British strata. Northwest France and the south of Ireland, on the other hand, offered them the hope of settling once and for all the validity of the analogues that De la Beche had cited repeatedly as evidence for the existence of genuine plant-bearing strata and anthracites far lower in the sequence than the true Carboniferous rocks. Only if those claims were decisively refuted could the potentially global validity of the Silurian system (and perhaps that of the Cambrian too) be put forward with full confidence.

Sedgwick approved of these plans, but warned that he could not get away as early in the year as Murchison wished, and that the pressure of his cathedral duties was such that the proposed trip might be the last serious fieldwork he would ever do. Murchison urged him however to "shake off the Norwichian trammels [and to] do that *something* more in field-geology without which your labours are incomplete and your general views *cannot be established.*" In fact, Sedgwick had no intention of foregoing this year's fieldwork; he said he preferred France and Ireland, but would go there if necessary on his own. For he objected to the way that his collaborator's plans were molded around the date of the British Association's summer meeting, right in the middle of the best time for fieldwork, and he refused to follow Murchison in being a "slave" to the Association.[54]

Even before hearing which option Sedgwick favored, however, Murchison had in fact dropped the other one: he feared his collaborator would not be available for a lengthy trip, and he considered Scandinavia too vast and desolate to tackle alone. Instead, he thought of a third option, at least for himself. This may have been prompted by the arrival in London of the latest

---

53. Murchison to Sedgwick, 19 January 1838 (ULE: RIM, Gen. 523/3; printed in Craig 1971, pp. 498–500). This important letter, in which Murchison also frankly revealed the poverty of his religious faith and the force of Sedgwick's example, was never sent.

54. Sedgwick to Murchison, 4 and 25 February 1839 (GSL: RIM, S11/150–51; the first dated in error 1838); Murchison to Sedgwick, 7 February 1839 (transcript in ULC: AS, IIID, 21).

issue of the *Annales des Mines*, containing a French translation of an important memoir by a student of von Buch's, the young Berlin geologist Ernst Beyrich (b. 1815). This would have alerted Murchison at once to the potential of the Rhineland for being "Silurianised." For Beyrich claimed that his careful study of the fossils in that region, and particularly the *Goniatites*, had enabled him to distinguish clearly between the *Kohlenkalkstein* or Mountain Limestone (e.g., in north Westphalia) on the one hand, and the true *Übergangskalkstein* or Transition limestone (e.g., in the Eifel and Nassau) on the other.[55] More specifically, however, Murchison received a letter from de Verneuil mentioning the definitely "Silurian fossils of the Eifel." Dumont had earlier inferred from his mapping that the Eifel limestones were equivalent to his lower limestone in Belgium, which, following Buckland's earlier suggestion, he had assigned to the Silurian; de Verneuil now supported that correlation on fossil evidence, as he had the previous summer with the equivalent limestone in the Boulonnais (§10.2).[56] A trip through Belgium and the Eifel, eastward across the Rhine and even as far as the Harz, would therefore have seemed an attractive proposition for Murchison's forthcoming fieldwork.

Hearing of this latest plan, and knowing that Murchison would be sure to visit Paris en route to collect all possible information from the French geologists, Sedgwick asked him to do the same for him too, so that he could go direct to Le Havre to begin his fieldwork in Normandy and Brittany. "Let us however meet if possible on the field of battle," he added; "of course the *old coal* question will be a *most prominent* point of our inquiry." For with De la Beche's *Report* in the public realm, it was more important than ever for them to test for themselves the validity of the French reports of coal in the Greywacke or Transition strata.[57]

"Like a weathercock," as he put it, Murchison promptly modified his plan once more in order to accommodate Sedgwick's proposal. He told him he himself would still go first to Belgium, and "floor that tract in a week" with Dumont and d'Omalius as his guides; instead of penetrating any further east, he would then just look briefly at the western edge of the Eifel. Returning to Paris before the French geologists left the capital for the summer, he would get their advice and then meet Sedgwick at Caen before they set out to study Normandy and Brittany together. "In two months we shall have *gutted* everything," he assured Sedgwick, and he himself could be back in England in time for the British Association. He proposed that they should then go together to the south of Ireland, where Griffith would "throw us in three weeks into every good cover," enabling them to check on the residual anomalies in that region, and Murchison could still be "home again for October shooting."

In this final plan, to which Sedgwick agreed, the most substantial work would be in northwest France, for this was the heartland of the analogies that

---

55. The original memoir (Beyrich 1837), which was Beyrich's doctoral dissertation at Berlin, was issued separately and probably not widely known outside Prussia. It was translated into French in two parts: the systematics of *Goniatites* and *Clymenia* in 1838, but the important discussion of their stratigraphical significance not until early 1839 in the first bi-monthly issue of *AM* for that year (Beyrich 1839a). The systematic part was retranslated into English shortly afterward, in the March and May 1839 issues of *ANH* (Beyrich 1839b).

56. De Verneuil's letter has not been located, but is reported in Murchison to Sedgwick, 28 February 1839 (transcript, dated in error 1835, in ULC: AS, IIID, 9).

57. Sedgwick to Murchison, 28 March 1839 (GSL: RIM, S11/153).

De la Beche had stressed in his own favor ever since the beginning of the Devon controversy. Murchison reported that he had "stuck like wax" to Buckland on a recent visit to Oxford in order to extract from him all possible advice about studying that region, which Buckland had explored in 1837 (§9.1). He told Sedgwick that on putting together Buckland's information and the published French literature, he considered "the case is pretty plain as to Silurian being strongly developed." But on the crucial question "whether the coal beds descend into this system" from the overlying Carboniferous strata *(terrain houiller)*, Buckland could only report that "all the Frenchmen swear it is below & in the older rocks" *(terrain de transition supérieur).*[58] The anomaly, so damaging to Murchison's interpretation of his Silurian system, thus remained to be resolved on the spot. Meanwhile, however, the threat nearer home had still to be faced.

58. Murchison to Sedgwick, 7 April 1839 (transcript in ULC: AS, IIID, 23).

Chapter Eleven

THE DEVONIAN SYSTEM REBORN

11.1 Old Red in Devon? (February and March 1839)

As soon as he bought a copy of De la Beche's *Report,* Murchison found that his opponent had indeed failed to acknowledge his geological debts. It was one thing however for Murchison to force De la Beche into some kind of apology over the issue of priority; it was quite another to devise a better interpretation of Devon geology.

Murchison may already have had a major alteration of Devon geology in mind when, toward the end of 1838, he had made his final revisions to *The Silurian System.* At that time he had hinted that the sharply distinct Silurian and Carboniferous faunas might be linked by some unrecognized intermediates; he had also elevated the Old Red Sandstone, which separated those faunas, into a "system" in its own right (§10.4). It is possible that those moves were intended to prepare the ground for a revival of the idea he had suggested to Sedgwick just two years earlier, namely, that the thick older strata of north Devon might be equivalent to the thick Old Red Sandstone of the Welsh Borderland and other parts of Britain, despite their striking differences in rock types and fossils. That conjecture had been ridiculed by Sedgwick, partly because he rightly suspected it was an ad hoc device to explain away the fossil plants that Harding had just reported (§8.2). Murchison had been highly unwilling to accept that those plants might come from Silurian or even older strata; although he had subsequently turned a blind eye to the anomaly, at least in public (§8.7), it had refused to go away.

276

Early in 1839, however, with *The Silurian System* out of the way, Murchison was at last obliged to look the anomaly in the face. His joint paper with Sedgwick on Devon geology had been read to the Geological Society some eighteen months earlier (§8.7), but it still had to be revised for full publication in the *Transactions*. In that form the anomaly could no longer be fudged. Sedgwick promised to get a second opinion on the plants, to add to Lindley's, from his Cambridge colleague, the botanist John Henslow (b. 1796).[1] But whatever Henslow's verdict, Murchison would have known that the Silurian could only be preserved as the system that predated all land plants if he could persuade his collaborator that the older Devon strata should be redated as equivalent to the Old Red Sandstone.

A revival of that idea, even if it risked Sedgwick's renewed derision, now had heightened explanatory attractions for Murchison. For as soon as he found time to study the *Report* closely, he was confronted once more with his opponent's impressively weighty evidence for the complete conformity between the Culm and the older strata, at least in north Devon. Murchison's failure to find field evidence for the major unconformity he had anticipated at this point in the sequence had been acutely embarrassing from the start (§7.2). His insistence that it was really there, despite appearances to the contrary, had progressively lost plausibility, as Austen and even Sedgwick had joined De la Beche (and Williams) on this crucial point, leaving Murchison with no supporters at all except "old Weaver." But if the older formations in Devon were not really Cambrian after all, but Old Red Sandstone in heavy disguise, Murchison would now be able not merely to concede the conformity below the Culm but to turn it into a powerful explanatory weapon. A complete conformity between Culm and older strata would become the natural expression of the transition from the Carboniferous group or system downward into what he now regarded as an equally important system, namely, the Old Red Sandstone. Harding's and Williams's plants in the older strata would become the anticipated traces of the flora of that "Old Red system," broadly similar to the plants of the Coal Measures but probably different in detail. Finally, the corresponding marine fossils from the older strata of Devon would become, by this new interpretation, the true record of the fauna of the Old Red Sandstone period; as such, they might bridge the otherwise anomalously abrupt disjunction between his Silurian fauna and that of the Mountain Limestone, smoothing it out into a gradual transition like those which he now claimed to see between the other major groups or systems.

To set against all its explanatory advantages, however, Murchison's revived but modified conjecture entailed the awkward task of explaining the striking contrasts between the Old Red Sandstone and its putative equivalent in Devonshire. Hugh Miller (b. 1802), the self-educated Scottish bank clerk who was supplying Agassiz with the finest Old Red Sandstone fish available, had asked Murchison the previous summer how he conceived the origin of that deposit. Murchison had replied that he thought the Old Red Sandstone was probably of marine origin, since in his own region of "Siluria" its lowest part contained fossils similar to those of the underlying Silurian.[2] But that

1. Sedgwick to Murchison, 4 February 1839 [dated in error 1838] (GSL: RIM, S11/150).
2. Miller to Murchison, 1 June 1838 (ULE: RIM, Gen. 1999/1/16); Murchison's reply is quoted in Geikie (1875, 1:259–60).

opinion made it all the more puzzling that the rest of the Old Red Sandstone in "Siluria," and the whole of it in Miller's native Highlands, contained little else than fossil fish, and was conspicuously lacking in any ordinary marine fossils, such as the mollusks, brachiopods, and corals that characterized the Silurian below and the Mountain Limestone above. To equate the Old Red Sandstone with the puzzling older strata in Devon, which did contain such fossils, therefore entailed a radical volte-face from Murchison's usual insistence on the primacy of fossil evidence. It involved equating two sets of strata that could *not* be identified by common "characteristic fossils." Furthermore, this move was all the more awkward for Murchison to propose, because it had just been suggested in print by his opponent De la Beche, albeit in the service of a different overall correlation of the Devon strata. Convinced of the ubiquity of lateral variation in both rock types and fossils, De la Beche had seen no problem with equating Austen's limestones, at least, with the Old Red Sandstone (§10.5). Now Murchison was being forced by the logic of his own interpretation to incorporate his opponent's suggestion, and thereby to abandon a central component of the rhetoric that had distinguished their positions hitherto. If the older strata of Devon were equivalent to the Old Red Sandstone, much that had been anomalous would in Murchison's view be resolved. But this could only be achieved at the cost of abandoning the full rigor of his insistence on fossil evidence, and of conceding the validity of his opponent's emphasis on the localized character of formations and their fossils.

Murchison's new conjecture received unexpected but timely support, when he heard from de Verneuil that "out of a lot of fossils sent by Austen [from Devon] seven or eight are certain Silurian fossils of the Eifel"; it is not clear whether de Verneuil acknowledged that it was Austen himself who had pointed out that affinity to him the previous year (§10.2). Phillips, Murchison reported, "seems to think they are of the same zoological epoch."[3] The comment suggests that Phillips, far from agreeing with de Verneuil that the Eifel fossils were truly Silurian, considered that both the Eifel *and* the south Devon fossils were distinct from either the Silurian *or* the Mountain Limestone faunas. In the light of that expert comment, however provisional, Murchison would have seen at once that the south Devon fossils were a promising candidate for the intermediate fauna he had predicted for the "Old Red System." All available collections of those fossils therefore became more important than ever. Austen's had already been scrutinized in detail, and Williams had provided fine specimens from north Devon; but only now was Hennah's equally important collection from the Plymouth limestone unpacked, more than two years after it had been sent to Sedgwick![4]

Murchison's brief allusion to de Verneuil's and Phillips's opinions may have been enough to alert Sedgwick to the suspicion that his collaborator was once again being tempted to equate the older Devon sequence with the Old Red Sandstone; or Murchison may have expounded his conjecture in another

3. De Verneuil's and Phillips's opinions are reported in Murchison to Sedgwick, 28 February 1839 (transcript, dated in error 1835, in ULC: AS, IIID, 9).

4. Hennah had sent the fossils after Sedgwick visited him in 1836 (§7.2), but apparently they lay unexamined among the mass of other boxes awaiting the completion of premises to house the rapidly expanding Woodwardian Museum at Cambridge. Hennah's patience was eventually rewarded with the tactful gift of a copy of *The Silurian System:* Hennah to Sedgwick, 20 October 1836, 28 October [1838], and 21 March 1839 (ULC: AS, IB, 70, 123, 178).

letter soon afterward. In any case, Sedgwick merely replied in haste, "the Devon fossils are a great puzzle: but I am as firm as ever— *No old red* in Devon."[5] His reasons for this opinion are not hard to guess. Without doubt he remained firm in his commitment to the supreme importance of field evidence, but in this respect nothing in the known sequence was now incompatible with Murchison's suggestion. Sedgwick had confirmed to his own (and Austen's) satisfaction that De la Beche was simply wrong to assert that the Culm underlay Austen's limestones (§9.7). With the elimination of that anomaly, the main features of the sequence in the whole of Devonshire were virtually agreed and settled. If the main part of the Culm were equivalent to the Coal Measures, and its basal black limestone to the Mountain Limestone, there could be no *structural* objection to the suggestion that the underlying sequence might include equivalents of the Old Red Sandstone.

What Sedgwick would have found unpalatable about his collaborator's suggestion, however, was that it threatened still further his own project for establishing his Cambrian system. Austen's proposal that some of the Devon limestones might be at least approximately of Mountain Limestone age, supported by the expert opinions of Sowerby and Lonsdale, had already robbed Sedgwick of the richest fossil faunas that might have served to characterize his Cambrian system more adequately than was possible in Wales itself. In his review of the older strata the previous summer, he had been obliged to retreat, as it were, to no more than a handful of Cornish localities (e.g., South Petherwin) as his only sources for the elusive Cambrian fauna (§9.8); his subsequent fieldwork had thrown even those into doubt (§9.9). If Murchison were now reviving his earlier suggestion that the whole of the north Devon sequence was of Old Red Sandstone age, then Sedgwick would have seen at once that there might be no justification for retaining *any* of the older strata of southwest England as Cambrian in age. Coupled with his continuing failure to find a distinctive Cambrian fauna in Wales, this might spell the end of any hope of defining the Cambrian system as clearly as Murchison had been able to define the Silurian.

Sedgwick planned to be in London the following week for a Council meeting at the Royal Society and promised to bring with him his copy of their Devon paper, so that he and Murchison could discuss these issues face to face. There is no record of what transpired when they met, but Murchison must have persuaded Sedgwick to abandon his opposition to the notion that most of the older rocks of Devon were equivalent to the Old Red Sandstone. For only ten days later he sent off an article to the editor of the *Philosophical Magazine*, under their joint names, in which that notion was put forward for the first time in public.[6] In the course of discussing this article, Murchison must also have persuaded Sedgwick that their reputations demanded a public criticism of De la Beche's failure to acknowledge the priority of their work. The article that Murchison wrote afterward, although it put forward a striking

5. Sedgwick to Murchison, 10 March 1839 (GSL: RIM, S11/153). This contains comments that suggest that it was a reply to an unlocated letter of Murchison's, later than that of 28 February (see above).

6. Ibid. Sedgwick certainly was in London for several days, for he attended the Council of GSL on 13 March (GSL: BP, CM 1/5, pp. 113) and dined with Greenough on 15 March: Greenough, diary for 1839 (UCL: GBG, 7/42); the joint article was dated 25 March (Sedgwick and Murchison 1839a, p. 260).

new interpretation of the geology of Devonshire, was at the same time a forthright attack on the author of the first *Report* of the Geological Survey.

11.2   The Devonian System Redefined (April 1839)

By writing for the *Philosophical Magazine*, rather than presenting a paper at the Geological Society, Murchison and Sedgwick ensured that their new interpretation of Devonshire would be published without delay and their priority thereby ensured. At the same time, they could criticize De la Beche's alleged plagiarism as forcefully as they liked, without the moderating influence of the Society's gentlemanly conventions. Their article on the "Classification of the Older Stratified Rocks of Devonshire and Cornwall" was published at the beginning of April, only about a week after it was sent to the editor. Although written by Murchison, Sedgwick had an opportunity to approve it beforehand and must have accepted joint responsibility for its contents.[7] It was a highly polemical document throughout, and not only where their criticisms of De la Beche became explicit. The article had two aims: to claim the authors' priority in establishing the correct structure of Devon and the correct place of the Culm strata there; and to propose a radically altered interpretation of the underlying (pre-Culm) strata.

On the first point, Murchison recalled how the original version of De la Beche's official map had been presented to the Geological Society in 1835, and put on sale, with *all* the older (pre-New Red Sandstone) strata colored uniformly as Greywacke. This established Sedgwick's and Murchison's priority in having subsequently distinguished the Culm as a separate formation, occupying a large part of central Devon, as first shown on the map they displayed at Bristol in 1836 (§7.3; fig. 7.4). Murchison was incensed that the revised version of De la Beche's map now reproduced that distinction (§10.5), but without acknowledgment of the source of the alteration. "A single sentence, a mere parenthesis *(if to the point)* would have satisfied us," he protested, but there had been none.

In a footnote he drew Phillips too into the fray. He implicitly criticized him for having cited De la Beche as his authority for coloring Devonshire in the same way on his little map of Britain (§10.3; fig. 10.2), and he mentioned that Phillips had promised to correct this "error" on its next edition. Murchison implied that De la Beche was simply incompetent to have failed to distinguish the Culm as a separate formation: even Harding, "though then very slightly acquainted with geological phenomena," had by his own efforts traced the boundary of the Culm in the north, distinguishing the black Culm limestone from the quite different limestones in the other strata. De la Beche, by contrast, had repeatedly confused the two, thereby failing to take even "the first step" toward determining the correct structure and sequence. "The truth is," he added bluntly, "that no one can make a correct section among slaty rocks till

---

7. Sedgwick and Murchison (1839a), published in April issue of *PMJS*, certainly very early in that month. Sedgwick alluded later to his opportunity to amend Murchison's text: Sedgwick to Murchison, 20 April 1839 (GSL: RIM, S11/156).

he learns to distinguish cleavage from stratification."[8] There is no evidence that De la Beche had in fact been misled in that way; but, on the other hand, his original interpretation of the structure and sequence, which he later conceded was mistaken, had unquestionably been due in part to his failure to distinguish the two sets of limestones (fig. 7.6).

Murchison contrasted De la Beche's original interpretation, by which the Culm was placed *below* some of the other formations in the north (§5.1), with his own and Sedgwick's dramatic revelation that the Culm was uppermost and lay in a huge trough in central Devon (§7.3). That interpretation was now illustrated with two new traverse sections, one of them showing clearly the symmetrical structure of the trough (fig. 11.1).[9] Murchison insisted that "the determination of the great *culm trough of Devon*," and the consequent establishment of the correct position of the Culm in the sequence, had been *"the key to the whole structure"* of the region, and that De la Beche should have acknowledged from whom he had borrowed that key. His failure to do so in his *Report*, Murchison asserted, had made the matter "now not merely a scientific, but also a moral question." For to add insult to injury, De la Beche had cited Williams's claim that *he* had been the first to report the trough structure in Devon (§9.1). Although the footnote was equivocally worded, Murchison was probably right to suspect that De la Beche had mentioned Williams's claim in order to undermine that of his opponents. Murchison protested that there was no evidence in the records of the Dublin meeting that Williams had made any such allusion to a trough structure; he relegated the matter to a footnote, "trusting that Mr. De la Beche is incapable of insinuating that which he knows to be incorrect."[10] That insinuation in turn could hardly go unchallenged.

When he dealt with the relation between the Culm and what were now established as still older strata, Murchison's polemical rewriting of history became staggering in its audacity. He referred to the lack of "any manifest discordancy" in the north as if it had been his own and Sedgwick's original opinion. He implied that this opinion had only been shaken temporarily by Weaver's report of an unconformity (§9.4), and then restored by Sedgwick's most recent fieldwork (§9.9). Likewise he mentioned that Sedgwick had found "what appeared an *unequivocal passage*" on the south during the same fieldwork, and he relegated to a footnote the admission that his collaborator had earlier claimed a clear unconformity in that area, dismissing it as "the exception and not the rule."[11] Murchison thus paid De la Beche back in his own ungenerous coinage: he conspicuously failed to give his opponent any credit or acknowledgment for having insisted all along on this complete conformity between the Culm and the other strata.

Murchison admitted however that "one of two things must follow" from his and Sedgwick's concession on this crucial point, reluctant and belated though it had been: "either the culm measures must be older, or the rocks of North Devon younger, than we had at first supposed" (in fact the comment

8. Sedgwick and Murchison (1839a, pp. 241, 255–57).
9. Fig. 11.1 is reproduced from ibid., p. 250.
10. Ibid., pp. 253–56.
11. Ibid., p. 242–43, 246.

Fig. 11.1. Murchison and Sedgwick's diagrammatic traverse sections from north Devon into north Cornwall, from their nominally joint article of April 1839, in which the term "Devonian System" was proposed as the equivalent of the Old Red Sandstone. The sections show a central trough of contorted "Culmiferous Rocks = Carboniferous System" (*a*), underlain conformably and symmetrically by "Slaty Rocks, Sandstone and Limestone = Old Red System" (*b*), the latter intruded by granite (*p*) to the south.

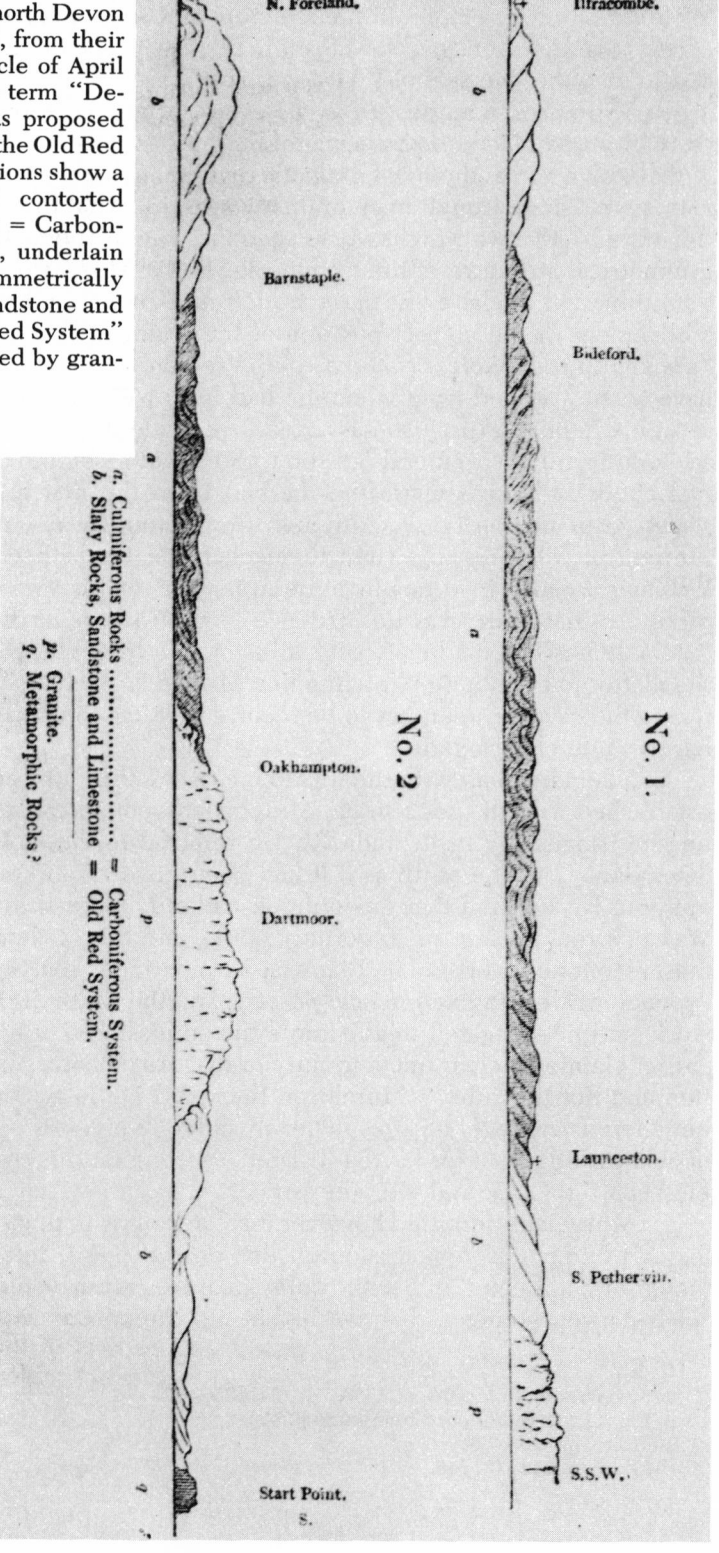

applied to south Devon too). He reviewed the fossil evidence for the age of the Culm and concluded that their original inference on that point had been correct. The plants that had precipitated the controversy proved, in their opinion, that the main part of the Culm was true Coal Measures, unusual in appearance but closely similar to those in Pembrokeshire; while some of the fossils in the black limestone near its base were of Mountain Limestone species.[12] There was no allusion to Sedgwick's earlier conjecture that the Culm might also extend downward to include equivalents of the Silurian (§9.8), presumably because Murchison had persuaded him that that interpretation had now become untenable in the light of the fossil evidence from the older strata. For it was purely on those grounds that Murchison now rejected the first option he had mentioned and chose the second instead. In their radical reevaluation of the age of the older strata, the fossil evidence was presented as preeminent.

Reviewing the improved fossil evidence from north Devon—much of it due to Williams's collecting!—Murchison now abandoned in public, as he had earlier in private (§9.4), the Lower Silurian affinities of the uppermost pre-Culm strata. Instead, he summarized the fossils from the whole sequence of thick formations as a mixture of brachiopod shells like those from the Mountain Limestone (*Spirifer, Productus*: see fig. 6.5, C, D); brachiopods and corals like those of the Upper Silurian (*Leptaena, Favosites*: see fig. 10.3, D, E); and even a mollusk like one from the base of the Old Red Sandstone (*Bellerophon*). There were also trilobites, "some of entirely new forms, and others approaching to Upper Silurian types"; and of course the now famous plants that Harding and Williams had discovered, here accepted on Lindley's authority as Carboniferous forms. As for south Devon, Murchison recalled that when Austen's fossils were discussed at the Geological Society, Lonsdale had considered that few could be "strictly identified" with Mountain Limestone species as Austen had claimed, and that many of the fossils were new and undescribed (§9.3). Murchison added that Lonsdale had also regarded some of Austen's corals as Silurian species, though that claim is not supported by the contemporary reports of the discussion. To add to all that evidence, Hennah's fine collection from Plymouth, when it was belatedly studied by Phillips and Sowerby, had proved to have the same curiously mixed character.[13]

For his new explanation of this mixed and anomalous fauna, Murchison turned the spotlight fully on to Lonsdale. He claimed that it was Lonsdale who had suggested that "the South Devon rocks would be found to occupy an intermediate place between the carboniferous and Silurian systems."[14] Of course it is possible that Lonsdale did in fact make that suggestion, but there is no contemporary evidence for it, and it has a suspiciously anachronistic tone. What Lonsdale had indeed suggested, judging from what was reported at the time, was only that Austen's fauna could not be much older than the Mountain Limestone; but that was a quite different claim (§9.3). At that time there had been no conceptual possibility that the Devon fauna could be "intermediate" between Carboniferous and Silurian, because the Carboniferous

---

12. Ibid., p. 243.
13. Ibid., pp. 243–47.
14. Ibid., p. 247.

was universally defined as including the Old Red Sandstone, and the faunas of the two systems were regarded—by Murchison and Lonsdale, among others—as almost totally distinct. Only later, when Murchison became convinced that the Old Red Sandstone deserved to be regarded as a system in its own right (§10.4), could Lonsdale's earlier remarks be retrospectively transformed in meaning and put to an explanatory use that no one at the time had foreseen.

Murchison pointed out that his own Silurian fauna, now at last fully described, coupled with the Mountain Limestone fauna that Phillips had already described so thoroughly, provided "for the first time, the means of placing the Devonian groups in their true order." He reviewed the Devon faunas—the word "Devonian" could here be read in its traditional topographical sense—and concluded that they were "all of characters intermediate between those which mark the Carboniferous and Silurian epochs." To take as an example the fossils belonging to one distinctive group of the mollusks, namely, the cephalopods, Murchison mentioned how the Devon rocks contained some that had hitherto been regarded as characteristically Carboniferous (*Goniatites:* fig. 6.5, B); some as mainly Silurian (*Orthoceratites:* fig. 10.3, B); and at least one that was unknown from either of those systems (*Clymenia:* fig. 9.3). The corals, brachiopods, and trilobites were of a similarly mixed character. Such evidence was persuasive indeed; but Murchison glossed over the fact that he himself had never until recently expected to find *any* intermediate fauna, or any intermediate system, between the Silurian and the Carboniferous, and that he had strongly emphasized their clear-cut contrast. With that striking volte-face skillfully concealed, however, he now drew the natural inference from his new interpretation of the Devon fauna. The article reached an italicized climax with the assertion *"that the oldest slaty and arenaceous rocks of Devon and Cornwall are the equivalents of the old red sandstone."*[15]

The striking character of the assertion can hardly be overestimated. At the Bristol meeting nearly three years before, Sedgwick and Murchison had proposed removing a large part of De la Beche's Devon Greywacke to the Coal Measures, and that proposal had since won growing support. Now the same authors proposed removing the rest of the Greywacke of southwest England—or, in their own terms, the Cambrian—to the Old Red Sandstone (fig. 11.1). What even they themselves had hitherto regarded as the Cambrian fauna was to be removed from *below* the Silurian system to a position *above* it, from before the Silurian "epoch" to a period that postdated it. Furthermore, the two systems that had hitherto seemed clear-cut, and sharply distinct in their faunas, were now to be separated by a new system with an intermediate fauna, which was to be equated in turn with a formation of quite different character in other regions (fig. 11.2).

Having baldly stated their proposal, Murchison skillfully pressed home its explanatory advantages. First and foremost, the "supposed difficulty" of the plants in the older strata was, he claimed, "at once obviated," though the uninitiated reader would hardly have guessed that it was Murchison himself, not De la Beche, who had found them anomalous in the first place (§8.2). With the older strata redated as equivalent to the Old Red Sandstone, the plants

15. Ibid., pp. 247–48.

Fig. 11.2. Diagrammatic columnar section to illustrate Murchison and Sedgwick's nominally joint "Devonian" interpretation of the sequence in Devonshire, published in April 1839. In contrast to all their earlier interpretations (except Sedgwick's most recent one: fig. 9.2, C), no unconformity or gap in the sequence was postulated. Also in contrast to all their earlier interpretations (except Murchison's fleeting conjecture in February 1837: fig. 8.1, C), most of the older strata were correlated with the Old Red Sandstone instead of the Cambrian. On the "standard" sequence used as a relative scale, the Old Red Sandstone (OR), now termed *Devonian*, is expanded into equality with the other systems to denote its new status, and the Carboniferous is redefined to exclude it. As in fig. 8.1, the large star denotes the position of the Culm plants; the small star, that of the older fossil plants found in strata previously attributed to the Lower Silurian.

became safely post-Silurian, and indeed little older than the Carboniferous in its newly narrowed sense; so Murchison's persistent claim that the Silurian predated any land plants or coal deposits could be salvaged after all. Second, Murchison's recently published prediction, that "zoological links to connect the *whole series*" from Silurian to Carboniferous would in due course be discovered, was triumphantly vindicated, though it had probably been inserted into *The Silurian System* almost at the last moment, as he began to anticipate the new interpretation (§10.4). "Instead of thinking ourselves rash and hasty" in proposing that interpretation, Murchison protested that he and his nominal co-author "would rather accuse ourselves of being tardy and overcautious." Having awarded Lonsdale all the credit for giving them the crucial clue, Murchison claimed that he and Sedgwick were "now surprised" that they themselves had taken so long to change their opinion on the age of the older strata. Only at this point did he concede rather grudgingly that Austen had provided the occasion for Lonsdale's clue by suggesting a Mountain Limestone age for the south Devon limestones (§9.3), thus reviving or repeating a still earlier comment of Greenough's to the same effect (§4.6). Likewise it was only here that he conceded that both Austen and De la Beche, as well as Lonsdale, had already suggested that the Old Red Sandstone was represented in Devon, even though De la Beche's suggestion was linked to a structural inference that Murchison considered untenable (§9.6, §10.5).[16]

 Presenting himself and Sedgwick as penitent sinners, Murchison protested that he and his co-author had "no desire to conceal the error into which we were first led by trusting too much to mineral characters." They claimed to have been, like other geologists, "deceived in attributing too high antiquity to strata having an antique lithological aspect and a slaty cleavage." Sedgwick's strong reasons for keeping a high antiquity for the strata, so that their fossils

16. Ibid., pp. 249, 258.

could help to characterize his Cambrian, was passed over in silence. Instead, Murchison presented them both—in tacit contrast to De la Beche—as reformed characters for whom "the day is now passed" when any but fossil criteria should be used in correlation. It was, he claimed, "a closer inspection of the organic remains" that had put them on the right track. Conversely, he argued that the striking contrast in rock types between most of the Devon rocks and the Old Red Sandstone was no legitimate objection: analogies elsewhere, such as those in the Oolitic group that had struck him so forcefully many years before (§4.1), demonstrated that "formations of the same epoch may have completely distinct mineral types."[17]

The effective conclusion of this outstandingly important article followed from that general claim. The Old Red Sandstone, which geologists had "been in the habit of regarding as the general type" for its epoch, was to be so no longer. The inference that that formation was equivalent to the older Devon strata, with their much richer fossil faunas, now demanded that a new name should replace the "old red system" to which Murchison had referred earlier in the article and in its illustration (fig. 11.1), as in his big book (§10.4). What was needed was a name that could be "applied, without any contradiction of terms, to rocks of every variety of mineral structure which contain the characteristic series of organic remains." At the end of the article that need was duly met: "we propose the term 'Devonian System' as that of all the great intermediate deposits between the Silurian and Carboniferous Systems." The fact that the authors had used that term at Bristol in a quite different sense, first as a "system" intercalated between the Cambrian and the Silurian, and then as a virtual synonym for Upper Cambrian (§7.3, fig. 7.4), was discreetly passed over. So was the inconvenient fact that the "characteristic series" of Devon fossils had *not* been found in the Old Red Sandstone itself, nor that formation's characteristic fossil fish in the rocks of Devon. There was also no allusion to the awkward fact that even the fundamental criterion of superposition could not be applied in any straightforward way to the "Devonian System" in its type area. For it was overlain by strata that were only questionably Carboniferous; and it was not demonstrably underlain by any undoubted Silurian strata, though Murchison did suggest tentatively that some of the oldest strata in Devon might prove to be Upper Silurian, doubtless in the hope of strengthening just this point.[18]

All such weaknesses in the nascent "Devonian" interpretation were swept out of sight in a final flood of persuasive prose. Murchison claimed that the full description of the Devon fossils, already being undertaken by Phillips for the Geological Survey, would complete "a regular descending order" in the older rocks. Geologists would henceforth assign their strata to "the three great systems which pass into each other, the *Carboniferous, Devonian* and *Silurian*." Whether the Cambrian would likewise prove to have "distinct typical fossils" was, Murchison admitted, still uncertain; he did not mention that it was the Devonian interpretation itself that had now eliminated Sedgwick's best chance of showing that it had. But at least above the Cambrian, "the dark and undefined aera of Greywacke" had been illuminated and clar-

---

17. Ibid., pp. 251–53. He mentioned in passing that they had in fact described *some* of the Devon rocks as *"much resembling"* the Old Red Sandstone (p. 252).
    18. Ibid., pp. 253, 257, 259.

ified, for Murchison proposed the three systems as groups of general validity. The Carboniferous had long been accepted internationally, and the Silurian was being recognized increasingly on the Continent (§10.2). So likewise Murchison now predicted that the "supposed absence" of the Old Red Sandstone beyond Britain would prove to be illusory, and that the "Devonian System" in its Devonshire guise would before long be discovered on the Continent too. Murchison urged that the "barbarous" word Greywacke, "serving as a shelter for ignorance, and paralyzing every effort for determining the succession of strata upon true principles," should therefore be abandoned once and for all.[19] With that covert stab at De la Beche's favorite term, Murchison brought this strikingly original but highly polemical article to a close.

11.3   First Reactions to the Devonian (April 1839)

The first reactions to this nominally joint article—or at least those Murchison thought worth reporting to Sedgwick—were concerned with their public attack on De la Beche rather than with their new interpretation of Devon geology. After gleaning the gossip at the Geological Society, Murchison reported that Lyell, predictably enough, was "delighted with our shot & says we have not hit hard enough," whereas Phillips "*dreads* the result as likely to be a fierce and acrimonious controversy." Lonsdale, on the other hand, "thinks we have done just enough," and expected no major row. "Nous verrons," Murchison commented skeptically; "if D[e la Beche] blazes up & agitates as an oppressed man, we have enough powder & shot in our tumbrils to sink him." Visiting Oxford shortly afterward to get Buckland's advice about their Continental fieldwork (§10.7), Murchison found the new president of the Society "a special pleader *in re* De la Beche," as he had been on earlier occasions (§8.5). But Murchison claimed he had forced Buckland to concede that their opponent "was *no gentleman* & had acted ill towards us."[20]

Buckland was not the only one, however, to criticize the two authors once again for their "fierce attack" on De la Beche at Bristol and since then in London. "Far too dead a set was made at him," commented Conybeare. He was shocked that his friend had shown "so much of the spiteful Cain," and indeed of the sinful "Old Adam"; he scolded Sedgwick for having let himself be "hurried forward by a judgement & feelings very inferior to your own." The "odium Geologicum" that Sedgwick had helped unleash was, Conybeare told him, even more distasteful than the "odium Theologicum" that he himself expected from the Oxford Tractarians in the wake of his Bampton lectures on critical patristic theology. He felt De la Beche's lack of acknowledgment was regrettable but understandable, not only because of the way he had been personally attacked but also because he, unlike the other gentlemen of geology, depended for his livelihood so directly on his public reputation. Furthermore, Conybeare defended him against the charge of incompetence, pointing out that *all* earlier geologists (including Conybeare

19. Ibid., pp. 253, 259, 260.
20. Murchison to Sedgwick, 2 and 7 April 1839 (defective transcripts in ULC: AS, IIID, 22–23).

himself) had made much the same mistake about the structure of Devonshire. That mistake was natural enough when the area was first being explored: not only was the Culm so contorted that the trough structure was difficult to detect, but in addition the then universal assumption that the line of granites marked the axis of oldest rocks made it natural to assume that the Culm would be older than the strata further to the north.

Conybeare was still skeptical about the correlation of the Culm with the Coal Measures, since the fossil evidence seemed equivocal, and he suspected the Culm would "turn out to be a local form of some middle portion of the Silurian group." The sharp distinction between the Culm and the rest of the "great grauwacke series," which Sedgwick and Murchison had stressed so forcefully, seemed to him a matter of less than first-rate significance. It is therefore not surprising that Conybeare was even less convinced by their new interpretation: "I can hardly believe you serious in referring the lowest Cambrian slates etc. to the old red," he told Sedgwick. But he realized that if it were correct, he and Sedgwick would have to make major alterations to their long-awaited joint sequel to Conybeare's classic *Outline* of British geology. The new interpretation was nothing if not a bold conjecture.[21]

Austen had been aware the previous summer how as a mere provincial he could expect to be "swallowed whole and digested" by the metropolitan geologists; "as you know," he had commented ruefully to De la Beche, "Sedgwick has a most voracious appetite and Murchison great powers of assimilation."[22] Sure enough, shortly before their new joint article was published, Austen heard that his own modest paper—which had in fact been the trigger for the new interpretation—had not been accepted for full publication in the *Transactions*, but that "any contributions will be absorbed by the authors of a paper which will be printed." When Sedgwick duly asked him to send a section showing his interpretation of the structure of his home area, Austen complied, but took the opportunity to comment on the vendetta against De la Beche. Like Conybeare, he criticized Murchison for his "personal attack" at Bristol, adding that a friend who had attended a subsequent Geological Society meeting had told him it seemed that "De la Beche was on his trial, Sedgwick and Murchison the prosecutors." Austen told Sedgwick that De la Beche regretted the quarrel "as far as you are concerned," since he was grateful to Sedgwick for having helped secure his Survey position (§5.8); by implication, Murchison's antagonism was quite another matter. "You may make of M. what you please," Austen commented to De la Beche a few days later, "but one is astonished that a man with a mind of the order of Sedgwick should come down so low" as in their latest attack: "the partnership," he predicted, "as it cannot bring up M., must end by sinking S. to the mortal level of his associate."

In Austen's opinion the affair was a disgrace to the science that Herschel had ranked next to astronomy in the "magnitude and sublimity of its objects," and which was "a science in which there is so much to be modified and corrected." He pointed out to Sedgwick that De la Beche had not been the

21. Conybeare to Sedgwick, 3 April 1839 (ULC: AS, IB, 188); Conybeare had not yet seen the article when he wrote. His lectures (1839) showed him—not for the first time—to be a Broad Churchman who was *au fait* with, and appreciative of, contemporary German biblical criticism.
22. Austen to De la Beche, 22 July 1838 (NMW: DLB).

only one to make mistakes, for Murchison's "original blunder" had been to attribute part of the north Devon sequence to the Silurian on quite inadequate evidence. Combined with their equally mistaken and obstinate denial of "the *passage* of one system into the other," this had placed alleged Coal Measures directly above alleged Silurian, and hence confused the whole correlation. Stuck in the "wilds" of rural Devonshire, Austen had not yet seen the published article; but he told De la Beche that when he read the summary Lonsdale had sent him of their "new view of the age of S. Devon, I cd. not but smile." For Austen saw at once that the proposed correlation knocked the bottom out of Sedgwick's hopes of emulating Murchison's *Silurian System* with a comparable volume on the Cambrian.[23]

Austen had sent a short note to the Geological Society, *before* he heard about the new interpretation, to establish publicly what he had been claiming privately for many months, namely, that the Great Limestone in his home area was roughly equivalent to the Mountain Limestone and the underlying "red arenaceous beds" to the Old Red Sandstone, the Plymouth limestone being still lower in the sequence (§9.6; fig. 9.2, A).[24] Although not identical to Sedgwick and Murchison's new interpretation, this was close enough to detract from their claim that Lonsdale was the only begetter of the new "Devonian" idea. It was therefore by no oversight that Austen's note was not read at the next meeting of the Geological Society, as he had been led to expect; for Murchison was determined to get in first with his own and Sedgwick's "very brief announcement" of the Devonian system. By publishing in the *Philosophical Magazine* they had bypassed the Geological Society; but a formal notice would have to be read there before they would be allowed to incorporate the Devonian idea into the revision of their long paper on Devon geology (§8.7). Even a short note about the new interpretation, however, could be expected to produce renewed argument about what had now become more important issues than ever.

11.4 Williams Provides a Peg for the Argument (April 1839)

In the event, Hamlet failed to appear and the play was postponed, as on a famous occasion more than two years earlier (§8.8). Sedgwick told his collaborator beforehand that he was too busy "hatching scholars in our College Hall" to get to London.[25] But the Devon controversy was too important to be shelved altogether, and a convenient peg for the argument was provided by a paper by Williams. It was the one that he had been planning back in the autumn of 1837 (§9.2); but in contrast to the similarly local papers that Austen and Weaver had written (§9.3, §9.4), Williams was too late for his to be read in the same session, and he did not submit it until the end of 1838. Its reading had then been postponed, owing to the author's illness. It was therefore a convenient coincidence that he was able to get to London at last, for the very

23. Austen to De la Beche, 19 March and 7 April 1839 (NMW: DLB); Austen to Sedgwick, 3 April 1839 (ULC: AS, IB, 187). The allusion was to Herschel (1830, p. 287), a sentiment quoted frequently by geologists.
24. Austen to De la Beche, 7 April 1839.
25. Sedgwick to Murchison, 4 April 1939 (GSL: RIM, S11/154).

first meeting after Sedgwick and Murchison's article had brought the Devon controversy once more into the limelight.[26]

"There is no fear of any controversy with T.H.D[e la Beche]," Sedgwick assured his collaborator beforehand; "a few angry words there may be, but that is all." Dealing with Williams was another matter, however, since they had implied in their article that he had been dishonest about his priority claim to the Culm trough. Murchison was embarrassed that he would now have to face on his own "a man to whom we have given the lie direct," and he asked Sedgwick's advice on how to handle the situation. "Pray *keep your temper,*" Sedgwick replied, knowing his collaborator's weakness in that respect; "it is most important for you to be *calm*; but *resolute* and *decisive. "* He rehearsed what he remembered of William's paper at the Dublin meeting, and claimed that Williams had not "even hinted at the existence of such a trough."[27]

As Sedgwick anticipated, Williams quoted from his Dublin paper (§6.2), in which he had noted how the strata in Devon dipped northward away from the Dartmoor granite *and* southward away from the granite of Lundy Island (see fig. 4.10). It is therefore likely that he had indeed regarded the large-scale structure as trough-like, a full year before his critics announced their independent discovery; but he had certainly failed to make the point explicit. Williams also quoted from the paper he had submitted for the Bristol meeting a year later (§7.3), in which he had mentioned more clearly the "great intermediate trough" of Culm; but that paper had not been read, so it hardly counted in the priority stakes.[28] Most of William's latest paper, however, was devoted to establishing the complete conformity between the Culm and the underlying strata on the north side of the trough. He cited De la Beche's favorite locality (Fremington Pill: fig. 8.2) as the best exposure of the gradual "passage," and displayed specimens to illustrate it. This evidence would have been highly effective in the debate if it had been presented a year earlier, since it would have directly contradicted Weaver's claim—which Murchison had welcomed enthusiastically—that there was an unconformity there (§9.4). But the point had been conceded in the meantime even by De la Beche's chief opponents, and it had lost all novelty value. Williams took the conformable relation as "establishing the grauwacke age of the Culm field beyond any further question"; but that confident assertion was just what the new Devonian interpretation had now brought back into question.[29]

As for the south side of the Culm trough, to which Williams had turned his attention since his appearance at Liverpool (§9.2), his comments would likewise have struck other geologists either as no longer novel or as implausible. He emphasized that here too the sequence was conformable, with fossil-bearing strata (at South Petherwin) to match those on the north. But De la Beche had emphasized this all along, and Sedgwick and Murchison had con-

26. Williams to Lonsdale, 17 and 23 May, 29 October, 15 December 1838; 1, 3, and 17 February, 15 March 1839 (GSL: BP, LR 3/334 and 341; 4/40, 86, 151, 251, 154, and 191); Williams (1839b), read at GSL on 10 April, the only GSL paper relevant to the Devonian controversy for which the original text has been located (SAS: DW, box II, unnumbered booklet). At its following meeting the Council agreed to William's request to withdraw it (GSL: BP, CM 1/5, p. 123); he planned to publish it himself as part of a larger work.

27. Sedgwick to Murchison, 4 and 9 April 1839 (GSL: RIM, S11/154–55); Murchison to Sedgwick, 7 April 1839 (transcript in ULC: AS, IIID, 23).

28. Williams, MS paper (n.26, pp. 1–2). The trough was "intermediate" probably in the sense that it lay *geographically* between the older rocks of north and south Devon.

29. Ibid., pp. 10–18; Williams to Lonsdale, 17 May 1838.

ceded the point more recently. In addition, however, Williams correlated *all* the strata of south Devon and north Cornwall, with a total thickness that he estimated at no less than 8½ miles, with just *one* (the uppermost) of his seven pre-Culm formations in the north.[30] At least some of those who heard his paper must then have concluded that the provincial geologist had overreached his own competence. Bilton reported to Phillips the next day that Williams's paper had been "long, jumbled, [and] confused," but full of useful local detail and illustrated by some fine fossils. Murchison reported to Sedgwick that although Williams's map and sections had been valuable, his observations had been "mingled with a 'quantum suff.' of absurdity."[31]

As expected, Williams's paper was used as a peg on which to hang a discussion of Sedgwick and Murchison's latest article, even though Williams had submitted his paper before it was published and had therefore made no reference to their new interpretation. As on the occasion at which the Devon controversy first erupted (§5.2), an account of the discussion recorded by a neutral participant, in this instance the Devonshire amateur William Bilton, can be used to check one that was far from impartial, namely Murchison's.[32]

De la Beche, who had taken the mail coach to London for the meeting, opened the discussion by claiming that he *had* mentioned his opponents' discovery of the Culm trough in his *Report;* he added disingenuously that he had simply stated what Williams had said in public, without intending to endorse his claim to priority. He asked Murchison to explain the "hard expressions" in their recent article, particularly the offensive words, "trusting that Mr De la Beche is incapable of insinuating that which he knows to be incorrect." But he spoke calmly and "in a very temperate & unobjectionable manner," and discreetly refrained from commenting on the work of his unreliable ally Williams. Since Sedgwick had, as Bilton put it, "taken care not to be present," Murchison was left to apologize for anything objectionable they had written in the heat of the moment, and he said it would have been more in line with their "strong desire to preserve harmony among the cultivators of geology" if the insinuating word "trusting" had been replaced by "believing." Northampton, who was now the president of the Royal Society, gave a noble seconding to this conciliatory gesture, and "it was voted nem. con. that the explanation was satisfactory & honourable on both sides." So "all ended à l'aimable," Murchison reported, though he added sourly that it would have done so anyway, "without any of the surpassing efforts of these good Samaritans."

This however merely cleared the decks for what those present regarded as a quite separate matter, namely, the geology of Devonshire. "Disavowing the least personality," as Murchison put it, he now "went 'the whole hog' touching the Devonian case." Bilton recounted how "he continued to assert most strongly his entire dissent from De la Beche's & Mr. W[illiams]'s views— & to be more than ever convinced that they [i.e., Sedgwick and himself] had at last hit the right nail on the head." As in the published article, Murchison

30. Ibid., pp. 27–37.

31. Bilton to Phillips, [11 April 1839] (UMO: JP, 1839/22); Murchison to Sedgwick, 11–12 April 1839 (defective transcript in ULC: AS, IIID, 24; extracts in Geikie 1875, 1:265–68). Apparently Williams displayed an updated version of the large-scale map he had shown at Liverpool (probably the sheets in SAS: DW, R/4).

32. Bilton to Phillips, 11 April 1839; Murchison to Sedgwick, 11–12 April 1839. The following reconstruction of this important discussion is based on these two letters.

gave Lonsdale all the credit for the new interpretation. But even this concession was not as generous as it may have seemed to the less knowledgeable among his hearers. For the Devonian interpretation rested far more on fossil evidence than on any fieldwork that Murchison (or Sedgwick) had done; only by praising Lonsdale's comments on Austen's fossils could he avoid having to acknowledge the major role that Austen himself had played. Austen, who was stuck in "the far west" and unable to be there, had to be kept out of the account at all costs, because he was now much too closely allied with De la Beche for Murchison's comfort. Since Lonsdale was popular in the Society and had been the object of much sympathy on account of his chronic ill-health, Murchison's gesture of generosity was a sure way of winning his audience's support.

De la Beche replied "in a very subdued tone"; according to Murchison "he did not place a veto on [any] one of my assertions," and he conceded their claim to "the originality of the culm-trough." He had probably decided to reserve judgment on the new interpretation of the older rocks until Phillips had got further with his study of the fossils; for the time being he merely said in his usual agnostic style that "it was immaterial to him what the things were called." Lyell, who spoke next, took evident pleasure in emphasizing the concession of priority that De la Beche had just made, and he inferred that his opponent was "by no means indisposed" to accept the latest interpretation too. Muchison thought Lyell advocated their ideas "very adroitly." Specifically, the Devonian interpretation was "most agreeable to him," above all because it appeared to "get rid of all the anomalies and difficulties (about plants and fossils)" which had indeed worried Lyell ever since the start of the controversy more than four years earlier. "Paddy Fitton," as Murchison nicknamed the Irish geologist, then "rose in great solemnity" to play his usual role as self-appointed guardian of the Society's gentlemanly standards. He urged "the propriety of restraining the too pungent expression of controversial writing," and he regretted that Murchison had not shown him the offending article before it was published. On the more substantive issue, Fitton probably approved the new interpretation, since, according to Murchison, he gave it "superlative praise" and emphasized the importance of what Lonsdale had done.

Greenough, as expected, "stuck up for the old names & old lights." By Murchison's account, he was "very ingeniously sophistical, tried to throw all into chaos, saw nothing new in our views, adhered to his old belief—greywacke for ever!—and sustained old Williams by casting fossil evidence overboard." According to Bilton's less biased account, what Greenough objected to was the establishment of new systems on purely local grounds, because it led to a "daily varying nomenclature" that could not be matched with the meaning of *Grauwacke* long established by "foreign Philosophes." This was a reasonable objection, since Murchison and Sedgwick had not yet demonstrated the validity of their Silurian and Cambrian, let alone their Devonian, outside the regions after which they were named. Greenough was justified in using this occasion, like many earlier ones, to castigate his colleagues for their insularity.

His international note was continued by George Featherstonhaugh (b. 1780), who claimed that "the great subdivisions of the old rocks" in the United States were "distinctly the same" as in Britain. De la Beche asked him whether the plants in the supposed Transition strata there were the same as in the overlying *"bituminous coal field"* of the productive Coal Measures, and scored a clear point when he was told that the fossils were probably distinct, as he believed they were in Devon. Following on from that reference to practical mining, a mine owner from south Wales who was prospecting in Devon said he could see none of the analogies that Murchison claimed; but Murchison retorted that he would need to trace his Welsh coal strata westward into Pembrokeshire before he would see them "put on such a greywacke aspect" that they looked like the Devon Culm.

Before the discussion was closed, Williams was given the last word and, according to Murchison, again "contended that his views were original & derived from no one," certainly not from his two chief critics. Judging from the notes he wrote in preparation for the meeting, Williams also complained that "these two Gentlemen cd. afford rather to encourage a Geologist struggling into existence as such, than to attempt to paralyze his efforts." He agreed that "no one can dispute the ingenuity of the hypothesis" they had proposed, but he found it implausible; Murchison told his collaborator afterward that Williams had "quizzed our accommodating disposition in altering our views to suit the fossils."[33] So the argument over the value of fossil evidence ended where it had been all along, with the provincial geologist as skeptical as his gentlemanly critics (or at least Murchison) were confident.

The discussion was brought to an end by the president, as custom required. According to Murchison, Buckland failed to give him and Sedgwick adequate credit for the discovery of the "coal trough" of Devon, "nor did he do justice to my Siluriana, without which, as you [Sedgwick] have justly said, no one could have started this hare." On the other hand, Murchison reported that Buckland was "particularly happy in assisting to demolish 'Greywacke' by bringing old Greenough up." For Buckland reckoned he had now seen the greywacke rock type in formations of so many different ages that "he did think with Conybeare that it was 'Jupiter quodcunque vides,' & agreed with us in the fitness of using it hereafter *entirely* as an adjective or expletive: Q.E.D." With Greywacke thus abandoned by the president, his acceptance of the new interpretation followed easily. According to Murchison, "he ended by *approving* highly of 'Devonian'—he now saw new light"; he gave Lonsdale the credit for it, and concluded with characteristic extravagance that henceforth there would be "two great names in English geology—W. Smith & W. Lonsdale!"

The meeting had gone on until the exceptionally late hour of half past eleven, and the customary tea and coffee upstairs had gone cold. Murchison told Sedgwick that the unbiased Warburton had thought the new interpretation "sound and unassailable." Warburton reckoned that "Buckland *had completely given in*, De la Beche was ready to do so, & Greenough alone held out, standing like a knight-errant upon his antiquas vias." Significantly, Murchison did not even mention the resistance of the provincial Williams, although it was his paper that had ostensibly been the subject of the meeting.

Summarizing the discussion the following day, Bilton told Phillips that the general opinion "with regard to the personal question, was, that De la

33. Williams, notes later than those dated 22 March 1839 (SAS: DW, box I, notebook 16). These lengthy notes are headed as being for "the meeting." They include the inference that, if the uppermost formation below the Culm were lowest Silurian (Llandeilo), as Murchison had originally suggested, then the Culm itself might be only slightly younger (Caradoc).

Beche had not mentioned S.'s & M.'s previous labours, as fully & honourably as he ought, but that the latters' paper was unjustifiably severe." Murchison reported to Sedgwick with his usual pugnacity that "the fight is over"; but he added that, "all things considered, we have come off remarkably well," since at least they had not been censured for their personal attack on De la Beche. As for their striking new interpretation of Devon, Murchison claimed that "our paper has *done the trick* & the recusants know it & feel it & sooner or later must bow to it." Bilton was somewhat less sanguine, for he told Phillips, "I do not think it can possibly be considered to be decided, but that it preponderates in favour of S. and M." In any case, it had certainly won an impressive measure of support at its very first exposure to expert debate. Bilton realized that Phillip's work on the fossils would now be crucial: "I do not altogether fancy you will find it easy," he told him, "to arbitrate satisfactorily between the contending opinions, & there will be no harm that you should first let matters settle a little."[34] That however was the very last thing that was likely to happen: far from being allowed to settle, the controversy had now been given a vigorous stir.

11.5   Consolidating the Devonian (April and May 1839)

Phillips had not been in London for the discussion that followed Williams's paper, but he could not keep out of the controversy. On the eve of the meeting De la Beche had asked him if it were true that he had promised Murchison to alter his little geological map, on which De la Beche had been cited as his authority for southwest England (§10.3). Phillips at once wrote a note to the editor of the *Philosophical Magazine*, protesting that his map had not contained the "error" Murchison had accused him of, for he *had* derived the geology of southwest England from De la Beche's then unpublished index map of the region. All he had borrowed from De la Beche's opponents was the Carboniferous attribution of the Culm, but this had long been in the public realm and needed no explicit acknowledgment on the legend of a map.

Murchison tried to persuade him not to publish this note; but Phillips protested to Sedgwick, "I *cannot withdraw it*, without injustice to De la Beche; the matter itself is of *no interest whatever to me personally*, but in the position of Mr De la Beche's scientific reputation, of much consequence (I think) to him." Sedgwick was full of remorse: "it was inconsiderate & unkind of us," he told Murchison, "to bring up Phillips by name, & I am very sorry I did not run my pen thro' the sentence." Alluding to Phillips's prospects of a permanent position with the Geological Survey, Sedgwick realized that "he must (more than any one) hate to have his name brought forward in our controversy." Phillips was indeed caught in the cross fire between his two powerful patrons, Murchison at the British Association and De la Beche at the Survey. Not surprisingly, he agreed to modify his note, and the tone of what was published was conciliatory. He referred to his critics' discovery of the true age of the Culm as a "very important change," and acknowledged

34. The advice was not disinterested, since Bilton hoped to persuade Phillips to accompany him to Norway that summer, now that Sedgwick and Murchison had decided against going to Scandinavia; he did not want Phillips to feel he must stay in England to see the controversy through: Bilton to Phillips, 1 and 11 April 1839 (UMO: JP 1839/17, 22).

the "originality and value" of the work that had led to their "new and remarkable views" on the age of the older strata, though he did not explicitly
accept the new interpretation.[35]

Murchison now realized that he and Sedgwick would have to publish
another note in the same journal, to withdraw their attacks on Phillips and
De la Beche, but at the same time to defend their criticism of Williams for
his unjustified claim to priority in the discovery of the Culm trough of Devon.
Such a note would also enable them to publicize their forthcoming fieldwork
in Brittany: "we can assure our friends that though various anomalous cases
on the continent are cited as impugning our views, that we have resolved to
encounter these Breton Knights forthwith & see what metal they are made
of." Alluding to their revision of their Devon memoir for the Geological Society's *Transactions*, Murchison told his collaborator, "we must offer our work
with a new face, or rather a new back side, to the Council, & if they kick it
we must pocket the affront." He foresaw that their critics on the Council might
raise procedural objections to printing a paper that would differ radically in
content from what had been read at the Society two years earlier, namely, in
the interpretation of the lower strata of Devon (its bottom or "back side"!).
Murchison considered that if such objections were made, it would be better
to publish a full account of their new interpretation elsewhere, rather than to
delay it any longer: "those may laugh who win."[36]

The day after the discussion of William's paper, Murchison had "a long
and satisfactory chat" with him at the Geological Society. "He really, poor
man, is not to blame," he told Sedgwick, "& though he started on a *wrong
scent*, & is a *coarse subject*, he has worked hard & well—twice as well as De
la Beche as regards N. Devon." He acknowledged that Williams's was "far
the *best* fossil collection we have ever had from north Devon," and he was
confident that the provincial geologist would "give way as soon as he is told
what to do by the fossil evidences." Having studied Williams's fossils with
Lonsdale, Murchison reported that "as far as we can yet see they confirm
better than ever the new view," and he promised to get drawings made of
"some of the best Devonian types" for them to take to the Continent. He sent
Williams's north Devon fossil plants to Cambridge, asking Sedgwick to get
Henslow to give a second opinion on them. Murchison hoped Henslow would
confirm Lindley's view that some species were different from any known from
the Coal Measures; "for if one or two such plants can be found in what the
French call their 'Transition coal fields' our case will be infinitely more complete." He added that "the existence of one or two true carboniferous species
is just what we would expect, together with a new form or two"; such a mixture
of plant species would match the mixture of animal fossils that had helped
define the new Devonian system. After a further's day work on Williams's
fossils, this time with Sowerby to help as well as Lonsdale, Murchison told
Sedgwick "our proofs are better than ever." He was already hard at work on
the short paper designed to establish their right to publish their new interpretation in the *Transactions*. "I am colouring a map of Devon & Cornwall

35. De la Beche to Phillips, 9 April 1839 (UMO: JP, uncat.); Phillips (1839), dated 11 April,
published in May issue of *PMJS*; Phillips to Sedgwick, 16 April 1839 (ULC: AS, IB, 192); Murchison to Sedgwick, 16 April 1839 (transcript in ULC: AS, IIID, 26); Sedgwick to Murchison, 20
April 1839 (GSL: RIM, S11/156).
36. Murchison to Sedgwick, 11–12 April 1839.

on the new entire scheme," he explained, "& thus the whole hog will be dressed harmoniously in Devonian sauce."[37]

Murchison contrasted Sedgwick's gouty indolence with his own frenetic activity: "I have been labouring myself into dyspepsia by working like a slave & a peace maker." In that unaccustomed role he duly wrote a note for the *Philosophical Magazine*, apologizing for their ungentlemanly insinuation against De la Beche, while maintaining their priority claim against Williams. It was also, he told Sedgwick, to "explain the beautiful diapason I have established in our joint names," namely the satisfying harmony of the Devonian interpretation. This time he took the precaution of getting Buckland's approval of what he had written, and he sent it to Sedgwick for his agreement too. In a striking reversal of their usual roles, however, Sedgwick told Murchison he had now become too conciliatory: "my impression is that the reader will think we have been in the wrong, & that under that feeling we have made a very *humble apology.*" He did not think any apology was called for and dissociated himself from Murchison's note: "I feel that De la Beche has treated us both unjustly, & that I still have much to complain of."

In the event Murchison did not wait for Sedgwick's reply, and the note was published under their joint names after all. Murchison asserted that William Smith deserved credit for having colored part of Devon as Old Red Sandstone a quarter-century earlier. Smith's map had in fact been very crude in that area, but the concession allowed Murchison to claim Smith once more as a legitimating forerunner. He maintained that he and Sedgwick had simply applied Smith's methods to Devon, using far better fossil evidence than had been available to Smith himself. Toward De la Beche he was duly conciliatory, maintaining that their quarrel had now been resolved. Toward Williams, who was described patronizingly as a diligent collector and "industrious observer," Murchison was more unyielding. He denied that Williams had postulated a trough structure at the Dublin meeting, and thereby rejected his claim to priority in its discovery. He conceded that Williams had indeed outlined such a structure in the paper he submitted for the Bristol meeting, but he pointed out that Williams had failed to attribute the Culm to the Coal Measures. Nonetheless, Murchison claimed that the priority issue had now been resolved amicably, and that the sequence of strata in Devon was "completely settled." All that remained to be discussed was their correlation. Murchison credited Lonsdale once again with having first suggested the new interpretation. Once the intermediate character of the Devon fossils was fully demonstrated, he concluded, all "lovers of truth" would be bound to "admit the necessity of so great and sweeping a reform," and would duly accept the *"Devonian System."* He added that he and Sedgwick planned to study the alleged Transition coalfields in France during the summer, and that they thought it "highly probable" that these would turn out to be Devonian too, so that the whole anomaly would finally be resolved.[38]

Williams too hurried into print with his own note for the *Philosophical Magazine*, referring to the Devonian interpretation as "the hypothesis of Mr Lonsdale." Once more he rejected the correlation of the Culm with the Coal Measures, and said he still believed it to be a true Transition group. But since

---

37. Ibid.; Murchison to Sedgwick, 16 April 1839.

38. Murchison to Sedgwick, 16 April 1839; Sedgwick and Murchison (1839b), dated 19 April, published in May issue of *PMJS;* Sedgwick to Murchison, 20 April 1839.

he used that term to refer to all strata up to and *including* the Old Red Sandstone, he was ready to accept the new correlation of the older strata of Devon provided that the Culm was *also* equated with the Old Red. The "rudimentary efforts of nature at a coal deposit," as seen in the Culm, would reflect what he termed a "first ascending within the carboniferous influence," or, in other words, an approach toward the true Carboniferous world. Williams offered "this flag of truce" to his opponents with his usual provincial obscurity of expression, but it did represent another distinct interpretation. In his own view it vindicated his (and De la Beche's) insistence of the complete conformity between the Culm and the older strata, but at the same time it conceded that part of that unbroken sequence might indeed be of Old Red Sandstone age. At least to that limited extent, even Williams had now made a certain concession toward the new interpretation.[39]

Before the three notes by Phillips, Murchison, and Williams were published, Murchison had arranged for his and Sedgwick's short paper on the new interpretation to be read at the very next meeting of the Geological Society. De la Beche was unable to come to London again so soon after his last visit, so Greenough undertook as usual to defend his interests. "The new go of Sedgwick and Murchison," De la Beche told him beforehand, "strikes me as a piece of ingenuity to get over difficulties which embarrassed their old story." Its ad hoc character was betrayed by the fact that it had only been produced after his *Report* was published, and not in Murchison's "Silurian tome." He also objected to the way Murchison repeatedly cited Pembrokeshire as an analogy to Devon, while keeping quiet about the geographically closer comparison with the Mendip Hills in Somerset. The normal Carboniferous formations there, he pointed out, were claimed to become "quite altered characters from a short confinement in the dark," where they were concealed by the younger rocks to the south, "while they would not change so long as they were permitted to roam freely in open day" over the rest of England and into south Wales (see fig. 10.2).

As on earlier occasions, however, De la Beche's most fundamental objection was that the Silurian "system" had no evident validity outside its original area. Even its alleged extension into Pembrokeshire seemed to him "a very forced thing" that would not stand up to critical scrutiny. "This Devon business," he was confident, "will do much to knock an undue extension of it on the head." If the Silurian were indeed just a useful regional category rather than being of worldwide validity, Murchison's boastful claims for it would soon prove to be exaggerated. "M. had better present the superbly bound copy of his tome to the Queen, as intended, as soon as he can," De la Beche commented, perhaps with a touch of jealousy; if the current rumors were true, Murchison would have to "push the negotiations for his Baronetcy briskly forward, or else by Jupiter I am afraid the thing will be too late." The latter-day chieftain of the Silures, Lord Caractacus—as Scrope had earlier suggested he should style himself, if elevated to the Tory peerage—might now fail to acquire even the lesser title of Sir Roderick![40]

39. Williams (1839a), dated 24 April, published in May issue of *PMJS*.

40. De la Beche to Greenough, 21 April 1839 (ULC: GBG, 63); De la Beche to Phillips, 21 April 1839 (UMO: JP, uncat.), closely similar in content but less outspoken; Scrope to Murchison, 5 November 1835 (GSL: RIM, S10/6). De la Beche had evidently not noticed Murchison's slight but significant hint in *The Silurian System* that an intermediate fauna to link Silurian and Carboniferous might be found (§10.4).

Sedgwick and Murchison's paper "on the classification of the older rocks of Devonshire and Cornwall," designed to legitimate the revision of their original paper, was read at the Geological Society a few days later. To emphasize that the change lay only in their broader correlations, and not at all in the structure or sequence, Murchison put up in the meeting room the same sections that they had displayed at Bristol in 1836. Their only major alteration was to "remove the lowest rocks from the Cambrian and Silurian systems to the old red," a change that was "founded on zoological evidence recently obtained." Furthermore, Henslow had reported just in time that the fossil plants from the lower strata in north Devon were "quite distinct from any known coal measures remains," thus confirming that the Devonian flora, like the fauna, was distinct. The striking contrast between the Devon rocks and the Old Red Sandstone in other regions was glossed over with the bland assertion that "the variation in composition of other formations within limited areas is equally great." Murchison mentioned "the previously ascertained regular sequence or passage" between the Culm and the older strata, without acknowledging that it was De la Beche who had originally ascertained it, or that both he himself and Sedgwick had until recently denied it vigorously. "Mineral character being no longer indicative of age," they asserted that a new system name was now needed; since "Devonshire affords the best type of the fossils of this intermediate system," they proposed "to substitute the term *Devonian* for old red sandstone." The magnitude of the claim being made for this term was clear from the hope they expressed that Continental geologists would soon recognize the new system by its fossils; they thus preempted any foreign claims to fuller or clearer representatives of the proposed intermediate system, by giving it an unmistakably English name before they had even looked for the same fauna abroad.[41]

Austen had submitted a paper earlier in the year, hoping to establish the priority of his own views before they were "swallowed" by Sedgwick and Murchison (§11.3). But Austen could wield no power from the depths of Devonshire, and his paper was held back until *after* his predatory seniors had had their say. He simply revised the sequence of older strata he had described a year and a half earlier (§9.3), confining himself again to southeast Devon. In the context of the continuing debate, the revision was mainly important for putting on public record what he had long been claiming in private, namely, that the strata underlying the Great Limestone in his area were equivalent to the Old Red Sandstone (§9.6).[42] But by delaying the reading of the paper, those with power in the Society had neatly transformed its apparent force. Instead of being seen as a priority claim to at least a part of the new Devonian interpretation, it was bound to appear simply as a local geologist's confirmation of what Sedgwick and Murchison had proposed in the preceding paper.

No direct account of the discussion that followed the reading of these papers has been preserved, but Bilton passed on to Phillips a second-hand account of what had been said. "Sedgwick expressed himself very strongly & eloquently in favour of the new Views of our N. Devon Grauwacke," he reported; "everyone appears to consider the last new light as much more

41. Sedgwick and Murchison (1839c), read at GSL on 24 April. Hennah's Plymouth fossils were apparently among the specimens displayed: Hennah to Sedgwick, 2 May 1839 (ULC: AS, IB, 196).

42. Austen (1839b), read at GSL on 24 April.

vraisemblable than the former." The comment shows not only that Sedgwick was perceived to be wholeheartedly in favor of the Devonian, and no longer merely a reluctant collaborator of Murchison's, but also that the general opinion of the meeting had been in its favor too. Bilton added on his own account that he thought Phillips would find the north Devon rocks "a tough morsel whenever you are called upon to crack them—& a still tougher job to reconcile all to your opinions whatever they may be."[43] Nonetheless it is clear that the new interpretation was gaining support with remarkable speed.

After the meeting Sedgwick got down once more to the revision of the paper that he and Murchison had presented almost two years earlier (§8.7). "We must try to support our *former system* and not change the names of the subordinate groups," he told Murchison; "all we do is to shift our horizon to fit the fossils." The Devonian could thus be presented as a striking innovation made necessary by the expertise of the fossil specialists, which in no way invalidated the complementary expertise of the field geologists. "Perhaps I shall be able to keep the descriptive part and the zoological part quite separate," Sedgwick suggested; "first describe the sections as geometrical phenomena; then the fossils." How soon this could be done remained uncertain, however, for Sedgwick's usual ailments were in full force: "a *curse* is on my *head,*" he told his collaborator, "I cannot sleep, or think, or write coherently."[44]

The Geological Society did not wait for Sedgwick's headache to clear. His and Murchison's latest paper, together with their original joint one, was sent off to Conybeare as referee. At first that overworked parson merely added a perfunctory postscript to his earlier report on Sedgwick's solo paper (§9.8); but when asked by the Council to give a clearer opinion, he recommended that the two joint papers should be published as they had been presented, "with their respective dates attached." Only the authors' "probable conjectures" on the Silurian and Cambrian dating of the older strata needed deletion from the paper they had written two years earlier; but Conybeare recommended that even this change should be duly noted in print and justified "with reference to the correct identifications as established by the Palaeontological evidence" given in the more recent paper.[45] That last comment indicates that Conybeare too already regarded the new interpretation as valid, or at least plausible.

Meanwhile Austen was understandably irritated at the way Sedgwick and Murchison were taking all the credit for the Devonian interpretation. When he next wrote to De la Beche, it was primarily to try once more to persuade him that he was mistaken in his interpretation of the field relations of the Culm in southeast Devon (§9.7). This however was now a marginal issue, in the sense that De la Beche was totally alone in arguing that the Culm was below and not above the limestones. But the fossil evidence in this case reminded Austen that "with respect to these older rocks of Devon, you know

43. Bilton to Phillips, 30 April 1839 (UMO: JP, 1839/21). Phillips's duties for BAAS had forced him to decline Bilton's invitation to accompany him (at Bilton's expense!) to Scandinavia; Bilton complained of BAAS's "very grievous tyranny" and said he would not regret its widely anticipated "early decay, if not decease."

44. Sedgwick to Murchison, 1 May 1839 (GSL: RIM, S11/157).

45. Conybeare to GSL, 13 and 24 May 1839 (GSL: BP, COM P4/2, pp. 3, 107); the paper was approved for publication on 5 June (GSL: BP, CM 1/5, pp. 137–38). The first report also recommended that Austen's 1837 paper (§9.3) should be further revised with Sedgwick's assistance, and that Weaver's paper (§9.4) had now been superseded. Austen's more recent paper was also returned to him to "revise and condense": Council of GSL, 22 May 1839 (GSL: BP, CM 1/5, p. 132).

that I have long held that they were of the age of the old Red, the uppermost portion being the equivalent of the Mountain Limestone; this M. and S. acknowledge, yet constantly speak of 'our' identification." Austen had indeed correlated some of the Devon rocks with the Old Red Sandstone before the Devonian interpretation had been devised (§9.6). But he seems not to have grasped the crucial point that Sedgwick and Murchison, in enlarging that correlation to take in almost *all* the pre-Culm strata in Devon, had at the same time eliminated the Mountain Limestone correlation that Austen had first suggested. For they had used Lonsdale's opinion that Austen's fauna was not exactly that of the Mountain Limestone as evidence that it was of a distinctly *different*—namely, a "Devonian"—age.

Nonetheless, Austen felt with good reason that it was his original suggestion, later confirmed by the fossil specialists, that had first made *any* such interpretation possible. "In fact Phillips and Lonsdale, from their very first inspection of the fossils, upset the trumpery about upper or lower Cambrian," he recalled to De la Beche; "but what could you expect but error, from identifications founded on 'lithological character'; it has led them into many scrapes, & will into many more—upon this, half of Silurianism is founded."[46] The criticism was not unfair: Murchison might lose no opportunity to proclaim the supremacy of fossil evidence, yet in practice he and Sedgwick had certainly been swayed more by the "antique" appearance of the rocks in Devon, when they first concluded that they were Cambrian (or, at youngest, Lower Silurian) in age (§7.2).

Austen was aggrieved that Sedgwick and Murchison had denied him credit even for the assembly of the relevant empirical evidence. He had heard that their forthcoming paper in the *Transactions* would be substantiated with *"numerous plates* of characteristic fossils by R.I.M. etc. & Prof. S.," yet many of the best of these had been due to Austen's collecting and fieldwork. "M. *was only a few days* in S. Devon before the Bristol meeting," he recalled; since then Austen himself had been "with Sedgwick on every occasion (in fact he always came here) so that I know exactly what he has seen." Nor would the older geologists acknowledge Austen's role with respect to the wider implications of the Devon fossils. In his most recent paper, Austen had correlated his south Devon limestones with some in the Rhineland: "this they have now taken up and are going there directly—the hint they attribute to Verneuil but he had it from me, with the very names of the fossils, which I sent him" (§10.2). "I am getting rather out of humour with this said Geolog. Soc. and shall have very little more to do with it," he told De la Beche; "it seems to be conducted for the benefit of a clique." At the age of thirty-one, he was about to move away from Devonshire altogether, to live near his father in Surrey. He still hoped Phillips would visit him before he left, and he even promised to delay the packing of his collection until the last moment. But clearly he sensed that his brief career as a valued local expert on the geology southeast Devon was coming to an end.[47] Just as Austen prepared to leave Devonshire, the metropolitan geologist he had criticized was beginning to explore the Continental correlation that Austen claimed *he* had been the first to suggest.

46. Austen to De la Beche, 20 May 1839 (NMW: DLB).
47. Ibid.; De la Beche to Phillips, 26 May 1839 (UMO: JP, uncat.). Austen's correlation of the Devon limestones with some in Germany is not mentioned in the brief official summary in *PGS*, but there is no reason to doubt that he made it.

Chapter Twelve

CONQUERING
THE CONTINENT

12.1   Murchison Opens the Rhineland Campaign
(May and June 1839)

With an unerring eye for the strategy of international persuasion, Murchison left London by steamer for Boulogne, less than three weeks after the "Devonian" paper was read at the Geological Society; he traveled straight to Paris to explain the new interpretation to the Société géologique. He had intended to go there anyway to get local information from the French geologists in preparation for his fieldwork in Belgium, the Eifel, and Brittany (§10.7). But the new Devonian interpretation gave his visit much greater significance. He duly consulted the French geologists, particularly de Verneuil, and he was lucky to find d'Omalius in Paris, too, to act as an informant about Belgian geology. But the most important event while he was there was the meeting of the Société, where he presented a copy of his and Sedgwick's article from the *Philosophical Magazine* and explained their new Devonian system. In response to criticism that it seemed poorly characterized, Murchison claimed that on the contrary the Devonian was as distinctive as his own Silurian and Sedgwick's Cambrian. In contrast to the French geologists' expectation that genuine *terrains* should be clearly distinguished by "séparations bien tranchées" (§10.2), Murchison now insisted that the strict application of that criterion would entail the absurdity of lumping *all* the English formations into a single system, because conformable "passages" were not the exception but the rule.[1]

1. Murchison, notebook "1839," notes for 15–25 May (GSL: RIM, N80, pp. 1–26); SGF meeting on 20 May (*BSGF*, 10:313–14).

Yet however confident he may have felt about the intermediate character of the Devonian fauna, Murchison was acutely aware that in Devonshire itself the field evidence for the intermediate position of the Devonian strata was very weak. There was a growing consensus that the Culm was overlying in position and Carboniferous in age, but no clear evidence that the "Devonian" strata were underlain by any Silurian. To strengthen the case for the Devonian system, Murchison would need to find much better evidence of superposition, linking strata that contained the Devonian fauna upward to the Carboniferous *and* downward to the Silurian. Since this was also important for establishing the wider validity of the Silurian itself, it was now in fact the primary goal of his forthcoming fieldwork.

At the time their joint paper was read at the Geological Society, Sedgwick had still been planning to work with Murchison in Brittany (§10.7), as he told Greenough when they dined together at the Athenaeum; and Élie de Beaumont had offered to help him with advice on that region.[2] But while Murchison was in Paris he sent Sedgwick a report which led them abruptly to abandon that plan. "I rejoice above measure at the news from Brittany," Sedgwick told his collaborator, and he agreed that they should work instead in the Rhineland. Judging from what Murchison later recalled, he had been studying the fossils from Brittany in the Paris museums and had concluded that the strata with coal seams in the *terrain de transition supérieur* were either Devonian or Carboniferous, but definitely not Silurian. Only some such firm conclusion would have led the British geologists to alter their plans so radically, for it would have convinced them that the alleged anomalies that De la Beche had cited so persistently could now be safely ignored, or at least shelved indefinitely. Conversely, Murchison's discussions in Paris may have strengthened his conviction that Belgium and the Rhineland would offer them better value for their time, by enabling them to see clear sequences of strata and fossils that might help consolidate the Silurian and Cambrian systems as well as the new Devonian. Anyway, Sedgwick agreed to meet Murchison in Bonn as soon as he was free, and meanwhile he worked hard on their Devonshire paper, consulting Lonsdale and Sowerby about the fossils.[3]

Murchison left Paris by mail coach and headed straight for Prussian territory. "Vive la Prusse, Down the French," exclaimed the Francophobe ex-soldier to himself as he crossed the frontier "clean out of the poisoned land, the curse of Europe." Safe on Prussian soil he found "order now at every step."[4] It can hardly be emphasized too strongly that the orderliness was not only civil or political; geologically too, Murchison was in no sense entering untamed territory. The older rocks of the Rhineland had been quite thoroughly explored and discussed, particularly in the massive four-volume compilation on *Das Gebirge in Rheinland-Westphalen*, edited in the 1820s by Jacob Nöggerath (b. 1788), the professor of mineralogy and mining at Bonn. He and his colleagues had described how the older rocks on both banks of the Rhine

2. Greenough, diary entry for 26 April 1839 (UCL: GBG, 7/42); Élie de Beaumont to Sedgwick, 30 April 1839 (UCL: AS, IF, 54).

3. Sedgwick to Murchison, 25 May and 4 June 1839 (GSL: RIM, S11/158–59). Murchison's letter from Paris has not been located; his retrospective remarks were made at the SGF field meeting in September 1839 (*BSGF* 10:418). The news of their change of plans must have spread rapidly, since Austen knew of it almost at once (§11.5).

4. Murchison, notebook "1839," p. 38.

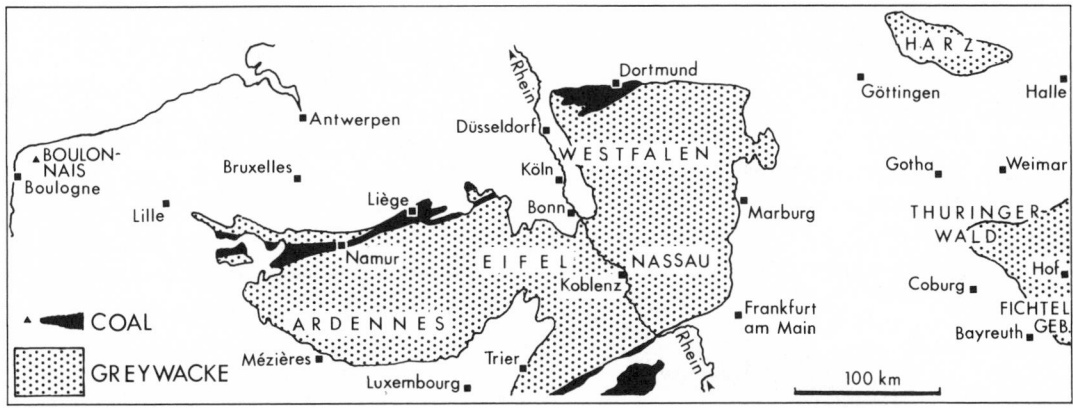

Fig. 12.1.   Outline map of parts of Germany, France, and Belgium to illustrate Murchison's and Sedgwick's fieldwork in the summer of 1839. The geology is based on von Dechen's geological map of Europe (1839), which was published while they were in Germany.

were lapped on all sides by Secondary strata (fig. 12.1). Disregarding the igneous rocks, most of that massif was composed of Greywacke or Transition strata, with many patches of Transition limestone, some with abundant fossils. But on the northern edge, in Westphalia, the Berlin mining geologist Heinrich von Dechen (b. 1800) had described a clearer sequence of distinctive formations with a folded structure not unlike what Dumont had later unraveled in Belgium. The formations ranged from the Coal Measures *(Steinkohlengebirge)* of the Ruhr coalfield, downward through sandstones, thin limestones and shales, to a massive limestone and so into the top of the main sequence of older *Grauwacke*. More recently, Beyrich's work (§10.7) had opened up the possibility of bringing order to the Greywacke itself by means of its fossils (particularly its cephalopod mollusks such as *Goniatites)*. Murchison would have had no chance of reaching even provisional correlations with Britain in a single season of fieldwork had he not been guided by published German memoirs such as these, and by the informal advice of the highly competent German geologists.[5]

    As soon as he reached Trier, Murchison called on Johann Steininger (b. 1794), the teacher of mathematics and physics at the Gymnasium, who had published a geological map of the Rhineland and knew the region well. He also studied the Eifel fossils in the local museum, but he intended to visit that area later in Sedgwick's company. Having decided to work meanwhile further east, he traveled directly to Koblenz and crossed the Rhine on to its right bank to begin his fieldwork there (this part of his itinerary can be followed on fig. 12.2). He worked first across the Taunus, where the rocks reminded him of the "S. Welsh System intermediate between the Silurian & Cambrian"; at Ems on the northern flank he noted that "they almost Silurianize, being there on the edge of a younger basin." After his arrival in Frankfurt he told his wife that "as I had no Silurian rocks (though sometimes very nearly approaching them) to traverse, this rapid section was enough." He noted, however, that his correlation was "but a guess, & is made before I see the key of

5. Von Dechen (1823), in Noggerath (1822–26); Beyrich (1839a); Wiegel (1973).

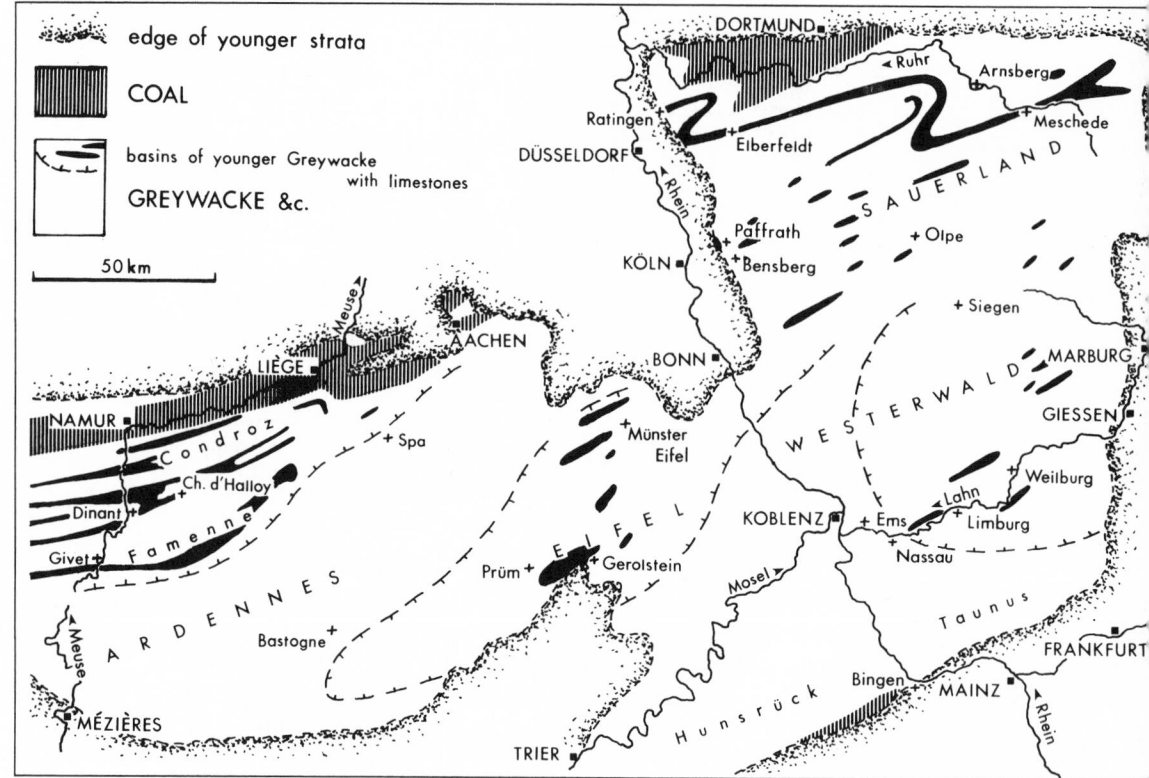

Fig. 12.2.   Map of the Rhineland and part of Belgium, on a larger scale than fig. 12.1, to illustrate Murchison's and Sedgwick's fieldwork in the summer of 1839. Most of the geology is based on von Dechen's map of Europe (1839), but the basins of younger Greywacke (*jüngere Grauwackenschiefer; terrain anthraxifère*) are derived from Steininger's map of the Eifel and Nassau (1822) and Dumont's report on the Eifel (1836).

the country, the limestones with goniatites etc." that Beyrich had described.[6] Still, for a start he was evidently anticipating that the Greywacke of the Rhineland would be somewhat like Wales, with his own Silurian overlying Sedgwick's ancient Cambrian rocks with complete conformity.

Having purchased a carriage in Frankfurt, which would enable him to cover the ground quickly without being dependent on public coaches, Murchison was tempted to travel at once to Bayreuth, to visit Count Georg zu Münster (b. 1776), the keen amateur geologist whose descriptions of fossil mollusks (*Goniatites* and *Clymenia*) from the Transition of the Fichtelgebirge had first suggested links with the Devon strata (§9.8); he told his wife "I believe there is much Devonian" there. But on second thoughts he decided "it is better to do *one thing* well than *two things* badly, & so I stick to the right bank of the Rhine." After getting local information from Hermann von Meyer (b. 1801), a civil servant in Frankfurt and the author of many publi-

6. Murchison, notebook, "1839," pp. 58, 61; Murchison to Charlotte Murchison, 29 May and 2 June 1839 (ULE: RIM); Beyrich (1839a). The area east of Ems and north of the Taunus was marked on Steininger's (1822) map as one of the basins of "jüngere Übergangsgebirge" (see fig. 12.2).

cations on German fossils, Murchison traversed back across the Taunus to the valley of the Lahn in Nassau. He was now in one of the basins of younger Greywacke marked on Steininger's map, and he clearly expected the fossils to be Silurian species. But in the event they seemed to tell a different story. "I am already more than half convinced that these limestones ... are our Devonians," he noted as he boated down the river; "there is nothing like them in the Silurian System, nor do they, as far as I can see, offer any Silurian fossils except a coral or two which are *equally* Devonian." That impression was probably reinforced soon afterward when he saw the fossils that Johann Sandberger, a schoolteacher and collector at Weilburg, had found in the local rocks.[7]

Leaving the Lahn valley, Murchison traversed rapidly northward across the Westerwald to Siegen, and thence to the upper valley of the Ruhr. "To a person bothering & losing himself in details," he told his wife, "the *geometry* of the country is puzzling as the same zones are repeated several times." But he was not one to be bothered, not least because the complex folded structure had already been sorted out for him on the published German maps. When he reached the Ruhr (at Meschede) he summarized for his wife what he had done.

> I have now gone clean across the region, & have looked into the zoological & mineralogical contents of each zone of rocks, as well as their geological relations, & what I have to say will surprise you. I do not believe there is a Silurian bed among them, & I am more than disposed to think that the whole is *DEVONIAN* (except perhaps the westward flanks). There are no Eifel fossils here.—The limestones are undistinguishable from those of Plymouth & North Devon & the organic remains are all of the same classes which occur in those rocks, *Goniatites*, large *Spirifers* etc.[8]

He anticipated that these Devonian strata would pass conformably under the Carboniferous strata of the Ruhr coalfield. He then hoped to confirm his provisional conclusions by traversing back to the south on another route, "so that when Sedgwick joins me I flatter myself that part of the campaign (& which *I* always thought would be *the key* to the whole thing) will be in my pocket, & I shall have swept the right bank of the Rhine." As a final fling, he added, "so much for unfortunate Grauwacke!" Just as he had long been fighting to replace that term in Britain, so now he hoped to show the inadequacy of the term on the German geologists' own territory. But for the moment he prudently warned his wife, "you need not boast too much of my geological hits, as some of them may fail."[9]

Sending his carriage on ahead, Murchison next worked his way on foot down the valley of the Ruhr. His tentative correlations culminated in a traverse across the Ruhr itself (at Schwerte) northward to Dortmund, summarized at the end of the day with a detailed colored section (fig. 12.3). In a straightforward sequence of steeply dipping strata, Murchison identified the outcrop of the "Great Mountain Limest[one] 2 miles wide," overlain by "Black or Culm

7. Murchison, notebook "1839," pp. 77, 85; Murchison to Charlotte Murchison, 9 June 1839 (ULE: RIM).
8. Murchison to Charlotte Murchison, 9 June 1839. For the fossils mentioned, see fig. 6.5, B, D.
9. Ibid.

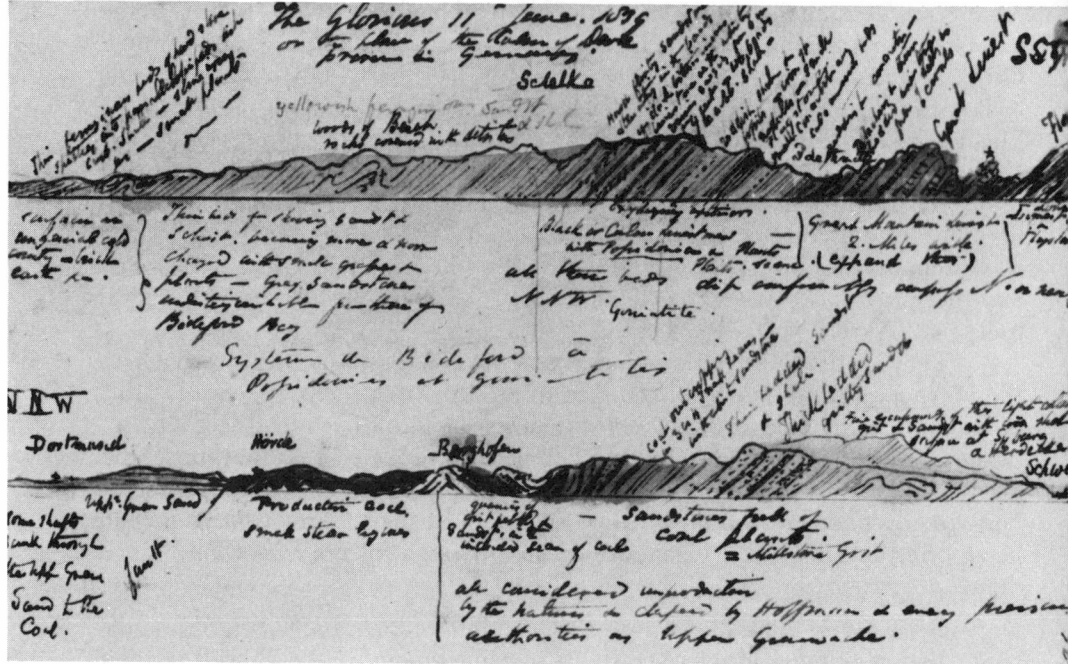

Fig. 12.3.   "The Glorious 11th of June 1839, or the place of the Culm of Devon proven in Germany": Murchison's traverse section (in two halves) across the Ruhr valley in Westphalia, northward to Dortmund on the edge of the Ruhr coalfield. The section shows what he regarded as the "Black or Culm limestone" (top center), sandwiched between "Great Mountain Limestone" (top right) and equivalents of the Millstone Grit and Coal Measures (bottom); the whole of the Culm was thereby "proven" to be Carboniferous in age.

limestones with *Posidonia*, Goniatite and Plants," followed by "Grey Sandstone undistinguishable from those of Bideford Bay." Then came "Sandstones full of coal plants = Millstone Grit," and finally the "Productive Coal [Measures]" on which the burgeoning industry of the Ruhr was being built. The most surprising aspect of these correlations was that Murchison followed von Dechen's recently published map of western Europe and identified the main limestone as Mountain Limestone *(Kohlenkalkstein)*, although he had made no more than a cursory study of its fossils. For Murchison that correlation had the great attraction of setting the Culm-like strata of Westphalia (von Dechen's *Kieselschiefer)* firmly *within* the Carboniferous. It is therefore no wonder that he titled his sketch section, in a tone of triumph, "The glorious 11th of June 1839, or the place of the Culm of Devon proven in Germany."[10]

Murchison also sketched a geological map of the area, clearly based on von Dechen's and Hoffmann's maps but with his own annotations added. On this the "Up[per] Grauwacke" below the productive coal strata became "Millstone Grit" or "Lower coal beds Anglice," and the "Old Grauwacke of Germans"(below von Dechen's Mountain Limestone) became, boldly, "old red

10. Murchison, notebook "1839," pp. 101, 106–8. fig. 12.3 is reproduced from: ibid., p. 108. The "glory" alluded to the traditional start of the grouse shooting season, a major date on Murchison's calendar.

or Devonian with bands of Limestone." As he traversed southward again, he marked "the great axis of this Region," where the oldest strata could be expected, as "Upper Silurian with Limestones." That attribution must have been based on Beyrich's opinion that the fossils in those limestones—which Murchison had not yet seen—were like those of the Eifel. This would have confirmed Murchison's nascent interpretation, by promising to disclose Silurian rocks *under* the Greywacke that he had boldly identified as Devonian. But as he approached Cologne and saw the limestones for himself (at Bensberg and Paffrath) he found their structural relations frustratingly obscure, and their fossils with "more the aspect of Devonian than Silurian."[11] At least for the time being, his newly discovered Devonian lacked any demonstrable Silurian base.

Nonetheless, on his arrival in Cologne Murchison felt sufficiently confident about his fieldwork to draw a "First Generalized Section of the older Formations" all the way from the Taunus to the Wesphalian coalfield (fig. 12.4). This duly showed the Culm limestone embedded in the middle of the Carboniferous, below which was a thick sequence of strata marked confidently as "Devonian." The core of the saddle he had just followed was now marked "Lower Devonians," and another further south in Nassau as "Plymouth limest.," with just a note that some rocks (at Ems) " *may be* upper Silurian."

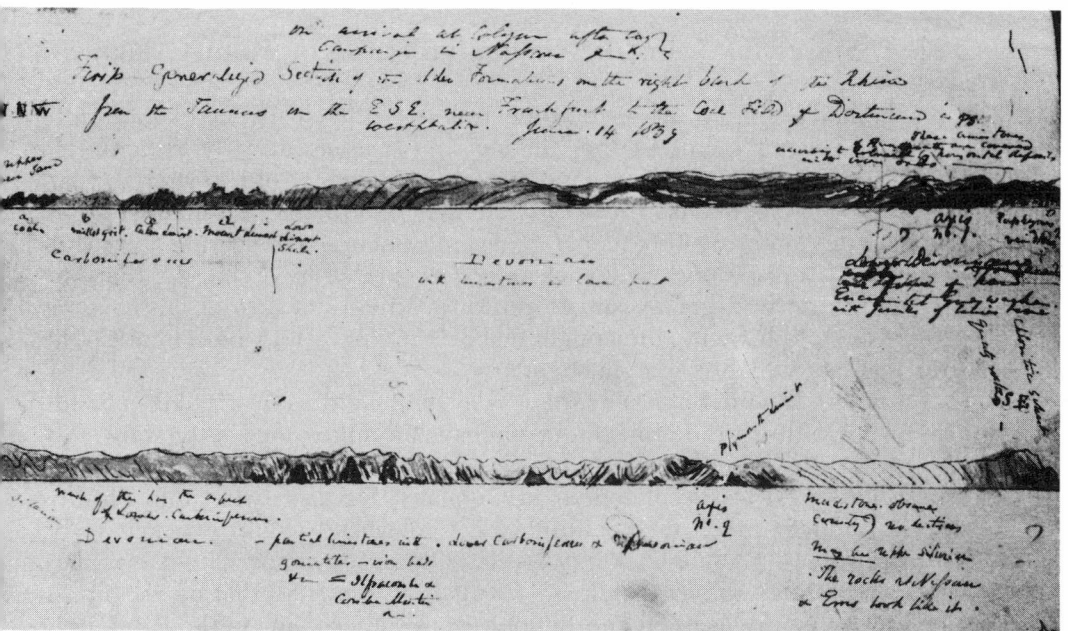

Fig. 12.4. Murchison's "First Generalized Section of the older Formations" in the Rhineland, drawn in June 1839. The Carboniferous (top left), with the putative Culm limestone in the middle, is condensed from sections such as the one reproduced in fig. 12.3. Almost all the thick underlying *Grauwacke* strata (top center and right, and lower section) are marked as "Devonian," with a note that a few might be Silurian (lower right).

11. Murchison, notebook "1839," pp. 207–8.

Thus almost *all* the older Greywacke east of the Rhine had been attributed to the Devonian, while the equivalents of the Culm were tucked neatly into the Carboniferous (fig. 12.5, A). His report to his wife was jubilant.

> I have been very successful & the cards played as I expected—indeed still better, for I obtained the direct proof of the black *culm limestones* of Devon forming the regular base of the Westphalia Coal field & resting *upon* the true Mountain Limestone which in turn reposes on true Devonian rocks!! Of my true Silurians I have seen but little . . . If I have my own way I shall not go near France again this season, at least not until the autumn after [the British Association meeting at] Birmingham. The mine I have opened here is *well worth ALL* our time & attention, particularly when coupled with the Hartz & the other transition tracts of N.W. Germany.[12]

Despite his failure to find any convincing Silurian beneath what he regarded as Devonian strata, this first phase of the trip had been as rewarding to Murchison as it was bound to seem disappointing to his collaborator. For with even the ancient-looking rocks of the Taunas turned into Devonian, Murchison was in effect repeating in the Rhineland what he had done in Devonshire, eliminating Sedgwick's hopes of finding a clear sequence of Cambrian rocks and fossils.

12.2   Sedgwick Joins the Campaign (June to August 1839)

Once he had attended the chapter meeting in Norwich that had kept him in England, Sedgwick crossed the Channel and traveled directly to Bonn. He arrived to find Murchison already there, drinking with "old Nöggerath" and with Karl von Oeynhausen (b. 1795), another Westphalian mining geologist, and doubtless interrogating them for local information. On hearing what Murchison had already done, Sedgwick agreed at once to spend most of the summer with him in Germany and to relegate Brittany to a brief tailpiece; as Murchison told his wife, they might make "a run of 3 or 4 weeks to settle the French affair, which is in a nutshell."[13]

Before leaving Bonn they spent a day with Goldfuss at his home outside the town. Goldfuss, the professor of zoology and mineralogy at the university and author of the lavishly illustrated *Petrefacta Germaniae,* was the owner of one of the finest fossil collections in Germany. He had earlier told Phillips that he thought much of the Rhineland Greywacke would turn out to be Silurian (§10.1). As Sedgwick recalled later, Murchison now found "so many Wenlock friends" among Goldfuss's fossils from the Eifel that he was firmly convinced that de Verneuil and Beyrich were right to equate the Eifel limestones with his own Upper Silurian (§10.7). Since Murchison had not found the distinctive Eifel fossils on the other side of the Rhine, this would have confirmed his provisional correlation by which most of the Greywacke on the right bank was to be regarded as Devonian. Leaving Bonn and returning to

---

12. Murchison to Charlotte Murchison, 14–15 June 1839 (ULE: RIM). Fig. 12.4 is reproduced from Murchison, notebook "1839," p. 209.
13. Murchison to Charlotte Murchison, 14–15 June 1839.

the Ruhr, his Westphalian "key" was accepted by Sedgwick without demur, while Murchison himself noted there was "nothing to add" to his earlier conclusions. Murchison had thus convinced Sedgwick that "the place of the Culm of Devon" had indeed been "proven in Germany."[14]

Soon afterward, as Murchison told his wife, "we bade adieu to our intimate friends the *Devonians* & coal fields without coal [i.e., Culm]," and they set out on a long circuit to the east (see fig. 12.1). First they traveled across Secondary country toward the Harz, the classic area for the German definition of the *Grauwacke*. Before starting work there, they got local information from Friedrich Hausmann (b. 1782), the professor of mineralogy and technology in Göttingen, and paid their respects to the eminent but now elderly Blumenbach. They identified Hausmann's Harz fossil as Devonian, and Sedgwick inferred that the massif was "evidently a compound of *Upper* & a small part of *Lower Devonian.*" When they reached the Harz (at Osterode) they were immediately convinced, as Murchison put it, that what had been "hitherto considered the *oldest* grauwacke of the Germans" was really Devonian and Culm-like Carboniferous. Sedgwick, he reported, "eats, drinks & digests like a Hercules," his chronic ailments being in abeyance, as usual when away from Cambridge; but the geological news was less encouraging. For although they believed they had fossil evidence for turning the Greywacke strata into Devonian, as in the Rhineland, "still they are here so thrown about & disjointed that there is no slight skill required in readjusting them."[15] They spent several days in the region, made the almost obligatory tourists' ascent of the Brocken, and found some Devonian fossils; but they were unable to "readjust" what they had seen into any clear sequence of formations. To that extent their time in the Harz was a failure.

Their next objective was to have been the Transition rocks in the Thuringerwald to the south. But in Gotha they found no local informant to help them, so after a brief stop in what Murchison noted scornfully as "the *Grand ducal village*" of Weimar, they made do with a single traverse to Coburg, finding slates with "all the character of the most ancient English rocks," namely those in Sedgwick's Cambrian of the Lake District, but no Silurian or Devonian fossils.[16]

Their last objective in the east was the massif of the Fichtelgebirge. Before going there, they visited Münster in Bayreuth, saw his outstanding fossil collection, and got advice on where to go; Murchison repaid the nobleman in a gentlemanly manner by presenting him with a copy of *The Silurian System*. On their brief trip to the Fichtelgebirge they duly found (near Hof) the fossils that Münster had described in his publications, particularly the cephalopod mollusks: *Orthoceratites, Goniatites,* and *Clymenia.* The last

---

14. Ibid.; Murchison, notebook "1839," pp. 210, 213; notebook "1839 book 2," pp. 1, 2, 10 (GSL: RIM, N81); Sedgwick, notebook "No. 1, 1839," pp. 1–11, dated 18–21 June (SMC: AS). The retrospective account is in Sedgwick to Murchison, n.d. (GSL: RIM, S11/180). Goldfuss's publication (1826–44), compiled in collaboration with Münster, had by this date covered the fossil corals, echinoderms, and some bivalve mollusks from German formations of all ages.

15. Murchison to Charlotte Murchison, 24 June and 1 July 1839 (ULE: RIM); Murchison, notebook "1839 book 2," pp. 27–42, 68; Sedgwick, notebook "No. 1, 1839," pp. 23–33, dated 23 June to 2 July.

16. Murchison, notebook "1839 book 2," pp. 105, 130–31. The distinguished Gotha geologists Ernst von Schlotheim (1764–1832) and Karl von Hoff (1771–1837) had died since Murchison's travels a decade earlier.

of these had earlier been suggested by Sedgwick as a characteristic fossil of the Cambrian (fig. 9.3); but now he joined Murchison in regarding this "Clymene series" as Devonian. As anticipated on that correlation, they found it was overlain by "Mount[ain] Limestone & Shale [with] true Carboniferous fossils"; but again they found no Silurian fossils lower in the sequence.[17] Like their sections in the Rhineland, this confirmed that the Devonian fauna lay below the Carboniferous, but failed to complete the proof by showing that it also lay above the Silurian fauna. Considering Murchison's confident claims for the universality of the Silurian, he must have been worried by now that it was proving so elusive in Germany.

Two long days of exhausting travel across Secondary country brought them back to Frankfurt. Murchison took Sedgwick over the Taunus, and then confirmed his first impression that the limestone at Ems contained Wenlock fossils. This made it the first convincing Silurian he had yet found in Germany, but its structural relations to the Greywacke were again frustratingly obscure. Traveling rapidly down the Rhine to Düsseldorf, they checked their earlier interpretation of the Westphalian sequence, met Beyrich at Siegen and saw some of his fossils, and found both Silurian (Wenlock) and Devonian fossils in the limestones near Cologne (at Bensberg and Paffrath).[18] But the notes that Sedgwick made during all this fieldwork were even briefer than usual, often dwindling to a bare itinerary, as if he had almost lost interest. For having once conceded that most of the Greywacke of Germany, judging by its fossils, must be Devonian, Sedgwick could no longer hope that it would help establish his Cambrian system on firmer foundations. The Devonian system had knocked the bottom out of the Cambrian as surely in Germany as in Devonshire.

After the two geologists reached Cologne again, Murchison summarized their results for his wife in a way that did not gloss over the crucial failure of their fieldwork so far.

> The only thing which annoys me in my work is, that although we have got excellent descending sections from the Coal Measures to the bottom of the Devonian or old Red System, into which *all* the *Grauwacke* of the right bank of the Rhine falls, still not a trace can I obtain of Ludlow, though the Wenlock appears on points, & thus we want [i.e., lack] the connection which exists in England.[19]

The putative Devonian of the Rhineland lay clearly in the expected position below the Carboniferous rocks, which included strata closely similar to the Culm of Devonshire; but it was not demonstrably in the position above the Silurian that the whole interpretation required (fig. 12.5, A). Murchison had only felt able to attribute a few patches of limestone to the Silurian on the basis of their fossils (at Ems and Bensberg); but even those rocks were not

17. Ibid., pp. 149–51; Sedgwick, notebook "No. 1, 1839," p. 49, dated 10 July; Murchison to Charlotte Murchison, 12–15 July 1839 (ULE: RIM). Sedgwick urged Munster to sell a duplicate collection of his fossils to some public museum; after his return to England he persuaded his own university to buy such a collection for the substantial sum of £500 (Clarke and Hughes 1890, 2:18–20).

18. Murchison, notebook "1839 book 2," pp. 163–64; Sedgwick, notebook "No. 1, 1839," pp. 53–63, dated 13–31 July; Murchison to Charlotte Murchison, 23 July 1839 (ULE: RIM). The new notebook that Murchison started soon after Frankfurt has been lost, and the rest of this phase of their fieldwork can only be reconstructed from his letters and Sedgwick's much more scrappy notes.

19. Murchison to Charlotte Murchison, 31 July 1839 (ULE: RIM).

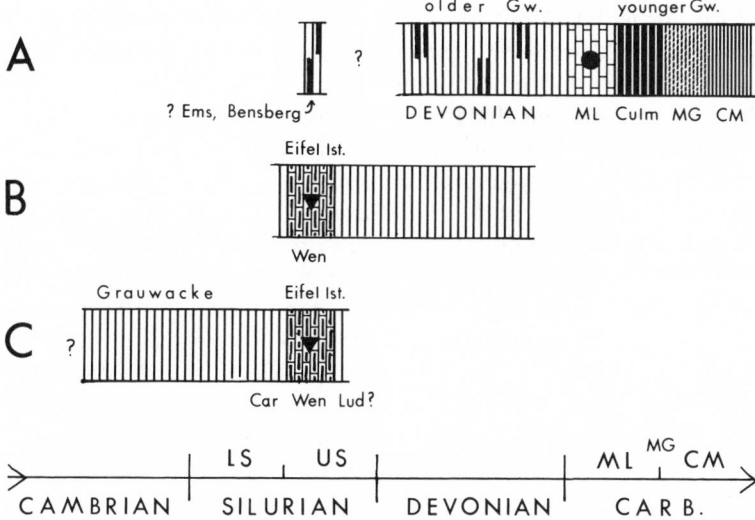

Fig. 12.5. Diagrammatic columnar sections to show Murchison and Sedgwick's first provisional correlation of the British sequence with the sequences (A) east of the Rhine, in Westphalia and Nassau (see fig. 12.4), and (B) west of the Rhine, in the Eifel, as a result of their fieldwork in the summer of 1839. The annotations below each section are Murchison's; the names used by German geologists are above. Murchison claimed that the Culm-like strata in Westphalia were clearly Carboniferous; that on both sides of the Rhine virtually all the German geologists' *Grauwacke* rocks were post-Silurian, thus closely paralleling his and Sedgwick's recent reinterpretation of Devonshire; and that the Devonian was closed off at the bottom by the Eifel limestone, which was attributed to the Upper Silurian (Wenlock). (C) shows the drastic revision of the Eifel, made necessary by Sedgwick's fieldwork in September 1839, when he readopted the structural interpretation of Dumont and the German geologists (see fig. 10.1): the *Grauwacke* west of the Rhine was then attributed to the Silurian and Cambrian, and the supposed Devonian lost any clear relation to the Silurian. The circular symbol marks the fauna of the putative Mountain Limestone; the triangular, that of the Eifel limestone; the short black bars denote other local limestones.

demonstrably below the alleged Devonian, and furthermore their fossils reminded him not of those at the top of the Silurian (Ludlow) but of those from a lower level (Wenlock).

Their final objective was therefore to search elsewhere for this crucial junction between the Devonian and the Silurian: "it is this," wrote Murchison in confident anticipation, "that we are to find in the Eifel and the Ardennes" (see fig. 12.2). Leaving Cologne, they first encountered the Eifel limestone (near Münster Eifel) apparently dipping conformably *under* the Greywacke. This would have seemed highly satisfactory, since Murchison was convinced by the fossils he had seen that the Eifel limestone was Upper Silurian (Wenlock); and the Greywacke looked similar to what he had attributed to the Devonian on the other side of the Rhine. It would have suggested that they had indeed found the elusive transition between the two systems (fig. 12.5, B). But this entailed the startling conclusion that *all* the Continental geologists who had studied the Eifel, such as Dumont, Steininger, and Beyrich, had been radically mistaken in their interpretation of the structure of the region. For they had all concluded that the fossil-rich limestones lay in troughs or basins, forming isolated outliers surrounded by older Greywacke (fig. 10.1; fig. 12.2), whereas Murchison's interpretation implied that the limestone must be in inliers surrounded by younger strata.

It is surprising that the British geologists were prepared even to consider the possibility that the structure of the Eifel had been so radically misinterpreted; the only plausible reason lies in the explanatory advantages that a complete inversion of the structure would bring. But as they continued their traverse across the region by way of Prüm, they must have felt increasing unease, as they saw more and more clearly that the Eifel limestone was lying on top of the Greywacke and not beneath it, and that their first sight of it had been an exception that could be attributed to a purely local overturning. As Sedgwick recalled later, by the time they reached Trier they were positively alarmed and had a sleepless night worrying about it. For if they were mistaken, the Eifel would consist only of Silurian strata with perhaps some Cambrian at the base, and would be useless for proving the Devonian case (fig. 12.5, C).[20]

That problem had to be left unsolved, at least by Murchison, who now had to return to England for the meeting of the British Association. From Trier he and Sedgwick traveled rapidly across Luxembourg into Belgium, but Murchison had to leave the Ardennes for his collaborator to study without him. When he got back to London, Murchison told Whewell of "the wonderful exploits of the Cambrian and Silurian knights" in finding their "followers" in the German Greywacke; but in fact they had found no Cambrian fossils for Sedgwick, and few Silurian ones for Murchison except in the Eifel. He did however refer to the regions they had studied as a "grand . . . Devonian field." To Phillips he was more explicit, making sure that his junior colleague in the Association knew their conclusions in advance of the meeting: "I am happy to tell you that the Devonian system now rests on a basis quite unmoveable, and that the coal-field of Devon will after this promulgation of our new data, never more be contested." Brimming with confidence, Murchison added that "even the sturdy Williams will be swept away!"[21]

12.3   Establishing the Devonian at Home
(June to August 1839)

Murchison had been out of England for three months, but the Devon controversy had not lapsed or been forgotten in his absence. Within a month of the publication of the article that launched the Devonian system, it had been adopted in the new edition of the *Wonders of Geology,* one of the most successful of the popular books on the science. The author, Gideon Mantell (b. 1790), a Sussex surgeon turned geological collector and lecturer, adopted Murchison's view that, with the new system, "the most ancient fossiliferous

20. Sedgwick to Murchison, 12 December 1839 and n.d. (GSL: RIM, S11/165, 180; on the dating of the second letter, see n. 77, below. Sedgwick's notes ("No. 1, 1839," p. 66, dated 1–10 August) are only a bare itinerary. The local overturning of the Eifel limestone at Münster Eifel was illustrated subsequently: Sedgwick and Murchison (1842, p. 277, fig. 12).

21. Sedgwick, notebook "No. 1, 1839," p. 68; Murchison to Phillips, 18 August 1839; Murchison to Whewell, 19 August 1839 (extracts from both letters in Geikie 1875, 1:278–79).

strata will constitute three great systems, which pass into each other, namely, the Carboniferous, Devonian and Silurian." But he failed to grasp that a *new* fauna was being claimed for the Devonian, and he mentioned only the few and peculiar fossils of the Old Red Sandstone.[22] The establishment of the Devonian would require more authoritative acceptance than this, but Mantell's inexpert adoption of the new system was at least a straw in the wind.

A more influential medium was the *Quarterly Review*. After Sedgwick declined Murchison's plea that he should use that periodical to defend their claim to priority over the interpretation of Devonshire (§10.6), the editor had persuaded Scrope to overcome his hesitations and to write the review after all. His essay covered not only *The Silurian System* but also De la Beche's *Report* and Lyell's *Elements;* as with other anonymous articles in the quarterlies, his authorship was no secret.[23] The main energies of "Pamphlet Scrope" had in the past years been directed more to political economy than to geology, and he praised De la Beche's chapter on economic geology as its most important part. He also stressed the economic value of Murchison's work, in that the recognition of Silurian strata could prevent useless prospecting for coal. That practical implication depended, of course, on Murchison's version of the directional theory of the history of the earth and of life; but Scrope was pleased to find Murchison, unlike Lyell, adopting his own earlier view "that the entire series of these changes [in organisms] from first to last were *progressive*, not *cyclical*, as some geologists are inclined to contend."

Such matters of high theory were relevant to the evaluation of Murchison's and De la Beche's work, because they impinged on the fundamental issue of correlation. Scrope argued that the Silurian fauna was further evidence that very general changes in organisms had accompanied the changes in global environments caused by the cooling earth. He thereby endorsed Murchison's prediction that the Silurian fauna would be found on a global scale (§10.4); conversely, he regretted De la Beche's evident reluctance to make any correlations outside his official region, and the consequent inconclusiveness of his *Report*. Scrope's essay was probably written before the Devonian interpretation was proposed, for he only mentioned in a footnote that, if confirmed, it would "go far to lessen the value of Mr De la Beche's survey." But apart from that one ominous note, the *Quarterly* had failed to savage the official surveyor's reputation as Murchison had hoped it would.

A month later, the *Edinburgh Review* also published a major geological essay. Written by Fitton, it was ostensibly a review of Lyell's *Elements*, but in fact it was largely devoted to claiming James Hutton as the inadequately acknowledged source of Lyell's distinctive approach. In the course of discussing Lyell's controversial reconstruction of the history of the earth, Fitton mentioned in passing that he saw much "propriety and fitness" in the new Devonian interpretation and was inclined to adopt it himself.[24] His reputation for sound judgment was such that the comment was bound to increase the plausibility of the Devonian system in the eyes of other geologists.

22. Mantell (1839, 2:610–14); the preface is dated 1 May. I am indebted to Dr D. R. Dean for drawing my attention to this change from the earlier editions.

23. Scrope (1839), published in June issue of *QR*.

24. Fitton (1839, pp. 424–25), published in July issue of *ER*. Lyell told Fitton afterward he was pleased Lonsdale's role had been recognized in print: Lyell to Fitton, 1 August 1839 (K. Lyell 1881, 2:47).

De la Beche, for his part, told Phillips in characteristic style that the truth would triumph in the end over unseemly personal squabbles about priority: "the public will trouble their heads very little whether it may proceed from Jack, Dick, Tom or Harry."[25] This was quite genuine, for he told Greenough soon afterward that he had been seeing how far he could accept his opponents' "new view of the Devon rocks." It was, he wrote, "decidedly a good hit and should be supported if it can," though its authors' "slapdash" style of fieldwork made him cautious about it. He claimed he was not surprised at the news already reaching England that Murchison had found the Mountain Limestone of Westphalia underlain by Greywacke equivalent to the Old Red Sandstone. As he reminded Greenough, "I have long contended against the old red sandstone of Herefordshire, etc. being taken as the type of the rocks of that age" (§10.5). He outlined the "bothery sectional work" in Devon that was difficult to square with the new interpretation, yet clearly he was now prepared to agree that at least *some* of the older strata might be of "Devonian" age. Indeed, although he avoided using that term, he went far toward accepting the general validity of Cambrian, Silurian, and Devonian. For he told Greenough that, if he himself were now making a geological map, he would show their outcrops by three distinct tints of a general Greywacke color. This was no casual remark, but a thinly veiled hint to Greenough that he would do well to adopt the three systems in the older rocks, or at least to color them separately on his forthcoming map of Britain. This could be done without having to adopt Murchison's and Sedgwick's names, for the systems could be shown simply as "the Upper, Central and Lower portions of one great whole," for which the time-honored term Greywacke could be retained.[26]

In his reply Greenough made no reference to what he planned to do with his map in this respect, but he put his finger on the explanatory problems posed by the new interpretation of Devonshire. He pointed out that in Devon the Culm, which he now accepted as equivalent to the Coal Measures, lay directly above strata that were being claimed as equivalent to the Old Red Sandstone. Sedgwick and Murchison had at last conceded that the two sets of Devon formations were strictly conformable, yet they rejected Austen's claim that some of the older ones were therefore likely to be equivalent to the Mountain Limestone. "This is not probable," Greenough exclaimed, adding with characteristic skepticism, "but alas! what is?" Like De la Beche he found it implausible to suppose that *all* the older rocks of southwest England, including even the metal-bearing rocks of Cornwall, could be as young as the Old Red Sandstone or—worse still—the Coal Measures. Yet he realized that once *any* of the Greywacke was claimed as the equivalent of those formations, the same attribution could hardly be denied to even the most ancient-looking rocks. "We seem hardly to have any other alternative, if we once quarrel with the wisdom of our ancestors," he told De la Beche; "well might the poet say 'Chaos is come again'."[27]

De la Beche suggested in response how the Devonian interpretation might be accepted *in part*, without going to the implausible lengths that

25. De la Beche to Phillips, 23 June 1839 (UMO: JP, uncat.).
26. De la Beche to Greenough, 16 July 1839 (ULC: GBG).
27. Greenough to De la Beche, 24 July 1839 (NMW: DLB); the poet was Shakespeare (*Othello*, III, iii, 90).

Greenough had foreseen. De la Beche said that he himself was adopting the new correlation with the Old Red Sandstone merely "for the sake of argument," but it is clear that he now found it plausible, at least for the *upper* part of the sequence of older strata in Devon. For he pointed out that, with an estimated thickness of 30,000 feet in north Devon, the pre-Culm strata were three times as thick as even the exceptionally massive Old Red Sandstone of the Welsh Borderland.[28] On the conventional rule-of-thumb correlation of thickness with time, this implied that the strata "much deeper in the series" in Devon might include equivalents of the Silurian and Cambrian as well. De la Beche had thus abandoned his idiosyncratic interpretation of the position of the Culm (fig. 9.2, B) and had adopted, at least tentatively, a scheme of correlation similar to Austen's (fig. 9.2, A).

Immediately after this exchange of correspondence, the *Philosophical Magazine* appeared with a further paper by Weaver on north Devon, designed to supplement what he had read at the Geological Society in January 1838 (§9.4). Weaver insisted once more, particularly in argument against Williams's recent criticism (§11.4), that there *was* an unconformity between the Culm and the older strata. He noted that Sedgwick and Murchison had changed their stance on this crucial point—a change that those geologists had glossed over and which now isolated Weaver from all the others involved. He argued, however, that to any "geologist of extended practical experience" it was well known that even a local unconformity was more significant than a general appearance of conformity, which could be a merely deceptive "parallelism." The unconformity at the base of the Culm in parts of *south* Devon was now well established; inverting Austen's judgment of its significance (§9.3), Weaver maintained in effect that this was the rule rather than the exception. He claimed that some small exposures in the north (near Barnstaple) supported his earlier contention that there was an unconformity there too.

The matter was even more important to Weaver than to the others involved, because, like the French geologists, he regarded unconformities as marking the truly *natural* divisions of the whole sequence of formations. In the case of Devonshire, he noted that the Carboniferous age of the Culm was now "generally admitted"; but the unconformity he claimed to find below the Culm then supported his view that the older strata must belong to a much older epoch than the Old Red Sandstone. He therefore rejected Sedgwick and Murchison's new interpretation as "singularly inappropriate": they were "throwing overboard all regard for the mineralogical distinctions" among the older rocks and being "carried away by hypothetical views" about piecemeal faunal change. But by stressing that he had held the same view of the Greywacke for a quarter-century, thereby treating it as one of the fruits of his "extended practical experience," his opinions could easily be dismissed as simply outdated. He was on stronger ground, however, when he asked for direct evidence that the Devonian was intermediate not only in its fossils but also in its position, since it was to search for just such evidence that Murchison had gone to the Continent.[29]

28. De la Beche to Greenough, 30 July 1839 (ULC: GBG, 65).
29. Weaver (1839, pp. 114, 116, 119n), published in August issue of *PMJS*. The *PGS* summary of his earlier paper (1838b) had been reprinted in June 1838 issue of *PMJS*.

Having rejected the Devonian or Old Red Sandstone attribution for the older Devon strata, Weaver refused to assign them instead to either the Cambrian or the Silurian, on the grounds that that very question embodied the dubious assumption that there was some universally valid order of strata within the Transition rocks. "Does such a sequence strictly prevail in nature, so as to admit of general application?" he asked rhetorically, and answered in the negative, alluding to the "tumultuary throes" that the earth's surface must have endured at that distant epoch. On the age of the older strata in Devon, "old Weaver" had thus defected from Murchison's camp, or more accurately, had remained on a campsite that Murchison had now abandoned; but on the possibility of making long-range correlations of these ancient strata in the same way that might be possible for the Secondary rocks, Weaver had retreated into a skeptical position close to Greenough's "chaos."[30]

Published on the eve of the British Association meeting, which he knew that illness would prevent him from attending, Weaver's paper served to remind geologists that the Devon controversy was still far from settled. It showed how there was still no consensus, even on such an apparently straightforward issue as the reality or otherwise of an unconformity at the base of the Culm. But most significantly it showed how the resolution of the controversy would affect the whole reconstruction of the early history of the earth. Murchison might be confident that fossils would in due course provide an unambiguous key to deciphering that history, but many other geologists were more skeptical. If they were to be convinced that all was not in fact chaotic beneath the Old Red Sandstone, they would have to be shown that clear sequences, with distinctive formations and fossils, could be recognized at least in some regions. Murchison had already done this for "Siluria"; but it was probably Fitton who at this point sent the editor of the *Philosophical Magazine* a translated summary of Dumont's work, which was published in the same issue as Weaver's paper. This not only brought the Belgian geologist's achievements more widely to the attention of the British, but also staked his claim to priority in advance of the Continental correlations that Murchison was expected to present to the British Association.[31]

When Murchison arrived back from the Continent, he had a week in London to help his wife with the "airy palace" they had just bought in Belgrave Square, which aptly symbolized the fashionable style of life to which they could now afford to become accustomed. He then took the newly opened railway to Birmingham for the British Association's meeting "among the Chartists." But in the event the threat of further political riots had little outward effect on the character of the meeting. In the geological Section, Buckland was president; De la Beche, Lyell, and Greenough were vice-presidents; and Darwin, in his first appearance at a meeting, was one of the secretaries. Since Williams and Austen were also present to read papers, most of those who had so far been involved in the Devon controversy were there, with only Sedgwick as a notable absentee. Phillips had to leave early, owing to the sudden death of his uncle, William Smith: geology had lost its father figure.[32]

30. Ibid., pp. 119–20.
31. Dumont (1839), published in August issue of *PMJS*, translated from Dumont (1838). Dumont's correlations had been adapted from what Buckland had suggested in 1835 (§6.3, §10.2).
32. Murchison to Charlotte Murchison, 23 and 31 July 1839 (ULE: RIM); Murchison to Yates, 18 August 1839 [dated June in error] (ULE: RIM, Dc.4, 101-3.2); *Ath.*, no. 618 (31 August 1839), pp. 645–46; Morrell and Thackray (1981, p. 252).

The committee had evidently planned one of the sessions to deal with Devon geology, for on the second full day De la Beche, Williams, and Austen spoke in turn on that subject in a hall crowded to capacity. De la Beche exhibited the revised version of the Survey's large-scale geological map of southwest England, which had not been completed in time for publication with the *Report* earlier in the year; it "excited much admiration" but no recorded discussion. Even if he had wished to alter the terminology on the map to take account of his qualified acceptance of the Devonian interpretation, he could hardly have done so, because the large-scale sheets had to remain consistent with the index map bound into the published *Report*. So the exhibited map would just have reminded the assembled geologists of De la Beche's public opinion that the "Carbonaceous rocks" of Devonshire lay conformably above "Grauwacke," neither group being firmly correlated with any particular strata elsewhere. Evidently Murchison kept quiet on this occasion about the lack of acknowledgment for which he had criticized De la Beche so bitterly earlier in the year (§10.6).

Williams then returned to the attack with a paper on south Devon and Cornwall. It was both a sequel to the paper on north Devon that he had presented to the Geological Society four months earlier (§11.4), and at the same time a response to Weaver's most recent article. Williams claimed once more that virtually all the older strata of southwest England constituted a single "grauwacke system," with its formations linked by gradual passages into "one great whole." With a good deal of rather confusing detail, Williams concluded that Weaver's "plea of unconformity" between the Culm rocks and the older strata was "abundantly negatived."[33] Like De la Beche's map, Williams's paper contained nothing new and merely confirmed that he remained a firm supporter of De la Beche, who indeed praised him publicly for his industry. There was certainly no hint that "the sturdy Williams" would shortly be "swept away," as Murchison had confidently predicted (§12.2).

Austen's paper was likewise more a matter of placing a viewpoint on public record than of reporting anything new. Ostensibly it dealt once more with the fossils of the south Devon limestones, but its covert purpose was to establish Austen's opinion about their correlation, in advance of Murchison's report on his Continental fieldwork. So Austen made a point of emphasizing the similarity of the Devon fossils with those in the Rhineland, since this was the hint he had given earlier to de Verneuil, which had then sent Murchison to Germany to look for himself (§10.2, §11.5). Austen attributed these limestones (and similar ones in northwest France) to a position "a little below our carboniferous series"; as on the occasion when his fossils were first discussed (§9.3), this implied a correlation with the very base of the Mountain Limestone or the uppermost part of the Old Red Sandstone. In the meanwhile, he had also made it clear that he regarded at least a part of the south Devon sequence as definitely equivalent to the Old Red (§9.6, §11.5). At Birmingham he might therefore have been expected to express public agreement with the new Devonian interpretation; but he was as much opposed as ever to such "local names, generalized." As in the paper read at the Newcastle meeting the previous year (§10.1), he argued that such terms had dubious theoretical over-

---

33. BAAS, session on 27 August 1839: *Ath.*, no. 618, pp. 660–61; *LG*, no. 1181 (7 September 1839), pp. 564–66; Williams (1840c).

tones: "they seem to assume the very prevalent but questionable notion, that, at various former periods, from the poles to the equator, certain uniform conditions and forms of animal and vegetable life universally prevailed."[34] Had Murchison's Silurian campaign been less imperialistic, Austen might have been willing to concede the plausibility of detecting that system, and the Devonian too, by their characteristic fossils, at least over areas such as western Europe. But by claiming the *global* uniformity of the Silurian fauna, Murchison had made the whole concept of ancient "systems" suspect in the eyes of those who, like Austen, doubted the view of earth history on which that putative uniformity was based.

Finally, before Buckland closed the session, Greenough displayed an advance copy of the new edition of his geological map, which was now at last almost ready for publication. A few weeks later, Murchison reminded him of "the exhibition of your map with Devon in a *new dress*."[35] This was almost certainly an allusion to the way Greenough had now colored the Culm as Carboniferous, thereby signaling publicly his acceptance of at least one part of the interpretation he had hitherto opposed. It is also likely that the "new dress" extended to the older strata, and that Greenough had shown them with the same color as the Old Red Sandstone.

By the end of that session there may well have been a general feeling that the Devonshire end of the problem was becoming stale, and that more interesting results could be expected from the work that Murchison and Sedgwick had been doing on the Continent. Two days later Murchison's interim report on their fieldwork was duly given pride of place, but not before he had underlined his global ambitions for the Silurian by displaying, among other exhibits, a collection of Silurian fossils from North America. He announced that he and Sedgwick would in due course be giving the Geological Society a paper arguing for the general adoption of the threefold sequence of systems: Silurian, Devonian, and Carboniferous. This would unambiguously stake their claim to the new Devonian system as an entity transcending its eponymous county and being potentially of worldwide validity. In the meantime, however, Murchison explained that his present paper would deal only with that part of their Continental work that had a bearing on the issue of the Devon Culm. He first summarized the traverses in Westphalia that had given him, a few weeks earlier, the "key" to the true position of the Culm rocks. He had found strata and fossils closely similar to those of the Culm limestone, clearly sandwiched between undoubted Coal Measures above (including the strata he equated with the Millstone Grit), and "unequivocal Mountain Limestone" below (§12.1). This showed clearly that the Culm really was Carboniferous in age, as he had argued from the very beginning of the controversy almost five years before; and it showed it more clearly than any British sections. As for the strata below the Mountain Limestone in Westphalia, Murchison just stated that they "fairly represent the British old red sandstone, or Devonian system." He admitted that the new term had not yet been adopted on the Continent, but argued that it would be found useful because these Devon-like strata occurred so widely on the Continent in the same position as the

34. *Ath.*, no. 618, p. 661; *LG*, no. 1181, p. 566; Austen (1840).
35. *LG*, no. 1181, p. 566; Murchison to Greenough, 22 October 1839 (ULC: GBG).

Old Red Sandstone in Britain.[36] But this formulation glossed over the unresolved weakness of his fieldwork; for he had not been able to find the Devonian in Germany demonstrably above any Silurian rocks, so that its equivalence to the Old Red had not yet been proved at all persuasively.

Greenough opened the discussion by stating that in the light of this new evidence he was now inclined to agree with Murchison's interpretation of the Devon Culm as truly Carboniferous in age, as indeed he had already shown it on his own map. De la Beche conceded likewise that the German sequence was the clearest yet available, and said he was "open to conviction on perfect evidence." Only Williams declared he was unconvinced, and Buckland said he was glad there was still one opponent to "the theory," implying that a little opposition was not unhealthy for its future development. All this referred however to the question of the age of the Culm, not to that of the age of the older strata of Devon. On the second point, Lyell claimed again that the credit for the Devonian interpretation belonged to Lonsdale, who "by an inspection of the fossils" had predicted the true age of the Devon strata. This not only served to obscure the earlier role of De la Beche's ally Austen, but also underlined the validity of Lyell's own theoretical scheme. For Lonsdale's inference, precisely because it had been based only on the fossils and not at all on fieldwork, could be used to confirm Lyell's view of the piecemeal character of faunal change. But the transitional character of the Devonian fauna would only vindicate Lyell's view if the transitional position of the Devonian strata could also be demonstrated; Murchison and Sedgwick had so far failed to do this, but that omission was not pointed out in the discussion.[37]

All in all, the meeting confirmed that the first part of Murchison's campaign of persuasion was developing satisfactorily: the consensus on the Carboniferous age of the Culm was consolidating with every month that passed. The fortunes of the second part of his argument were much less certain: there was quite widespread goodwill toward the striking idea that most of the older strata in Devonshire—and perhaps much of the Greywacke on the Continent— might be no older than the Old Red Sandstone, but more work was evidently needed to make it persuasive. Murchison returned to the Continent, after only three weeks in England, hoping to find that missing evidence.

12.4   The Devonian Vanishes on the Continent (August and September 1839)

Meanwhile Sedgwick had been hard at work in Belgium. After Murchison left him at Namur to return to England, d'Omalius took him to his château in the Condroz, and thence on an excursion in the Famenne (see fig. 12.2). It was, he told Murchison later, "among your Silurian friends—The system repeated again and again in undulations." He fully accepted the correlations

36. Murchison (1840a), read at BAAS on 29 August 1839; *Ath.*, no. 620 (14 September 1839), pp. 702–3; *LG*, no. 1180 (31 August 1839), p. 550; *LG*, no. 1183 (21 September 1839), pp. 600–601.

37. *Ath.*, no. 620, pp. 702–3.

that Dumont (following Buckland) had already made in print (§10.2), and correlated Dumont's lower limestone group with Murchison's Upper Silurian (Wenlock). Since the overlying strata (Dumont's upper sandstone group) passed upward without a break into unmistakable Mountain Limestone (Dumont's upper limestone group), they were, Sedgwick noted, "very naturally supposed by Mr Dumont to represent both the Ludlow [i.e., uppermost Silurian] & the O[ld] R[ed]." But this was a far cry from Murchison's vision of a major "Devonian System" as important as the Silurian and the Carboniferous. Sedgwick's notes give no hint that he shared that vision, and he used the term "Devonian" only in a geographical sense.[38]

Returning to Namur, Sedgwick studied the fossil collection of Philippe Cauchy (b. 1795), the professor of mineralogy at the Athenaeum there, and identified what he believed to be Silurian (Caradoc and Wenlock) species. He and d'Omalius were then joined by Dumont. Demonstrating the research that had launched him on his career a few years earlier, Dumont took Sedgwick up the valley of the Meuse and showed him the fine sections of folded strata that he had expounded to Buckland, Greenough, and the French geologists in 1835 (§6.3).[39] After they returned to Namur, Sedgwick summarized what he had seen, in a letter that Murchison must have found waiting for him in London after the Birmingham meeting. The terms he used show once again his complete acceptance of Dumont's correlations.

> The scenery is exquisitely beautiful, & the geology of its kind, glorious. 1st. Coal measures crumpled into a kind of basin. 2nd. Mountain Lime[stone], Ludlow, Wenlock all *inverted* & *sur-mounted* by the Caradoc. Then a complicated system of saddles & basins extending to Dinant & Givet on the French frontier [see fig. 6.4] . . . . On the whole I have no doubt now that the Ardennes are Cambrian.[40]

Murchison must have been pleased by this news of the Silurian in Belgium, but also disturbed at the lack of any reference to "Devonian" rocks. Even more alarming would have been Sedgwick's report that "Dumont is most positive that the Eifel is a basin [see fig. 10.1]."[41] For Murchison would have seen at once that this put at risk their own earlier interpretation of the Eifel, and with it their tentative identification of the base of the Devonian system (§12.2). Sedgwick must by now have realized that Dumont's flair for unraveling regions of folded strata was such that his opinion could not be lightly ignored. Accepting (with Dumont and de Verneuil) that the Eifel limestone was Upper Silurian (Wenlock), he and Murchison had assigned the apparently overlying strata to the Devonian (fig. 12.5, B). But if the true sequence were exactly the opposite, as Dumont insisted, these "Devonian" rocks would be pre-Wenlock and therefore Lower Silurian, while the lowest rocks of all (the eastward extension of the *terrain ardoisier*) would belong down in the Cambrian (fig. 12.5, C).

"As for myself, I am in a great puzzle," Sedgwick admitted to his collaborator. "Of one thing I am sure. That we ought never to have left the Rhine

---

38. Sedgwick, notes dated 16 and 17 August 1839 (SMC: AS, notebook "No. 1, 1839," pp. 73–75); Sedgwick to Murchison, 25 August 1839 (GSL: RIM, S11/160).

39. Sedgwick, notes dated 18–23 August 1839 (notebook "No. 1, 1839," pp. 76–88).

40. Sedgwick to Murchison, 25 August 1839.

41. Ibid.

till we had scoured both its banks including the Eifel & the Silurians of Liège."
To repair this omission, Sedgwick decided to return to the Rhineland while
waiting for Murchison to rejoin him. Since Dumont was unable to leave his
survey work to show him over the Eifel from the Belgian side, Sedgwick
traveled straight back to Bonn and approached the Eifel from the east. Having
left Murchison's carriage in Liège for him to collect later, Sedgwick had to
do his fieldwork on foot, but this was no unusual hardship for him. He walked
through the Eifel on a long traverse by Prüm in the west, being "half eaten
with fleas" at one overnight inn and arriving "in a miserable wet condition"
at another. He reached Gerolstein "dead tired and worried," and recorded the
cause of the worry: "as far as the red sandstone is concerned (Caradoc) there
appears to me no doubt, and the L[ime] S[tone] is a basin or trough." So
Dumont and the German geologists were definitely correct after all: the rocks
that he and Murchison had ascribed to the "Devonian" on their first traverse
of the Eifel only a month before now seemed to be Lower Silurian.

Returning to the Rhine at Koblenz, Sedgwick then traversed up the
valley as far as Bingen, probably to study the most ancient-looking rocks of
the region (Taunus and Hunsrück) where they were cut through by the river
gorge, in the hope of retrieving them at least for his Cambrian. Finally, he
returned downstream to Bonn, awaiting Murchison's arrival in a mood of
profound skepticism about the whole Devonian interpretation.[42] For if their
supposed "Devonian" in the Eifel were demonstrably overlain by an Upper
Silurian (Wenlock) limestone, and if it were therefore Lower Silurian and
even Cambrian in age, then their similar "Devonian" rocks to the east of the
Rhine became equally questionable and might need to be reassigned likewise
to the Silurian and Cambrian.

Meanwhile Murchison had returned to the Continent, but not to join
Sedgwick in northwest France as they had earlier intended, nor to take his
wife on the quiet seaside holiday he had half-promised her. Instead, before
rejoining Sedgwick in the Rhineland, he attended the annual field meeting
of the Société géologique, which this year was held conveniently at Boulogne.
The twenty members present included not only Frenchmen such as Prévost
and de Verneuil, but also four geologists from England—Buckland, Fitton,
Greenough, and Murchison himself; there were also no fewer than fifty-seven
nonmembers present, including several more Englishmen. A geological *en-
tente cordiale* was celebrated with the election of Fitton as president of the
réunion; he in turn invited the presidents of the two societies (Prévost and
Buckland) to sit beside him in the formal evening sessions.[43]

After a first day's fieldwork on the Secondary formations, de Verneuil
outlined how he planned to show the party the older rocks of the Bas Boulonnais inlier, which he had described at the Porrentruy meeting the previous
year (§10.2). The position of the inlier and its tiny coalfield, right on the line
of the concealed extension into France of the great Belgian coalfield (see fig.
12.1), was enough to explain the intense local interest in the meeting. At the

42. Ibid.; Sedgwick, notes dated 30 August to 18 September 1839 (SMC: AS, notebook "XXXIII
1839," pp. 1–31).
43. Murchison to Charlotte Murchison, 23 July 1839; SGF meeting at Boulogne, 8–11 September 1839 (*BSGF* 10:385–87). Fitton (1827) had been the first to propose correlations between
the Boulonnais and England.

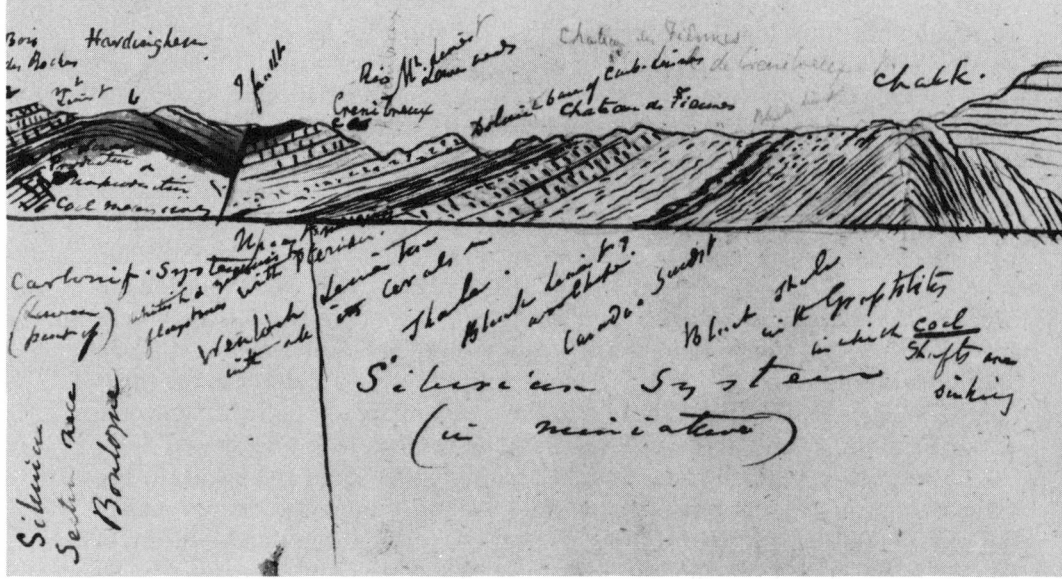

Fig. 12.6.   Murchison's section of the inlier of older strata near Boulogne, sketched during the field meeting of the Société géologique in September 1839. The Carboniferous (left) is shown directly and conformably overlying the "Silurian System (in miniature)," with no Old Red Sandstone or Devonian between; the whole sequence is overlain by the Chalk (right), with a major unconformity at its base. Note the comment (right) that the oldest strata (Lower Silurian) were being explored for coal.

end of the first day on the older rocks, Prévost summarized how they had seen the "système carbonifère," including the workable coal seams as well as limestones, conformably overlying the rocks that de Verneuil had identified by their fossils as being Silurian. Murchison had publicly endorsed that Silurian correlation during the day, commenting on the close similarity between the lower Boulonnais limestone and those of Belgium and the Eifel. Prévost commented that in the Boulonnais there was thus no Old Red Sandstone between Carboniferous and Silurian. The next day this point was further underlined, when the uppermost Silurian formation ("psammite du Ludlow rock"), complete with an appropriate characteristic fossil found by Murchison (*Bellerophon globatus*), was seen to be conformably overlain by the basal limestone of the Carboniferous. As Murchison summarized the matter in a section sketched in his notebook, the Carboniferous was thus directly overlying the "Silurian System (in miniature)," which was complete from top to bottom—Ludlow, Wenlock, Caradoc, Llandeilo (fig. 12.6).[44]

   That evening Murchison was invited to comment on what the party had seen. Greenough had brought from Birmingham the advance copy of his map, and Murchison had already praised it generously before the French geologists. He now used it to summarize the three great systems of older strata established

44. *BSGF* 10:397–402. Fig. 12.6 is reproduced from Murchison, notebook "1839 Bk 4" (BGS: RIM, GSM, 1/127). The lowest strata (with graptolites) were not named as Llandeilo, but they had the right rock type and fossils, and they lay below what Murchison marked as Caradoc sandstone.

in Britain, each with a characteristic fossil assemblage. Sedgwick's poorly characterized Cambrian was not mentioned, but the new *"système dévonien"* of southwest England was carefully explained as the equivalent of the Old Red Sandstone. When he turned to the local French sequence, however, Murchison simply accepted the Silurian correlations that de Verneuil had proposed, noting only that the limestone attributed to the Upper Silurian contained fossils more like those of the Eifel than the classic English localities (Wenlock, Dudley). He left completely without explanation the apparent absence of either Old Red Sandstone or "Devonian" in the Boulonnais, yet he made no comment to suggest that he had any doubt about the conformable relation between the Carboniferous and the Silurian there.[45]

Murchison thus abandoned, or glossed over, the wider implications of his recent proposal of a Devonian system (§11.2), and reverted to his earlier interpretation of the European sequence, as enshrined in *The Silurian System* (§10.4). He interpreted the Boulonnais as if rocks with the distinctive Silurian fauna were to be expected *immediately* below those with Carboniferous fossils, and as if the absence of the Old Red Sandstone between were no more than a minor local feature. It is possible that Sedgwick's recent letter from Belgium had made him uneasy about the validity of his earlier bold identification of the Devonian in the Rhineland, and that he decided to keep a discreet silence about the Devonian until he had checked that work again. But his confident identificaton of the whole Silurian sequence in the Boulonnais suggests otherwise. In fact a clue is given by the note on his section that *"coal* shafts are sinking" in the black shale at the base of the Silurian there (fig. 12.6, right). For only if the older strata were maintained as Silurian could Murchison use them to demonstrate the value of fossil evidence in the hunt for coal. This became clear in the somewhat egocentric account of his talk that he wrote to his wife the following day.

> It so happens that owing to my having more knowledge of the older rocks than other geologists here, I have been obliged to become a sort of Cicerone & orator, & yesterday evening in the great Library, the Mayor of Boulogne & many French present, I delivered myself of an hour of Silurianism [and] explained the relations of the old rocks of the country. The effect of my discourse was to destroy the coal boring mania in rocks of Silurian age. They have a poor little coal field here which lies low in the Carboniferous Limestone group, & this being immediately recumbent on my Silurian schists and shales they have ... been poking, at great expense & with the money of unfortunate shareholders, into my Stygian abysses.[46]

Once again, Murchison's interpretation of the Silurian as a period that had predated any landplants whatever, and therefore any coal deposits, was being put directly to economic use.

In the subsequent discussion, this point was challenged by de Verneuil, who claimed that the anthracites of Brittany and Normandy—so long emphasized by De la Beche—were in the Upper Silurian, directly above "Caradoc sandstone" with large Silurian trilobites. Murchison admitted he had not yet

---

45. *BSGF* 10: 395, 412–18.
46. Murchison to Charlotte Murchison, 19 September 1839 (ULE: RIM).

been to Brittany to see this for himself, but he claimed instead that the fossils he had seen in Paris earlier in the year implied that the anthracites were more probably Devonian or even Carboniferous (§12.1).[47] So the Devonian system was mentioned after all in a Continental context, but only because it helped undermine the plausibility of the Transition dating of the anthracites of northwest France. Indeed, Murchison ended his remarks by saying that he hoped his forthcoming fieldwork with Sedgwick in the Rhineland would establish on firmer grounds his view of the Silurian as a system that could be relied on as a baseline in the search for coal.

Murchison left Boulogne with a major prize: de Verneuil had agreed to join him for the next phase of his fieldwork. This was important, because the French geologist's "intimate knowledge of species" would enable them to make provisional but authoritative correlations on the basis of fossil evidence while still in the field, without waiting for the specimens they collected to be studied by fossil specialists in London or Paris. From Boulogne the two geologists traveled straight to Liège, where they retrieved Murchison's carriage, and then traversed across the Eifel to Bonn. It is possible that Murchison met Dumont while passing through Liège, or maybe he had just taken note of Sedgwick's warning. In any case, the many sections he sketched in the Eifel prove that he now adopted Dumont's view of the structure of that region (fig. 10.1), although it was directly contrary to his own earlier interpretation. He showed the overall structure as a series of troughs; following de Verneuil's opinion, he marked the Eifel limestone as "Wenlock limestone," and the lower strata as, for example, "Caradoc passing up into Wenlock." He now interpreted the Eifel as a whole as "showing Cambrians and Silurians all in their proper places & clearing away many difficulties" (fig. 12.5, C).[48] That tacit withdrawal of his (and Sedgwick's) earlier interpretation left the Eifel useless for establishing the elusive base of the Devonian in the Rhineland, but the setback did not lessen Murchison's confidence that the Devonian system could still be put on firmer foundations on the other side of the Rhine.

12.5   A Triumvirate in the Rhineland (September and October 1839)

When Murchison and de Verneuil reached Bonn, they found Sedgwick waiting for them. Sedgwick recalled later how he told Murchison "point blank" that in his opinion their earlier interpretation of the structure of the Eifel had been mistaken; and that with that correction "the *devonians* of the Rhenish provinces, on both banks [of the Rhine], became sub-Wenlock," or in other words Lower Silurian or even Cambrian. "Of course we both saw the tremendous consequences which followed," he recalled, for in his opinion this knocked the bottom out of their case for *any* Devonian strata in the Rhineland. But Murchison was unperturbed: "I find Sedgwick much bothered & disconcerted about many essential geological points & much disposed to go into a 'chaotic' state," he reported to his wife the next day; "but I hope we shall put

47. *BSGF* 10:417–18.
48. Murchison, notebook "1839 Book 5," pp. 1–11 (GSL: RIM, N82).

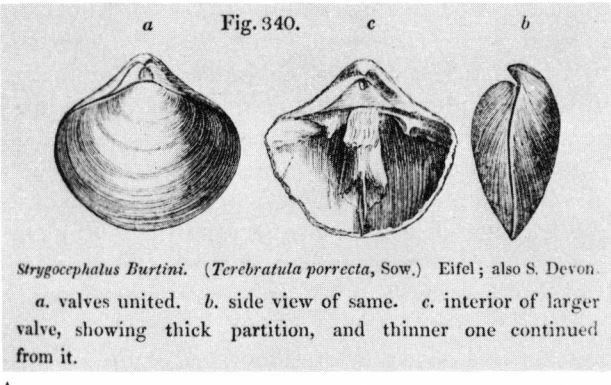

Fig. 340.

Strygocephalus Burtini. (*Terebratula porrecta*, Sow.) Eifel; also S. Devon.

a. valves united.   *b.* side view of same.   *c.* interior of larger valve, showing thick partition, and thinner one continued from it.

A

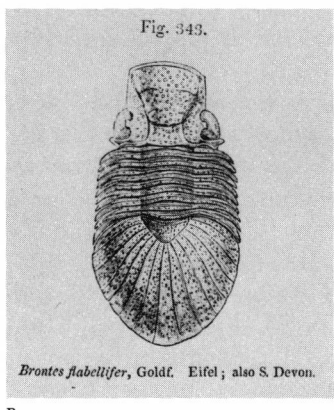

Fig. 343.

Brontes flabellifer, Goldf. Eifel; also S. Devon.

B

Fig. 12.7. The brachiopod *Strygocephalus* (A) and the trilobite *Brontes* (B): two distinctive fossils regarded as characteristically "Devonian" by Murchison and de Verneuil during their fieldwork in the Rhineland in the autumn of 1839. The illustrations are from the second edition of Lyell's *Elements of Geology* (1842).

up our horses & come to some clear general conclusions, in spite of the apparent hotch potch of this volcanized country [i.e., the Eifel] in which he has got *basketted* (but this entre nous)." After a day's indoor work on the sections to be published with their long-delayed paper on Devonshire, they began their fieldwork in a "valiant triumvirate" with de Verneuil.[49]

Taking the carriage to Cologne, they headed for the edge of the Westphalian coalfield, evidently in order to revise the traverses on which Murchison's interim report to the British Association had been based. But the effect of the new member of the party was apparent almost at once, and not only in the increased frequency and specificity of Murchison's and Sedgwick's notes on the fossils they collected. As soon as they reached (near Elberfeldt) the outcrop of what Murchison had previously termed the "Great Mountain Limestone," de Verneuil confirmed Sedgwick's suspicion that his collaborator had made a major mistake in that correlation. They found it contained corals like those of the Eifel limestone, and the highly distinctive brachiopod shell *Strygocephalus* (fig. 12.7, A), which they had previously seen in one of the older limestones near Cologne (at Paffrath). As Sedgwick noted, the formation was therefore clearly "a lower L[ime] S[tone] and *not mountain.*" Murchison

49. Sedgwick to Murchison, n.d. [mid-December 1839] (see n. 77); Murchison to Charlotte Murchison, 19 September 1839. To be "basketted" was slang derived from cockfighting: the phrase implied that Sedgwick's abandonment of the Devonian was an interpretaion that would be defeated, and that he would then be unable to pay the "debts" arising from his wager on it.

drew a revised version of part of his "Glorious 11th of June" section near
Dortmund (fig. 12.3), but the triumphant tone was now muted. The Devonian
had almost vanished. All that he had regarded as Devonian earlier in the
summer was reassigned to the Silurian; the overlying "Great Mountain Limestone" was renamed "*Strygocephalus* Limestone," with some Silurian corals
"but not one Silurian shell"; and the strata still higher in the sequence, which
so closely resembled the Culm limestone and associated rocks in Devonshire,
were thereby deprived of the "proven" Carboniferous dating that had made
the earlier occasion seem so glorious.[50] Murchison had dismissed Sedgwick's
worries arising from the Eifel, but now he himself must have been deeply
worried. For the structure of Westphalia was more straightforward, and the
revision had been made necessary by his own favored method, namely, by de
Verneuil's on-the-spot assessment of the fossils. While the new interpretation
extended the sway of the Silurian, it simultaneously called into radical doubt
the validity of the Devonian interpretation. Sedgwick was not the only one
to have got "basketted" in the Rhineland.

The three geologists turned next to Nassau to review the sequences
there. From Koblenz they traversed up the valley of the Lahn and then made
a loop to the north across the Westerwald. Murchison summarized the sequence there by distinguishing two main groups of strata: above, a "Système
à Goniatites" which he correlated with the "Strygocephalus Zone," noting
that its fossils were more "like Devonian than Silurian"; and below, a "Quaere
Infra Devonian" group, marked as an "expansion of System overlying Silurian
proper." Back in the Lahn valley, a further study of Sandberger's collection
at Weilburg prompted him to conclude that "in no part of Westphalia or Nassau
are the lower Limestones etc. unequivocally like those of Wenlock." Notes
such as these suggest that Murchison still regarded at least part of the pre-
Carboniferous sequence as somehow younger than the "Silurian proper," but
his adoption of the epithet "Devonian" was now both cautious and oblique.[51]

De Verneuil had to leave them at this point to return to Paris, but when
the two British geologists got back to Koblenz, Murchison drew—at a distance,
as it were—a "General Succession in North Westphalia," which summarized
his latest interpretation (fig. 12.8). At the top of the sequence, the "Millstone
Grit & Coal" were unchanged from his earlier work, but below that point all
the correlations were drastically altered. At the bottom, all the strata he had
earlier attributed to the Devonian were now marked with "Silurian fossils"
and labeled as "Upper Silurian Sandstones & Hard Greywacke" (the latter
was also marked "Caradoc?"). In between those limits lay the Culm-like strata
and his former "Great Mountain Limestone." With the removal of that limestone from the Carboniferous, a new equivalent for the Mountain Limestone
had to be found at a higher level. Murchison now inferred that they, like the
German geologists, had earlier confused two quite distinct limestones in Westphalia: the *Strygocephalus* Limestone and what he now believed was the *real*
Mountain Limestone at a higher level (at Ratingen). So on this sketch section
he marked some "Black Posidonia Limestones," quite high among the Culm-

50. Sedgwick, notebook "XXXIII 1839," p. 33, dated 21 September; Murchison, notebook
"1839 Book 5," pp. 17–25.
51. Murchison, notebook "1839 Book 5," pp. 46, 54–55; Sedgwick, notebook "XXXIII 1839,"
p. 42, dated 3 October.

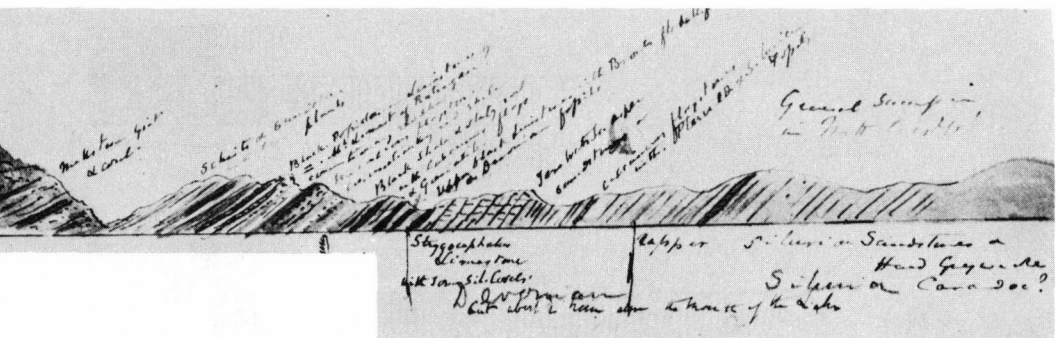

Fig. 12.8. Murchison's revised section of the sequence in Westphalia, sketched in October 1839 and embodying a radical revision of his interpretation in June (fig. 12.3). The thick Devonian at the bottom became Silurian (right), and the earlier Devonian was implicitly reduced to no more than the "*Strygocephalus* limestone"—his former Mountain Limestone—and some overlying strata with *Brontes* (center). Murchison *later* divided this section into three parts (separated by vertical strokes), annotating them as "Silurian" (right) and "Devonian" (center), with the Carboniferous (left) taken for granted.

like strata, as " = Mt. Limest. of Ratingen?" The Carboniferous was then implicitly confined to the strata from about that point upward. But below that point in the sequence, and above what he now attributed to the Silurian, lay the intermediate strata about which he had become so hesitant and confused: the *Strygocephalus* Limestone, "with some Sil[urian] corals," and the overlying black shales and limestones, with the distinctive trilobite *Brontes* (fig. 12.7, B) and other "Devonian fossils."[52] That last attribution was as near as he would now come to claiming the presence of a "Devonian System" between the Carboniferous above and the Silurian below. The claim was not only oblique; the implicit Devonian was also greatly reduced from its former status in the whole sequence, merely taking over the lower part of what Murchison had earlier regarded as the Carboniferous strata (fig. 12.9, A2; compare fig. 12.5, A).

Before returning to Westphalia to check this interpretation, Sedgwick and Murchison made a last brief excursion into the Eifel, boating back down the Mosel to study the rocks on the banks of the river. Murchison thought them Silurian, while Sedgwick equated them with the supposedly Cambrian *terrain ardoisier* of the Ardennes; but the difference of opinion was only relevant to the broader issues in that it showed again that both geologists had now abandoned their earlier Devonian interpretation of at least that part of the Eifel. Returning to Bonn, Murchison sold his carriage, and he and Sedgwick took a steamer down the Rhine to Düsseldorf for their last fieldwork in Germany. "There are two limestones in the country we mean to visit, . . . and we want to separate them," Sedgwick told Whewell, as he sent word of his imminent return to Cambridge. On the edge of the Westphalian coalfield they duly confirmed that what they now regarded as the real Mountain Limestone (at Ratingen) was distinctly higher in the sequence than the "Lower or Stry-

<hr />

52. Fig. 12.7 is reproduced from Lyell (1841a, 2:155, 157). Fig. 12.8 is from Murchison, notebook "1839, Book 5," p. 64.

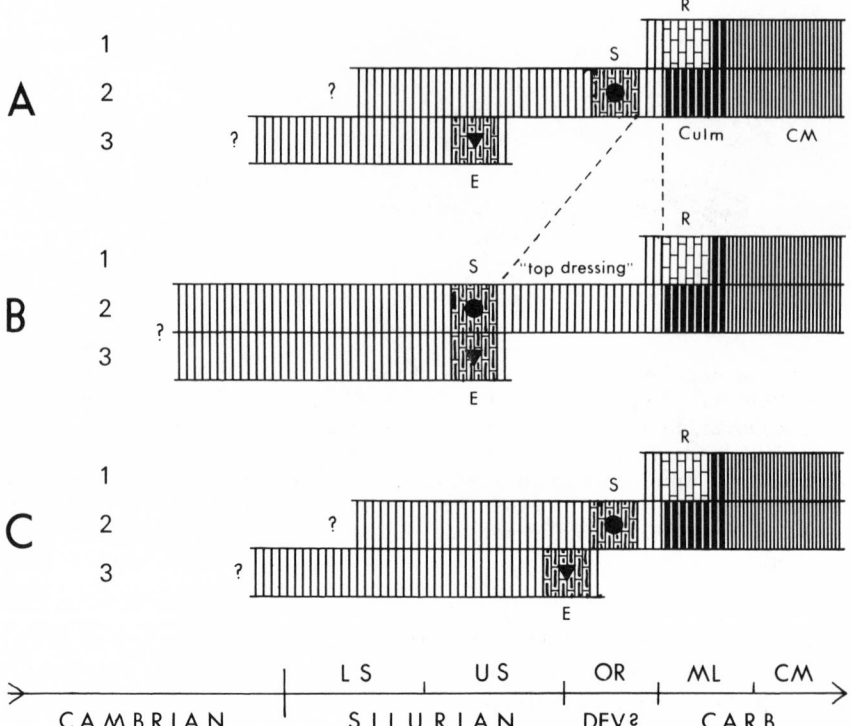

Fig. 12.9. Diagrammatic columnar sections to illustrate Sedgwick's and Murchison's interpretations of the Rhineland sequences during the weeks immediately after their joint fieldwork there. In each correlation, (1) denotes the sequence around Ratingen in north Westphalia, (2) the main Westphalian sequence, and (3) the sequence in the Eifel. On the "standard" scale the Devonian or Old Red is reduced to its earlier size, to reflect the new doubts of both geologists about its status as a major system. Murchison and Sedgwick agreed that the limestone at Ratingen (R) was the true Mountain Lime-stone (ML), equivalent to the black Culm-like limestones further east, and that the Eifel limestone (E: fauna marked by black triangles) was Upper Silurian (US). But Murchison (correlation A) now regarded the *Strygocephalus* limestone (S: fauna marked by black circles)—his former Mountain Limestone—as *Devonian* (see fig. 12.8). Sedg-wick (correlation B) insisted, on the contrary, that on fossil evidence it must be close to the Eifel limestone, and thereby reduced the Devonian still further to some thin strata forming a mere "top dressing" to the Silurian. Later, in the winter of 1839, Murchison (correlation C) conceded that those two faunas were indeed similar, but put *both* limestones into the Devonian (the Eifel one at the boundary with the Silurian). This move began the revival of his claim that the Devonian should be regarded as a major system with a distinctive fauna transitional between the Silurian and the Carboniferous.

gocephalus Limestone," which Sedgwick firmly referred to as "Wenlock" (fig. 12.9, B1, B2).[53] Since at least some of the Culm-like strata were above the upper limestone, this retrieved some of the glory of "the Glorious 11th of June" by restoring Murchison's evidence for the Carboniferous age of the Culm. But it was no longer clear what strata, if any, Sedgwick was prepared to retain as Devonian.

53. Murchison, notebook "1839 Book 5," pp. 77, 92; Murchison to Charlotte Murchison, 8 and 10 October 1839 (ULE: RIM); Sedgwick to Whewell, 11 October 1839 (TCC: WW, Add. Ms. a. 213²⁴).

The whole sequence in Westphalia now seemed to Sedgwick to be closely similar to that in Belgium, with Carboniferous strata lying directly and conformably above Silurian strata, each system including a distinct limestone. Probably as a result of his arguing for that analogy, the two geologists abruptly altered their plan for their return journey. Instead of continuing by steamer down the Rhine to Rotterdam and thence back to England, they took a coach to Liège and traveled on to Namur to give Murchison a chance to see the Belgian sequence for himself. In a near concession of defeat over the Devonian, Murchison drew one Belgian section en route as "Infra Carboniferous strata," although he did mark some of those strata as "Devonian Limest. of Plymouth." But when d'Omalius had taken them through the Famenne and down the now celebrated traverse of the Meuse, Murchison meekly accepted all the conventional Silurian correlations: below the "Carboniferous Series" he duly marked in order the Ludlow, Wenlock, and Caradoc. It was left to Sedgwick to draw the awkward conclusion: between the uppermost Silurian ("Upper Ludlow") and the unmistakable Mountain Limestone (Dumont's upper limestone group) there was "no interruption indicated by the sections, or any appearance to mark the want of the *Old red system.*"[54] The Devonian had virtually vanished.

Leaving d'Omalius at Namur, Sedgwick and Murchison traveled by coach to Antwerp to board a steamer for London. Murchison had been on the Continent for five months (not counting his brief return to England for the British Association); Sedgwick had spent four months in continuous fieldwork, and thanks to fresh air, exercise, and plenty of Rhine wine he had no ailments to complain of. In the course of their long expedition, Murchison had turned most of the ancient Greywacke of the Rhineland into Devonian, only to find himself forced by the fossil evidence to turn much of it back into Silurian, leaving his confidence in the Devonian precarious if not collapsed. Sedgwick had seen his potential Cambrian annexed by the Devonian but later at least partially restored. He had totally lost confidence in the Devonian interpretation, of which he had been the nominal co-author only six months earlier. But whatever their differences, it would have been clear to both geologists that their best hope of resolving the Devonian problem, after almost five years of controversy, lay packed inside the boxes they had been sending back to London. At least for a time, the field geologists would have to yield pride of place to the fossil specialists.

12.6   A Postmortem on the Devonian (October to December 1839)

When the two geologists disembarked in London, Sedgwick had to return to Cambridge at once to begin his lectures, while Murchison took their fossils to Lonsdale at the Geological Society. But before joining his wife in the country, Murchison wrote an urgent letter that reflected their disarray over the Devonian interpretation. "I hasten to prevent your directing the colourer to 'ruddle' Devonshire," he told Greenough, alluding to the reddish color

54. Murchison, notebook "1839 Book 5," pp. 97, 99; Sedgwick, notebook "XXXIII 1839", p. 48, dated 17 October; Murchison to Charlotte Murchison, 12 October 1839 (ULE: RIM).

conventionally used on geological maps to depict the Old Red Sandstone. He assured Greenough there was "no ground *whatever* for abandoning one jot" of their interpretation of the Culm as Carboniferous; that aspect required no alteration from the draft map that Greenough had displayed at Birmingham (§12.3). But "*Germanic* evidences" had now made Murchison doubt "the 'Old Red' equivalent of Devon & Cornwall" sufficiently to urge Greenough to hold up the final preparation of his map.[55]

Murchison explained how he and Sedgwick had concluded that on all the German geological maps the true Mountain Limestone (of Ratingen) had been confused with the *Strygocephalus* Limestone lower in the sequence, similar in rock type but distinct in fossils. "It is a long yarn to spin," he told Greenough, but he revealed that he and Sedgwick were now in serious disagreement (fig. 12.9, A, B).

> You will ask what is this "Strygocephalus Limestone." If I *by myself* answered—the reply would be "Plymouth & South Devon Limestone (with which its fossils & corals most closely agree), that limestone being intermediate between the Carboniferous & Silurian Systems." Ergo in *my* view there would still be in this [Westphalian] Series an equivalent for the Devonian [i.e., Devonshire] *Lower Rocks* in the place to which we assigned them [fig. 12.9, A]. But Sedgwick is *not* (at present) of this mind & seems disposed to abandon the case altogether as to the existence of rocks of such age in Westphalia *or* in Devon. He merges the Strygocephalus black Limestone with the *Wenlock* [Upper Silurian], to which I cannot subscribe [fig. 12.9, B].

Although Murchison thus claimed to retain his confidence in the Devonian interpretation, he was sufficiently shaken by Sedgwick's defection to tell Greenough he would have to decide for himself after hearing "the full rationale" when they met: "I shall indeed be truly sorry to be the means of leading you into any error."[56] Devonshire was back in a state of uncertainty and confusion.

"The Devonian & the Silurian will not I hope induce you to delay," Greenough had been told by Jameson only the previous week, but after so many delays another would make little difference. Jameson might protest from Edinburgh that "the Devonian system is a riddle for us in this quarter," and complain that "the geognostical characters & relations of the rocks" had not received sufficient attention.[57] But in quarters less provincial than Jameson's Wernerian Society, the Devonian remained a crucially important if controversial interpretation. In fact, Sedgwick's and Murchison's latest fieldwork had been concerned as much with the structural relations and rock types of the putative Devonian strata as with their fossils.

Unlike Greenough and indeed Murchison, Sedgwick could not bide his time but had to declare his position on the Devonian controversy as soon as he returned to Cambridge. Having announced the previous year that he intended to test the Devonian interpretation in Brittany, he now felt bound

55. Murchison to Greenough, 22 October 1839 (ULC: GBG, 1281).
56. Ibid.
57. Jameson to Greenough, 14 October 1839 (ULC: GBG, 1033). "I cannot now understand your Devonian and other novelties," he added a few weeks later: Jameson to Greenough, 15 December 1839 (ULC: GBG, 1034).

to explain to the "two Heads of Colleges—many Masters of Arts—and about 60 undergraduates" in his audience how his fieldwork in Belgium and the Rhineland had altered his opinions. "I told them plainly that I gave up the Devonian case," he reported to Murchison, "& that I considered the whole of the old rocks of Devon & Cornwall (excepting the Culms) as inferior to the Dudley Limestone," that is, as Lower Silurian and Cambrian. Sedgwick had thus reverted to almost exactly the interpretation he had held before the Devonian was proposed. But he assured Murchison that no harm would be done if he were to change his mind again, since the lectures would not be published. He added that he had told his audience "that the matter was still *sub judice* & that our fossils had not been examined."[58]

"I don't think we can get any *Devonian* (in our old sense of the word) out of the Westphalian sec[tio]n," Sedgwick warned his collaborator. But it is striking that after his lengthy fieldwork with Murchison and their triumvirate with de Verneuil, he gave far more weight to fossil evidence than ever before. "The Eifel horizon is too firm to be shifted," he insisted, meaning that de Verneuil's opinion of the Upper Silurian (Wenlock) affinities of its fauna would have to be accepted (fig. 12.9, B3). He doubted if Murchison could persuade anyone else that the Eifel limestone was Silurian but the *Strygocephalus* limestone Devonian (fig. 12.9, A), because the two faunas were so similar. He suggested that at most the Westphalian limestone might be uppermost Silurian (Ludlow): but then, he pointed out, "all you would have for Devonian will be pinched into a few bands of Goniatite slates" between the *Strygocephalus* limestone and the Culm-like equivalents of the Mountain Limestone (fig. 12.9, B2). "By such a supposition we gain nothing," he concluded: "is our Devonian & Cornish case worth fighting for?"[59]

"Sedgwick has been telling us *his* German story," Whewell reported to Murchison after attending the lecture; "I want to hear yours," he added, evidently suspecting that the Devonian might still be retrieved. Whewell was amused to hear that Greenough had "struck his flag a few months too soon" by giving the Devon rocks the same color as the Old Red Sandstone before the case was confirmed. "Well for you," he told Murchison, "or he would have buried you under a grauwacke tombstone for the rest of your days."[60] Greenough's map was indeed likely to be a monument so authoritative that it would embalm geological opinion for a generation, but Whewell was evidently unaware that Murchison himself had advised Greenough to delay the coloring until the matter was clearer. Devonshire might yet turn out to be an ancient Greywacke tombstone.

Murchison's doubts about the Devonian are also reflected in the way he was now talking about his summer's fieldwork. He told Bilton—who was passing through London after collecting fossils from the older rocks of Norway—that he had been "very successful in applying Silurian tests to the Rhine—& Belgium," but there was no mention of the Devonian.[61] Even before Phillips heard that report, Murchison had already told him directly about his

58. Sedgwick to Murchison, 28 October 1839 (GSL: RIM, S11/164).
59. Ibid.
60. Whewell to Murchison, [early] November 1839 (transcript in TCC: WW, O.15.47[283]); the original is not in GSL.
61. Bilton to Phillips, 15 November 1839 (UMO: JP, 1839/48).

problems; and his junior colleague agreed tactfully with "the natural, just & proper perplexity which you feel & ought to feel, . . . regarding the true & *general* dovetailing of [the] Silurian & Devonian *aeras.* " Phillips was uneasy about the apparently casual way that Dumont and other geologists invoked "the reversaling of large masses of strata," for he saw that this could lead to circular reasoning about the sequence of fossils: "it is indeed a *radical* cure of a difficulty, by throwing away the premises which lead to a plaguey conclusion."[62] For the time being, Phillips was evidently as skeptical about the Devonian interpretation as Murchison had become.

Whatever problems the Devonian might be raising, however, there is no doubt that, as *The Silurian System* got into the hands of Continental geologists, they were increasingly inclined to support Murchison's and Sedgwick's efforts to establish *some* such sequence of major systems internationally. When de Verneuil left them in the Rhineland, for example, Murchison had given him a letter for Élie de Beaumont, summarizing what they had done. In his reply the French geologist praised their work "en cherchant à classer les couches de transition avec la même régularité que les couches secondaires." He recognized the importance of their attempts "à dechiffrer ce vieux chaos" on both sides of the Channel, and he hoped de Verneuil's work on the fossils of the "ci-devant terrains primitifs" would give France a worthy share in that enterprise.[63] Clearly Élie de Beaumont had no doubt about its feasibility: the "chaos" of the older strata was there to be deciphered, not accepted as insuperable.

12.7   An Interlude on Devonshire (September to January 1839)

Meanwhile at least one geologist continued to argue about the Devon problem in terms of local detail rather than grand internatonal correlations. Austen had moved to Surrey during the summer and expected to have little more to do with Devonshire geology (§11.5). De la Beche had moved the headquarters of his survey to south Wales, and now that his *Report* and its associated maps were fully published he too expected little further involvement with southwest England. Phillips had failed to go to Devon during the summer, and De la Beche had to implore him not to give up his projected work on its fossils.[64] But Williams continued to fight a lone battle for the interpretation that even De la Beche and Greenough had publicly abandoned at the Birmingham meeting (§12.3).

Not content to wait several months for the report of the British Association to be published, Williams sent the *Philosophical Magazine* an expanded version of what he had read, only a few days after the meeting ended. The crucial question of the conformity below the Culm in north Devon had now turned into an argument about competence in the field. Williams scorn-

62. Phillips to Murchison, 7 December 1839 (GSL: RIM, P14/15); Murchison's letter has not been located in UMO.
63. Élie de Beaumont to Murchison, 31 October 1839 (ULC: AS, IF, 78).
64. De la Beche to Phillips, 23 June and 18 September 1839 (UMO: JP, uncat.).

fully rejected Weaver's instances of unconformiity in north Devon (§12.3), dismissing one of them, for example, as being "in such a decomposing loose condition, and so concealed by herbage, tangled briar and brushwood, that I should invariably reject it from my field-book, as utterly inconclusive." On this point Williams was in good company, for Weaver was now alone in claiming there was an uncomformity there. But on the question of the age of the Culm itself, it was Williams who was now on his own, since he continued to insist that the Culm was much older than the Carboniferous. Characteristically, he remarked that as support for this position ebbed away, he himself became more confident in it: "I fearlessly challenge the banded world of geologists to disprove it," he concluded.[65] A few days later, while doing some fieldwork on the southern edge of the Culm area, Williams sent a postscript to his article, adding that he had now seen "such an accumulation of concurrent testimony, radiating from different sources to a common centre," that he was more than ever convinced that Sedgwick and Murchison were mistaken about the Culm. He concluded provocatively that "the merest tyro in geology might take his stand" on what could be seen in one small area (around Chudleigh), and yet "defeat the assaults of a thousand adversaries."[66] When Murchison read this after his return to London, he would have seen that "the sturdy Williams" had fight in him yet.

What Williams did not reveal in his article, however, was that his continuing insistence on the ancient date of the Culm had just been reinforced by a radical revision of his interpretation of the south side of the Culm area. Partly as a result of seeing the exposures that had earlier led De la Beche to a similar conclusion (§9.7), Williams now inferred that the rocks immediately south of the Culm *overlay* the Culm itself (fig.12.10). For example, the strata near Launceston (at South Petherwin) were *not*, he now thought, a repetition of those bordering the Culm on the north, brought up to the surface by the trough structure of central Devon, but were rather a part of a vast formation *younger* than the Culm. This reversal of the order of superposition was decisive. "It shows, to demonstration, the true position of the Culm field," he noted confidently: the Culm and plant-bearing beds of Devons[hire] are thus separated from the upper great Coal field [i.e., the Coal Measures] by the entire [thickness] of the Slates of Cornwall, by the Old Red Sandstone proper, and by the M[ountai]n Limestone."[67] All the older rocks of southwest England were thus to remain more firmly than ever in the Greywacke.

On his return to his Somerset rectory, Williams wrote a paper incorporating this new interpretation and sent it to the Geological Society. Though the officers may have been getting tired of the obstinate provincial, they scheduled it to be read at the last meeting but one in 1839; like his paper earlier in the year, it would at least provide a peg for a general discussion of the Devon problem (§11.4). Shortly beforehand, Williams asked Phillips's opinion of "the O.R.S. theory" or Devonian interpretation; but he now used fossil evidence to make his point against it, asking if Phillips agreed that some of the south Devon corals were Murchison's Upper Silurian species. If those

65. Williams (1839c), dated 9 September, published in October issue of *PMJS*.
66. Ibid., postscript dated 19 September.
67. Williams, notebook for "August 1839 to April 1840," notes dated 21 September and 10 October 1839 (SAS: DW, box I, notebook 17).

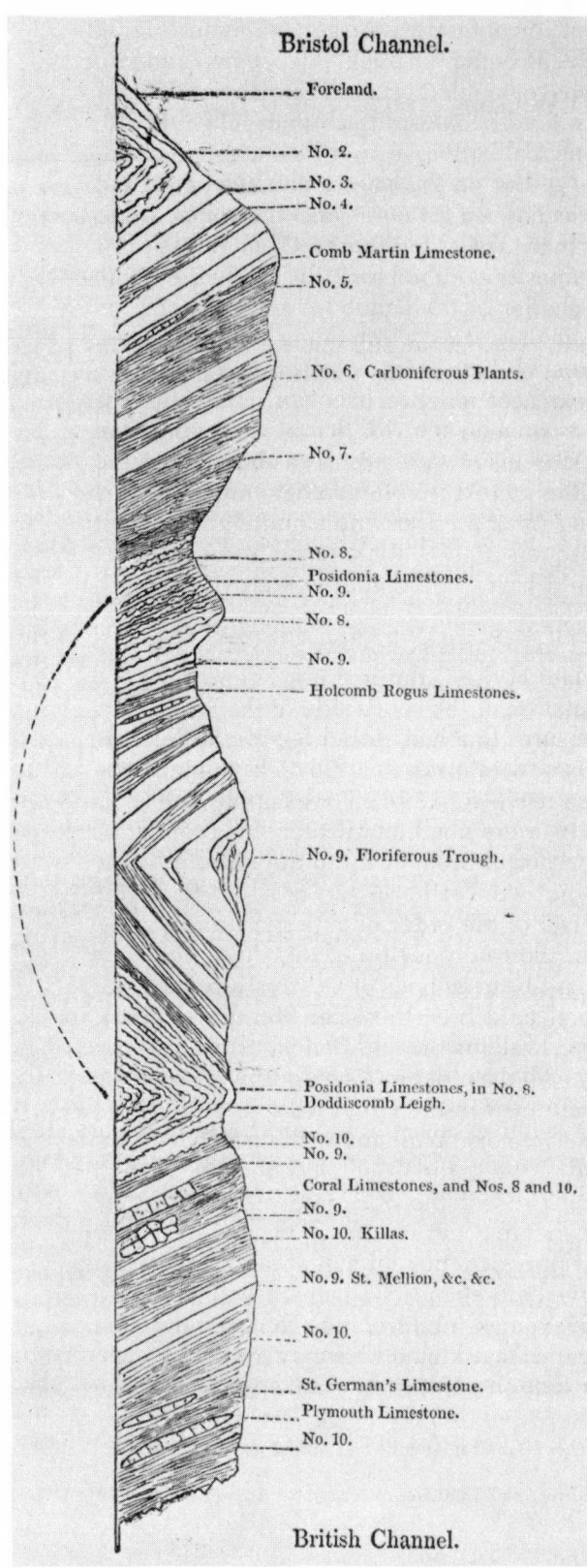

Fig. 12.10. Williams's diagrammatic traverse section of Devonshire, published in January 1840, showing the *saddle* of Culm or "Posidonia Limestones" (center right) that he claimed to have detected on the south of the Culm trough (center). This implied that most of the south Devon and Cornish strata (right, No. 10) were *above* and therefore younger than the Culm (No. 9). This radical reinterpretation of the structure supported his continuing insistence that the Culm must be of ancient, Greywacke, date.

limestones were indeed Upper Silurian, then Williams concluded that on his new interpretation of the sequence the Culm was "as likely to be ab[ou]t the age of the Caradoc Sands[tone]s as anything else."[68] By implication the north Devon strata, which he regarded as even older, would also be Lower Silurian or even Cambrian.

Darwin, as secretary, read the paper at the Geological Society the following week.[69] Williams criticized "the proposed law of Mr W. Smith," but only because he considered that its rigorous application would entail "a final and universal extinction of species" between every separate formation. Williams insisted that he himself believed such extinctions had only happened "at very distinct epochs," between which the apparent changes in fossils were to be attributed to the migration of organisms and to local changes in environment. In this way he could neatly circumvent Phillips's claim that the fossil shells in the Culm limestone were of Carboniferous species, and reconcile it with his own Lower Silurian dating of the Culm. De la Beche and others had earlier used a similar argument, but its plausibility was wearing thinner with every year that passed. Furthermore, any support Williams might have had from some of the more prominent geologists, for his rejection of Smith's simplistic use of fossils, would have been forfeited by the evident archaism of his higher-level theorizing, as when he revealed that he believed some characteristically Carboniferous species might still be alive somewhere in the world. Finally, those geologists would not have been pleased to hear Williams's concluding remark that the age of the Devon rocks ought not to be settled "by reference to the structure of a foreign district." That chauvinistic criticism of Sedgwick and Murchison's German research was not welcome in the cosmopolitan ambiance of the Geological Society.

Austen reported afterward that Williams had failed to treat the problem "Zoologically," though he himself had tried to do so during the discussion, and he hoped Phillips would exploit that approach more thoroughly. Murchison told Sedgwick more bluntly that "Parson Williams gave us another diatribe on Devon last meeting & I was compelled to speak." But what he criticized mainly was Williams's new interpretation of the structure of Devon, not his dating of the rocks. "The blundering Welshman has now got into a more extraordinary blunder than ever," Murchison explained, referring to Williams's inversion of the sequence to the south of the Culm trough: "I think I have settled the Parson's hash to the satisfaction of every real geologist present, but enough of this dull dog."[70]

When the *Athenaeum* published the Geological Society's official summary of his paper a few days later, Williams was incensed that it unfairly made him out to be a blundering reactionary in the matter of fossil evidence. He therefore wrote another article for the *Philosophical Magazine* to set the record straight. He challenged the "gentlemen" of the Society to face the implications

68. Williams to Phillips, 28 November 1839 (UMO: JP, 1839/50).

69. Williams (1840b), read at GSL on 4 December 1839. The Council agreed to his request to withdraw the paper afterward (GSL: BP, CM 1/5, p. 168).

70. Austen to Phillips, 8 December 1839 (UMO: JP, 1839/54); Murchison to Sedgwick, 8 December 1839 (defective transcript in ULC: AS, IIID, 27). Austen's comments on the fossils in the south Devon limestones are recorded by Darwin, note dated 4 December 1839 (ULC: CD, 205. 10.70); I am indebted to Dr David Kohn for telling me of this note. To settle Williams's hash was slang for beating him into silence.

of the field evidence that supported his new interpretation of the structure of
Devon: "to doubt it is to doubt the plainest evidence of the senses." He
recalled how, the previous summer, he had found that crucial evidence (near
Chudleigh): "the scales fell from my eyes—every difficulty and apparent
anomaly vanished as if by magic." The traverse section that illustrated his
article (fig. 12.10) showed clearly how this "master key" had resolved his
problems. On the south side of the Culm trough, where all were now agreed
that the black Culm limestone emerged from beneath the Culm proper, Williams claimed to have detected a saddle, such that that limestone then dipped
southward again under a repeat of the Culm proper, which was then followed
upward by the *whole* of the thick sequence of south Devon and Cornwall,
with the Plymouth limestone near the top.[71]

That such a deviant interpretation could be put forward at this stage
was due above all to the nature of the field evidence, which other geologists
considered far less decisive than Williams claimed. In southeast Devon, where
the structure was unquestionably complex and the rock exposures widely
scattered across a gentle landscape of fields and woods, no "master key" could
be unambiguously persuasive. "He stands alone," one observer commented
to Phillips after Williams's article was published; "like a fighting Cock on his
own field, [he is] going to fight you all together."[72] But with the evidence in
Devonshire so indecisive, it was now clear that Williams's gentlemanly critics
might leave him to crow unchallenged on his chosen territory, while they
themselves transferred their fight wholly to the wider arena of the Continent.

12.8   The Devonian Redivivus (December 1839 to February 1840)

Having crushed Williams's implausible interpretation of the structure of Devon
and Cornwall when the provincial geologist's paper was read, Murchison had
taken the opportunity to report "viva voce to the Society" how, despite Sedgwick's defection, he himself had now regained confidence in the Devonian
interpretation for both England and Germany. "I am fully prepared to stand
out for the perfect parallelism of the two countries," he told his collaborator,
"& to contend as of old that the Devonshire fossilif[erous] strata do really
occupy . . . the intermediate position & must therefore be on the Old Red
[Sandstone] parallel."[73] The Devonian was showing life again after all.

The evidence that had gradually brought Murchison "back to the *status
ante bellum*" was that of the fossils or "pièces justificatives" that he and
Sedgwick had brought back from the Continent. In the weeks since their
return, while Sedgwick was busy lecturing in Cambridge, Murchison had
been able to discuss their specimens at leisure with Lonsdale and Sowerby
at the Geological Society. He summarized their conclusions in a long and
detailed letter that was explicitly intended to bring Sedgwick back to his

71. Williams (1840a), dated 16 December 1839, published in January 1840 issue of *PMJS*. Fig.
12.10 is reproduced from ibid., p. 60. Williams to Phillips, 26 December 1839 (UMO: JP, 1839/
58).
72. Sanders to Phillips, 1 February 1840 (UMO: JP, 1840/5).
73. Murchison to Sedgwick, 8 December 1839.

"former true path" of the Devonian interpretation.[74] "Thank God! I now see daylight again," Murchison told his collaborator; "all our follies proceeded from our attending to these cursed mineralogists and gentlemen who deal in 'symétrie de position,' whose doctrines will now, I bless my stars, go by the board." But in fact his letter shows that it was not only the arguments of geologists such as Dumont that had made them doubt the Devonian interpretation; they had also been swayed by fossil specialists such as de Verneuil. It was not such a simple struggle between the forces of darkness and light, between the criteria of structure and rock type and the criterion of fossils, as Murchison polemically portrayed it.

Dealing first with the Belgian sequence, Murchison claimed that "the fundamental and cardinal point" that they had both accepted too uncritically was that the formations underlying the Mountain Limestone (Dumont's upper limestone group) were "true equivalents of the Silurian System." But in fact Dumont had borrowed that correlation from Buckland, who had based it on a first impression of the fossil evidence (§6.3; fig.6.3, B); and until *The Silurian System* was almost completed, Murchison himself had anticipated that Carboniferous formations would be directly underlain by Silurian formations in just this way (§10.4). However, he now reported that the fossils in the strata immediately under the Mountain Limestone of Belgium (i.e., in Dumont's upper sandstone group) were not Silurian species at all, but those found in precisely the same position in Britain (i.e., in Phillips's "transition group"). "You will see as soon as anyone," Murchison told his collaborator bluntly, "that the grand fulcrum of your ratiocination is gone." In the next formation down the sequence (Dumont's lower limestone), the fossils were likewise not Silurian, apart from some corals also known from Devonshire; Murchison inferred that it was what Sedgwick, before his recent defection, "*would* have called Devonian, i.e. of a type intermediate between the Carboniferous & Silurian." Murchison expected this revived Devonian interpretation to make his collaborator "fidget or jump," but he urged that it was "a very rational solution" in view of the complete conformity of these formations below the Belgian Carboniferous.

Turning next to Westphalia and Nassau, "the Classic country for my money," Murchison had regained his original confidence that it "would settle the whole question" and prove that the Devonian was sandwiched between Carboniferous and Silurian, more clearly than in Belgium. For the *Strygocephalus* limestone had now turned out to be "*IN EVERY ONE* of its fossils South Devon, & not Silurian anglice in any one save 2 or 3 corals which are common to S. Devon & Silurian." He added that "the shells are all distinct— they approach near to Carboniferous types & are *one & all* common in the S. Devon limestones!" Lying conformably above this formation, now unquestionably "Devonian," were the black Culm-like limestones, which Murchison said he had "not the *smallest* doubt are ours of Barnstaple" in north Devon, and was "now pretty well convinced" were also the equivalent of the true Mountain Limestone nearer the Rhine (at Ratingen). The position of the Devonian below the Carboniferous was thus secure once more. Lying conformably below the *Strygocephalus* limestone was "the fossiliferous Greywacke" of Westphalia and Nassau, which Murchison had first assigned to the Devonian

74. Ibid.

but later to the Silurian (§12.1, §12.5); he now reported that it contained Upper Silurian fossils. "In fine I revert to my original creed," he concluded, though in fact it was to a much more modest version. The Devonian had contracted to just the *Strygocephalus* limestone and the strata immediately above it, and the Silurian had expanded correspondingly (fig.12.9, C1, C2). But this reestablishment of the Silurian to the east of the Rhine did now give Murchison the firm base for the Devonian that had earlier been so elusive (§12.2). It was probably about the time he wrote this letter that he went back over his field notebooks and confidently marked many of his earlier sketch sections to show what he now believed was the sequence from Silurian through Devonian into Carboniferous (see fig. 12.8).

Dealing thirdly with the Eifel to the west of the Rhine, Murchison reported another "radical error," which if uncorrected "would have brought us back into Old Chaos." This was that they had assumed, following de Verneuil, that the Eifel limestone was Upper Silurian (Wenlock). "It is no such thing!" Murchison exclaimed; but nor were its fossils quite the same as those of the south Devon limestone or of the newly "Devonian" limestones of Belgium and Westphalia. "It is Devonian (Lowest) plus Silurian," he inferred; "or in other words it is (if reference to the established type in England be worth any thing) a member of transition from one system to the other, in which there are *many more Devonian* than Silurian fossils, but not *exclusively* Devonian." This inference was a striking expression of Murchison's renewed confidence in the Lyellian model of piecemeal faunal change, on which the Devonian interpretation had originally been based. For since the Eifel limestone lay almost at the top of its local sequence (fig. 10.1), there could be no direct *field* evidence that it was "Lowest" Devonian and lower than the more definitely Devonian limestones elsewhere (fig. 12.9, C3). On the other hand, Murchison did now support his inference by claiming that the fossils in the underlying strata were "all *Upper Silurian* & positively not Lower Silurian" species.

Finally, Murchison reported that the Harz fossils were also either Devonian or Carboniferous, while the sections in the Fichtelgebirge were "most decisive," in that below the Carboniferous strata was "a fine Devonian Zone with all the fossils & corals which nothing can gainsay & which no man alive shall ever place in parallel with *my* Silurians." All in all, the cumulative evidence of the fossils from all five regions put the Devonian interpretation, or at least a modest version of it, on to much firmer foundations. "I pledge my life," Murchison assured his collaborator, "that if plain facts be laid before plain geologists there will be no escape from my present induction." He was, he told Sedgwick, "once more redivivus, although you had well-nigh killed me."

Sedgwick may have been a plain geologist, but he did not find Murchison's conclusion persuasive. Although he hoped to be able to discuss the matter face to face at the Geological Society the following week, it was far too important to wait till then, and he wrote back with a detailed response to Murchison's arguments. "The Westphalian sections are very good because they tell a good story," he admitted, "but for classification I think the left Bank of the Rhine better because of the Eifel limestone." He recalled that that limestone contained many Upper Silurian fossils (e.g. the trilobite *Calymene* and the cephalopod *Phragmoceras*: fig. 10.3, A, C), which told "a plain story which Buckland, Goldfuss, Dumont & ourselves all read one way"; and the Eifel strata below the limestone also contained Lower Silurian fossils (e.g., the trilobite *Trinucleus*). Sedgwick found it highly implausible that these Silurian genera should "struggle up into a *terra incognita* of so called *old red*," as Murchison's latest interpretation implied; "I don't believe a word of it." He reminded Murchison that "all the Geologists of the Continent" had equated the Eifel limestone with the lower Belgian limestone, from which Dumont had collected "some very firm Eifel fossils." In both regions, on "the old view," the underlying Lower Silurian (Caradoc) strata "pass downwards into slates with few fossils & finally with no fossils" (the Belgian *terrain ardoisier*), which Sedgwick evidently still regarded as the Cambrian. "From my general views of the structure of the Rhenish provinces I therefore don't budge a step," he added firmly.[75]

"Don't suppose I shall not rejoice to get a little Devonian, or rather old red, top dressing," Sedgwick assured his collaborator; "it would delight me to find it & the more the better." But it is clear that his interpretation would only leave room for a thin "top dressing" to what he considered to be Silurian limestones, immediately under the Carboniferous formations (fig. 12.9 B). For example, he accepted readily that the strata immediately below the Mountain Limestone of Belgium were what Murchison had termed "sub-Carboniferous"; but he was surprised his collaborator had found no Upper Silurian (Ludlow) fossils in the strata between those rocks and Dumont's lower limestone group below. This indicates that Sedgwick, like Dumont, expected the equivalents of the Old Red Sandstone to be squeezed into no more than a fraction of one of the Belgian formations (the upper sandstone group: see fig. 6.3, B). It is therefore not surprising that Sedgwick now revealed that he had also lost confidence in the Devonian interpretation of the south Devon limestone itself: "I deny that it is old red," he told Murchison. "I deny that its corals are intermediate between the Silurian & Carboniferous polypes," he added, arrogating Lonsdale's expertise to himself; but he did admit that "the shells are the only difficulty and a great one they certainly present." A more weighty point, however, was that the Devon limestone was "overlaid by slate rocks thicker than all your Silurians put together"; for this was incompatible with the minimal or "top dressing" version of the Devonian interpretation that was all Sedgwick would now concede.[76]

Shortly afterward, Sedgwick told Murchison he could not meet him in London after all. "Only knock up the Eifel case & I am your man," he assured him, recalling how Goldfuss's fossils at Bonn had convinced Murchison himself that the Eifel limestone was Upper Silurian (§12.2). "All that you & I have wrangled about since," he pointed out, was Murchison's claim that the *Strygocephalus* limestone of Westphalia should be "split off from the Eifel," as Devonian and Silurian, respectively (fig. 12.9, A). Sedgwick continued to

75. Sedgwick to Murchison, 12 December 1839 (GSL: RIM, S11/165).
76. Ibid. Perhaps inadvertently, Sedgwick mentioned one new piece of evidence that reinforced Murchison's more ambitious version, namely that Ansted had found Lower Silurian (Caradoc) fossils in the Fichtelgebirge, below the putative Devonian (Münster's goniatite-bearing strata), which was thus provided with its previously elusive base (§12.2). This was not mentioned in the paper Ansted presented shortly afterward, which was confined to Bohemia: Ansted (1840), read at GSL on 8 January.

believe they were "in *one system,* " even if not "on the same *exact line*" (fig. 12.9, B); and Murchison had now virtually conceded this (fig. 12.9, C). What remained to be agreed was *which* system, Devonian or Silurian.[77]

Murchison waited over a month before replying, in the vain hope that they would be able to discuss the problem face to face. When at last he wrote again, he claimed it was now "impossible to look at the fossil evidences" from all the Continental regions, and "not to allow the existence of a group intermediate between the Carboniferous & Silurian"—a view which, he reported, de Verneuil too was now inclined to adopt. Putting past mistakes behind them, Murchison pleaded with Sedgwick to consider "how we can set ourselves straight & *take* and *keep* the true geological line" of the Devonian interpretation; otherwise, he warned, "all will again become Chaos & years may elapse before the right road is hit upon." With their Devonshire paper about to appear at long last in published form, Murchison was already drafting an introduction to their projected paper on their Continental work. "I have in my mind's eye how *I* would treat the subject on my own bottom," he told Sedgwick, "but what a much better bottom will the case stand upon if you on your honest judgement arrive at the same conclusions!" He claimed that "the sequence and passage" from Silurian through Devonian to Carboniferous was the same in Brittany—which in the event they had failed to visit—as in Belgium and the Rhineland; and he concluded that "the cases will all stand or all fall together, & the question is one of the utmost geological importance."[78]

Murchison's proposed revision of Dumont's work had now brought him into conflict with Fitton. It was probably Fitton who had arranged the previous summer for Dumont's correlations to be published in English (§12.3). More recently he had proposed Dumont for the next Wollaston medal, implying that Murchison had not given the Belgian geologist the credit he deserved. Murchison protested to Sedgwick that he did appreciate Dumont's sequence, but that the Belgian's correlations had been based only on "mineralogical" criteria. Dumont's own lists of fossils were "quite enough to prove that these rocks cannot be Silurian," he claimed; and "never were passages & sequences more perfect than from his carboniferous limestone downwards." Murchison therefore insisted that the true interpretation of the Belgian sequence was his own achievement and not Dumont's at all: "when we find the zoological passage equally good—in short all the evidence one way, why are we in doubt?"[79]

Meanwhile Sedgwick was preparing the last of his lectures. He saw "no difficulty with the British sequence," since the *order* of the formations in Devon was no longer controversial. But "on the Continental Equivalents I shall speak with *proper caution,* " he told his collaborator; "I must state the case with all its difficulties, & so let it rest—till time has set all right." Since

77. Sedgwick to Murchison, n.d. [mid-December 1839] (GSL: RIM, S11/180). By internal evidence, this important but undated letter appears to be a reply to a missing letter from Murchison, replying in turn to Sedgwick's letter of 12 December. It was certainly written before the end of the year, since Sedgwick regrets that they cannot meet on "Wednesday" (probably the GSL meeting on 18 December) and adds "we can hit again in January." The dating is important because the letter records a crucial stage in the development of Sedgwick's views.

78. Murchison to Sedgwick, 23 January 1840 (defective transcript in ULC: AS, IIID, 28).

79. Ibid.; Dumont to Murchison, 11 January 1840 (GSL: RIM, D27/1); Council of GSL, 5 February 1840 (GSL: BP, CM 1/5, p. 186).

the lectures would again prevent them from meeting in London, he set out in another long letter the reasons why, despite all Murchison's rhetoric, he was still skeptical about the Devonian interpretation. "That there is a Devonian series I doubt not," he admitted, "but what is its extent?" That Dumont's upper sandstone group, *"at least a portion of it,* is coeval with our old red seems certain," he conceded; but this was "exactly Dumont's view of the case" and no discovery of theirs.[80] Anyway, that thin Devonian top-dressing was quite different from Murchison's far more ambitious Devonian system, which took in *all* Dumont's formations between the Mountain Limestone and the *terrain ardoisier.*

"Here I don't see my way," Sedgwick admitted; if the lower limestone in Belgium contained many Upper Silurian fossils, as he thought it did, he feared that "we shall never be able to shake our Continental neighbours out of their present notions" based on Dumont's correlations. Those might indeed have been "founded on mineral analogies," Sedgwick conceded; but "supposing him wrong, and his Silurians to be our Devonians, what follows?" In Sedgwick's view, the Devonian of Belgium would then be suspiciously like the Silurian of Britain; if Murchison's Devonian correlation were adopted nonetheless, their problems would still not be over, "for where are your Silurians?" To call the lowest Belgian strata Silurian would be "hypothetical, because there are no fossils," Sedgwick pointed out; and without fossils to identify the strata, "there is an end of your Silurian types, for the general purpose of Continental comparison & subdivision." Sedgwick's solicitude for his collaborator's system was tacitly tinged, however, with apprehension about his own; for if Dumont's Silurian were reassigned to the Devonian, the underlying *terrain ardoisier* would have to be assigned to the Silurian, and Sedgwick might be left with no Cambrian at all. He therefore tried to forestall any move by Murchison to redefine the Silurian to denote "all the fossiliferous rocks below the Devonian"; to do so, he pointed out, would deprive the Silurian of "all its value as a term of distinct classification."[81]

These criticisms were cogent indeed, though Sedgwick claimed they were merely an attempt "to state the whole difficulties we have to fight against and reconcile." But it is clear that now, in striking contrast to his views earlier in the controversy, he expected that what would "set all right" in the end would be the fossil evidence. "You have the whip hand of me with the fossils before you," he pointed out, implying that he might well come round into agreement when he had seen them for himself. As for Dumont, he thought Murchison had referred to the Belgian "far too contemptuously," and that his groups were "quite natural" and his local sequence "as good as it can be." Fitton had written Sedgwick a long letter, "partly apologetic—as if he thought I should be jealous of Dumont's claim"; but Sedgwick had replied bluntly that he knew Dumont's work at firsthand much better than Fitton, and that he would certainly support the proposal to give him the Wollaston medal.[82]

Saving the full force of his rhetoric till they met face to face, Murchison only sent his collaborator a short note in reply to this long letter, to report one further piece of evidence without delay. Sowerby and Lonsdale had now

80. Sedgwick to Murchison, 27 January 1840 (GSL: RIM, S11/151).
81. Ibid.
82. Ibid. Fitton's letter has not been located.

identified the "Silurian" fossils Buckland had collected on his traverse of the Meuse in 1835 (§6.3). "The characteristic Corals are *pure* Devonian," Murchison reported, and the "shells" were just the kinds they had anticipated as "filling up the intervals" between Silurian and Carboniferous. "Dumont's Silurian is thus utterly annihilated root & branch," he concluded, leaving it to Sedgwick to draw the corresponding inference that the Devonian was now all set to be triumphant.[83] Their intensive exchange of correspondence may not have brought Sedgwick round completely, but it had certainly left him more ready to return to the full Devonian interpretation that he had nominally co-authored ten months earlier. It remained to be seen, however, whether that interpretation would become the focus for a new and wider consensus, or remain as an implausible minority opinion.

83. Murchison to Sedgwick, 1 February 1840 (defective transcript in ULC: AS, IIID, 29).

Chapter Thirteen

CREATING A CONSENSUS

13.1   Devonshire in a New Dress (February 1840)

At the Geological Society the following week, Greenough presented, displayed, and explained the long-awaited revised edition of his great geological map of England and Wales. He congratulated himself that, after all the intensive research of the previous twenty years, his map had needed so few major alterations. But as on the advance copy he had shown at Birmingham and Boulogne the previous summer (§12.3, §12.4), Devonshire was depicted, as Murchison had put it, "in a new dress." Indeed, although there were many minor corrections and improvements to geological boundaries all over the map, only the older rocks had needed any major change, and only Devonshire and Cornwall had needed radical revision.[1]

Greenough had colored the main part of the Culm as Coal Measures, and its basal black limestones as Mountain Limestone. Like Phillips a year and a half earlier (fig. 10.2), he thereby explicitly accepted the whole of the Culm as Carboniferous, as Sedgwick and Murchison had first claimed at Bristol in 1836 (§7.3). But the "new dress" of Devonshire extended to the older rocks too, for Greenough had colored them the same as the Old Red Sandstone

---

1. Greenough's map (1840a), dated 1 November 1839, but not in fact published until January 1840 (an advertisement appended to *PGS* 3 no. 66, with summaries of meetings up to 8 January, refers to the map as "nearly ready"). Greenough presented the map and read a paper about it (Greenough 1840c) at GSL on 5 February; the *Memoir* to accompany the map (Greenough 1840b) was evidently not published until after that meeting.

elsewhere, thereby implicitly accepting Sedgwick and Murchison's more recent Devonian interpretation. He had not "ruddled" them with a reddish tint,
however, as Murchison had expected, but had followed De la Beche's suggestion that he should show the Cambrian, Silurian, and Devonian by three
related tints of the same Greywacke color (§12.3). This was no trivial carto-
graphical detail, but a visual feature pregnant with theoretical significance.

Greenough claimed that the "false colouring" of Devonshire on his
original map had only arisen "from imperfect information"; but that "Baconian" assertion was belied by his reference to the "animated discussions" on
the Devon problem over the past years. In fact he had only accepted the
Devonian interpretation by giving it an important theoretical twist. Rather
than coloring the older strata of Devonshire as Old Red Sandstone, he had
depicted the Old Red Sandstone "as an *integrant member of the Greywacke
formation.*" He thereby joined De la Beche in claiming that it was the Old
Red Sandstone that was atypical of its period, not the Greywacke rocks of
Devon and the Continent. But he refused to adopt the term "Devonian," on
the grounds that that word should be reserved for its traditional topographical
usage. This was somewhat specious, for he conceded that "in Siluria the
divisions first established by Mr. Murchison have broken up the unity of the
quondam greywacke formation," and he adopted the term "Silurian" for the
middle division of the Greywacke, even though it too was in origin a topographical term.[2]

By a characteristic piece of sophistry, Greenough turned to his own
advantage the fact that Murchison now regarded the Harz as Devonian. For
he claimed that, since the term *Grauwacke* had been "first applied in the
Harz, and . . . adopted by Werner & by his pupils spread all over the world,"
it had precedence over the term Devonian, if indeed the correlation between
the two regions were valid. However, he conceded that the greywacke rock
type could be found among strata of many different periods, and that the
geology of the Harz was obscure. So he proposed that the term Greywacke
should be dropped altogether, as Murchison had been urging since early in
the controversy. But instead of adopting the term Devonian for the Greywacke
above the Silurian, and the term Cambrian for the Greywacke below it, Greenough reused another old miners' term and called them both "Killas."[3] The
effect of these terminological maneuvers was to let Greenough have his cake
and eat it. He adopted the major divisions of the pre-Carboniferous strata that
Sedgwick and Murchison had proposed, but instead of terming them in descending order the Devonian, Silurian, and Cambrian, they appeared on his
map as Upper Killas, Silurian, and Lower Killas. (The Silurian was presumably
too well established by now to be rejected, even by Greenough.)

It is a mark of the perceived importance of "this *vexata quaestio*," as
Greenough called it, that he devoted most of his paper to the Geological
Society, and most of the short explanatory *Memoir* that was published shortly
afterward, to the discussion of the Devon controversy. Although his treatment
of Sedgwick's and Murchison's work was less than generous, there can be no

2. Greenough (1840b, pp. ii–iv).

3. Ibid., p. iv; Greenough, note beginning "The term Greywacke . . . ," n.d. (UCL: GBG, 16/
9, loose sheet), almost certainly a note for his speech at GSL. He had used the term *Killas* in a
different sense on the original edition of the map (§4.6).

doubt that he had completely accepted both the Carboniferous age of the Devon Culm and the Old Red Sandstone or Devonian age of the older rocks of that county. Furthermore, Murchison had given him a summary of his Devonian interpretation of the Westphalian sequence, and Greenough quoted this with full approval.[4] Ironically, therefore, the geologist whom Murchison regarded as the most die-hard reactionary had accepted the Devonian interpretation in its fullest international form, even before Murchison's most trusted collaborator!

After Greenough had introduced his map, Buckland exercised his prerogative and forbade any discussion of it, probably to avoid another outbreak of personal squabbling over priority and credit. But Murchison, in what he termed afterward a *"spontaneous act,"* asked the president's permission to move a vote of thanks, and praised the *"great merits"* of the map. He wrote to Greenough the next day, however, to express how he was hurt by Greenough's *"new* method of explaining *why* you have coloured Devon & Cornwall as *Carboniferous* & *Old Red."* He complained that, in contrast to what Greenough had generously said at Birmingham, he and Sedgwick had not received due credit as the authors of the most radical changes on the map. Greenough had mentioned Phillips, De la Beche, and even MacCulloch, "yet not a word of those who first propounded the *new* view which you have adopted after stoutly *opposing it* up to last summer." The most striking component of that new view was of course the Devonian interpretation, but Murchison claimed to be indifferent about the name: "the great fact is that there does exist—and we we were the first to propound it—a system intercalated between the Carboniferous & Silurian Systems," with a deceptively *"antique* impress" in Devonshire and a characteristic assemblage of fossils. This was the outstanding achievement for which Murchison demanded full credit, precisely because it had initially been regarded as implausible or even absurd. "It therefore delights me to see you (our father) taking our view at the eleventh hour," he concluded, in a backhanded compliment to the first president of the Society; and he asked only that Greenough should openly acknowledge their originality.[5]

Greenough was unmoved. On the attribution of the Culm to the Coal Measures, he told Murchison, "your conclusion was true, your premises were unsound"; as at the start of the controversy (§5.2), he rejected Murchison's dogmatic assertion "that plants found within a true coal field could not be found any where else." On the coloring of the older rocks of Devon, he turned the priority tables on his critic, and claimed that *he* had not changed his interpretation, since he had always equated them with the true and original *Grauwacke* of the Harz, which Murchison now termed Devonian. Greenough disclaimed any intention of appropriating for himself the credit for the changes on his map; but he deftly denied it to Murchison, by claiming that it was Harding's fossils and Phillips's identification of them that had first convinced him that the Culm was Carboniferous, while it was Lonsdale's opinions on Austen's fossils that had shown him that the older rocks of Devon were Upper Killas or Devonian. Reprovingly, he told Murchison that "no praise is worth

---

4. Ibid., p. viii; Greenough, note beginning "if further evidence . . . ," n.d. (UCL: GBG, 16/9, loose sheet), bearing on verso an undated memorandum in Murchison's hand, almost certainly used by Greenough for his paper at GSL.

5. Murchison to Greenough, 6 February 1840 (ULC: GBG).

having that is not strictly just"; but his own treatment of the issue put a rather narrow justice far above generosity.[6]

Murchison was incensed at Greenough's intransigence: in his response the velvet gloves came off. "You announce the Devon changes as the greatest in the Map," he complained, "& yet you swamp the authors of the change." What angered him most was that Greenough himself had been so prominent among the earlier critics of both components of that change. "So *horribly* did it grate on *all your ears*," Murchison recalled, referring to the interpretation of the Culm as Carboniferous, "that we were pronounced all but mad," and it was only "after years of battling" that it had won some support. Again he protested that he had no objection if Greenough wished to "throw overboard" the term Devonian; "but pray do not endeavour to do it by a side wind." He added, however, that Greenough should give his reasons for choosing "Upper Killas" instead: "some may side with you, some with us, but we do not abandon the term [Devonian]." He concluded with a scarcely veiled threat: "if you do not interpolate our names, my honest opinion is that you will do yourself a disservice & be sorry for it hereafter."[7]

The threat behind the veil was disclosed the next day, when Murchison reported to Sedgwick what had happened at and since the meeting. "The *most public method*" might have to be used to expose "the pirates," he told his collaborator, if Greenough did not insert an adequate acknowledgment into his *Memoir* and if Buckland did not do likewise in his forthcoming anniversary address. "I would myself write a statement for the English, French & German Journals," he proposed, "that would have our President & Greenough in no enviable position." He was disturbed, however, that he and Sedgwick were unable to "take a strong & united line on the whole subject," since Sedgwick was still skeptical about the Devonian interpretation that was one important component of the alleged piracy. Yet "something must be done soon," he insisted, "for Paddy Fitton has so stirred the cauldron that we must go ahead." Significantly, Murchison had taken the precaution of seconding Fitton's proposal that Dumont should be awarded the Wollaston medal, when it was finally approved by the Council of the Geological Society just after Greenough presented his map.[8] That would at least forestall any criticism that Murchison was claiming all the credit for working out the Belgian sequence, even though he was boasting how he had "annihilated" Dumont's Silurian correlations.

13.2   A Geological Enclosure Act (February and March 1840)

Dumont had pride of place at the anniversary meeting of the Geological Society a few days later, when his friend Fitton accepted the Wollaston medal on his behalf. The reason given in Buckland's eulogy for the Society's "tardy

6. Greenough to Murchison, n.d. [between 6 and 9 February 1840] (original not located; Greenough's copy in ULC: GBG).

7. Murchison to Greenough, 9 February 1840 (ULC: GBG); Murchison, note beginning "Bottom of p. 3," n.d. (UCL: GBG, 16/9, loose sheet), referring to Greenough's *Memoir* and probably sent with this letter.

8. Murchison to Sedgwick, 10 February 1840 (transcript in ULC: AS, IIID, 30); Council of GSL, 22 January and 5 February 1840 (GSL: BP, CM 1/5, pp. 181, 186).

recognition" of Dumont's work was that the intense folding of the strata de-picted in his original memoir had seemed "unusually complex and improb-able" to English geologists. They had therefore doubted its reality until they had seen it, as Greenough and Buckland himself had in 1835 (§6.3), and Sedgwick and Murchison more recently (§12.4, §12.5).[9] But the award seemed even more tardy than it might have, because it had to be given exclusively for Dumont's original memoir of 1832 and not for his more recent papers that had brought that work up-to-date. For only in that way could the Society acknowledge Dumont's brilliant structural analysis of the Belgian rocks and his unraveling of the sequence there, while avoiding having to endorse his more recent correlations between the Belgian sequence and the British, about which Murchison at least was now highly skeptical.

After the award had been made, Buckland delivered his presidential address. In his review of "the Home Department of Positive Geology," he put *"the Devonian system"* into first place; and his generous treatment of Sedg-wick's and Murchison's originality must have made Murchison abandon his threat to criticize him in the scientific press. Lonsdale was given the credit for making the suggestion, which "was not at first appreciated," on which the new system was based. But Sedgwick and Murchison were fully acknowl-edged as the eventual authors of "the greatest [change] ever made at one time in the classification of our English formations"; they were praised for the way they had extended it, "with a boldness which does credit to their sagacity," even to the ancient-looking, metal-bearing rocks of Cornwall. Buckland noted how De la Beche, and more recently Greenough, had adopted the same group-ing of the rocks, though under different names (§10.5, §13.1); he tactfully agreed with Greenough that no single term could adequately characterize formations as diverse as the Old Red Sandstone and the rocks of Devonshire. Nevertheless, he did adopt the term Devonian, and he emphasized how its recognition explained at last why the Old Red Sandstone had seemed so puzzlingly absent on the Continent.[10]

In an eirenical gesture of compromise, Buckland suggested that the term Greywacke should be retained in its time-honored sense, to denote *all* the Transition strata below the Carboniferous (the latter now redefined to exclude the Old Red Sandstone), but that it should be divided into Devonian, Silurian, and Cambrian. This proposal followed the interpretation embodied in Greenough's new map, while allowing Sedgwick and Murchison the names of all the three systems they had defined. Buckland gave that threefold division the highest praise possible, by comparing it to the systematic order that the Secondary strata had received at the hands of William Smith, whose *éloge* he delivered at the end of his address. Combined with the analogous work that Lyell and some of the French geologists had been doing on the Tertiary strata in the same years as the Devonian controversy, the whole sequence of for-mations was becoming more orderly with every year that passed, and the

---

9. Anniversary meeting of GSL, 21 February 1840, award of Wollaston medal (*PGS* 3:206–7). The cash prize from the same fund was awarded to Sowerby as a subsidy for his *Mineral Con-chology* (ibid., pp. 208–9).

10. Buckland (1840, pp. 224–26), read at GSL on 21 February. He had asked Williams for permission to cite his unpublished discovery of fossil plants in the Cornish metal-bearing slates: Buckland to Williams, 15 February 1840 (SAS: DW, box I, enclosed in notebook 17).

history of the earth and of life ever clearer. In a vivid and appropriate metaphor, Buckland concluded that with the establishment of the three systems of older rocks, "we are, as it were, extending the progressive operations of a general inclosure act over the great common field of geology." The "chaos" of an undifferentiated Greywacke was being banished to the margins of the science; the fossil-based methods of the deceased "father of English geology" were triumphant.[11]

Greenough made no comparable gesture to assuage Murchison's indignation. The published versions of his Geological Society paper and the *Memoir* accompanying his map both mentioned Sedgwick and Murchison by name, but only alongside many others and with no special emphasis.[12] But as on earlier occasions Murchison's bark was worse than his bite, and he dropped his plan to accuse Greenough of plagiarism.

Among those present to hear Buckland's address was Sedgwick, making his first visit to the Geological Society since returning from the Continent exactly four months earlier. During that time Murchison had repeatedly bemoaned his inability to win his collaborator back into full adherence to the Devonian interpretation, because Sedgwick had not been able to see the fossil evidence for himself. But while he was in London for the anniversary meeting, Sedgwick did at last see the fossils they had both collected in Belgium and the Rhineland, duly identified by Lonsdale and Sowerby. According to Murchison—Sedgwick did not record his own reaction at firsthand—the comparison of the fossils with those from Devonshire convinced him that the Devonian interpretation was valid after all. Sedgwick's support gave Murchison the signal he had been waiting for, as he told Phillips, "to open the campaign" to establish the Devonian as a distinctive system that could be recognized all over western Europe, sandwiched between the Silurian and the Carboniferous. "I am now highly delighted in having insisted on the 'Old Red' as a system," he added, "and on my prophecy of what it would turn out in fossils," alluding to his cryptic hints in *The Silurian System* (§10.4). With Buckland and Greenough both adopting the Devonian interpretation, he concluded confidently that "at last all is settled as to the great boundaries."[13]

In fact persuasive work was still needed in order to settle anything, as Murchison tacitly conceded by his talk of a coming "campaign." His letter to Phillips was itself an opening shot; when he reported that Phillips's own identifications of Austen's fossils added to "a list already too strong to admit of any doubt" about the reality of the Devonian, it was Phillips's doubts he was covertly trying to dispel. As part of the persuasive process, he had to amend the record of his own past opinions. He asked Phillips—addressing him now as secretary of the British Association rather than as a fossil specialist—if he could alter the summary of his paper from the Birmingham meeting (§12.3) before it was published in the Association's annual *Report*. Almost certainly he wanted to strengthen his argument for the Devonian on the Continent in the light of his work since the meeting took place. More serious, however, was the amendment needed to the record of the subsequent

11. Buckland (1840, pp. 227–29); the Tertiary research is reviewed on pp. 233–37; the *éloge* on Smith is on pp. 248–54.
12. Greenough (1840b, p. iii; 1840c, p. 181).
13. Murchison to Phillips, 25 February 1840 (extract in Geikie 1875, 1:286–87).

meeting at Boulogne, where he had firmly identified Silurian formations and fossils immediately below the Carboniferous (§12.4). Once his confidence in the Devonian interpretation had revived, as his other Continental fossils were identified after his return to England (§12.8), Murchison realized he must have made a major mistake in the Boulonnais, and he asked the curator of the museum in Boulogne to send him some fossils to check the point. "They clearly Devonianize it," he admitted to Phillips. He claimed that "without looking for fossils" he had accepted on trust that Dumont and de Verneuil were right in "Silurianizing" the rocks below the Carboniferous.[14] But here he was rewriting history: the record of the meeting and his own field notes show that he himself had had no hesitation in identifying the fossils as Silurian, and that he had thereby put a stop to the hunt for coal in those rocks (§12.4).

Murchison resolved to go to Paris without delay, to further his campaign by explaining the Devonian interpretation at firsthand to the French geologists, correcting his mistake over the Boulonnais and turning it into an instance of the Devonian system on French soil. That persuasive task became more urgent when de Verneuil told him about the paper *he* intended to present to the Société géologique, even before Murchison could get there. For de Verneuil revealed that he had not in fact accepted the Devonian interpretation in its full form; as Murchison reported to Sedgwick, the French geologist wanted to treat the Devonian only as "an upper member" of the Silurian. When he read his paper in Paris a week later, de Verneuil showed he was mainly concerned to argue for the global validity of just two great systems of older strata, the Carboniferous and the Silurian. All he did with the Devonian interpretation was to claim that, if and when his British collaborators proved the equivalence of the Old Red Sandstone and the Devonshire strata, that group would have to be shifted from the Carboniferous to the Silurian side of a great and natural divide, to become a newly defined "Silurien supérieur." In a wide-ranging review, de Verneuil argued that both great systems could now be recognized throughout Europe, in North and South America, South Africa, and even Australasia; and that they were sharply distinct in their fossil assemblages.[15]

These views were not unlike those Murchison himself had espoused, until the final stages of writing *The Silurian System* (§10.4); and de Verneuil set them in a wider framework with which Murchison could partly agree. Like Murchison, de Verneuil attributed the apparent global uniformity of the faunas of these ancient epochs to a climatic homogeneity much greater than at the present day. But he regarded that long development from homogeneity to heterogeneity as having been punctuated by occasional "grandes révolutions" in conditions, each causing a corresponding "renouvellement des espèces." As de Verneuil acknowledged, the theory was Élie de Beaumont's by origin; but Murchison, having shifted into a much more Lyellian or gradualist position since his adoption of the Devonian interpretation, would have none of it. "I hold *stoutly* to the doctrine of old Lonsdale," he told Sedgwick, "which from all I see is coming out stronger than ever." For de Verneuil's treatment of the Devonian could not be reconciled with the opinion of all the English

14. Ibid.
15. Murchison to Sedgwick, 24 February 1840 (ULC: AS, ID, 26); de Verneuil (1840), read at SGF on 2 March.

fossil specialists, that that fauna contained many Carboniferous species as well as those with Silurian affinities.[16] The full Devonian interpretation, with its emphasis on the *gradual* faunal transitions from Silurian to Devonian and from Devonian to Carboniferous, was incompatible with what de Verneuil had proposed; he had still to be won round after all.

Murchison fared better with de Verneuil's mentor, Élie de Beaumont, to whom he wrote in advance of his visit to explain his latest views on the Devonian interpretation. Although Élie de Beaumont remained skeptical about the gradual faunal transitions that Murchison claimed, there is no doubt that, unlike de Verneuil, he was inclined to accept the Devonian as a full-blown system or *terrain* on a par with the Silurian and the Carboniferous. Murchison's new interpretation of the Rhineland, he told him, matched his own observations there, and had, he thought, "de grandes chances de stabilité"; and he found the proposed changes in the Belgian correlation plausible. Having come to appreciate "la puissance de métamorphisme," he was prepared to believe that even the Cornish rocks, as ancient in appearance as any, might belong to the "terrain dévonien." He promised that he and Dufrénoy would look again at Brittany in the light of Murchison's new ideas; he anticipated that the Devonian might prove to be even more widespread in France than the Silurian.[17] De Verneuil might still be a half-hearted and unreliable ally, but Murchison had evidently won for the Devonian interpretation the far more weighty support of Élie de Beaumont. To add to that good news, he had just received two letters from Berlin which revealed that the new interpretation had also won the support of two leading Prussian men of science. The creation of a fully international consensus, among those whose opinions mattered most, was now well under way.

Alexander von Humboldt (b. 1769), the most outstanding of Prussian men of science and one of the most distinguished in all Europe, wrote with lavish if belated praise for *The Silurian System*, a copy of which Murchison had sent him while he was in Germany. In the early 1820s, Humboldt's *Essai géognostique* had been a seminal work on the principles of stratigraphical geology; although his scientific interests and obligations had widened immensely since that time, he may not have been exaggerating much when he claimed to have read every line of Murchison's book. He also told him he was now waiting impatiently for Sedgwick's companion volume on the Cambrian, "le spectre qui se montre au dessous de votre Système Silurien"; and he expressed full approval of both names as "sonores et flexibles." But what is most striking is that he had already grasped the British geologists' proposal of the Devonian system, that he appreciated how it elevated the Old Red Sandstone into a quite new significance, and that he recognized how fruitful the threefold systems might be in German geology. Far from resenting Murchison's intended suppression of the Germanic term *Grauwacke*, this most cosmopolitan of scientists welcomed the use of British terms in geology, alongside, for example, the Arabic and Greek terms long used by astronomers, as "une nouvelle marque de la communauté intellectuelle des peuples civil-

---

16. Murchison to Sedgwick, 24 February 1840; de Verneuil (1840); Élie de Beaumont (1829–30).

17. Élie de Beaumont to Murchison, 5 March 1840 (GSL: RIM, E4/10).

isés."[18] Murchison and Sedgwick could hardly have wished for more striking international recognition of their geological Enclosure Act on the older rocks of the earth's crust.

Murchison's second letter from the Prussian capital, likewise thanking him for a copy of his big book, was from von Buch. As Humboldt admitted, his colleague's appreciation of the work would be even more informed than his own. Although von Buch's praise was as unstinted as Humboldt's, he did also sound a note of caution about any attempt to recognize the finer divisions of the Silurian outside its original region: "de vouloir les faire entrer dans une case de la série établie," he told Murchison, "me parait vouloir l'étendre dans le lit de Procrust." Not knowing of Murchison's latest conclusions, he assumed he would somehow insert the *Strygocephalus* limestone into his "Série Velche" or Silurian.[19]

All this, however, referred only to the finer divisions; about the broader systems von Buch had no hesitations. An English oyster might not taste quite the same as a Holstein oyster, as he put it, yet they were the same species. He himself had been perhaps the first geologist on the Continent to use the Silurian and Cambrian to bring provisional order to the Greywacke there, soon after those systems were defined (§6.3). Now he revealed that he had adopted the Devonian with equal enthusiasm. When he saw Murchison's illustrations of the Old Red Sandstone fish *Holoptychius* (fig. 13.1), he had recognized at once that its scales were those long since known from sandstones in the Russian Baltic province of Livonia, associated with limestones con-

Fig. 13.1.   The fossil fish *Holoptychius*, as illustrated in Murchison's *Silurian System* (1839). Its distinctive bony scales proved to be a valuable marker for the Old Red Sandstone throughout Europe, even if there were no well-preserved specimens of the whole fish. This specimen was 28 inches (72 cm) long: the large size and complex structure of many of these oldest known vertebrates were widely regarded as powerful evidence against theories of transmutation or organic evolution.

18. Humboldt to Murchison, 26 February 1840 (GSL: RIM, H32/1; more legible contemporary transcript in ULE: RIM, Gen. 523/4.45). Humboldt's *Essai* (1823) had been published the same year in German and English translations.

19. Von Buch to Murchison, 23 February 1840 (GSL: RIM, B33/1; a reliable but abridged transcript is printed in Geikie (1875; 1:291–92n).

taining shells and corals. He now set out for Murchison a provisional classification of all the formations that outcropped over a vast area from Riga to Novgorod, adopting a tripartite scheme in which the "*Système dévonien*" was firmly sandwiched between the "*Système carbonifère*" above and the "*Système silurien*" below.[20] Murchison could hardly have asked for more exciting news. Not only did von Buch, like Humboldt, accept the full Devonian interpretation on an international level. He had also pointed out to Murchison a region in which it might be decisively confirmed, not only by an unambiguous order of superposition in a region of unfolded strata, but also by a direct association of Old Red Sandstone fish with Devonian shells and corals. If the Baltic provinces of Russia fulfilled the promise suggested by von Buch, the Devonian controversy would be over.

Finally, Murchison received from Neuchâtel a small but potentially decisive piece of news which fitted neatly into what von Buch had just suggested. Murchison had sent Agassiz various fossil fish to identify and had explained how his Continental fieldwork had in his opinion reinforced the Devonian interpretation of the Old Red Sandstone. But in particular he had asked Agassiz to identify a single large fish scale that d'Omalius had found in the older Belgian rocks. Agassiz replied at once to tell him the result: "le Poisson de Mr d'Omalius que vous placez de le Dévonien appartient au genre Holoptychius, dont je me connois aucune trace de le Silurien."[21] Not only did this support Murchison's Devonian attribution for what Dumont had termed the Silurian of Belgium; far more important, it was the first authoritative report of a distinctive Old Red Sandstone fossil among formations full of "Devonian" shells and corals. It was the first positive evidence that those two disparate faunas, which Murchison had so implausibly correlated when he first devised the Devonian interpretation, really had existed contemporaneously at that distant epoch. Von Buch had hinted that the two faunas might be found together in the Baltic provinces, but that still had to be confirmed. D'Omalius's fish scale was only a single specimen, but for the Devonian interpretation it was no ordinary straw in the wind.

13.3   Setting the Record Straight (March to May 1840)

Before Murchison could get to Paris to further the Devonian campaign among the French geologists, Fitton gave the caldron another stir at home. "Paddy Fitton . . . says that we ought to put Lonsdale more in the foreground," Murchison reported to Sedgwick. Having successfully put Dumont in the limelight as the author of the first clear sequence of Transition formations—a sequence published in full several years before Murchison's—Fitton was now championing the Geological Society's modest and overworked librarian as the author of the crucial suggestion that had made Murchison's Devonian interpretation possible. Murchison protested that he and Sedgwick had already credited Lonsdale fully and generously in their memoir on Devonshire which was now

20. Ibid. Fig. 13.1 is reproduced from Murchison (1839, pl. 2 bis, fig. 1).
21. Agassiz to Murchison, 4 March 1840 (GSL: RIM, A2/6). The Belgian fish scale was subsequently illustrated by Agassiz (1844, pl. 24, fig. 11).

at last nearing publication. But just as he had forestalled criticism by supporting the award to Dumont, so Murchison was now careful to concur with what Fitton proposed, though to Sedgwick he was highly scornful about "the well meaning but occasionally over-meddling deeds of our warm-hearted & potato-headed friend." Fitton was too influential to be ignored, even though Murchison found him "petulant and ignorant" about the affair. Fitton's proposal was that Lonsdale himself should put on record his own role in the genesis of the Devonian interpretation, at the very next meeting of the Geological Society. It is a sure sign of the importance of the affair that, when Lonsdale's paper was postponed, probably because he needed more than a bare week to write it, Murchison put off his visit to Paris to make sure he could be present to hear it read.[22]

In his paper, Lonsdale summarized how the south Devon fossils, with their apparent affinities to both the Mountain Limestone fauna and the fossils of the Transition, had earlier led to conflicting and uncertain interpretations by, among others, De la Beche and Phillips. But he conceded that before the Silurian fauna had been fully studied that confusion had been unavoidable. He recalled that once the Silurian fauna was available as a standard of comparison, the decisive clue had been Austen's collection from southeast Devon, which he first saw in December 1837 (§9.3). For he had then agreed with Sowerby that the shells looked Carboniferous, while perceiving in the corals a Silurian affinity; other fossils were not known elsewhere, at least in Britain. It was this mixed and anomalous character of the Devon fauna, he claimed, that had first led him to suggest that the rocks might be equivalent to the Old Red Sandstone.[23]

There is no strictly contemporary evidence that Lonsdale had made the suggestion in quite the form that seemed so neat in retrospect. Had he stated it at the time in terms of a correlation with the Old Red Sandstone, it is inconceivable that such a striking conjecture would have gone unrecorded. The available evidence implies that what he had really suggested, when Austen's fossils were discussed, was only that they were not precisely those of the Mountain Limestone, as both Sowerby and Austen himself believed at that time, but that they must be somewhat older (§9.3). Only in retrospect, after Murchison began to regard the sharp contrast between the Carboniferous and Silurian faunas as anomalous rather than natural, did the Devon fossils become plausible candidates for a strictly intermediate fauna. Lonsdale's "historical précis of the course of events *in re* Devoniana," as Murchison had termed it beforehand, thus embodied a subtle rewriting of history. Lonsdale was not being dishonest: no more than anyone else could he easily project himself back into an earlier phase of the controversy, when a "Devonian System" with an intermediate fauna had been literally inconceivable. History apart, however, Lonsdale did give an important and up-to-date list of the south Devon fauna. He cited sixty-two species (in thirty-five genera) in categories that displayed its mixed character: thirteen species (in six genera) were also known from the Silurian; thirty-eight species (in twenty-four genera) were

22. Murchison to Sedgwick, 2 March 1840 (ULC: AS, ID, 26a); Fitton to Murchison, 3 and [5] March 1840 (ULE: RIM, Gen. 523/4, 14, 15).
23. Lonsdale (1840a), read at GSL on 25 March.

known—within Britain—only from the Devonian rocks themselves; and eleven species (in five genera) were also known from the Carboniferous.[24]

Lonsdale may have been, as Murchison put it, "a noble creature & unconscious of such jealousies"; but his performance at the Geological Society was certainly used by others to promote their own causes. Those, led by Fitton, who considered that Murchison was claiming far too much of the credit for unraveling the older rocks and their faunas, were now determined to force him to share the limelight with Lonsdale. Only that gathering resistance to Murchison's overweaning ambition can account for the unprecedented speed with which Lonsdale's paper was rushed into published form. At the first meeting after it was read, the Council assigned it as usual to a referee, but then, in a most unusual way, accepted the referee's report orally and on the spot; at the following meeting a ballot was taken as usual, but then, most unusually, the paper was ordered to be printed immediately and out of turn. This highly irregular procedure ensured that it would be published in the Society's *Transactions* in the same issue as Sedgwick and Murchison's Devonian interpretation of Devonshire, and hence that it would be seen as modifying that paper's claims to originality.[25]

By the time those decisions were taken, however, Murchison was in France. Landing in Boulogne, he took the opportunity to check his new interpretation once more on the fossils in the local museum before taking a coach to Paris. Élie de Beaumont had mentioned to him beforehand that Baron Alexander von Meyendorf (b. 1798) was visiting Paris in preparation for an extensive official journey through the Russian empire to collect information on, among other things, its mines and mineral resources; and that he would be prepared to undertake scientific commissions, as it were, for foreign savants. Murchison was just in time to meet Meyendorf before he left Paris, and promptly got his agreement that he should join him in person for that "Russian Campaign" in the coming summer. Since de Verneuil also agreed to go, Murchison would have a chance of seeing for himself the undisturbed ancient strata in northern Russia, which von Buch had just suggested might contain decisive evidence to extend the Silurian system and to clinch the Devonian case once and for all. Meyendorf's official position would smooth their way through the imperial bureaucracy; and de Verneuil's expertise with fossils would make it possible, as in the Rhineland, to correlate the strata with reasonable certainty while still on the spot.[26]

First, however, Murchison had to persuade the French geologists of the validity of the Devonian interpretation, not least in the Boulonnais. "Both Dumont and the French may naturally look with some jealousy on the correction of *their views*," Fitton had warned him on the eve of his departure. Élie de Beaumont was not at the meeting of the Société géologique to hear Murchison read his paper: he had told him beforehand that he never attended

---

24. Lonsdale (1840b), list of fossils on pp. 737–38.
25. Council of GSL, 8 and 29 April 1840 (GSL: BP, CM 1/5, pp. 203–7); the referee was probably Fitton. The Devonshire paper (Sedgwick and Murchison 1840b) was in an issue of *TGS* originally scheduled for publication on 21 February, but delayed—partly by the printing of Lonsdale's paper—until 17 June (GSL: BP, CM 1/5, pp. 169, 230).
26. Murchison, notebook "France and Normandy 1840," notes for 1–3 April (GSL: RIM, N83, pp. 1–8); Murchison to Charlotte Murchison, 4 April 1840 (ULE: RIM); Élie de Beaumont to Murchison, 5 March 1840 (GSL: RIM, E4/10).

now, because he considered its discussions had become as futile as crackpot claims to perpetual motion.[27] With that important exception, however, there was a good assembly of French geologists, together with Nöggerath from Bonn.

Murchison's paper on "les *roches dévoniennes*" was ostensibly on the Boulonnais, but in fact he reviewed the genesis and development of the Devonian interpretation in some detail, duly giving Lonsdale the now routine if somewhat mythical credit for the initial suggestion. He summarized the previous summer's fieldwork in the Rhineland, exhibiting a sketch map to show the outcrops of the two limestones (now identified as Carboniferous and Devonian), the clear separation of which had been the clue to the true sequence there (§12.5). A brief review of the Belgian strata, regarded likewise as "dévoniens et *non siluriens*" except at the base of the sequence, then led to Murchison's similar reinterpretation of the Boulonnais. Here he retained the lowest formation ("schistes à graptolites") for the Silurian, and thereby claimed that the Devonian—the rest of the erstwhile Silurian—was clearly sandwiched between Silurian and Carboniferous, just as in Belgium and the Rhineland. He praised Dumont's work in all respects except for his correlations, which he dismissed because they were not adequately based on fossil criteria. As for de Verneuil, Murchison rejected what his past and future companion had put forward only two meetings earlier, and insisted that there was no sharp distinction between the Silurian and Carboniferous faunas, and that the Devonian fauna was truly intermediate. The Devonian interpretation was thus fully launched on the French scene.[28]

Although Élie de Beaumont was personally helpful and friendly to him, Murchison soon saw that the French geologist's influential theory of occasional "révolutions du globe" was blocking any full acceptance of the Devonian interpretation in France. "Theory 1st & facts *afterwards* is their rule," he noted privately; "it is clear that with French ideas the Devonian will scarcely be admitted as a *terrain* but only as an 'étage supérieur' of the Silurian System." For he saw that Élie de Beaumont's theory predisposed the French to expect that any genuine *terrain*—or in Murchison's terms, a genuine system—would be bounded at top and bottom by unconformities, or at the very least by sharp changes in fossil assemblages. This had been Murchison's own view up to a late stage in the composition of *The Silurian System* (§10.4); but now he fought against it with all the zeal of a convert. He insisted that the Devonian counted as a true *terrain* by its great thickness, by its "many peculiar types [of fossils] found neither in Carboniferous or Silurian," but also by the many other fossils "showing passages into & connections with the Systems above & below it." He noted that this new view of gradual faunal changes "does not suit the French school of séparations tranchées," which de Verneuil had exemplified in his recent paper. "But this will not do!" he exclaimed, for even de Verneuil himself had had to admit that one of the putatively Devonian limestones in the Rhineland (at Paffrath) contained sev-

---

27. Fitton to Murchison, 30 [March] 1840 (ULE: RIM, Gen. 1999/1/6; the month reads like "Nov.," but only March fits the contents); Élie de Beaumont to Murchison, 5 March 1840.

28. Murchison (1840b), read at SGF on 6 April; Murchison to Charlotte Murchison, [10] April 1840 (ULE: RIM). The sketch map mentioned in the paper was probably a topographical map of 1830 in four sheets, with the geology added in watercolors and MS annotations (BGS: RIM); the Mountain Limestone of Belgium and Westphalia is shown clearly distinct from the various Devonian limestones.

eral shells *"undistinguishable* from Carboniferous [species]." So although he was pleased that Dufrénoy conceded that there was probably some Devonian in Brittany, Murchison was scornful about the general level of the discussion of his paper, which left him feeling some sympathy with Élie de Beaumont's attitude toward the Société géologique.[29]

Before returning to England, Murchison did some fieldwork at long last on the older rocks of northwest France. But what had been planned a year earlier as the main work for a whole summer season (§10.7) had now shrunk into a mere three days in one corner of Normandy en route to a steamer from Le Havre to Portsmouth. Normandy and Brittany had first been shelved and were now skimped. For the significance of the region in the strategy of Murchison's (and Sedgwick's) research on the older systems had steadily declined as it became clear that other regions offered the chance of less ambiguous results. Murchison certainly considered he had identified Silurian rocks near Caen, underlain by "Cambrian" rocks without fossils; and he may also have found them overlain by what he regarded as Devonian strata. But in any case he saw enough to convince him that the region was too complex in its structure and too obscure to be worth a more thorough study.[30] The reports of anthracite in the Greywacke of northwest France, which De la Beche had invoked repeatedly ever since the start of the controversy (§5.3), were not to be refuted directly after all. They could safely be ignored, for Murchison was now more confident than ever that the anthracites lay either among Devonian strata or in the Carboniferous, and certainly not in the Silurian or Cambrian. Normandy and Brittany were therefore no longer a threat to his view of the Silurian as a period that had predated all land plants and all coal deposits. He could now pursue his Devonian campaign without distraction, carrying the heart of the persuasive argument from the Rhineland to Russia.

While his collaborator was away in France, Sedgwick worked hard on their joint paper describing their Continental results, and it was ready for the first meeting of the Geological Society after Murchison returned. It set out for the public record the sequences they had worked out in Germany and Belgium (§12.1–§12.5), with the correlations that Murchison had proposed after their return (§12.8), and with which Sedgwick had eventually agreed. But the overall purpose of the paper was stated clearly at the outset: it was to establish the validity of the Devonian system, "not merely on plausible arguments derived from its suite of fossils, but also on the more direct evidence of natural sections."[31] This encapsulated the condition on which Sedgwick was now content to be its co-author. For he had been persuaded in retrospect that the Devonian interpretation did not depend—as it had done when first propounded for Devonshire—only on an unproven assumption about gradual faunal change. Instead, the distinctive Devonian fauna could now be *seen* to be sandwiched between the Silurian and the Carboniferous. While satisfying

29. Printed discussion of Murchison (1840b), in *BSGF* 11:256–57; a fuller though hardly impartial record, including conversations after the formal discussion was over, is in Murchison, notebook "France & Normandy 1840," pp. 9–15.

30. Murchison, notebook "France & Normandy 1840," notes dated 25–29 April 1840, pp. 51–63. His record of finding the Silurian "surmounted by Devonian rocks" is only in his MS book "Wanderings in Russia I," written in 1865 on the basis of this and later notebooks (GSL: RIM, J8).

31. Sedgwick and Murchison (1840a, p. 300), read at GSL on 13 and 27 May.

Sedgwick's geological scruples, however, the formulation also embodied a covert counterattack on the pretensions of Lonsdale's recent paper. Murchison and Sedgwick valued Lonsdale for his outstanding expertise with fossils and had no personal quarrel with him. But their new joint paper on Germany and Belgium was tacitly designed to show that, contrary to the claims being made on Lonsdale's behalf, his priority lay only in one component of the Devonian interpretation, namely, in the recognition of its mixed and transitional fauna, and not in the field evidence that was needed to clinch its validity.

Sedgwick and Murchison's paper reviewed the regions of their field work in the most effective order, beginning with Westphalia and continuing with Belgium; the more problematical Eifel was left to the last, and the Harz and Fichtelgebirge relegated to a virtual appendix. In their main summary, they claimed to have delineated "a great and uninterrupted series of formations," which was "in general accordance with the British series." The boundaries of the major systems were said, however, to be marked by "no great mineralogical interruptions," let alone unconformities. There was likewise "no want of continuity among the groups of the great palaeozoic series of animal forms," as it was now called, extending Sedgwick's earlier term to include the faunas of *all* the systems from Cambrian to Carboniferous. This was a decisive rejection of any definition that depended on "séparations tranchées" between successive *terrains*. In defiance of French sensibilities, the paper concluded with the firm claim "that the Devonian system is therefore a natural system," both by its distinctive fauna and by its "manifest position in the German and Belgian sequences." Since the Old Red Sandstone of Britain had exactly the same position, the final claim was that "the Devonian system . . . is contemporaneous with, and representative of, this old red sandstone."[32] Just as the plausible claim to the Devonian on the grounds of the fauna in Devonshire had now been confirmed by the direct evidence of sequences in Germany and Belgium, so Murchison and Sedgwick anticipated that the plausible equivalence of the Devonian and the Old Red Sandstone would be confirmed in Russia by the direct evidence of a mixture of their fossils.

This long paper was read in the customary manner at two successive meetings of the Geological Society. Murchison left Sedgwick to deal with the second occasion, for he had to start for Russia the week after the first. "The case is I think pretty well in hand," Sedgwick assured him before he left. He still had reservations about Murchison's correlations for the Eifel rocks, but on the main points their collaboration was once more secure. There is no record of any discussion after the paper was read; "nobody made any fight," Sedgwick reported later.[33] One of De la Beche's friends, writing after the first half had been read, referred to it sarcastically as "the song of joy chanted by the Devonian rocks on the occasion of their at last getting into their proper places." The mock-heroic rendering of that song may well express a rather general feeling about the Devonian campaign, as the Napoleon of geology left London to carry it into Russia.

32. Ibid., p. 309; Sedgwick to Murchison, 2 [–5] April 1840 (GSL: RIM, S11/169).
33. Sedgwick to Murchison, 18 and [19?] May and 25 July 1840 (GSL: RIM, S11/171–73); the second letter is dated only "Tuesday," but was evidently written not long before Murchison's departure on Friday 22 May.

Thrice fortunate Devonians to have met with protectors who have rescued you from the evil company to which mortal ignorance had condemned you—powerful friends who have taken you by the hand and segregated you from the villainous companions [i.e., Greywacke, Killas] whose very names proclaim their *low* origin. Rejoice, Oh Devonians! and sing the praises of your benefactors in sweet Silurian strains! Your fame is about to be proclaimed from Archangel to Cornwall—'da Scilla al Tanai, Dall'uno all'altro Mar.'[34]

13.4   Resolving Russia (May to September 1840)

"When I return," Murchison boasted publicly before he left London, "you shall have the Carboniferous, Devonian & Silurian *en grand*."[35] To give geologists that threefold simplicity in persuasive form, he had first to gather all the information he could about Russian geology. Having sailed with de Verneuil to Hamburg, he first made a detour to Berlin for just that purpose. They bought Russian maps and studied Russian fossils in the university museum and in von Buch's private collection. Murchison was flattered by the attentions of "Der Humboldt," and his social appetite was satisfied by being placed above a Prince and a Duke at the dinner given by Humboldt in honor of the two foreign geologists. But by far the most important news that Murchison sent back to Sedgwick was that "our classification is already quite accepted by the Prussian geologists." This referred not only to the Silurian and Cambrian; "the Devonian is also greeted by all as a capital thing done & a *good name*, & they can now see every where proofs of it which they did not see before." Humboldt and von Buch confirmed their acceptance of it, which they had indicated by letter earlier in the year (§13.2). Murchison also persuaded von Dechen that the new interpretation was valid in Westphalia, and that geologist promised to provide a geological map to illustrate the full version of the paper that had just been presented in London. Von Dechen also showed Murchison his own and von Buch's maps and sections of Bohemia and Silesia, which demonstrated that there too the "true Mountain Limestone" was underlain by "true Devonian [with] loads of Clymenes" and that in turn by "older Greywacke, my Silurian. Q.E.D." This "Euclidean" proof was enough to make Murchison decide on the spot to take his wife to that part of central Europe for the next summer's fieldwork.

Not least among events in Berlin, however, was a visit to Christian Ehrenberg (b. 1795), the professor of medicine at the university. Murchison was able to see at firsthand the outstanding quality of Ehrenberg's already celebrated microscopical researches. In particular, he was duly convinced that living species of "infusoria" (i.e., single-celled animals) were indeed to be found in the fossil state as far down the sequence as the Chalk, thereby

34. Charters to De la Beche, 23 May 1840 (NMW: DLB). The inferred allusion to terms such as Greywacke and Killas makes sense of the pun: they were low in the geological sequence but also, as miners' terms, low in social origin as seen from a gentlemanly perspective. The quotation was from Alessandro Manzoni's celebration of Napoleon's triumphant campaigning: "Dall'Alpi alle Piramidi,/ Dal Manzannare al Reno,/ Di quel securo il fulmine/ Tenea dietro al baleno;/ Scoppiò da Scilla al Tanai,/ Dall'uno all'altro Mar" (*Il Cinque Maggio*, 25–30).

35. Quoted in Weaver to Williams, 2 July 1840 (SAS: DW, box I, enclosed in notebook 17).

weakening the supposed *séparation tranchée* or sharp faunal boundary between the Secondary and Tertiary strata. That Murchison put his Devonian
campaign into just that theoretical framework is clear from his comment to
Sedgwick immediately after describing Ehrenberg's work: "therefore I am
already convinced that we shall ride a winning horse [on] easy ground, *Carb.
Dev.* et *Sil.*"[36]

The next lap of that geological race was to start at St Petersburg. But
instead of tackling the long overland journey eastward through the Baltic
provinces, Murchison and de Verneuil doubled back to the west and took a
steamer from Lübeck. Passing Öland and Gotland, Murchison noted that the
first island "belongs entirely to me, being throughout Silurian," and he longed
to check on the second; he had not yet visited either, but his confident attribution reflects the routine way the Silurian was already being generally adopted
(see fig. 13.2). Their arrival in the Russian capital was facilitated by the official
character of Meyendorf's expedition; they met the young naturalist Count
Alexander von Keyserling (b. 1815), who was to accompany them, and also
several local geologists. On studying the collections in the Academy of Sciences and the School of Mines, Murchison soon decided that "the most useful
fossil" would be his *Holoptychius* (fig. 13.1): "my Scottish fish make a clear
horizon of old Red, & we have also Silurian & lots of Carboniferous." His
global ambitions for the Silurian were encouraged by seeing what he and de
Verneuil regarded as characteristic Silurian fossils from far-flung parts of the
Russian empire: not only from the Urals, but also from the Altai on the borders
of China, from Novaya Zemlya in the Arctic, and even from Alaska.[37]

As in Belgium and the Rhineland, the foreign geologists would not of
course be working in geological *terra incognita*, except in the more remote
areas; what they had brought to Russia was their firsthand experience of
analogous formations and fossils in the more thoroughly studied regions of
western Europe. When the expedition was assembled, Murchison, de Verneuil, and the Russians left St Petersburg and set out toward Archangel on
the White Sea. Exposures were few and far between, for the country was
covered with bogs on top of thick "diluvial" debris, and the only natural
exposures were where rivers had cut down to the solid rock. But one of the
first of these exposures made up in importance what they lacked in frequency.
On the banks of the Volkov near Lake Ladoga they found a section of almost
horizontal strata that, as Murchison told his wife, "enabled me to connect the
Devonian more clearly with the Silurian." Not only was that relation clearer
than in Belgium and the Rhineland. Far more important, this Russian "Devonian" was found to be "uniting within a very small compass the fishes of
the old Red Sandst. of Scotland! & the shells of Devon & the Eifel!"[38] The

36. Murchison to Sedgwick, 30 May 1840 (defective transcript in ULC: AS, IIID, 31); Murchison to Charlotte Murchison, 28 May 1840 (ULE: RIM); Ehrenberg (1839). Murchison's notebook from the first part of his trip to Russia is lost; his "Wanderings in Russia I," written in 1865
from it and other sources (GSL: RIM, J8), incorporates many retrospective comments and cannot
be relied on.

37. Murchison to Charlotte Murchison, 2[–5] June 1840 (ULE: RIM); Eichwald (1825, 1840);
Pander (1830); Thackray (1978a).

38. Murchison to Charlotte Murchison, 27 June 1840 (ULE: RIM); Murchison to Sedgwick,
1[–7] Setpember 1840 (defective transcript in ULC: AS, IIID, 32). The corresponding note in
"Wanderings in Russia I" may be an authentic transcript from his lost field notebook, but is
immediately followed by a comment that is certainly retrospective (GSL: RIM, J8, pp. 32–33).

faunal mixture that von Buch had reported from the Baltic provinces further west had duly been found; d'Omalius's Old Red Sandstone fish scale from the Devonian of Belgium had proved a true straw in the wind (§13.2).

The remainder of their journey to Archangel was by comparison an anticlimax, though Murchison found plenty of "our old friend the Old Red Sandstone," with its upper part capped by a limestone with Carboniferous fossils, which completed the sequence from Silurian into Carboniferous.[39] But the exposures were so widely scattered that, although the strata were nearly horizontal and undisturbed, the sequence could not always be pieced together with confidence. Nonetheless, the experience of seeing these ancient formations looking as recent as the younger Secondary strata of England was for Murchison a final and decisive proof that such appearances counted for nothing in geology: only the fossils could reveal the true history of the earth.

From Archangel they turned south again, following the Dvina upstream and then cutting across country to the Volga, which they followed down to Nijni Novgorod; from there they turned west in a wide loop to Moscow. These vast distances, over which the geologists often separated from the slower party under Meyendorf, yielded little of geological interest; but on the return journey to St Petersburg, Murchison and de Verneuil did find in the Valdai Hills some exposures that fully confirmed the sequence they had inferred at the start of their ten-week tour.[40]

One of Murchison's final notes summarized what was by far the most important result of the 4500-mile expedition: "Q.E.D.—the mixture of Devonian shells of Devonshire with fishes of the Highlands of Scotland demonstrates that Sedgwick and myself were right in identifying the Scottish Old Red with the Devonshire rocks, under the name of Devonian." By comparison, he believed, it would be "merely a work of detail" to trace the same sequence of older formations into the east of Russia; "the great points are fixed," he noted, "and if I could see the Ural, I should then get them all into order." To illustrate the order he had already inferred, he marked a topographical map of western Russia and the Baltic with three broad bands of color (fig. 13.2). The colors covered areas far beyond what he had seen at firsthand, and were doubtless derived in part from published descriptions of these vast spreads of almost horizontal strata. What was novel, however, was that Murchison had confirmed the hint von Buch had sent him that spring (§13.2): sandwiched between strata with Silurian fossils and those with Carboniferous fossils were *both* limestones with Devonian shells *and* sandstones and marls with Old Red Sandstone fish. "The Q.E.D. is accomplished," he noted with evident delight.[41]

On board a Baltic steamer on his way home, Murchison wrote to tell Sedgwick how the Russian sequence from Silurian through Devonian (now

39. Murchison, "Wanderings in Russia I," pp. 63, 65.

40. Murchison to Charlotte Murchison, 12 August 1840 (ULE: RIM). Murchison's field notebook "2. Russia 1840" covers only the journey from Archangel to Nijni Novgorod, with a few later notes at the back (GSL: RIM, N84), but it can be supplemented with the retrospective account in "Wanderings in Russia II' (GSL: RIM, J9).

41. Murchison, "Wanderings in Russia II," pp. 177–78; these quotations are probably authentic transcripts from a lost field notebook that covered the last third of the trip. Fig. 13.2 is traced from Murchison's copy of Arrowsmith's *Russia and Poland* (1832), with his "First sketch of Russia, 1840" added in MS (BGS: RIM). One of his probable sources was the geological sketch map in Strangways (1822).

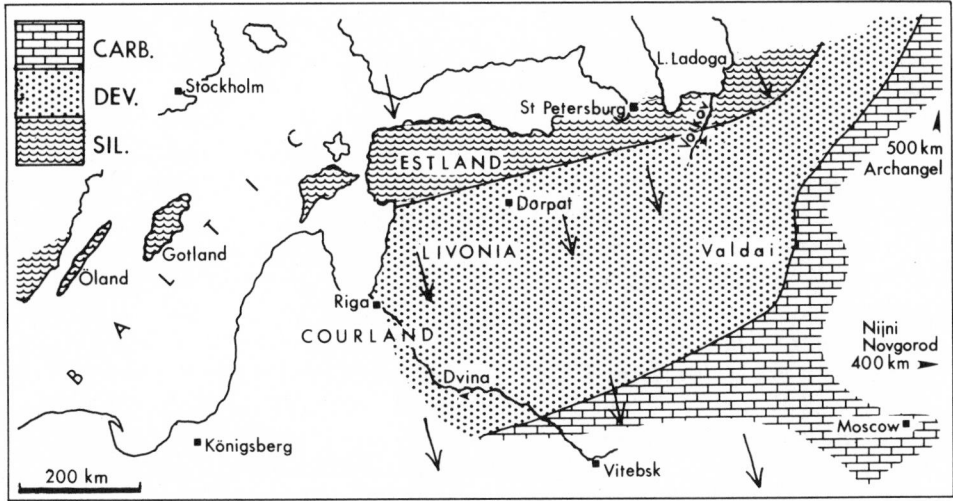

Fig. 13.2. Murchison's geological map of western Russia and the Baltic, sketched after his fieldwork in the summer of 1840, to show his interpretation of the very gently dipping strata as a sequence of Silurian, Devonian, and Carboniferous rocks. The annotations (not copied here) recorded the association of Old Red Sandstone fish with Devonian fossil shells in what was here firmly established as the Devonian system.

clearly equivalent to the Old Red Sandstone) into Carboniferous provided "the trinitarian proof" that demonstrated "the truth of Devonianism." Anxious as ever to forestall any backsliding by his collaborator, Murchison told Sedgwick he hoped this "splendid and unanswerable confirmation of our views" would banish his ailments and make him "Adamus redivivus." But Sedgwick's duties were keeping him in Norwich, and Murchison had to prepare to fight for their Devonianism without his support at the forthcoming meeting of the British Association.[42]

13.5   Devonshire on the Sidelines (May to September 1840)

The Devonian controversy had from the start been recognized as a matter of international significance for geology, but in its earlier phases it had naturally been focused on the detailed interpretation of Devonshire. Once the Devonian interpretation of the older rocks had been put forward, however, its proponents had been obliged to seek justification for it outside England. For by that time it was evident that the rocks of Devonshire itself were too complex in structure, and the exposures too obscure, to yield persuasive and unambiguous evidence either for or against the new interpretation. Murchison, Sedgwick, and de Verneuil—solo or in various combinations—had since discovered what they took to be just such persuasive evidence in Belgium, the Rhineland, and Russia for the establishment of the Devonian system. Devonshire was thereby left on the sidelines of the debate (§12.7). Work on the

42. Murchison to Sedgwick, 1[–7] September 1840; Sedgwick to Murchison, 13 September 1840 (GSL: RIM, S11/174). The trinitarian reference alluded both to the threefold systems and to Sedgwick's Cambridge college.

strata and fossils of Devonshire did not come to a halt, of course; but its role changed from being the spearhead of research to merely supplying ancillary support for arguments generated elsewhere.

Ironically, a thorough and systematic study of the fossil evidence from Devonshire, which had been so urgently needed in the earlier phases of the controversy, finally got under way just as its broader significance declined. Phillips had been identifying Devonshire collections for De la Beche throughout 1839, but only on an ad hoc basis. De la Beche had secured government money to enable Phillips to travel in Devonshire and collect the fossils in situ, but only in the spring of 1840 had Phillips at last been able to go. Once he was there, De la Beche had asked if he would be interested in some more permanent way of working for the Geological Survey, since his competence in the field evidently matched his expertise in the museum. Phillips was eager to spend more time on research and had planned to resign from his curatorial position in York and his lecturing position in London. Meanwhile, as a skillful collector and with the help of all the local geologists, he had quickly amassed larger collections of Devon fossils than any he had yet seen. Despite all that new material, De la Beche urged him to have his report on the fossils completed in time for what he sarcastically termed the "Wisdom meeting" in Glasgow, where it could be displayed as valuable publicity for the Survey before it was submitted to the Ordnance authorities for printing at government expense.[43]

In all his progress reports to De la Beche, Phillips showed by his fossil identifications that he continued to regard the Culm as unquestionably Carboniferous, its black limestones having many unmistakably Mountain Limestone fossils. As for the older rocks, on the other hand, Phillips was much less certain. Early on in his fieldwork he had suggested to De la Beche that southwest Wales ought to provide the clue to Devonshire, implying perhaps that some of the older rocks of Devon might be as old as the Silurian, as De la Beche maintained. But later his collecting convinced him that, in any case, north and south Devon were broadly similar in their fossils, which had at least some Carboniferous affinities. In the new edition of his little *Index Geological Map* of Britain, published at about this time, Phillips used a special tint for the older rocks of southwest England, explicitly to avoid having to decide on their age prematurely. It is clear, however, that like De la Beche he found the Devonian interpretation quite plausible. But whereas De la Beche expressed radical doubts about the validity of the Silurian, on the grounds that it could not be separated from the Cambrian in Wales, Phillips showed himself much less skeptical. When De la Beche told him scornfully how he had heard that supposed Silurian and Carboniferous fossils had been found together in a single quarry in North America, Phillips corrected the report in such a way as to make it clear that he accepted those systems on a global scale, at least provisionally, and he implicitly left room for the Devonian too.[44]

43. Phillips to De la Beche, 22 February, 6 April, and 30 September 1840 (NMW: DLB); De la Beche to Phillips, 26 February, 25 March, 10 April, 3 May, 2 July, and 4 October 1840 (UMO: JP, uncat.); Phillips's report was finished almost on time, and he got his fee of £250.

44. Phillips to De la Beche, 28 March, 6, 20, and 27 April, 10 May, and 16 June 1840 (NMW: DLB); De la Beche to Phillips, 3, 14, and 18 May 1840 (UMO: JP, uncat.); Phillips (1840a), the "fifth" edition of the map in Douglas and Edmonds (1950).

In an article for the popular serial *Penny Cyclopedia*, Phillips put the matter into an even wider context by arguing that there had been "one system of life," recorded as a broadly similar assemblage of fossils, extending all the way from Sedgwick's Lower Cambrian at least as far as the base of the Old Red Sandstone. But he suggested that this "Palaeozoic Series"—here he adopted Sedgwick's term—might need to be extended up through the Carboniferous as far as the New Red Sandstone as a result of Sedgwick and Murchison's new intermediate (or "supposed" Devonian) fauna linking the Silurian with the Carboniferous. So the Devonian, at least provisionally, provided the crucial link that unified this newly defined *"Palaeozoic"* era in the history of life; Phillips suggested that the corresponding terms "Mesozoic" and "Kainozoic" would then be appropriate to characterize the quite different faunas and floras of the rest of the Secondary and the Tertiary, respectively.[45] Phillips might not yet have joined the growing international consensus on the Devonian system, but he was clearly open to conversion.

Weaver, on the other hand, was not. Back in the spring, before Sedgwick and Murchison produced their paper on Belgium and Germany at the Geological Society (§13.3), Weaver had tried to upstage them with the first installment of a long article in the *Philosophical Magazine*. It was primarily an attack on his long-standing Irish rival, Griffith, who had proposed changes in the geology of the south of Ireland in response to the Devonian interpretation of Devonshire. But Weaver took his dense and tortuous argument to the Continent too; he insisted once more that the older rocks of Devon were clearly "older Transition," judging by the way their fossils (e.g., *Clymenia*) matched some from the Transition rocks of Germany. But Sedgwick and Murchison had already claimed that many of those Transition rocks were really Devonian. Weaver therefore suggested as an alternative that the Devon rocks might belong *between* the Silurian and the Old Red Sandstone, but urged that their correlation should be deferred until the fossils were fully described. That was perhaps intended as a partial concession toward the possible validity of the Devonian interpretation; but it would have seemed extremely implausible to other geologists, since the completely conformable passage from Silurian into Old Red Sandstone in the Welsh Borderland implied that there could be *no* significant formation of intermediate age elsewhere. There was tacit agreement that Weaver could henceforth be ignored.[46]

"My dear friend Mr. Weaver, tho' very likely in Ireland, Belgium &c. is kilt [i.e., beaten] in Devonsh. and handles myself & it very gingerly," wrote his Somerset neighbor Williams in one of the few recorded reactions to Weaver's article. Williams himself was undeterred by his own increasingly isolated position on the correlation of the Devon strata. "I will bide my time," he had noted earlier in the year, confident that he would be vindicated eventually and his gentlemanly critics silenced; "you have all snubbed me—and I will blow you all up as *surely* as Guy Fawkes & his comp[anio]ns *did contrive* to

---

45. Phillips (1840b); although the names were new, Phillips's threefold division had figured in his earlier article in the same serial on "Geology" (Phillips 1838b). The new term for the Tertiary became "Cainozoic" in his later writings.

46. Weaver (1840), published in April, May, and June issues of *PMJS*; Weaver to Williams, 2 July 1840 (SAS: DW, box I, enclosed in notebook 17).

blow the King & Parliament up alive."[47] But as least in one important respect, Williams had been influenced in spite of himself by his opponents' approach; for the focus of his attention, like Sedgwick's, had now turned to fossils, despite his earlier scorn for their value in geology. He planned to publish a book of his own about them, and somewhat tactlessly asked Phillips to help him with their identification. When Phillips declined to work on a project that would compete with his own, Williams got Sowerby's assistance instead. Sowerby reported that Lonsdale welcomed Williams's increased attention to fossils as being most likely to help resolve the problems of Devon geology. "I believe the Devonian or old red classification to be good," Lonsdale told Phillips; "labourers" like Williams would help decide the matter one way or the other, for "the truth must come out, oppose it who will." But Williams himself was less open-minded: "at all events," he told Phillips, "I'll give the old red sandstone hypothesis a good kick at Glasgow."[48] As the geologists converged on the Scottish city for the British Association's meeting, it was clear, at least to Phillips, that Williams would not let Murchison's grandiose conclusions go unchallenged.

13.6   From Moscow to New York (September 1840 to April 1841)

At the Glasgow meeting of the British Association, most of the leading geologists were present as usual, with only Sedgwick as a conspicuous exception. In his capacity as general secretary, Murchison had suggested that local pride would be flattered if the presidents of the Sections were Scotsmen; in the case of the geological Section, the appointment of Lyell neatly fulfilled this condition while keeping the position in the hands of the London gentlemen of science. Among other activities of the Section, a day's excursion to the Isle of Arran, already celebrated for its spectacular and compact geology, brought the leading figures amicably together at the head of a party some 200 strong.[49] Only after the Association had tactfully observed the Sabbath with Scottish inactivity, however, did the geological Section really come to life. In their eagerly awaited "brief outline" of their Russian research, Murchison and de Verneuil described how the remarkably undisturbed Silurian strata near St Petersburg, identified by their rich fossil fauna, were conformably overlain by vast spreads of the "Old Red or Devonian System," and those in turn by limestones with unmistakably Carboniferous fossils. This was illustrated with a traverse section from St Petersburg to Moscow (fig. 13.3). Agassiz had arrived in Glasgow just in time to confirm that the fossil fish in the middle of those three systems were indeed Old Red Sandstone forms ((*Holoptychius, Coccosteus*). "The evidences now found in Russia," claimed the authors, or more

47. Williams to Phillips, 27 May 1840 (UMO: JP, 1840/24); Williams, undated notes written on Buckland to Williams, 15 February 1840, and probably intended for a future speech at GSL.

48. Williams to Phillips, 27 May and 30 June 1840 (UMO: JP, 1840/24, 30); Phillips to De la Beche, 16 June 1840; Lonsdale to Phillips, 22 August 1840 (UMO: JP, 1840/51); Sowerby to Williams, 28 August 1840 (SAS: DW, box I, enclosed in notebook 17).

49. *Ath.*, no. 673 (19 September 1840), p. 730; no. 674 (26 September), p. 744; no. 675 (3 October), p. 777; Morrell and Thackray (1981, pp. 216, 219).

Fig. 13.3.   Traverse section from St. Petersburg to Moscow, to show the place of the "Devonian System" (center) clearly sandwiched between the Silurian (left) and the Carboniferous (right) in a sequence of undisturbed strata. The section is based on the one displayed at the meeting of the British Association in Glasgow in September 1840 to illustrate Murchison and de Verneuil's first report on their travels in Russia.

THE ATHENÆUM

S.S.E.

Sand and grit (a). Shale of oolite (b).

Carb. limes. Thin coal beds at bottom.

Moscow.

Much drift.

Devonian System.

Waldai Hills.

Silurian System.

Northern drift.

N.N.W.

St. Petersburgh.

Mr. LYELL dwelt on the importance of Mr. Murchison's observations in removing many doubts as to the new arrangement of some of the British formations. Some doubts had continued to exist of the identity of the age of the beds containing fishes in Herefordshire and the north of Scotland, with those limestones containing corals and other organic remains in Devonshire, but these had been now set at rest by the observations made in Russia. We see, at the present day, different formations in the process of creation in the various seas on the surface of the globe, and these must contain distinct organic remains according to the locality: the same circumstances must have taken place at a former period, and thus the old theory of universal formations become exploded. From the almost perfect horizontality of the Russian strata, they offered the phenomena of formations existing now as they had been deposited, and undisturbed by igneous rocks. Still a general rising of the land had been proved; Baron Wrangel had observed it in the northern parts of the empire: or, if not a general rising, it could be proved that there had been an oscillation, as in some places there had been depressions. Mr. Lyell referred to his own observation of similar phenomena in Scandinavia. He instanced, also, his discovery of Arctic Sea shells at Uddavalla, 200 feet above the present sea level: also in Sweden at an elevation of 600 feet. The erratic blocks he regarded as more modern than the recent Arctic shell beds, as near Stockholm he had found blocks, at 200 feet elevation, lying on shells similar to those now existing in the Baltic—a sea of brackish water.

Mr. CRAIG read a paper 'On the Coal Formations of the West of Scotland.'—This communication was of considerable practical value, and of much local interest; but an intelligible abstract could not be presented to the general reader without reference to maps and sections. Mr. Craig mentioned that the coal formations of the West of Scotland occupy a space of 3,000 square miles. He alluded to some of the superficial geological features, especially to the drift, which is found only on the hills and slopes, the low grounds being quite devoid of it; this may have been caused both by denudation and by outbursts from below, and the drift seems to have come from the north-west. Under the drift and subjacent beds of clay and sand, recent marine shells are often seen, at various elevations, in one place 360 feet above the present sea level. Mr. Craig described the coal for-

probably Murchison, "leave not a possibility of doubt as to the old red and Devonian systems of rocks being identical"; and he asserted firmly that, after years of controversy, "this question is for ever set at rest." Lyell as president stressed the outstanding importance of this conclusion and backed Murchison with the assurance that any remaining doubts about the Devonian interpretation were indeed "now set at rest." Any doubters present held their fire, and there was no further discussion.[50]

Later in the same session, however, these conclusions *"en grand"* were neatly contrasted with the limited vision of a mere provincial, when Williams was duly given his chance of "a good kick" at the Devonian. This he did in his usual forthright style, but with no new arguments. Implicitly he did perhaps accept what Murchison had claimed for Russia, because he emphasized by contrast that in Devonshire there were *no* Old Red Sandstone fossils, let alone the fish whose geographical range was now so spectacularly extended. Repeating his earlier claim that the Culm passed *under* the ancient-looking rocks of Cornwall (§12.7), Williams insisted that the Culm could not be Carboniferous, nor the older rocks equivalent to the Old Red Sandstone, unless his opponents accepted the reductio ad absurdum that the true Coal Measures lay under the Old Red. It is surely significant that, if there was any discussion, it was not thought worth recording. Like Weaver, Williams could now be safely ignored.[51]

The following day, the Devonian issue and even Murchison's Russian travels were upstaged by Agassiz's striking prediction that the traces of vanished glaciers he had detected in the Alps would also be found in the Scottish Highlands. The day after that, Murchison read a paper on the Old Red Sandstone and its fish in the same region, making use of the fossils that Hugh Miller had begun to write about on a popular level in his newspaper *The Witness*; and De la Beche suggested an environmental reason why shelly organisms apparently could not live in the areas where the Old Red Sandstone was deposited. But otherwise the Devonian controversy did not surface again, and as the geologists dispersed they may well have felt that it was almost over. If they were to maintain their reputation for pugnacious argument, Agassiz's sensational theory of a global Ice Age seemed a promising candidate for its successor.[52]

"I am planning the Ural on one hand, and the Alleghenies on the other," Murchison had told Sedgwick on his return from Russia, "for nothing short of Continental masses will now suit my palate." After reading about the paper

50. BAAS, session on 21 September 1840: *Ath.*, no. 676 (10 October 1840), pp. 800–801, from which fig. 13.3 is reproduced; Murchison and de Verneuil (1841a). Lyell had taken care to keep up-to-date with the Devonian problem: Lyell, notebooks for 1840 (KHK: CL, notebook 80, pp. 45, 50, 58; notebook 83, pp. 44–48).

51. *Ath.*, no. 676, p. 801; Williams (1841).

52. BAAS, sessions on 22 and 23 September 1840: *Ath.*, 677 (17 October 1840), p. 824; no. 678 (24 October), p. 846; Murchison to Sedgwick, 26 September 1840 (extracts in Geikie 1875, 1:306–8); Lyell to Miller, 21 October 1840 (NLS: HM, Ms. 7528, f. 30); [Herries-]Davies (1968; 1969, pp. 273–76). On Agassiz's glacial theory, which did indeed become a new focus of controversy from this moment onward, see more generally [Herries-]Davies (1969, chap. 8), and Rudwick (1970b). *The Witness* was the organ of the "anti-intrusionist" party in the Church of Scotland; Miller had recently been appointed its editor. The first of his articles, with their anti-transmutationist interpretation of the Old Red Sandstone fish, was published on 9 September; they were later expanded into book form (Miller 1841).

Murchison had given in Glasgow, Conybeare wrote to urge him to go next to the United States to explore the "virgin" spreads of ancient strata there, which might go far to extend the validity of the Palaeozoic systems to a global level. Although he was no longer active in field work, except near his home in east Devon, Conybeare's judgment was still highly respected by other leading geologists. His evident acceptance of the Devonian system, in the wake of Murchison's Russian research, therefore strengthened the growing consensus substantially.[53]

Early in 1841 Murchison received welcome news that Humboldt and von Buch fully accepted his Russian conclusions and wished him well with his plans for the Urals. Likewise von Dechen showed that his conversion to the Devonian interpretation the previous summer (§13.4) had not been a figment of Murchison's wishful thinking. He praised "the great Silurian and Cambrian, and Devonian labours of the famous geologists of England," and told Sedgwick he fully approved of their new interpretation of his own earlier work in the Rhineland. But in what was becoming for Sedgwick an all too familiar move, von Dechen extended the Devonian and Silurian to cover even the ancient-looking rocks (of the Taunus and Hunsrück) that Sedgwick had hoped to salvage for the Cambrian. For von Dechen attributed the lack of fossils in those rocks simply to metamorphic action: "they are eaten by the plutonic powers of the depth," as he put it. But if the Cambrian was in von Dechen's opinion still insecure, the Devonian was certainly well established. He expected it to take its place, along with the other older systems, as one of "a limited number of subdivisions very easy to recognize, which are not in every locality developed in the same manner." In other words, the systems would prove valid and recognizable on a global scale, however much their constituent formations and even their precise fossil contents varied locally. With the establishment of the Devonian to link the Silurian to the younger systems, von Dechen considered that the earlier chapters of the history of the earth and of life were well on the way to being as clearly understood as the later.[54]

That grand conclusion had in fact been anticipated, just before von Dechen wrote, when Buckland delivered his last anniversary address to the Geological Society in London. As in the previous year (§13.2), the Devonian issue was given pride of place in his review of "Positive Geology." He summarized how Sedgwick's and Murchison's work in Belgium, the Rhineland, and Russia had "confirmed the palaeozoic classification of the Carboniferous, Devonian and Silurian Systems," and he predicted that it would soon be extended to America. Meanwhile, Lonsdale's work on the Devonian fossils had, he asserted, added an "instructive lesson" in stratigraphical geology; for it had proved that, in the absence of direct evidence from superposition, fossils were "the surest and safest criterion" of relative age. The "paramount value of Palaeontology"—the term now beginning to be adopted in English—was

---

53. Murchison to Sedgwick, 1[–7] September 1840; Conybeare to Murchison, n.d. [? October 1840] (GSL: RIM, C17/10), probably written soon after the account of Murchison's paper was published in *Ath.* on 10 October.

54. Humboldt to Murchison, 12 January 1841 (ULE: RIM, Gen. 523/4, 46); von Dechen to Sedgwick, 4 March 1841 (ULC: AS, IF, 83). Münster (1840) was unconvinced, but his opinion on such a matter of theory counted for much less.

thus underlined by a president whose recent work had of course been in just that branch of the science.[55]

After this advance publicity, Murchison and de Verneuil's paper was an anticlimax when it was read at the Geological Society the following month. It simply repeated in a more formal way what they had announced in Glasgow six months earlier, stressing how the characteristic Devonian shells were "found in Russia *in the same beds with the fossil fishes of the old red sandstone* of the British Isles." In addition, the unaltered condition of these ancient strata in Russia proved that a "crystalline and hardened state" was *not* a necessary concomitant of great age in rocks, thereby demonstrating conclusively that fossils alone were a reliable criterion for long-range correlation. No discussion was recorded afterward in the correspondence of those present, and probably there was none. For the conclusions of the paper were now accepted by almost all whose opinions counted; the "Devonian System" was already passing into quite routine use, for example, in the arrangement of the Society's museum.[56]

Finally, at the very next meeting after the reading of the Russian paper was concluded, a paper by Murchison was read in his absence, summarizing a sequence of formations and a list of fossils from New York State, which had been sent by James Hall (b. 1811) of that state's geological survey. Murchison used this material skillfully to the advantage of his own overall interpretation, identifying Hall's formations by their fossils as the equivalents of "the old red sandstone or Devonian system" and of the Silurian. In particular, he claimed that as in Russia the sandstones with Old Red Sandstone fish *(Holoptychius, Coccosteus)* were closely associated with strata containing Devonian shells. He was in fact drawing on a growing body of published descriptions of sequences in North America, in some of which correlations with Europe had already been proposed. But this was the first overt extension to America of *all* the new Palaeozoic systems, including the Devonian, to be advocated from the European end.[57] With those systems in increasingly routine use all the way from western Russia to eastern North America, and with the Devonian in particular accepted by most of the geologists whose opinions counted, the Devonian controversy had shrunk, at least in Murchison's view, into little more than a mopping-up operation.

55. Buckland (1841, pp. 481–87), read at GSL on 19 February but not published until June or July.

56. Murchison and de Verneuil (1841b), read at GSL on 10 and 24 March; Murchison to Sedgwick, 30 March 1841 (transcript in ULC: AS, IIID, 33); report of Museum Committee of GSL for 1840, 19 February 1841 (GSL: BP, OM 1/9; *PGS* 3:368). Sedgwick later refereed the paper and approved it for publication in *TGS:* Sedgwick to GSL, 3 November 1841 (GSL: BP, COM, P4/2, p. 131).

57. Murchison (1841a), read at GSL on 7 April; Conrad (1838, 1840a, b). On the New York State survey, see Hendrickson (1961) and Schneer (1969); on earlier work on the same strata, see Wells (1963).

Chapter Fourteen

THE DEVONIAN NATURALIZED

14.1 Mopping up Devonshire (April to August 1841)

With only one or two exceptions, those who still rejected the Devonian interpretation of Devonshire itself were geologists whose opinions were now considered to be of only marginal significance. Weaver had lost credibility when he continued to insist that there was field evidence for an unconformity at the base of the Culm, after Sedgwick and even Murchison found themselves obliged to concede to De la Beche (and Williams) that the sequence was perfectly conformable almost everywhere (§12.3). Williams had likewise lost credibility when he claimed field evidence in favor of a new structural interpretation that others found highly implausible, namely, that the Culm was below, and older than, all the rocks of south Devon and Cornwall (§12.7). Weaver, now 67 years old, was treated in effect as superannuated; Williams was relegated willy-nilly to the role of a diligent "laborer" or local fossil collector.

Austen had not lost credibility, but circumstances had forced him to retreat to the sidelines of the Devonian controversy. After he moved his home from Devonshire to Surrey in the summer of 1839 (§11.5), his more frequent attendance at the Geological Society in London was no compensation for his inability to contribute to the debate by further fieldwork in Devon, although he did take a keen interest in Phillips's opinions on his own outstanding collection of fossils. "Devon geology seems still a *vexata quaestio* as [to] age, order of superposition &c.," he had remarked after he read about Williams's paper at Glasgow; but it was the comment of an outsider, and he showed little

369

appreciation of how the controversy had developed in the nine months since Greenough had used that phrase to characterize it (§13.1).[1] Sending some more south Devon fossils a few months later, just after the paper on Russia had been read in London, Austen mentioned to Phillips that "you do not, I understand, admit the 'Carboniferous character' which I imagined many of these fossils exhibited." When Phillips told him he did agree, although he interpreted the affinity differently, Austin was pleased to have his own original view vindicated, at least indirectly. Although he thought de Verneuil had been "talked into" the new interpretation by Murchison's overbearing rhetoric, Austen himself must by now have been inclined to adopt it, for he felt aggrieved that Lonsdale had not publicly given him any credit for his role in its genesis (§13.3). He claimed once more that it was *he* who had first pointed out to de Verneuil that the Eifel fossils were like those of south Devon, and that it was that hint that had first led Murchison to think of trying to solve the Devon problem by going to the Rhineland (§10.2, §10.7).[2]

Austen was scornful of "such a foolish name as Devonian," adding that "Silurian is just as bad"; but this was only because he felt "Devonian" should refer to Devonshire, while "the Silurian period belongs to [Roman] history, not to Geology." To avoid the ambiguities of such topographical terms, he suggested to De le Beche an alternative scheme of nomenclature for the whole stratigraphical sequence, based explicitly on the model of Lyell's Tertiary periods. In this scheme, the Carboniferous would be renamed the "superior primary." But below that was the "medio-primary," sandwiched unambiguously between the Carboniferous and the Silurian, and clearly a synonym for Devonian. So Austen, like Greenough (§13.1), had now accepted the Devonian system after all, in all but name. The form and timing of his proposal on nomenclature make it likely that he did so as a direct result of hearing Murchison and de Verneuil's Russian paper read in London and being convinced by their arguments. Certainly he also saw at about the same time the illustrations that de Verneuil and his Parisian collaborator, the Vicomte Étienne d'Archiac (b. 1802), had submitted to the Geological Society to accompany their palaeontological appendix to Sedgwick and Murchison's paper on the Rhineland. Austen would have seen at once that many of those fossils resembled his own from south Devon, while the main paper had presented good evidence that they came from strata below those with the Carboniferous fauna.[3] In any case, whatever the exact evidence that he found persuasive, Austen had now in effect joined the general consensus that the Devonian interpretation was the true solution to the Devon controversy.

By the time the leading British geologists had had the chance to reflect on Murchison's Russian research, only two of them had failed to endorse the Devonian interpretation unambiguously. De la Beche, like Austen, had in

---

1. Austen to Phillips, n.d. [? October 1840] (UMO: JP, 1841/73), probably written soon after reading report of Williams's paper at BAAS, published in *Ath.* on 10 October; Austen to De la Beche, 27 January 1841 (NMW: DLB).

2. Austen to Phillips, 2 and 18 April 1841 (UMO: JP, uncat.).

3. Austen to [De la Beche], n.d. [? April or May 1841] (NMW: DLB), an incomplete letter written after the publication of the review of *The Silurian System* in the April 1841 issue of *ER* (Fitton 1841), but before the BAAS meeting in July. De Verneuil and d'Archiac's appendix was not officially received at GSL until 15 December, but the plates for it must have been there by early April, when Austen saw them: Council of GSL, 15 December 1841 (GSL: BP, CM 1/5); Austen to Phillips, 2 April 1841.

effect retreated geographically, for the publication of his *Report* and his revised maps of Devon early in 1839 had spelt the end of his official responsibilities for southwest England. He was now deeply involved with his survey in south Wales, which presented equally difficult but different problems. His recorded remarks at the Glasgow meeting (§13.6), combined with other comments since the publication of his *Report*, suggest however that he did accept the Devonian interpretation, at least in a qualified way. He had stated in his *Report* that he saw nothing implausible about the putative equivalence of the Old Red Sandstone with Greywacke rocks in other regions (§10.5), so he may have been persuaded by Murchison's claim to have found both types of strata, with their appropriate fossils, juxtaposed in Russia. But his characteristically noncommittal policy toward nomenclature and correlation allowed him simply to sit on the fence about the ages of the Culm and the still older strata of Devon. He may have remained ultimately skeptical about the possibility of identifying Murchisonian "systems" on any wide scale; certainly he avoided having to commit himself publicly on the correlation of what he defiantly continued to term the Greywacke.

De la Beche's agnostic ambiguity about correlation left only Phillips to make up his mind. "I begin to be anxious about Murchison's *Old Red fossils* in Russia," he had told De la Beche soon after the Glasgow meeting, referring to the fossil shells rather than the fish. "Do they prove what he fancies?" he asked; *"are the things below them Silurian really!?"* Having not yet seen the fossils, he had wanted De la Beche to find out for him what the fossil specialists really thought about them; but when he himself was next able to visit London, he had probably been convinced. As he completed his official report on the Devon fossils toward the end of 1840, he had told De la Beche he thought local formation names better for descriptive purposes than the broader but more controversial system names, like Devonian; but as in the case of Austen, this does not necessarily show that he rejected the idea behind the name.[4] When in the spring of 1841 he refereed Sedgwick and Murchison's Rhineland paper for the Geological Society, he recommended it for publication and praised their "sagacity and resolute perseverance." But he discreetly avoided giving any opinion on the authors' chief claim, namely, to have located the Devonian system in Germany between the Carboniferous and the Silurian. He stressed only the great importance of their attempted correlation with Devonshire, "since the determination of the changes produced in deposits of a given age, by circumstances related to geographical position, is become of essential value in theoretical & practical geology."[5] Although this was clumsily expressed, Phillips, like Austen, clearly appreciated that the crucial problem was that of *recognizing* major units such as systems over long distances, since— by analogy with the present day—much environmental variation, and hence great variety of rock types and fossils, were to be expected between one region and another.

Soon after writing that report, Phillips took up the official appointment that De la Beche had secured for him with the Geological Survey. Although burdened with discharging William Smith's debts and with providing for his widow, Phillips was in no position to grumble at his modest salary of £300 a

---

4. Phillips to De la Beche, 20 October and 20 November 1840 (NMW: DLB).
5. Phillips to GSL, 19 March 1841 (GSL: BP, COM P4.2, p. 208).

year. When toward midsummer the printing of his Devon report was at last complete, he promptly turned his attention to south Wales, the subject of his next report; and his practical involvement with the problems of Devonshire was all set to weaken as rapidly as De la Beche's had done.[6] But first he returned to do three months of final fieldwork in Devon. Shortly beforehand he told De la Beche about the conclusions embodied in his report. His "general view of the analogies of the Devon fossils" was expressed in a simple sequence. The Devon and Cornish fossils older than the Culm were assigned to a position *between* what he termed "Lower Palaeozoic" and "Upper Palaeozoic." What exactly this meant was not revealed, but he whetted De la Beche's appetite for the published work by telling him cryptically, "I think you will be amused by the arithmetic at all events."[7]

Phillips's *Palaeozoic Fossils* was published shortly afterward. As the full title made plain, the bulk of the book was devoted to the straightforward textual description and lithographic illustration of the fossils that he and many of the local amateur collectors had amassed. No fewer than 275 species were described, which, Phillips claimed cautiously, was sufficient for "a partial removal of the veil which has so long obscured the age and affinities of the strata of Devon and Cornwall."[8] In a brief concluding section, he explained why rock types, at least in this region, were deceptive as criteria of age, so that fossils were indispensable. He mentioned the increasing acceptance of "three great systems" in the classification of strata generally. This referred not to Murchison's "trinity"of Silurian, Devonian, and Carboniferous but to his own much broader categories: the Palaeozoic, Mesozoic, and Cainozoic (§13.5). Phillips divided the "Palaeozoic Strata," which of course included *all* the older (pre-New Red Sandstone) rocks of southwest England, into three great groups. His summary to De la Beche was now made plain. The "Upper Palaeozoic" comprised all the Carboniferous, together with some strata in the north of England overlying the Coal Measures ("Magnesian Limestone formation"). The "Lower Palaeozoic" comprised "Transition Strata"—elsewhere identified as the Silurian—and "Primary Strata." In between was the "Middle Palaeozoic," represented by the rocks of south Devon and the Eifel. The term "Devonian" was used only in its traditional topographical sense; but it was now clear at last that Phillips had stepped off the fence, and that, like Greenough, Austen, and perhaps De la Beche, he had accepted the Devonian interpretation of Devonshire in all but name.[9]

The "arithmetic" to which Phillips had alluded beforehand was now revealed as a quantitative or "statistical" analysis of the Devon faunas in relation to the Silurian and Carboniferous. Unlike the metropolitan gentlemen of geology, Phillips acknowledged the value of Austen's theoretical reflections as well as his abilities as a collector; he praised Austen for the way he had emphasized the necessity of basing correlations on whole assemblages of

6. De la Beche to Ordnance, 27 January 1841 (NMW: TS, II, pp. 356–57); the request to employ Phillips from 1 April was approved on 17 February (ibid., pp. 359, 364–65, 386); Phillips to De la Beche, 20 October 1840, and 18 and 26 June 1841 (NMW: DLB).

7. Phillips to De la Beche, 21 and 26 June 1841 (NMW: DLB).

8. Phillips (1841, preface, pp. v–xii), published by 30 July (*Ath.*, no. 720, p. 626).

9. Phillips (1841, pp. 155–82), "Notices and Inferences." He had got his Eifel fossils not only from Sedgwick and Murchison but also directly from Goldfuss: Phillips to De la Beche, 4 December 1840 (NMW: DLB).

fossils, rather than on a few supposedly "characteristic" species. But the main model for Phillips's analysis was Lyell's earlier treatment of the Tertiary faunas. Phillips's proposed Palaeozoic, Mesozoic, and Cainozoic recalled the classically based terms (Eocene, Miocene, Pliocene) that Lyell had used to subdivide the Tertiary, or what Phillips now termed the Cainozoic. Phillips's ambition was to subdivide the Palaeozoic likewise into smaller units that would be based purely on the "statistics" of a slowly changing faunal composition.

Like Lyell, Phillips took as his units the *species* identified in the fossil faunas. Like Lyell again, he was well aware that taxonomic lumpers and splitters might identify different numbers of "species" in any given collection. But just as Lyell had applied a uniform standard by getting the Parisian palaeontologist Paul Deshayes (b. 1797) to do all his identifications for him, so Phillips got round the same problem by relying throughout on his own identifications. To compare with his 275 species from Devon and Cornwall, he had 336 species from the Silurian and 420 from the Carboniferous. Not only did his detailed tabular analysis show that the Devon fossils *as a whole* were clearly intermediate between Silurian and Carboniferous, as Lonsdale had demonstrated the previous year from much smaller collections (§13.3). Further than that, Phillips claimed to be able to divide the fossils of Devon and Cornwall into three *subordinate* groups on the basis of a gradually changing faunal composition. In inferred chronological order, the fauna of the south Devon limestones (e. g., Tor Bay and Plymouth) was nearest to that of the Silurian; then came the faunas of the upper part of the north Devon sequence (e. g., Barnstaple); and finally those of the north Cornish strata (e. g., South Petherwin), which had the strongest affinities to the true Carboniferous faunas.[10]

Phillips had thus not merely accepted the Devonian interpretation but had taken it much further than Murchison or Lonsdale. He regarded the Devonian rocks and fossils not so much as a clearly defined "system" flanked by equally well-defined Silurian and Carboniferous systems, but rather as a series of faunas that reflected an imperceptibly gradual process of piecemeal faunal change. Only by the analysis of large numbers of species could the distorting effects of local *ecological* variations be ironed out, as it were, to reveal the overall effects of *temporal* change. Phillips had declined to adopt the term "Devonian system," probably because he felt it failed to express the essentially palaeontological basis of any adequate definition of the corresponding *period*. But he had nonetheless provided the Devonian interpretation with a masterly vindication in terms of the fossil faunas of Devonshire itself.

Phillips's book was published just in time for the meeting of the British Association, which this year took place for the first time in southwest England itself, under the general presidency of Whewell. The previous year the leading geologists had wanted Conybeare to preside over their Section in Plymouth, probably because he was the only member of the inner circle of the Geological Society who actually lived in Devonshire. But Conybeare had declined and suggested De la Beche instead, perhaps as a way of hinting that the official geological surveyor deserved greater recognition.[11] Anyway, De la

10. Phillips (1841), the "statistical" tables are on pp. 141–54. Lyell's were in Lyell (1830–33, 3: appendix); for their significance, see Rudwick (1978).

11. Murchison to De la Beche, 9 and 20 March 1841 (NMW: DLB); Murchison to Sedgwick, 8 April 1841 (transcript in ULC: AS, IIID, 35).

Beche had duly been appointed president by the Council, and he came from south Wales for the occasion. Conybeare acted as a vice-president, together with the other leading geologist who was a Devonian by birth, namely Buckland; also present was Sedgwick, who combined the meeting with a few days' fieldwork among the Devonian rocks; while Phillips was there once more in his official capacity. But Lyell had just left for the United States, and Murchison was somewhere in the wilds of the Urals, missing a meeting for the very first time. Perhaps because the ranks of the leading geologists were rather depleted, three local amateurs who had long been active in the Devonian debate were condescendingly admitted for the first time to the committee of the Section: the argumentative Williams, the venerable Hennah—now 85 years old—and Austen, who had returned to Devon to visit his old friends.[12]

At the first session of the geological Section, a paper on some strata in north Wales gave rise to a wide-ranging discussion. De la Beche pointed out that "general principles were involved," because the strata, lying unconformably below the Mountain Limestone, resembled the Old Red Sandstone but had Upper Silurian fossils. Phillips urged caution, in view of the possibility that purely local or ecological factors might have affected the fossil fauna, and Sedgwick agreed. When Williams asked provocatively about the Devonian interpretation, Sedgwick admitted that geologists on the Continent "would doubtless pack the old red with the Silurian," implying that they had not yet accepted the Devonian as a full-blown system; he emphasized that the "ultimate arrangement of large groups" would need a study of the relevant fossils on a worldwide scale.[13]

In another session, the fossil collections of various local geologists were likewise discussed, in the usual manner, almost exclusively by the dominant leaders of the Section. Edward Moore and another Plymouth amateur displayed some of their local fossils, but it was Phillips who gave his opinion that one specimen of fish scales might perhaps represent the Old Red Sandstone genus *Holoptychius*. He recognized that alternatively the scales might belong to a Carboniferous genus (*Palaeoniscus*), and Williams suspected the specimen might have drifted from a passing Welsh coal boat. But if it were genuine it would be the first faint hint of direct evidence for the equivalence of the Old Red Sandstone and Devonian in Devonshire itself.[14] Harding showed his latest fossils from some of the lowest strata in north Devon (near Combe Martin), but it was Sedgwick who identified them as uppermost Silurian (Upper Ludlow) forms, and emphasized that there was "no good line of separation" between the Silurian and Devonian in that sequence. Sedgwick gave a highly unhistorical reconstruction of his own drastic change of opinion on the age of the strata of north Devon. More generally, however, he gave Lonsdale the credit for first showing that the Devonian fossils were intermediate between the Silurian and the Carboniferous, "the exact lines of

12. *Ath.*, no. 718 (31 July 1841), p. 579; no. 719 (7 August 1841), p. 597: Sedgwick, notebook "No. 35, 1841" (SMC: AS).

13. BAAS, session on 29 July 1841: *Ath.*, no. 719, pp. 497–99; Bowman (1842). John Bowman (b. 1785) was a Wrexham banker, and an amateur geologist and naturalist, to whom Murchison had suggested some research on Denbighshire strata.

14. BAAS, session on 3 August 1841 (the second amateur was S. P. Pratt): *Ath.*, no. 722 (28 August 1841), p. 674; Williams to Phillips, 3 August 1841 (UMO: JP, 1841/37).

demarcation between these systems not being finally settled"; and he alluded to the similar problems that he and Murchison were now having over the boundary between the Silurian and the Cambrian. "Certain typical forms appear and disappear, one series melting into another," he explained, in a way that Phillips—or Lyell—would have applauded; although this made the boundaries more difficult to draw, he insisted that it "does not make them less real." In other words, he tacitly claimed the validity of the systems that he and the absent Murchison had defined, even though he had to concede to Phillips and the absent Lyell that the boundaries of those systems were inescapably arbitrary, at least in faunal terms. "Stand to your guns," Murchison had ordered him beforehand; "the types are clear & distinct, & birds of *passage* are not to frighten us."[15] Sedgwick had done his best to obey those orders.

Finally, when another local amateur displayed his collection of 150 species of "Devonian fossils," Conybeare objected to that term on the usual grounds that it was a topographical word; he proposed "Epi-Silurian"instead, implying perhaps that his adherence to the full Devonian interpretation was less than wholehearted. But it was left to Phillips to place the amateur's collection in the context of his own recent research; he summarized the changing character of the faunal assemblages when traced through "the Devonian rocks," and claimed that the fossil evidence for their intermediate position was now "as free from imperfection as any which could be obtained." Putting that solution of the Devonian controversy into a still wider framework, Phillips took the opportunity to urge the adoption of his own "three great classes" of Palaeozoic, Mesozoic, and Cainozoic. It was "a classification depending wholly on general views of the associations of organic life," or in other words, on his highly ecological conception of piecemeal faunal change; yet he pointed out that it could be combined harmoniously with the more conventional descriptions of the sequences of formations.[16]

Phillips's "luminous address," which according to one reporter was "listened to with marked and almost anxious attention" and applauded heartily, was "a fit finale to a brilliant campaign" by that geologist among the fossil-bearing strata of Devon and Cornwall, and a rousing conclusion to the geological Section's discussions in Plymouth.[17] But it also marked Phillips's final emergence from any subservient or ancillary role into full recognition as one of the leading geologists in England, with as much authority in his theoretical interpretations as in his fossil identifications and practical fieldwork. It may not be fanciful to link his personal triumph at Plymouth with the absence thousands of miles away of the two geologists who, in different ways, had unavoidably and perhaps unconsciously cramped his style hitherto. By contrast, the meeting also marked a further stage in Williams's relegation to the margins of the science. He circulated among the assembled geologists a prospectus for his projected book on the geology of southwest England; but evidently there was little positive response, since nothing more was heard of the proposal. "I wish he could publish his facts and a good map, and omit his

15. *Ath.*, no. 722, p. 674; Murchison to Sedgwick, 8 April 1841. On the problems of the Cambrian-Silurian boundary, see Secord (1986).

16. *Ath.*, no. 722, pp. 674–75. This collection had been made by J. C. Bellamy.

17. *LG*, no. 1284 (28 August 1841), p. 560.

classification," commented Lonsdale; but that modest role was one that Williams had long since rejected.[18]

With Williams no longer taken seriously, and Weaver faded from the scene, the triumph of the Devonian interpretation of Devonshire was now almost complete. Some geologists might for various reasons be unwilling to use the term Devonian in its new stratigraphical and temporal senses. But almost all were now agreed that the older (pre-Culm) rocks of southwest England were far less ancient than had been assumed when the controversy began, and that they were at least roughly equivalent to the Old Red Sandstone elsewhere in Britain. The full significance of that deceptively local conclusion remained to be established on a wider stage.

14.2   From the Urals to Niagara (April to October 1841)

Murchison had left London in the spring, soon after the reading of his and de Verneuil's paper on Russia. As in the previous year, he planned to gather information, sound out opinion, and promote his own views in both Paris and Berlin before setting out for his second and more ambitious tour of Russia. From Paris he reported to Sedgwick on the progress of the Devonian interpretation and sought to strengthen his collaborator's resolve in defending it. "Whatever dubiety may shroud the minds of some of our countrymen," he assured him, "the thing is already *quite done as to the Continent.*" He reported that de Verneuil was now fully convinced, and that Élie de Beaumont and Dufrénoy had engraved the terms Devonian, Silurian, and Cambrian on their long-delayed geological map of France. Dumont's colleague Laurent de Koninck (b. 1809), a palaeontologist and chemist from Liège, was also in Paris, partly in connection with his forthcoming monograph on the Carboniferous fossils of Belgium; Murchison reported to Sedgwick that de Koninck too was convinced by their Devonian interpretation of the Belgian sequence. But Murchison was not leaving the expansion of the consensus to chance, for he instructed his wife to send offprints of his Russian paper to all the key geologists in Britain and the United States. "Pray read attentively the long abstract I made of the Russian case," he told Sedgwick with particular insistence; "for there the Devonian matter is *completely settled.*[19] His second Russian campaign was thereby implicitly characterized as no more than a mopping-up operation, at least as far as the Devonian interpretation was concerned.

Joined by de Verneuil, who was once more to be his palaeontological aide, Murchison traveled direct from Paris to Berlin, stopping there only briefly to renew his contacts with Humboldt and the Prussian geologists. This time he planned to travel overland through the Baltic provinces to St Petersburg (see fig. 13.2) in order to study for himself, however briefly, the sequence of

18. The prospectus is mentioned in Williams (1842b, p. 27n); also Lonsdale to Phillips, 31 August 1841 (UMO: JP, 1841/45). Substantial MS drafts, almost certainly for his book, are in SAS: DW, boxes II, III.

19. Murchison to Charlotte Murchison, 2 and [c. 8] April 1841 (ULE: RIM); Murchison to Sedgwick, 8 April 1841 (transcript in ULC: AS, IIID, 35); Koninck (1842–44). The American recipients of Murchison's Russian paper were to be Silliman at Yale, and Hall, Conrad, and Lardner Vanuxem (b. 1792) of the New York survey.

almost horizontal Palaeozoic strata that he had seen the previous year further to the east (§13.4). After crossing the Russian frontier, almost their first section in Courland showed satisfactorily "a succession from the Silurian to the Red Devonians." On reaching Riga they met up with Keyserling, who had been finding fossil fish in the Old Red Sandstone on the banks of the Dvina, and who was again to be their traveling companion. Together they visited Christian Pander (b. 1794), a member of a wealthy banking family and a well-known naturalist, whose published descriptions of fossils from his outstanding collection had first alerted Murchison to the presence of Silurian strata in the region. Both Pander and Hermann Asmus, the professor of zoology in Dorpat, whom they visited later, had found massive fossil fragments which they were reconstructing into what Murchison noted as "gigantic fishes of the old red."[20] All this added little to Murchison's general conclusions, however, although it was satisfactory confirmation of the previous year's work.

The geologists spent some time in St Petersburg, studying the fossil collections in the School of Mines and preparing for their major expedition to the Urals. In the midst of a round of fashionable parties and balls matching the splendor of the imperial court, Murchison was gratified to be presented to Czar Nicholas himself; later he had a private conversation with him. He told the Czar it was *"physically impossible"* for coal deposits to be found in the north of Russia, and that the Donets coalfield in the south should be exploited instead.[21] His confident prediction was once again the product of his interpretation of the Silurian system and, now, the Devonian too. He had concluded the previous year that the sequence of Secondary-like strata in northern Russia was in fact Silurian, Devonian, and, at the top, the equivalents of the Mountain Limestone. He was therefore convinced that there could be no significant Coal Measures anywhere in the region; and his concept of the Devonian and Silurian periods, as recording the very dawn of land plants, ruled out the possibility of any coal deposits in those older strata. Conversely, although he had not yet visited the Donets coalfield, museum specimens convinced him that it was in true Carboniferous rocks, which he believed to represent the first, if not the only, significant epoch of coal formation in the history of the earth. Such were the economic fruits of the fossil-based geology he had worked so hard to promote.

Murchison's advice to the Czar was only the first instance of the way the Devonian interpretation had now been integrated into his geological work so fully that it could be used as a matter of routine. From St Petersburg the party traveled first to Moscow and then, at the end of May, on to Kazan, where the expedition to the Urals was to begin. Passing through Perm, the geologists confirmed that a great thickness of strata was sandwiched between the Carboniferous and a younger limestone that they judged by its fossils to be equivalent to the uppermost "Palaeozoic" formation in western Europe (*Zechstein* or Magnesian Limestone). But as they approached the Urals, such undisturbed formations were replaced by strongly folded rocks. What is striking about the

---

20. Murchison to Charlotte Murchison, 23 April and 1 May 1841 (ULE: RIM); Murchison, notebook "1841, Vol. I," pp. 47, 54–68 (GSL: RIM, N85); Pander 1830. The "gigantic fishes" were *Asterolepis* (Agassiz 1844, pls. 30, 32).

21. Murchison to Charlotte Murchison, 6 and 25 May 1841 (ULE: RIM); Murchison, notebook "1841, Vol. I," pp. 115–50, amplified and partly transcribed in the 1860s in "Wanderings in Russia II" (GSL: RIM, J9).

notes that Murchison made in the subsequent weeks, as they crossed the mountains on no fewer than seven routes, is that he repeatedly identified "the equivalent of the Devonian," or "the Old Red Sandstone or Devonian," simply by a handful of key fossils (e. g., the trilobite *Brontes*; fig. 12.7, B), and the Silurian likewise. The Urals, he told his wife, were "our Devonians & Silurians—with nothing left for poor Sedgwick's Cambrian!!" Rarely was the structure clear enough to prove a definite sequence; often the rocks were highly disturbed or even altered by metamorphism. But the Devonian was no longer on trial; it had passed, at least for Murchison, into routine and unquestioned use.[22]

The geological exploration of the Urals, amicably combined with the Russians' official survey of mineral resources, was the climax of Murchison's fieldwork. The sheer unfamiliarity of the country, the exhilaration of rough and even tough travel through remote and roadless terrain, the excitement of reaching the edge of the Siberian plain, and the novelty of being entertained by Kirghiz nomads: such experiences combined to make the summer more than memorable for Murchison. As he drank a solitary toast to the British Association to mark the start of the Plymouth meeting, he doubtless felt some satisfaction that he could leave residual British problems to others, for his own horizons were now far too broad to be confined to Devonshire and "Siluria."

Leaving the Urals at the end of July, the party traveled westward across the Volga and the Don to study the Donets coalfield, which the geologists confirmed as truly Carboniferous. Returning northward toward Moscow, they found near Orel some fine exposures of soft yellow limestone, which Murchison noted as "leaving no doubt that the whole group represents the Zechstein," making it younger than the Carboniferous. But soon afterward they discovered fossil shells and fish in these strata, including what he termed "my own dear *Holopt[ychius] nobilissimus*" (fig. 13.1), which made him change his mind abruptly. "Devonian System proved," he noted tersely. Although "undistinguishable from our Magnesian Limestone" in its rock type, its fossils had once again saved him from gross error. "This discovery is of *great importance*," he noted; "this axis along the middle region of Russia is more than I expected." For such a broad strip of Devonian rocks confirmed his confident prediction to the Czar that the productive coal strata of the south were completely cut off from what was now revealed to be a basin of Carboniferous strata (but without significant coal) around Moscow. The Devonian had proved its value in very practical use.[23]

As soon as the party reached Moscow and the 13,000-mile expedition was over, Murchison summarized their principal conclusions in an article for the journal of the naturalists' society there. In St Petersburg he submitted an official report to the government and explained their results personally to the Czar; he and de Verneuil were both awarded the Cross of St Ann (second class) for their pains—though de Verneuil's lacked the diamonds! On his return to London, Murchison sent the *Philosophical Magazine* an English version

22. Murchison to Charlotte Murchison, 21 June and 12 July 1841 (ULE: RIM); Murchison, notebooks "Vol. 3, 1841," pp. 38, 100, 132–34, 156; "Vol. 4, Ekaterinbourg to Orenbourg," p. 72; "Vol. 6, Russia 1841," pp. 41–42 (GSL: RIM, N87, N88, N90); "Wanderings in Russia, III, IV, V" (GSL: RIM, J10–12).

23. Murchison to Charlotte Murchison, 25 September 1841 (RIM: ULE); Murchison, notebook "VIII," pp. 161–74, and "Wanderings in Russia VI" (GSL: RIM, N92, J13).

of the article he had left in Moscow. The Urals were not "primitive," he reported, but largely composed of strongly folded and even metamorphosed Silurian, Devonian, and Carboniferous rocks; the Donets coalfield was truly Carboniferous, but in a basin quite separate from that of Moscow. His most striking announcement, however, concerned the formations the geologists had found overlying the Carboniferous strata around Perm to the west of the Urals. Murchison termed them "*Permian*"; he claimed not only that their animal and plant fossils were intermediate between those of the Carboniferous and the New Red Sandstone (or "Trias" of German geologists), but also that their position and thickness qualified the Permian to be regarded as a separate major system.[24] This new Permian system had no direct connection with Murchison's campaign to establish the Silurian and the Devonian, but it did complete his conquest of almost the whole of the "Palaeozoic" strata, in Phillips's extended definition of that term. The Permian system was now staked out above the Carboniferous, while the Devonian and the Silurian were firmly established below, and the Silurian fossils were edging Sedgwick's Cambrian into oblivion. Murchison's set of major systems for the older rocks, each marked by distinctive characteristic fossils, was now in place; it remained only for him to press them into general use.

That persuasive process, by which the geological world might be convinced that these systems were natural units of global validity, had been set in motion, even before Murchison returned to England, through the activities of Lyell. All through the spring and early summer of 1841, Lyell was busy with two closely related projects: the revision of his *Elements of Geology*, and the preparation of the Lowell lectures he had been invited to give in Boston in the autumn. A rough outline of headings for one lecture included "old red" between Silurian and Carboniferous; but then, probably after reading Murchison and de Verneuil's paper on Russia (§13.6), Lyell amended it to "Devonian or old red." Later he made a careful study of Sedgwick and Murchison's paper on Devonshire, and he began to use the term "Devonian" routinely in his notes of the frequent discussions he had with Lonsdale about the Palaeozoic fossils. The *Elements* was completed in July, shortly before he and his wife sailed to the United States; when it was published not long afterward, Lyell's adoption of the Devonian interpretation was at last fully in the public realm.[25]

In striking contrast to the first edition that he had published three years earlier, Lyell's treatment of the "Cambrian Group" was now drastically curtailed, and he doubted if it had a sufficiently distinctive fossil assemblage to be maintained as a separate category at all. That opinion was partly the result of Murchison's and Sedgwick's growing problems over the demarcation of their respective systems in Wales; but in Lyell's treatment it was due just as much to the removal from the Cambrian of the fossil-bearing rocks of southwest England, which he had used in the first edition as its chief exemplar (§9.8). The engraving of the mollusk *Clymenia* that he had used to illustrate the Cambrian fauna (fig. 9.3) was now reused to illustrate the *Devonian* fauna,

24. Murchison (1841b), dated 5 November and published in December issue of *PMJS*.

25. Lyell, notebook 86, December 1840 to February 1841, pp. 17, 53; notebook 87, February to June 1841, pp. 46–47, 68–70, 91, 95, between dated entries for 16 April and 3 June (KHK: CL); Lyell (1841a), dedication (to Fitton) dated 10 July.

along with new engravings of several other characteristic fossils (fig. 12.7). For after the chapters dealing with the Carboniferous, and before that on the "Primary Fossiliferous Strata" (including the Silurian and residual Cambrian), Lyell had inserted a new chapter on the "Old Red Sandstone or Devonian Group."[26] Though he preferred the term "group" to "system," Lyell's adoption of the Devonian interpretation was thus complete in all essentials. He had accepted the equivalence of the Old Red Sandstone with the shell-bearing "Devonian" strata, not least on the evidence from Russia; and he had treated all those strata as the record of a period that was as important in the history of the earth and of life as the Silurian before it and the Carboniferous after it.

Lyell intended to make good use of his generous invitation from the Bostonians and planned to stay in the United States for about a year and to travel extensively. He had many objects in mind for his fieldwork there, and hoped in particular to study the extensive Tertiary deposits. But he also wanted to see for himself the spectacular development of undisturbed Palaeozoic strata that the geologists of the New York State survey had begun to describe (§13.6). In preparation for his visit, he not only toured the Welsh Borderland to see Murchison's Silurian at firsthand but also made a close study of the published work of Timothy Conrad (b. 1803), who since 1837 had been official Palaeontologist on the New York survey.[27]

Lyell and his wife finally left Liverpool by steamship a few days before the Plymouth meeting of the British Association, and sailed to Boston. They traveled by railway and steamboat to New Haven to visit Benjamin Silliman (b. 1789), the country's most respected geologist, at Yale. Continuing the journey by steamboat, Lyell called at New York and then traveled up the Hudson to Albany, the state capital and the headquarters of its geological survey. There he met Hall, who took him westward through the region of almost horizontal strata that he had described for Murchison earlier in the year (§13.6) and that his colleague Conrad regarded as Silurian. Lyell's first substantial fieldwork in America was thus on the older Palaeozoic rocks. At Niagara his attention was diverted by what he regarded as spectacular evidence for the vast timescale involved in the erosion of the falls. Somewhat bemused, if not bewildered, by the lack of deference shown to him by the egalitarian rural Americans, the gentlemanly Lyell then returned to Albany by a different route along the border with Pennsylvania. There he resumed his stratigraphical fieldwork with a study of the Old Red Sandstone strata that the American geologists had described overlying the Silurian, and he followed that sequence up into the coal strata on the northern edge of the great Pennsylvania coalfield.[28]

Like Sedgwick and Murchison in the Rhineland two years earlier, Lyell was in no sense doing original research: he was merely seeing for himself

26. Lyell (1841a, 2:154–58, 176–77).

27. Lyell, notebook 88, June and July 1841, pp. 1, 49, 93 (KHK: CL); Conrad (1838, 1840a, b); Hendrickson (1961); Schneer (1969).

28. Lyell to Horner, 26 August and 21 September 1841; Lyell to Mantell, 29 October 1841 (K. Lyell 1881, 2:54–59). Lyell's American tour comes within the purview of the unpublished sequel to the first volume (1972) of Professor Leonard Wilson's biography, and I have not seen the relevant field notebooks at KHK; Lyell's American fieldwork cannot therefore be described and analyzed here as thoroughly as Murchison's in Russia. The account he wrote much later (Lyell 1845, 1:1–76), although useful in general terms, cannot be relied on for evidence of his immediate impressions and interpretations.

what American geologists had already explored with great thoroughness. What he was able to add to their work, of course, was a firsthand comparison with similar rocks and fossils in Europe. Later in the autumn, Conrad showed him the Tertiary strata of New Jersey, and Henry Rogers the central part of the Pennsylvania coalfield. In the latter area, he was able to see for himself the overwhelming evidence for equating the anthracitic coal with the productive Coal Measures of western Europe, a similarity that had earlier undermined the plausibility of claims that analogues such as the Devonshire Culm were of Transition or Greywacke age. While he was in Pennsylvania, Lyell was chiefly interested in the American evidence for the mode of formation of coal and in the spectacular folding of the strata in the Appalachians generally. But he had now seen, both in the folded region and in undisturbed condition further north, an American sequence of Silurian, "Old Red," and Carboniferous strata that quite closely matched the European.[29]

The day after he returned to Boston, Lyell wrote Fitton a long account of what he had seen in some ten weeks' travel and fieldwork. As he doubtless intended, the letter was read at the Geological Society later in the autumn, and his first impressions of American geology thereby became widely known in England. His account covered the work that has just been summarized, and much more besides; only one further point needs particular comment. In his letter to Fitton, Lyell specifically mentioned that the American formations sandwiched between the Silurian and the Carboniferous included not only Old Red Sandstone strata with the usual fossil fish, but also shales with "Devonian fossils." So Lyell now shared Murchison's opinion that the Devonian interpretation could be sustained in its fullest form in the eastern United States, with evidence closely comparable to that in Russia (§13.6). Only three months later, Conrad showed that he too shared that opinion: in his paper "On the Silurian and Devonian Systems of the United States," read in Philadelphia early in 1842, he claimed the American sequence as the "most perfect series of the older Palaeozoic rocks" anywhere in the world, and summarized the fossil evidence for its correlation with Europe. Since Conrad had almost certainly discussed this with him beforehand, Lyell was able to explain confidently to his lecture audience of some 2000 people in Boston, when he dealt with the record of earlier periods in the history of the earth and of life, that the sequence of Silurian, Devonian, and Carboniferous strata was now soundly established in both the Old and New Worlds.[30]

14.3   Securing Murchison's Monument (November 1841 to February 1842)

Soon after Murchison returned from Russia, Phillips wrote to explain that in his report on the Devon fossils, which had been published during Murchison's absence abroad, he had proposed a new way of subdividing the Palaeozoic

29. Lyell (1845, 1:77–100). On the interpretation of the folding in the Appalachians, see Greene (1982, chap. 5).

30. Lyell (1841b), a letter to Fitton dated 15 October, read at GSL on 17 November; Conrad (1842), read at the Academy of Natural Sciences on 18 January; Lyell (1845, 1:106–23). Conrad assigned to the Devonian the Ithaca, Chemung, and Blossburg formations of the Erie group at the top of the "New York System," and commented on the many Devonshire species in the Chemung and on the Old Red Sandstone fish *Holoptychius* in the Blossburg.

formations (§14.1). As he recalled later, he had "pointed out to him the necessity of regarding 'Silurian,' 'Cambrian' &c. Systems as *local* and limited
generalisations." Securely employed with the Survey, Phillips may well have
felt he could now afford to be less deferential toward his earlier patron; in
any case, he told Murchison bluntly that his systems, though valuable locally,
were "not to be put in place of larger generalisations less dependent on
physical peculiarities of land & sea." Only his own method of analyzing whole
assemblages of fossils in minute detail, Phillips implied, could be an adequate
basis for more satisfactory major categories. By contrast, Murchison's method
of identifying his systems, even in the field, by a few supposedly "characteristic" fossils, was doomed to lead sooner or later to spurious results, because
any such handful of species was likely to have been confined by local environmental conditions.[31]

Murchison was incensed that his systems should thus be challenged
by the younger geologist, and he promptly broke off all contact with him.
When he replied to Lyell's American news with an account of his own work
in Russia, Murchison told Lyell about his new Permian system, and about the
"Q.E.D." of his discovery of the Devonian in central Russia. But he was most
concerned about Phillips's unwelcome criticism of the theoretical basis for
all these systems in the Palaeozoic. "I hope you will not desert me," Murchison
pleaded, "but stick to the nomenclature first resolved upon from fair geological
work." He was anxious that Lyell, having pioneered in the Tertiary the "statistical" methods that Phillips had now exploited in his analysis of the Devon
fossils, might drop Murchison's systems and follow Phillips in adopting a new
and purely fossil-based nomenclature for the older strata (§14.1). "His effort
to subdivide our Palaeozoics by new names is really to avoid the word Devonian & so gratify De la Beche," Murchison claimed, unable or unwilling
to conceive that Phillips might also have had reasons of scientific substance.[32]

Lyell's report to Fitton about his American fieldwork had been read at
the Geological Society immediately after Sedgwick presented a supplement
to his earlier general *Synopsis* of all the older rocks of Britain. Just as Lyell
had had to make major changes in his elementary book in response to the
Devonian interpretation, so Sedgwick had to do likewise at the more elevated
level of this important synthesis of recent research. Three years earlier, his
original *Synopsis* had been full of concealed uncertainties about the age of
the older rocks of Devon and their wider implications (§9.8). With the establishment of the *"Devonian system"* these problems could now be resolved.
Sedgwick pointed out that from the New Red Sandstone down to the Lower
Silurian in Britain "there is one continuous unbroken sequence in which no
term is wanting." Since the Old Red Sandstone manifestly filled the"whole
intermediate position" between Silurian and Carboniferous strata, he argued
that any formations with fossils that were similarly intermediate *must* be "on
the parallel of *some part* of the old red sandstone." Sedgwick therefore concluded that "the argument for the true place of the Devonian system is com-

31. Phillips's correspondence with Murchison on this point has not been located, but it is
summarized in Phillips to De la Beche, 14 December 1841 (NMW: DLB).
32. Phillips to De la Beche, 24 March 1842 (NMW: DLB); Murchison to Lyell, 1 December
1841 (APS: CL); Murchison's first summary of his Russian work had just been published in
English in the December issue of *PMJS* (Murchison 1841b).

plete." He had greater difficulties with his own Cambrian rocks, since he had to admit that their demarcation from the Lower Silurian was far from clear. But he maintained that his "Upper Cambrian" and "Lower Cambrian" were still important and natural "groups" for the strata that contained the very earliest fossils, and he revived the term "Cumbrian" for still older strata with no traces of animal life. Whatever the problems of those pre-Silurian strata, however, Sedgwick had now publicly adopted the Devonian as a major group between Silurian and Carboniferous, without any prop of co-authorship with Murchison.[33]

In the subsequent discussion, the propriety of these terms was argued over by Greenough, Murchison, and Fitton, but Sedgwick was later reported as saying "they might call the rocks what they pleased."[34] Unlike Murchison, he felt no special personal attachment to the terms he himself had coined, but he *was* concerned that whatever terms came into general use should reflect the natural boundaries suggested by the fossil record itself. Despite his lecturing duties in Cambridge and "an enormous & suffocating correspondence," Sedgwick later found time to write to Murchison to explain why he was so unsatisfied with the major categories that his collaborator proposed. "There is a quasi Devonian system," he argued, with particular reference to Belgium and the Rhineland, "but it blends with the upper Silurian." This comment did not indicate, however, that he had reverted to his earlier view that the Devonian was a mere "top-dressing" to the Silurian (§12.8). What he was now urging was the recognition of a new major "natural division" below the Carboniferous to include the Devonian *and* the upper Silurian; he claimed to be indifferent about its name, suggesting that either "Devonian" or "Silurian" (in modified senses) would be appropriate. The lowest "natural division" would then include the Lower Silurian *and* the Cambrian, and could likewise be termed either "Silurian" or "Cambrian" (in modified senses). Doubtless these proposals were in part a tacit bid to retrieve a palaeontological basis for his Cambrian, after the loss of what was now recognized as the Devonian fauna of southwest England, and in the face of the encroachment of Murchison's Lower Silurian fauna into Cambrian territory in Wales.[35] But Sedgwick's suggestions for the reclassification of the older Palaeozoic formations indicate that his adherence to the Devonian as a *major* unit was unaffected by his recognition of its ultimately conventional status.

"Sedgwick has fairly given up the Cambrian as a Zoological group," Murchison reported triumphantly to Lyell, having no intention of ceding any Lower Silurian strata or fossils to an enlarged Cambrian. By contrast, his own systems were secure: as he told Lyell, he had "the satisfaction of knowing that Silurian & Devonian are engraved on the geological Maps of Élie de Beaumont," and in increasing use elsewhere in Europe and America.[36] It was

33. Sedgwick (1841), read at GSL on 3 and 17 November.

34. The prohibition of public reporting of GSL discussions was flouted on this occasion by Charles Moxon, the enterprising editor of a new journal, who printed Sedgwick's remark and was subsequently reprimanded by the Council: *Geologist* 2:131; Council of GSL, 15 June 1842 (GSL:BP, CM 1/5, p. 385).

35. Sedgwick to Murchison, n.d. [? late November], wmk 1841 (GSL: RIM, S11/187), probably written soon after his "Supplement" (1841) was read at GSL on 3 and 17 November, and shortly before he went to Norwich to begin his usual period of duty (December and January) at the cathedral. See Secord (1986).

36. Murchison to Lyell, 1 December 1841.

perhaps by an unconscious slip of pen or memory that Murchison omitted to mention that what he had seen in Élie de Beaumont's study some months earlier had been an advance copy of the French map with Devonian, Silurian, *and Cambrian* engraved on the key (§14.2).

That is what he duly found on the presentation copy of the map that reached the Geological Society a few days later, accompanied by a massive volume of *Explication*. Dufrénoy and Élie de Beaumont had retained the tripartite division of the pre-Carboniferous *terrain de transition*, which had been used on the map that had been widely available to geologists for several years in unpublished form. But at a fairly late stage, probably after Murchison read his paper on "les roches dévoniennes" in Paris in April 1840 (§13.3), the French geologists had added, beside the *supérieur, moyen* and *inférieur* divisions of the *terrain de transition*, the words "Système dévonien, Vieux Grès rouge des Anglais," "Système silurien," and "Système cambrien," respectively. Murchison's (and Sedgwick's!) nomenclature was thus engraved on what was now perhaps the finest geological map in the world.[37]

The introductory chapter to the *Explication* revealed, however, that the French geologists had combined their acceptance of one aspect of the Devonian interpretation with a rejection of another. Their "tableau général" split an idealized columnar section of the whole sequence of formations into a series of discrete chunks, each representing a *terrain* with characteristic fossils. These were separated by a series of gaps, each representing a *système* in the French sense, or in other words an epoch of folding and mountain building with a characteristic compass orientation. The French geologists' acceptance of the Devonian and other Pre-Carboniferous systems (in the English sense) was thus combined with a continuing adherence to Élie de Beaumont's high-level theory of occasional *révolutions*.[38] As Élie de Beaumont had told Murchison and De la Beche early in the Devonian controversy, the three divisions of the *terrain de transition* were indeed separated by major unconformities in Brittany (§5.5). But although the French geologists followed Dumont in describing the Belgian sequence as completely conformable, from the *terrain houiller* down into the ancient *terrain ardoisier*, they still tacitly took Brittany as their model. So their acceptance of the Devonian and Old Red Sandstone as a major *terrain* between the Carboniferous and the Silurian was *not* combined with Murchison's insistence that the boundaries of the Devonian system were often completely conformable and its fossils of an essentially transitional character. Nonetheless, the published map of France and its *Explication* did finally eliminate the anthracites of Brittany from being in ancient Greywacke, as De la Beche had repeatedly claimed in the earlier phases of the Devonian controversy. They were indeed in the *terrain de transition supérieur*, but that was now identified as Devonian.[39] Murchison

---

37. Dufrénoy and Élie de Beaumont (1841a, b), presented to GSL on 15 December (*PGS* 3:588–89), and to ASP on 20 December (*CRAS* 13:1131–34).

38. Dufrénoy and Élie de Beaumont (1841b); the "tableau général des formations" is on pp. 58–60.

39. Ibid., pp. 221–32. Brittany (pp. 176–239) was by Dufrénoy, the Ardennes (pp. 240–64) by Élie de Beaumont, and the long chapter on the *terrain houiller* (pp. 499–786) was by both authors. Élie de Beaumont still equated the Culm of Devon with his *terrain de transition supérieur* or Devonian, *not* with the *terrain houiller* or Carboniferous: Élie de Beaumont to De la Beche, 15 February 1841 (NMW: DLB).

could therefore take comfort that the last of the alleged instances of Greywacke coal had been toppled from its position as a counter example to his characterization of the Silurian as a period that had predated any land plants whatever. The mopping up of the Devonian controversy was almost complete.

By this time, Murchison must have been expressing in public his displeasure at Phillips's challenge to his own scheme of systems for the older rocks, for De la Beche warned Phillips about the "coming storm." Undeterred, Phillips set about tackling Murchison's collaborator instead. "I think it is very evident that we must endeavour to unsystematize our minds again," he told Sedgwick, punning on the word system; he considered it would be better "to keep for some time, *in doubt,* the exact *lines* of demarcation which best suit the soft *shades* of mother nature." Phillips praised Murchison's Silurian for its "noble assemblage" of fossils; but Murchison would not have been pleased to hear from Sedgwick that the younger geologist still wanted "the whole basis & superstructure" of his systems "to undergo a new & searching analysis."[40] Although Murchison had rejected the French geologists' "séparations tranchées" and emphasized the transitional character of the Devonian fauna, he had now been effectively upstaged by Phillips's more radically Lyellian emphasis on "the soft *shades* of mother nature," which made *any* boundaries arbitrary and problematical.

At the anniversary meeting of the Geological Society, Murchison announced the award of the Wollaston medal to von Buch, and De la Beche as foreign secretary received it on his behalf. As expected, Murchison's address gave pride of place to "Palaeozoic Geology," and he used the occasion to insist on the value of his own nomenclature. He based it firmly on regarding fossils as "the language of nature," a usage that was buttressed with the customary invocation of "our distinguished leader William Smith." Murchison therefore rejected Sedgwick's Cambrian on the grounds that it was synonymous with his own Lower Silurian, since the Cambrian no longer had any distinctive fossils of its own. The validity of the Silurian was emphasized as usual by reference to the wide geographical range of its fossils; he told Sedgwick afterward that he regarded the Lower Silurian fauna as the oldest of all, thus depriving his erstwhile collaborator of *any* role in plotting the origins of life on the earth. With Lyell skimming the cream off North America, Murchison had evidently abandoned his own campaigning ambitions in that direction and was turning his gaze eastward again, for he told his audience, "I despair not of planting the Silurian standard on the wall of China." As for the Devonian, Murchison recounted what was now the established myth of its genesis in such a way that it too supported his emphasis on "the language of nature". It was, he claimed, "merely by seeing the letters of the alphabet spread out before him in a cabinet" that Lonsdale had solved the Devonian problem.[41]

The anticipated storm burst, however, when Murchison dealt with Phillips's report on the fossils of Devonshire. He attacked Phillips's proposed classification of the Palaeozoic into Lower, Middle, and Upper divisions (§14.1)

40. Phillips to De la Beche, 14 December 1841; Phillips to Sedgwick, 24 January 1842 (ULC: AS, ID, 120a).

41. Murchison (1842, pp. 640–46), read at GSL on 18 February (the award to von Buch is on pp. 633–35); Murchison to Sedgwick, 25 February 1842 (transcript in ULC: AS, IIID, 36). Von Buch fully endorsed his Russian research: von Buch to Murchison, 20 February 1842 (GSL: RIM, B33/7).

because it "suppressed" and "submerged" his own systems. "The perpetuity of a name affixed to any group of rocks through his original research," he insisted, "is the highest distinction to which any working geologist can aspire." Leaving his audience in no doubt about which geologist he had in mind, he added, "it is in truth his monument." He criticized Phillips for extending the "Palaeozoic" upward to include the formations he himself had just named Permian, on the grounds that the existence of the earliest fossil reptiles ("Saurians") in those strata linked them clearly with still younger formations, rather than with older ones. Likewise, he pointed out, at the very top of the Silurian "a new era is announced by the presence of the earliest Vertebrata," in the form of the Old Red Sandstone fish. With that kind of criterion for defining major boundaries, Muchison's strictures on any purely fossil-based classification were cogent enough: "all nomenclature founded *solely* upon our present knowledge of the distribution of animal and vegetable life," he claimed, "must be liable to change with every new important discovery." That disadvantage did not apply with equal force, however, to Phillips's "statistical" method of defining major categories by the *overall* composition of their faunas and floras; but Murchison showed no sign of having grasped what Phillips had proposed. Instead, he insisted on the priority of his own topographically defined systems, and he gave De la Beche an unmistakable hint that he expected the Geological Survey to adopt those systems as a matter of official policy on all its future maps and memoirs. Murchison was determined not to be robbed of his monument.[42]

14.4   Murchison's Apotheosis (December 1841 to November 1842)

One of the cornerstones of Murchison's monument, namely, his joint paper with Sedgwick on the sequences in the Rhineland and Belgium, had been read at the Geological Society in May 1840 (§13.3), and Phillips had subsequently recommended it for full publication in the *Transactions*. But it was only in the spring of 1842 that it was at last printed; although it was not published until late in that year, it was available beforehand to all the leading geologists in London. Embellished with fine colored sections, and a detailed map derived from von Dechen and other German sources, the paper at last put in the public realm the full evidence for the authors' Devonian interpretation of the Continental sequences. What the map and sections showed, more clearly than the paper that had been read at the Society, was that the strata assigned to the Devonian were now restricted to those directly underlying the Carboniferous of Westphalia and Belgium, together with the limestone patches of the Eifel and Nassau. Beneath those rocks was a vast mass of Greywacke, *all* of which was marked on the map as Silurian. The Cambrian had thus been completely eliminated (except for the lower part of the *terrain ardoisier* in Belgium), and the Devonian had after all become little more than a "top dressing" to the Silurian. The Devonian was thus firmly sandwiched between Carboniferous and Silurian, but at the cost of greatly reducing its

42. Murchison (1842, pp. 646–49).

relative importance in the sequence, compared with Murchison's emphemeral first interpretation (§12.1).[43]

About the importance of the fossil evidence, on the other hand, there could be no doubt. Appended to the paper was a detailed description of many of the most significant Devonian fossils from the region, with fine lithographed illustrations. When this appendix reached the Geological Society from Paris, Phillips at once recognized "the Palaeozoic species of Devon" among them, and he told De la Beche that the list of species would be very useful "if analysed in my way." In fact, de Verneuil and d'Archiac did analyze the fossils, if not in quantitative or "statistical" terms, then at least in terms of a sequence of faunas that had changed gradually in composition. Presenting a set of the illustrations to the Société géologique in advance of their publication in London, they summarized the faunas of "les trois systèmes paléozoiques," claiming to see a gradual overall increase in the number and diversity of the species as the faunas were traced upward, or forward through the history of the earth. Together with the text submitted to the Geological Society, their comments proved that de Verneuil had now definitely abandoned his earlier insistence on "séparations tranchées" between the successive systems, and that he had replaced it with Murchison's view of gradual faunal transitions, if not quite with Phillips's more radically Lyellian interpretation of faunal change.[44]

Only two days after de Verneuil and d'Archiac commented on the Devonian fossils in Paris, Murchison had a further opportunity to insist in London on the validity of his systems, when his and de Verneuil's joint paper on their second Russian expedition was read at the Geological Society. This merely amplified what he had published in the *Philosophical Magazine* after his return to England (§14.2); but he was able to emphasize again how their traverse of the Baltic provinces had confirmed the validity of the "old red or Devonian series," by proving the close association of Old Red Sandstone fish with Devonian shells. A month later, the reading of their paper on the Urals— this time with Keyserling added as a third author—showed the Devonian in routine use as a system that could be recognized by its characteristic fossils and by its position between Silurian and Carboniferous, despite strong folding and even some metamorphic alteration. Summarizing the whole of their Russian research, Murchison pointed to the wider significance of the vast spreads of almost horizontal strata west of the Urals. Forestalling any rival claim that Lyell or the Americans might make, he maintained that the Russian geology was "the best proof which has yet been obtained in any part of the world of *the same extent*" to show the character of the processes of organic change. The Russian strata proved not only that "distinct forms of animal life were successively created and entombed in each succeeding deposit," but also that "the successive obliteration of these classes" had not been caused by volcanic eruptions or by any "great physical disturbances of the strata."[45] Murchison

43. Sedgwick and Murchison (1842), published in *TGS* on 1 November. The key to the map (pl. 24) allotted the same color but different numbers to the Silurian and the Cambrian, but on the map itself *no* number denoting the Cambrian was printed anywhere!

44. D'Archiac and de Verneuil (1842a), read at SGF on 4 April; d'Archiac and de Verneuil (1842b), received at GSL on 15 December 1841 and published in *TGS* on 1 November 1842; Phillips to De la Beche, 14 December 1841.

45. Murchison and de Verneuil (1842), read at GSL on 6 and 20 April; Murchison, de Verneuil, and von Keyserling (1842, p. 752), read at GSL on 18 May.

and de Verneuil thus insisted jointly that Russia disproved Élie de Beaumont's theory of occasional physical *révolutions* as the cause of extinction.

In the context of such grand conclusions, the residual problems of Devonshire now seemed increasingly marginal, if not irrelevant. Williams had continued his lone fight against the tide of opinion; another long diatribe against Sedgwick and Murchison, full of "plausible reasons and positive proofs" against the Devonian interpretation, had appeared in the *Philosophical Magazine* not long before the 1842 anniversary meeting. But neither Murchison nor any other leading geologist paid it any attention. In fact this was just what Williams himself had predicted: he anticipated that his many years of "patient and cautious and rigid investigation" would "go for nothing," and that his conclusions would be considered "not worth refuting." Among his many objections to the Devonian interpretation were some that his opponents would have found awkwardly cogent, if they had bothered to read the article. In particular, Williams pointed out that no plausible explanation had yet been produced to account for the completed disparity between the Old Red Sandstone and the alleged "Devonian" strata within England itself—even between the Mendip Hills behind his rectory and the Quantock Hills he could see in the distance—whereas the Old Red Sandstone itself had now been traced by Murchison with almost unaltered character all the way to central Russia. But Williams forfeited any credibility he might have gained from such arguments, by his continued insistence that all the Greywacke of south Devon and Cornwall was younger than the Culm (§12.7). He assured Phillips, at least half seriously, that he would prove the point "by swearing it before a Justice and sending my friends the affidavit." But that unorthodox mode of scientific persuasion did not alter the opinion of all the other geologists familiar with the region, that Williams had made a gross mistake in the structural interpretation of an area with poor and scattered exposures, and that he had got the sequence radically wrong. Williams did not give up completely, and he continued to argue in print for his view of the "true position" of the Devon rocks. But he made no further attempt to present his work at the Geological Society or to convince the gentlemanly geologists who dominated it.[46]

The other provincial geologist who had been equally active earlier in the controversy was preparing his swan song on Devonshire at about the same time. After reading many papers at the Geological Society over the past years, and having them all rejected by referees as too insubstantial or incomplete to warrant full publication, Austen had at last been encouraged to consolidate them all into a single regional memoir on his former home area of southeast Devon. His paper, which was scheduled for the same issue of the *Transactions* as the papers on the Rhineland, was worthy but dull. Austen had perhaps lost interest in Devon geology, and the paper contained few traces of his earlier imaginative theorizing about the original conditions in which the rocks had been deposited and in which the fossil organisms had lived. He rejected the term "Devonian" once more, because he felt it was still needed as a topographical term; but his approval of Lonsdale's inference on the transitional

46. Williams (1842a, b), published in February and July issues of *PMJS*; Williams to Phillips, 24 April 1842 (UMO: JP, 1842/9); Williams (1844, 1846).

character of the south Devon fossils shows that Austen too had now adopted the Devonian interpretation in all but name.[47]

The other geologist still active in Devonshire, though no mere provincial, was Phillips. Even after his report was published, he had continued to work on fossils sent to him by Devon collectors such as Harding. He also continued to stress the importance of Devonshire itself, and therefore of his own and De la Beche's work there, in comparison to the more exotic foreign regions. "We shall learn more about *Devonianism* by North Devon examinations than by any other researches, I feel sure," he told De la Beche. Certainly they were both more aware than Murchison of the theoretical problems of correlation, and they discussed between themselves the "laws of possible variation in mineral deposits of a given geological age." That question of local variation in the physical environment and in the corresponding forms of life was quite general, but it was not for nothing that Phillips alluded in this context to the "Cambro-Siluro-Devono-&c.ian problem." For in all those three systems, but most acutely in the Devonian and Old Red Sandstone, the problem was to account for local contrasts in rock types and fossils in circumstances that suggested that the variations were more a function of original environments than of relative ages. As for Murchison's attack on Phillips over the subdivision of the Palaeozoic, its impact was muted by Phillips's absence from the anniversary meeting, and it was not until much later in the year that he read the address in its printed form. He then told his Survey colleague that he had written to Murchison, protesting "against the 'judgement' of the Pres. of the Geol. Soc. *in Re Palaeozoica*." But Murchison must have been conciliatory, for Phillips did not in fact appeal to the public against the president's verdict.[48]

In the summer of 1842 "the twelfth assemblage of the Longheads," as one reporter called the British Association meeting, was held in Manchester and was the occasion for a further public airing of the residual issues in the Devonian controversy. Murchison added the presidency of the geological Section to his similar position in the Geological Society, and most of the other leading British geologists (except Lyell) were also present. Several American geologists, and Keyserling from Russia, gave the sessions an international flavor that compensated Murchison a little for the failure of his earlier proposal that such national meetings should have been replaced this year by a pan-European scientific assembly in Frankfurt.[49]

On the second day of ordinary sessions, Murchison summarized a paper on the geology of the United States beyond the Appalachians, which had recently been sent to the Geological Society; and this provided a convenient

47. Austen (1842, pp. 465–66), published in *TGS* on 1 November. His most recent paper (§11.5) had been approved for printing after more than two years' delay, but no further action was apparently taken until even later: Council of GSL, 30 June 1841 (GSL, BP, CM 1/5, p. 314); Austen to Phillips, 27 February 1842 (UMO: JP, 1842/4). No leading member of GSL would have experienced, or tolerated, such treatment.

48. Harding to Phillips, 28 August, 19 September, and 27 October 1841 (UMO: JP, 1841/44, 49.1, 57.1); Phillips to De la Beche, 4 February, 23 May, 21 August, and 2 September 1842 (NMW: DLB). Phillips continued to regard north Devon as "an area of greater interest to philosophical geology than almost any other": Phillips to De la Beche, 21 December 1842 (NMW: DLB).

49. *LG*, no. 1327 (25 June 1841), p. 438; *Ath.*, no. 766 (2 July 1842), p. 591. On the plan for a meeting in Frankfurt, see Morrell and Thackray (1981, pp. 383–84).

peg for the leading British geologists to comment on the problems of inter-continental correlation among the older rocks. Phillips firmly maintained the leading position he had won for himself in Plymouth the previous year in Murchison's absence, and he opened the discussion with his customary note of caution about the simplistic use of fossils, "the evidence being not of *time*, but of *circumstance*." Sedgwick agreed with that cautious approach, but he urged that a tentative extrapolation could be made from the well-known European sequences and commented on the "remarkable analogy" across the Atlantic. De la Beche then came out at last with an unambiguous acceptance of the Devonian interpretation: he expressed confidence that at least the larger groups of formations would be found to be valid throughout the world, and he backed the provisional adoption of the terms Carboniferous, Devonian, and Silurian. His reward after the meeting was a belated personal gift of a copy of *The Silurian System:* after nearly eight years of almost unremitting hostility, Mr Murchison had made his peace with the newly knighted Sir Henry.[50]

Later in the same session, a progress report from Agassiz on his research on Old Red Sandstone fish, although read in his absence, brought the Devonian even more firmly to the center of the stage. Summarizing some eight years' work on many fine collections, Agassiz reported that all the species and most of the genera of the fish were known only from the Old Red Sandstone itself. He claimed that the remarkable diversity and elaboration of the fish in this, almost their earliest known occurrence, was a "great argument against the theory of the successive transformation of species and the descent of all living organized beings from a small number of primitive forms" (see fig. 13.1). He "startled his compeers," however, with the further claim that each such formation had a completely distinct fauna, sharply divided from those older and younger, bearing witness to frequent episodes of sudden extinction and new creation of species. Opening the discussion, Murchison agreed that Agassiz's results could be regarded as "disproving the transmutation of species, in passing from one stratum to another." But Agassiz's other theoretical claim was, as one reporter put it, "at once repudiated by many of the most eminent geologists in Manchester." Murchison, Sedgwick, De la Beche, and Phillips all spoke up in turn, insisting that many species of fossils had much longer ranges in time than Agassiz allowed.[51] There might be no sympathy for any Lamarckian fantasy of the transmutation of species, but at least among leading British geologists there was now a virtual consensus that faunal change had taken place in a piecemeal fashion. Excepting only Lyell's own research on the Tertiary faunas, that Lyellian concept had become persuasive and widespread chiefly as a result of the Devonian interpretation. For the Devonian had broken down the almost complete contrast between what had previously seemed to be two distinct major faunas, the Silurian and the Carboniferous, and had replaced it with a nuanced transition that was now taken as a model for the whole fossil record.

50. BAAS, session on 24 June 1842: *Ath.*, no. 767 (9 July), p. 615; the paper (see Murchison 1843a) was by David Dale Owen (b. 1807), the state geologist in Indiana and a son of the social reformer Robert Owen. De la Beche to Murchison, 31 July 1842 (ULE: RIM, Gen. 1425/389). De la Beche's knighthood, an official recognition of his work in building up the Geological Survey, gave him a title before Murchison, who had hoped to get a baronetcy on the strength of *The Silurian System* (§11.5).

51. *Ath.*, no. 767, pp. 615–16; *LG*, no. 1329 (9 July 1842), p. 481; Agassiz (1843).

Despite the rejection of Agassiz's theoretical claims about catastrophic faunal change, however, there was no doubt about the importance of what his research revealed about the early history of the vertebrates, the highest group of animal life. A letter he sent to Manchester the day after his paper was read probably reached Murchison just too late for him to report its contents to the meeting. One of his immediate purposes was to report on some fossil fish collected by Pander, whom Murchison had visited on his traverse of the Baltic provinces the previous summer (§14.2). But his broader conclusions would have come as no surprise: Agassiz claimed that his study of the fossil fish "à l'origine des choses" would throw light on "les circonstances dans lesquelles les êtres organisés ont été creés." The broader theoretical claim was not new; but Murchison's establishment of the position of the Old Red Sandstone in both Russia and Britain, and its correlation with the more ordinary fossils of the Devonian, had given its remarkable fish quite new significance.[52]

The day after Agassiz's report was read, a paper by Griffith on the Carboniferous fossils of Ireland showed how Phillips's application of the Lyellian "percentage system" to the Devonian faunas was already being adopted by others. Griffith used a percentage analysis to argue that the fauna of the lower part of the Mountain Limestone in Ireland was unlike that in England, and should be correlated instead with some of the upper Devonian strata in north Devon. In the discussion, Sedgwick conceded this point, not least in the light of his own fieldwork in Ireland, but evidently he did not regard this as upsetting the validity of the Devonian itself. De la Beche, in a clear expression of support for his colleague Phillips, urged that "the gradual change of organic life, from the lowest deposits upwards, could alone indicate with certainty the exact age of any deposit." Phillips himself was skeptical about the contrast Griffith had asserted between the English and Irish strata, but he agreed that the north Devon strata were likely to be some of the youngest of that epoch. All in all, the discussion showed how the use of the Devonian in stratigraphical geology had now settled down onto a level of local detail.[53] The Devonian might be gaining more significance than ever for the understanding of the history of life, particularly as the record of the earliest period at which vertebrate animals and land plants had flourished; but as a category in the description of strata and their fossils its use was becoming routine.

It was therefore appropriate that Murchison, as president of the geological Section, should have been invited to give the evening lecture on the final day of the meeting, on the spectacular achievements of geology in general and on his own Russian research in particular (fig. 2.4). The Manchester meeting thus reached its climax with the apotheosis of Murchison as the hero of Palaeozoic geology.[54]

---

52. Agassiz to Murchison, 25 June 1842 (HLH: LA, bMS. Am. 1419, 51); Murchison to Agassiz, 21 May and 13 June 1842 (ibid., 501–2). Agassiz had already developed his theoretical position more fully in a lecture in Neuchâtel in 1841, which was published in English translation after the Manchester meeting, in the October 1842 issue of *ENPJ* (Agassiz 1842). Miller (1841) had already used the Old Red Sandstone fish on a more popular level in a similar argument against transmutation.

53. BAAS, session on 25 June 1842: *Ath.*, no. 768 (16 July), pp. 640–41; *LG*, no. 1331 (23 July), pp. 511–12; Griffith (1843).

54. BAAS, evening lecture on 28 June 1842: Murchison (1843b).

14.5   Completing the Consensus (1842 to 1847)

The Devonian controversy had no termination as sharp or striking as its first eruption. Not only did the controversy subside gradually into a consensus, but the Devonian itself passed slowly into routine use among geologists almost everywhere.[55] In the summer of 1843 Sedgwick conceded formally at the Geological Society what he had mentioned informally at the Manchester meeting the previous year, namely, that the Devonian deserved to be treated as one of the "primary divisions" of the Palaeozoic rocks, having a fauna that shaded into those of the Carboniferous above and the Silurian below. In a paper concerned principally with the much older rocks of north Wales, Sedgwick mentioned that it was the Russian and American sequences that had brought him at last to that full acceptance of the Devonian interpretation, of which he had originally been a reluctant co-author. But the triumph of the Devonian had spelled the demise of his Cambrian, at least as a distinctive period in the history of life on earth. He could only salvage his project by annexing Murchison's Lower Silurian to his former Cambrian, as he had already proposed in private, and by terming that oldest division the "great protozoic group." His subsequent argument with his erstwhile collaborator over the definition of the Cambrian in relation to the Silurian became more acrimonious and emotional with the passage of time; but unlike the Devonian controversy, it never engaged the attention of any large or diverse group. It remained largely confined to the two chief protagonists and their immediate supporters and clients.[56] It ceased to have even an indirect impact on the routine use of the Devonian system and period.

The British Association meeting at Cork later that summer showed, even more clearly than the Manchester meeting the previous year, that the use of the Devonian was now routine among British and Irish geologists. Griffith presented a major paper on "the old red or Devonian system" in relation to the Carboniferous of Ireland, but the discussion merely showed that any disagreement was now confined to correlations of relatively local significance. Phillips did however use the occasion to express his conviction that the Devonian strata of Devonshire, the Rhineland, and Russia contained fossils "sufficiently characteristic in themselves to constitute a distinct system." In a later session, when the amateur collector Charles Peach (b.1800) reported fragments of fossil fish from the older rocks of Cornwall, Murchison commented that the discovery confirmed his and Sedgwick's original Devonian interpretation "in a very remarkable manner"; and Phillips tacitly agreed, since he considered the fossils were of Old Red Sandstone forms. At least a little direct fossil evidence for the equivalence of the Old Red Sandstone with the more normal Devonian had now been found in southwest England

---

55. Perhaps the most striking sign of the virtual end of the controversy is the rapid decline in the frequency and content of references to the Devonian problem in the *correspondence* of the participants.

56. Sedgwick (1843), read at GSL on 21 June. For the further development of the argument between Murchison and Sedgwick, see Secord (1986).

itself. The Devonian interpretation in England was in that respect no longer dependent on analogy with Murchison's much clearer case in Russia.[57]

Murchison consolidated the use of his systems among British geologists by incorporating them into a useful small-scale geological map of Britain. Competing in the same market as Phillips's earlier map (§10.3), Murchison's was published by the Society for the Diffusion of Useful Knowledge just in time for its author to display it at the meeting in Cork. Based largely on the revised edition of Greenough's great map (§13.1), but with unambiguous identification of the older systems, Murchison's map used the same color and symbol for the Old Red Sandstone of the Welsh Borderland and the older rocks of north and south Devon, and both were firmly labeled "Devonian." As on Greenough's map, the Culm limestones were correlated with the Mountain Limestone and the main body of the Culm with the Coal Measures. In Wales, the Cambrian was eliminated altogether; but in the key the "Palaeozoic" was adopted in Phillips's sense, stretching from the "Primary" through Silurian, Devonian, and Carboniferous as far as the new "Permian" at the top. Phillips later adopted virtually the same classification on a new edition of his own small-scale map, so the controversy on that score was over.[58]

By far the most important new research on Devonian rocks and fossils was that of German geologists working in the Rhineland. In particular, Ferdinand Römer (b. 1818), who had studied palaeontology in Berlin, undertook a major new study of the Rhineland sequences and their fossils, enlarging on Beyrich's earlier work. In his memoir on *Das Rheinische Übergangsgebirge* (1844), Römer argued from detailed fossil evidence that almost *all* the strata below the unquestioned Carboniferous of Westphalia belonged to the Devonian, as indeed they had in Murchison's first and ephemeral interpretation (§12.1). The Silurian was thus virtually eliminated from the Rhineland, but in compensation the Devonian was enlarged from what had become little more than a "top dressing" to the Silurian into what was clearly a major system in its own right. Within this greatly expanded Devonian sequence Römer distinguished several successive faunas, most of which could be matched with those Phillips had described from various parts of Devonshire itself (§14.1). Römer's memoir initiated a phase of intensive research in which the Rhineland, together with Belgium, became de facto the "type" area for the Devonian system. For a time, the term *Rheinisch* competed with *Devonisch* among German-speaking geologists; but the English name was soon adopted, since it had historical precedence and was already in general use. Nevertheless, it was in the Rhineland and Belgium that the full sequence of Devonian faunas was worked out thoroughly and the subdivisions of the Devonian system gradually defined.[59]

57. BAAS, sessions on 17 and 21 August 1843: *Ath.*, no. 827 (2 September), pp. 800–801; no. 828 (9 September), p. 827; Griffith (1844); Peach (1844). As at Manchester, Williams was among those present but his contribution was marginal.

58. BAAS, session on 22 August 1843: *Ath.*, no. 829 (16 September) p. 849; Murchison (1843c). Phillips's map (1846) is the so-called seventh edition of Douglas and Edmonds (1950); the Devonian was not shown on the key as a system as thick as the Silurian and Carboniferous, probably because to do so would have entailed a complete reengraving.

59. Römer (1844); the Belgian sequence was revised by Dumont (1848). Another important young recruit to Rhineland geology was Guido Sandberger (b. 1821), the elder son of the Weilburg amateur whose collection Murchison had found valuable while he was in Nassau (§12.1, §12.2):

Meanwhile Agassiz had continued his exploration of what he termed "un monde nouveau," the world of the earliest fish. Although many geologists were dismayed at his preoccupation with the intense controversy over his glacial theory, Agassiz assured Murchison late in 1842 that he had not abandoned his palaeontological research and announced that he was writing a separate monograph on the Old Red Sandstone fish. He confirmed the identity of many of the species in Russia and Britain, but also stressed the differences in the faunas, which he interpreted as the result of deposition in Devonian seas with little direct connection. As the work advanced, he stressed the theoretical importance of his evidence that the organic diversity among the fish had been as great in the immeasurably remote Devonian period as in the modern Mediterranean. For he claimed that this refuted the transmutationists' argument that the earliest representatives of any major group would have been few in number and undifferentiated in structure. Agassiz continued to be isolated in his idiosyncratic claim that each formation contained an independently created set of species; but his more general argument against transmutation was widely shared, and his masterly monograph on the fish of the Old Red Sandstone "ou Système Dévonien" (1844) was taken as authoritative evidence in that increasingly acrimonious argument.[60]

This is not the place to trace the course of the controversy over transmutation or, as it later came to be termed, organic evolution. The outcome of the Devonian controversy, particularly in the recognition of the earliest records of vertebrates and land plants and in the support it gave to the idea of slow and piecemeal faunal change, provided important ammunition for all sides in the evolutionary debate. But this was essentially a matter of *using* for wider purposes what was already consensually accepted as reliable knowledge. Indeed it was precisely that consensus that gave the evidence its perceived authority. Rather than attempting to trace the widening use of the Devonian, it is therefore more appropriate at this point to conclude the narrative account of how that consensus was formed.[61]

While Keyserling and de Verneuil were in England in 1842, they and Murchison had agreed to take up the hint they had been given the previous year in St Petersburg, that the imperial government would subsidize the full publication of their Russian research. Their papers to the Geological Society were therefore withdrawn, and before the end of the year Murchison had sent

---

Sandberger (1842). Bronn's *Index palaeontologicus* (1848–49) is a good example of a general work in German that used "Devonisch" in a routine manner as a major category. The later subdivision of the Devonian in the Rhineland and Belgium is summarized in Ziegler (1979).

60. Agassiz to Murchison, 16 and 18 November 1842 (HLH: LA, bMS. Am. 1419, 54 and 55). Agassiz (1844), preface dated August; the introduction (pp. ix–xxxvi) merely amplified what he had already published in English translation in *RBAAS* (1843), but the superb colored illustrations gave an exceptionally vivid impression of these large and spectacular fossils.

61. Chambers's anonymous *Vestiges* (1844), the most popular essay on transmutation, was rebutted in a scathing review by Sedgwick (1845) and later in Miller's similarly popular *Footprints* (1847). For an account of these polemics, see Gillispie (1951, chap. 6). Hodge (1972) analyzes the contrast between Chambers's theory and Darwin's; the latter was almost exactly contemporary but remained private until much later (Ospovat 1981). Chambers did not use the term "Devonian," but made great use of the "Era of the Old Red Sandstone" in his naively transmutationist interpretation of the fossil record; he suggested, for example (pp. 68–69), that the earliest fish (e.g., the Old Red Sandstone *Cephalaspis*) had been derived from "crustacean" ancestors (e.g., the Silurian trilobite *Asaphus*)! On the later debate among men of science about the progressionist interpretation of the fossil record, see Bowler (1976).

a first chapter to the printer in London. The fossils had already been farmed out to specialists, as they had been for *The Silurian System*. But Murchison was not content to complete the book without a season of field work in central Europe in 1843 and another in Scandinavia in 1844. The latter provided him with what he regarded as conclusive proof that the Silurian fauna was the earliest of all in the fossil record—thus finally eliminating Sedgwick's Cambrian. At the same time, Keyserling was extending their earlier survey to remote areas of northeast Russia. Only in the summer of 1845 were advance copies of the first, strictly geological volume of *The Geology of Russia* ready for Murchison to take to St Petersburg to present to the Czar and to members of his government. After that there were further revisions and amendments before the work at last went on sale at the beginning of 1846.[62]

As with *The Silurian System* just six years earlier, Murchison's geological *magnum opus* on Russia contained no surprises by the time of its belated publication. On the other hand, the descriptions and illustrations of the fossils, in the separate palaeontological volume edited in Paris by de Verneuil, were indispensable for further research on the Palaeozoic strata anywhere in the world. In the geological volume, Murchison inverted the conventional order of description that he, like most other geologists, had used for his earlier book. For "having learnt to decipher the very first letters in the long records of animal life," Murchison considered that geologists could now "assume a more distinct position as historians" and describe the successive systems in their true chronological order. With the unforeseen resolution of the Devonian controversy, which had finally linked the ancient Silurian to the better-known Carboniferous and still younger strata, he claimed that "the hard and indelible register, as preserved for our inspection in the great book of ancient Nature, is at length interpreted and read off with clearness and precision."[63]

It was the Russian evidence of Old Red Sandstone fish in close association with Devonshire shells, Murchison recalled, that had "entirely dispelled any doubts" about the Devonian interpretation (§13.4). Like all other geologists who had engaged in the controversy, he remained convinced that the fish, just as much as the shelly organisms and corals, had been marine organisms, being in Russia "cohabitants of many parts of the same great sea." The striking contrast between the two Devonian faunas was related to the frequent difference in the sediments enclosing the fossils: Murchison argued that, as at the present day, the fish had lived on sandy bottoms and the shelly organisms and corals in muddy or limey conditions. Despite the great diversity of the sediments, the "Devonian types" of fossils were, he maintained, essentially uniform; and he concluded, doubtless with the full agreement of de Verneuil, that "we have no hesitation in adhering to the word 'Devonian,' and in urging geologists to follow our example." The Devonian controversy, they implied, could now be regarded as closed.[64]

That conclusion was no longer either presumptuous or premature. When at the end of 1845 Élie de Beaumont chose stratigraphical geology as the

---

62. Murchison, de Verneuil, and Keyserling (1845a, b). The complex publishing history is summarized in Thackray (1978a); there were German and Russian translations of the first (geological) volume by 1848.

63. Murchison, de Verneuil, and Keyserling (1845a, introduction, p. 9*).

64. Ibid., p. 40; chap. 4, "Devonian or Old Red System," pp. 63–68.

theme for his next course at the Collège de France, he acknowledged the outstanding interest of the older strata by allocating no fewer than three-quarters of his lectures to the "paléozoique." Alongside the conventional French term *terrain,* he now used the term *système* in its newer, English, sense as a virtual synonym. Murchison sent him a copy of the first volume of *Geology of Russia* after the course had begun; shortly afterward, Élie de Beaumont outlined the whole stratigraphical record, from the "terrains tertiaires" back to the "système cambrien," and accepted Murchison's and de Verneuil's conclusion that "le vieux grès rouge et le système dévonien sont parfaitement le prolongement l'un à l'autre." He also treated the Devonian as a set of formations of comparable importance to the other Palaeozoic *terrains* and as a comparable period of earth history. But most important of all, Élie de Beaumont now accepted the essential continuity of the systems or *terrains.* In some regions they might be separated by unconformities that were causally related to distinct *systèmes* (in the French sense) of folding and elevation; but the Russian research had proved that such "séparations tranchées" were not universal. Likewise, Élie de Beaumont now joined de Verneuil in a much more gradualistic view of faunal change, citing as a particular example the faunal transitions from Silurian into Devonian and from Devonian into Carboniferous, and urging the importance of plotting the ranges of individual species even *within* a single system such as the Devonian. With the full adherence of the influential Élie de Beaumont, the Devonian interpretation was now as well established in France as it was in Britain, Germany, and Russia.[65]

At about the same time, De la Beche wrote a massive book-length review of the Geological Survey's work in south Wales and southwest England. He stressed throughout the importance of interpreting local sequences of strata and fossils in terms of their original environments. Within that characteristic framework, he used the "Old Red Sandstone and Devonian Rocks" as the finest exemplar of the temporal equivalence of formations and fossils of strikingly contrasted character. De la Beche's acceptance of the Devonian interpretation, which he had indicated informally at the British Association meeting in 1842 (§14.4), was thus put unambiguously on record in his own inaugural contribution to the Survey's new series of *Memoirs.*[66]

Murchison himself had failed to reach "the walls of China"; but even before the publication of *The Geology of Russia,* Devonian rocks had been reported from as far away as the Altai. To establish the Devonian in the other direction, beyond the Atlantic, was all that remained. In the introduction to his part of *The Geology of Russia,* Murchison—doubtless with de Verneuil's approval—had mentioned how the Silurian had long been recognized in New York State, and that "a true Devonian system" below the Carboniferous of Pennsylvania had been confirmed by Lyell's visit (§14.2). In fact, Conrad had begun to use the Devonian routinely in his interpretation of the New York

---

65. Élie de Beaumont, lectures at Collège de France, 20 December 1845 to 27 June 1846 (ASP: LEB, box 11). The initial review of the Devonian was in lecture 13 (31 January 1846); from lecture 35 (16 May 1846) onward, the course was largely a summary of Murchison's Russian work. The general remarks on faunal change were in the concluding lecture 46. Although converted to a gradualistic view of faunal change, Élie de Beaumont continued to champion a paroxysmal theory of mountain building: Greene (1982, chap. 3).

66. De la Beche (1846, pp. 50–105).

sequence as early as 1842, but the "New York system" of American geologists embodied, according to Murchison and de Verneuil, equivalents of both the Silurian and the Devonian. De Verneuil therefore visited the United States to study the American sequences at firsthand. He returned from his trip full of praise for the detailed work of the American geologists, but critical of their attempts to correlate their sequences with Europe. His modestly titled "Note sur le parallélisme des roches des dépôts paléozoiques,"read at the Société géologique in the spring of 1847, was in fact a major review of the North American sequences and their fossils. De Verneuil correlated the various formations with the European sequences on the basis of a far more thorough comparison of the fossils than had previously been possible. In particular, he attributed a much more substantial part of the sequence in New York State to the Devonian rather than the Silurian. Since the strata there were undisturbed and the fossils abundant and well preserved, New York became as important for the detailed understanding of the Devonian in America as Belgium and the Rhineland were in Europe.[67]

With de Verneuil's authoritative correlations of Devonian strata, by their fossils, all the way from the Urals to the United States, this narrative of the Devonian controversy can conveniently—if arbitrarily—be brought to a close. The Devonian system, and the corresponding period in the history of the earth and of life, had now been *naturalized,* in both senses of the word. By the later 1840s the term Devonian had virtually lost its originally English connotations. It had been naturalized in the sense of being adopted by geologists in every scientific nation for the description of their own strata and fossils, and Devonshire was no longer even regarded as the "type" area for establishing its meaning in case of doubt. But furthermore, it was becoming so deeply embedded in routine geological practice that the complex story of its slow, hesitant, tortuous, and above all human construction was being forgotten. The Devonian had now been naturalized, in the sense that it was fast becoming an unproblematic fact of *nature.*

67. Tchihatcheff (1845); Murchison, de Verneuil, and Keyserling (1845a, 1:4*); Conrad (1842); de Verneuil (1847), read at SGF on 19 April. De Verneuil assigned to the Devonian the upper part of the American geologists' Helderberg group, as well as the whole of the Erie group and the overlying Catskill or Old Red Sandstone.

Part Three

THE ACTION
ANALYZED

Chapter Fifteen

THE SHAPE
OF THE
CONTROVERSY

15.1 A First Approximation

The Devonian controversy has now been traced in detail, in terms that the participants themselves could have recognized and understood, all the way from its background in the early 1830s, through its first eruption into the public realm late in 1834, to the subsiding of the argument into a virtual consensus in the mid-1840s. The controversy has been followed from its origins in a disagreement ostensibly confined to the interpretation of a few fossils from strata in one corner of Devonshire, all the way to the establishment and international use of a new and potentially global system or major grouping of strata, representing a newly distinguished major period in the history of the earth and of life. The controversy is epitomized in the change of meaning and usage of its central term. When it began, the word "Devonian" was applied by geologists, in simple extension of its everyday usage, to any rocks or fossils found in Devonshire. By the time the controversy subsided, the same word was used internationally in a new and strictly technical sense: to denote any rocks or fossils that had originated during a specific period in the history of the earth, and indeed to denote any other events attributed to that period, in whatever part of the globe they might be found or inferred.

The remainder of this book is an analysis of some features of the Devonian controversy, using descriptive and analytical categories as different as may be needed from those used by the participants. The goal of this analysis is to explore the implications of this exceptionally well-documented historical case study for our understanding of the shaping of scientific knowledge, past

401

and present. The moment for rescue from the perhaps overwhelming flood of detailed immediacies has at last arrived. Retrospective reflection in the light of the known outcome of the controversy becomes at this point permissible, indeed essential. The scientific natives are now safely out of earshot (§1.6).

In the course of the detailed narrative, successive or alternative interpretations of the rocks and fossils in Devonshire itself have frequently been summarized in the form of columnar sections laid on their side to represent conceptions of temporal sequence. Leaving aside for the moment the identity of those who conceived or adopted these schemes of interpretation, their bewildering variety can be condensed into three major categories. The codeword *GRE* (an abbreviation of "Greywacke") will be used in this analysis to denote those interpretative schemes in which *all* the older (pre-New Red Sandstone) strata of Devonshire (a) were regarded as forming an unbroken sequence of strata, and (b) were attributed to the Greywacke or Transition, or at youngest to the Old Red Sandstone. The codeword *COA* (an abbreviation of "Coal Measures") will denote those schemes (a) in which the Culm strata of Devonshire, or at least those with fossil plants, were attributed to the Coal Measures, even though (b) this involved postulating a major gap in the sequence, and hence implicitly a major unconformity, separating those Culm rocks from the still older strata. Third and last, the codeword *DEV* (an abbreviation of "Devonian") will denote those schemes in which (a) all these strata were regarded—as in GRE—as forming an unbroken sequence, but (b) the plant-bearing Culm strata were attributed—as in COA—to the Coal Measures (or at oldest to the Mountain Limestone), while (c) a significant part of the still older strata were regarded as equivalent to the Old Red Sandstone.

For a first approximation or rough outline of the shape of the controversy, it is sufficient to note two successive variants of GRE and two of COA (fig. 15.1). In what will be termed *GRE.1*, the Culm and its fossil plants were considered to be somewhere in the middle of the unbroken sequence, all of which was attributed to the older or lower Greywacke (LG). In *GRE.2* the Culm was placed at the top of the sequence, the upper part of which was extended upward into the Upper Greywacke or Silurian. In *COA.1*, all the Culm strata above the putative gap or unconformity were attributed to the Coal Measures (CM). In *COA.2* a lower part (including particularly the black Culm limestone) was distinguished from the main or plant-bearing part of the Culm and was attributed to the Mountain Limestone (ML), thus extending the upper part of the sequence downward toward the lower.

Once these four variants are placed in the temporal order in which they were proposed, with DEV later than them all, the relations between them are immediately apparent (fig. 15.1). GRE.1 and COA.1 were at the greatest *theoretical distance* apart. GRE.2 was a modification of GRE.1 which had the effect of lessening that distance, because it brought the Culm plants somewhat closer to the date (CM) to which they were attributed on COA.1, while retaining the unbroken character of the sequence, or complete conformity between the Culm and the other strata. COA.2 was a modification of COA.1 which likewise had the effect of lessening the theoretical distance from GRE, because it brought the Culm sequence somewhat closer to the still older strata, and hence reduced the magnitude of the putative gap or unconformity that separated them. DEV, on the other hand, combined features from both GRE.2

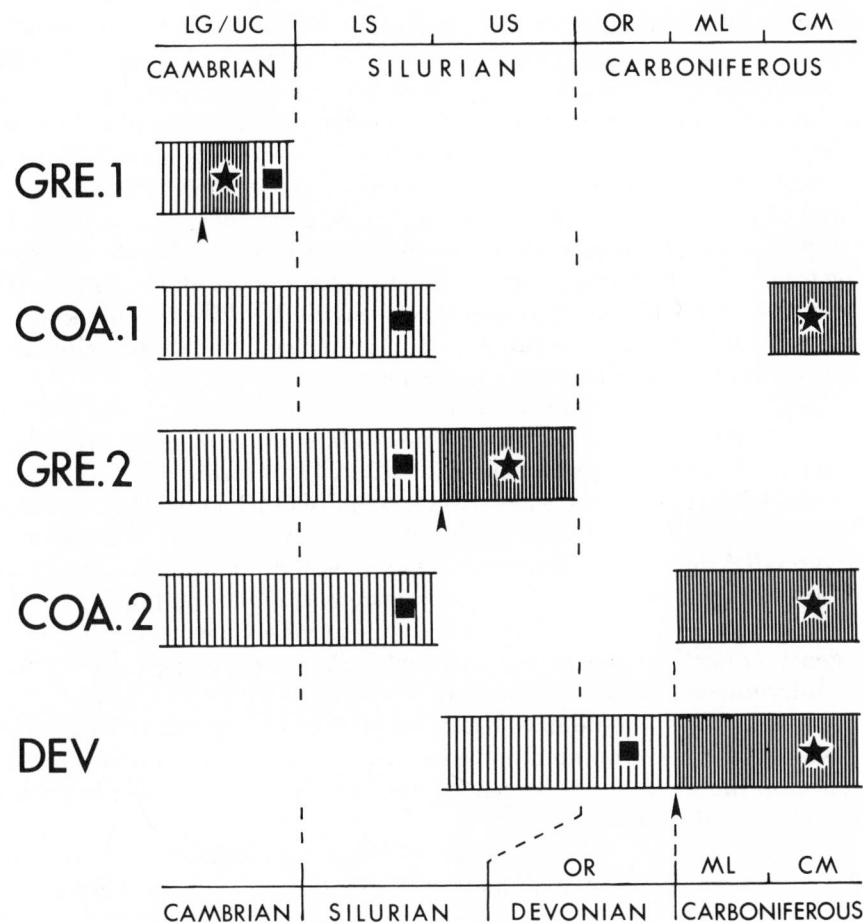

Fig. 15.1. Diagrammatic summary of five successive major schemes for the interpretation of the older strata of Devonshire. The "standard" sequence derived from other regions, which acted as a relative timescale, is shown at top and bottom, representing (respectively) versions before and after the controversy, the latter incorporating the enlargement of the Old Red Sandstone (OR) into a new "Devonian system." On the five columnar sections the distinctive Culm rocks are heavily shaded, and the position of their fossil plants is denoted by a star symbol. The other strata, regarded as *older* (except in GRE.1), are lightly shaded; the square symbol denotes the position of what were regarded as the *youngest* of their rich fossil faunas (e.g., around Tor Bay and Barnstaple). The arrowhead symbols mark the claimed conformity at the base of the Culm; gaps in the sequence represent the claimed unconformity at that point.

and COA.2. From GRE.2 it retained the unbroken character of the sequence, or complete conformity between the Culm and the older strata, and brought those strata still further forward on the relative timescale. From COA.2 it retained the Carboniferous dating of the Culm. Conversely, DEV rejected or abandoned the Greywacke dating of the Culm plants in GRE.2, but also the gap or unconformity postulated in COA.2.

The relations of these five main interpretative schemes for Devon geology can now be used to express at a first approximation the theoretical dynamics of the controversy. On figure 15.2, the horizontal axis represents the (as yet unquantified) timescale of the controversy, and the vertical axis an (unquantifiable) impression of theoretical distance. The thick arrows are lines of *interpretative development*. They show first the slight lessening of theoretical distance that was entailed in the development of GRE.2 and COA.2 from their respective earlier variants. The thick arrows leading *to* GRE.1 and COA.1 represent the precedents and theoretical expectations that gave both GRE and COA a preexistence, as it were, even before they were formulated explicitly to explain the specific problems of Devonshire. The thick arrow leading from COA.2 to DEV is highly tilted to represent the much more drastic interpretative development that that move entailed, while the thick arrow leading onward *from* DEV represents its further consolidation, validation, and extension to regions far beyond Devonshire.

The five main schemes were linked, however, not only by lines of interpretative development, but also by lines of *interpretative pressure*, which are shown by the thin arrows on figure 15.2. COA.1 was formulated in *reaction* to GRE.1. GRE.2 represented a tacit *concession* to COA.1, and COA.2 likewise to GRE.2. Above all, DEV embodied a decisive though tacit concession to the whole GRE development, in that it incorporated the GRE insistence on the unbroken character of the Devon sequence and abandoned the COA postulate of a gap or unconformity. But it is unquestionably a matter of historical contingency, rather than logical necessity, that DEV was developed from COA.2; under different circumstances it might have been developed from GRE.2 instead. In that case it would have had to embody a corresponding concession to the whole COA development, in that it would have incorporated the COA insistence on the Coal Measures age of the Culm plants and abandoned the GRE claim to their Greywacke dating.

The theoretical field was in effect divided into three distinct *interpretative domains*, separated by two sharp *interpretative boundaries*. These terms

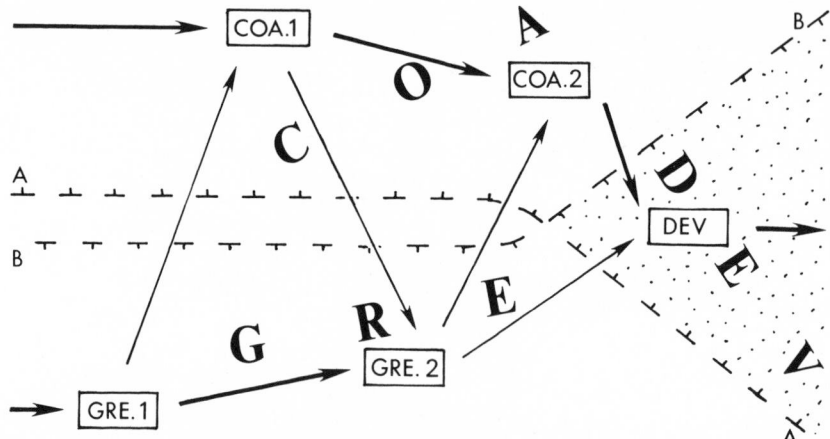

Fig. 15.2.   Schematic chart of the development of the Devonian controversy from the first formulation of interpretation GRE.1 to the establishment of interpretation DEV (see fig. 15.1). The horizontal axis represents historical time, the vertical axis *theoretical distance*. Thick arrows are lines of *interpretative development*; thin arrows are lines of *interpretative pressure*. Three *interpretative domains*, GRE, COA, and DEV, are bounded by two *interpretative boundaries*, A and B; the novel and unforeseen DEV domain is stippled.

will explicate the sense in which DEV was no mere compromise between GRE and COA. In the earlier phases of the controversy, GRE and COA were not only far apart in theoretical distance, but they also faced one another across a frontier that was perceived as a pair of irreconcilable battle lines. The banners flying over the battle lines, as it were, read in effect "No Coal Measures plants in the Greywacke!" on the COA side (boundary A on fig. 15.2), and "No gap in the sequence of Devonshire strata!" on the GRE side (boundary B). The battle lines were perceived as being in direct confrontation, with no intervening ground on which any compromise position could possibly be established. The two slogans that encapsulated the initial confrontation between GRE and COA were not *logically* opposed, but they seemed incompatible owing to the much wider theoretical implications carried in their train. In the course of the controversy, however, that perception changed, and DEV was eventually formulated within a *new* interpretative domain that no participant had foreseen as a possibility when the controversy began. What made that domain at first conceivable, then plausible, and finally persuasive—as represented on figure 15.2 by the progressive widening of DEV—was that it was *not* created by a widening of the untenable no-man's-land between the earlier embattled positions. Instead, the battle lines defended by GRE and COA, having initially faced each other in opposition, filtered silently through each other, as it were, until they faced outward, leaving at their rear a domain defended by them *both*. It was in this new and unforeseen domain that DEV was at first formulated and later developed and consolidated, becoming the focus for the virtual consensus that brought the controversy to an end.

DEV was undeniably a compromise between GRE and COA in certain formal respects; but when analyzed as a historical or temporal development, it is clear that it was no *mere* compromise. It was a significantly novel construction. Furthermore, that unforeseen novelty was a product of a dialectical exchange that modified *both* the opposed positions from which it was derived. Even at this crude first approximation, therefore, the Devonian controversy provides a general conclusion of much wider validity for the understanding of scientific knowledge. Whether DEV be characterized as consensual—which it was unquestionably, at the very least—or as fruitful, successful, reliable, valid, or even correct or true, it was clearly *not* established by the straightforward victory of one interpretation or theory over its rival or competitor. *Neither* of the initial alternatives (GRE and COA) had a monopoly in the requirements for victory, whether those requirements be taken to include natural truth, political clout, or anything else. The development of a successful interpretation evidently required the interaction of *both* those initial alternatives, and it resulted in claimed knowledge that was unforeseen, unexpected, and above all *novel*.

15.2   A Coarse-Grained Analysis

The three-letter codewords GRE, COA, and DEV have been chosen in deliberate stylistic contrast to philosophical analyses in terms of theories T_1, $T_2, \ldots, T_n$. Nonetheless they are still uncomfortably reminiscent of such analyses, which are so formal and abstract that they make little or no contact with the actualities of scientific research. This can be remedied, however, by transcending the neat but somewhat spurious simplicities of the foregoing first approximation, which has now served its purpose as an initial orientation. It can be replaced by an analysis that, while still relatively "coarse-grained," does greater justice to the complexities of the real arguments, and that begins to bring the arguing human actors into view. First, however, it is necessary to plot and classify a greater variety of the interpretative schemes that were proposed to explain the puzzling features of Devonshire geology. They still fall into three fairly distinct groups, which can continue to be termed GRE, COA, and DEV, although the formal definitions of those categories require slight modifications to accommodate some aberrant variants. Figure 15.3 shows all the important variants known to have been proposed during the Devonian controversy, reduced to the simplest form possible.

In *GRE.1a* (hitherto described as GRE.1) the Culm and its fossil plants were placed somewhere in the middle of an unbroken sequence of Lower Greywacke (LG). In *GRE.1b* the Culm was switched to the top of the sequence as a result of a revised interpretation of the *structure* of the region, but the dating of the sequence as a whole was unchanged. *GRE.2* (as already described) retained the same sequence as GRE.1b, but stretched it toward the younger end of the relative timescale (to the right on fig. 15.3), the Culm being attributed to the Upper Silurian (US) and the pre-Culm strata being taken to include some Lower Silurian (LS). *GRE.3a* continued the stretching of the sequence as a whole, the youngest strata being attributed to the Old Red (OR), but the Culm was switched back into the middle of the sequence, somewhat as in GRE.1a and likewise on account of a new interpretation of the structure of the region. In respect of overall dating, *GRE.3b* was a similar modification of GRE.2, but the Culm was kept at the top and equated with the Old Red.

COA.1 (as already described) attributed all the pre-Culm strata to the Lower Greywacke and all the Culm to the Coal Measures (CM), postulating a huge gap and hence a major unconformity between them. *COA.2a* extended the pre-Culm sequence to include some Lower Silurian, whereas *COA.2b* (previously described as COA.2) also reduced the gap from the other end by attributing the Culm limestones below the plant-bearing strata to the Mountain Limestone (ML). (In one variant of COA.2b the Lower Silurian dating was abandoned.) *COA.3* retained from COA.2a a Coal Measures dating for the upper or plant-bearing part of the Culm, but it was aberrant in that the putative unconformity was shifted to a new position *within* the Culm, the lower part of which was assigned to the Upper Silurian (and perhaps also to the Old Red) and regarded as overlying the older strata conformably. *COA.4* retained from COA.2a the unconformity below the Culm, regarded as Coal Measures in age, but it too was aberrant in that the older strata were attributed to the Old Red (with a little Silurian too).

DEV.1 borrowed from GRE the conformable relation between the older strata and the Culm, adopted from COA.2a the dating of the older strata, but modified the dating of the Culm to some lower part of the Carboniferous (probably the Mountain Limestone and Old Red). *DEV.2a* adopted much the same dating for the older strata, and also the conformity between them and the Culm, but stretched the older strata far forward, as far as the Old Red or

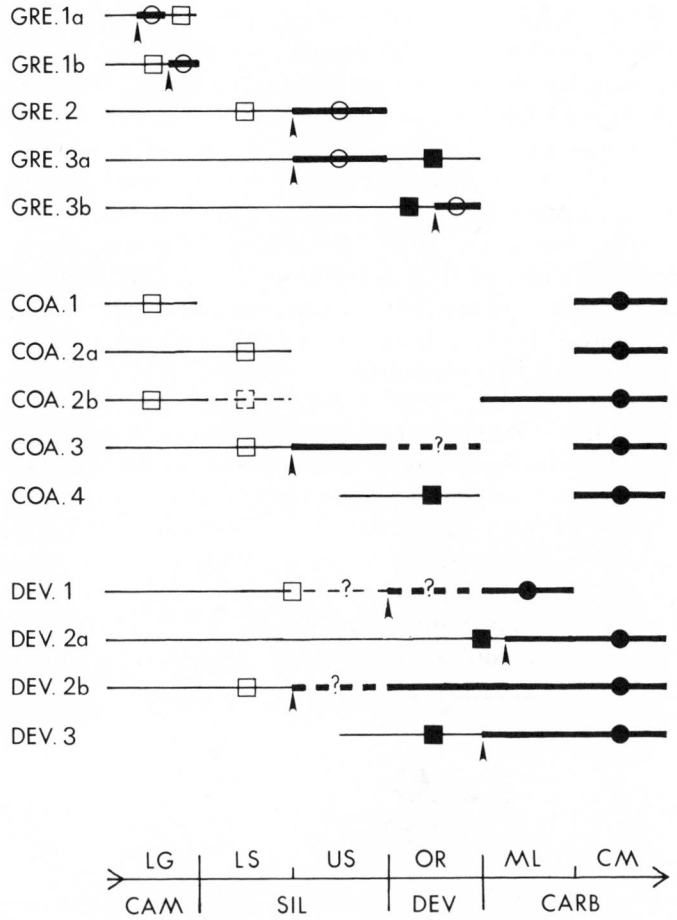

Fig. 15.3.   Diagrammatic summary of fourteen alternative schemes for the interpretation of the older strata of Devonshire, amplifying Fig. 15.1. For each scheme an unbroken sequence of strata is represented by a horizontal line, the thicker lines denoting the distinctive Culm strata, plotted against a simplified "standard" sequence or relative timescale. As in fig. 15.1, square symbols represent the dating of the youngest of the rich fossil faunas in the older strata; circular symbols (replacing the stars in fig. 15.1) represent the dating of the fossil plants of the Culm. On this diagram, *black* circles show Carboniferous datings of the Culm plants, and *black* squares show Old Red (or Devonian) datings of the youngest of the fossil faunas in the older strata. Arrowhead symbols point to the claimed conformity at the base of the Culm; claimed unconformities are marked by gaps in the sequence. The various schemes are arranged here in a roughly logical order, and only incidentally in temporal order. For the identification of the schemes in terms of the narrative, see the text.

even the base of the Mountain Limestone. *DEV.2b* was similar in the overall range attributed to the whole unbroken sequence, but instead stretched the Culm strata far backward from the Coal Measures, tentatively as far as the Silurian. Finally, *DEV.3* was similar to DEV.2a, but as in COA.4 *most* of the older strata were attributed to the Old Red—which was then renamed the Devonian—rather than only the youngest part of that sequence.

All these interpretative schemes can now be plotted on a schematic chart of the development of the Devonian controversy, which brings the first approximation on figure 15.2 somewhat closer to the real complexities of the argument. Figure 15.4 is similar in form, but *all* the arrows now represent lines of interpretative development (not of interpretative pressure); they represent the routes taken in the *derivation* of new interpretations. *Boundary A* and *boundary B* appear again as in figure 15.2; but three other important boundaries now need to be added. *Boundary C* encloses those interpretations—namely DEV.2a and DEV.3, but also GRE.3a, GRE.3b, and COA.4—in which at least the younger part of the pre-Culm sequence was equated with the Old Red. *Boundary D* encloses DEV.3, in which the "Devonian," equivalent to the Old Red elsewhere, was regarded as a major system on a par with, for example, the whole of the Silurian, rather than being treated merely as the oldest subdivision of the Carboniferous. A position *ORS* is also shown enclosed by this boundary, to denote the interpretation of the Old Red itself as a major system in regions other than Devonshire. Finally, *boundary E* separates GRE.1a and GRE.3a from all other interpretations, in that they alone placed the Culm somewhere in the middle of the sequence rather than at the top.

Several conclusions, which may be valid far beyond the particular example of the Devonian controversy, can be drawn at once from the "coarse-

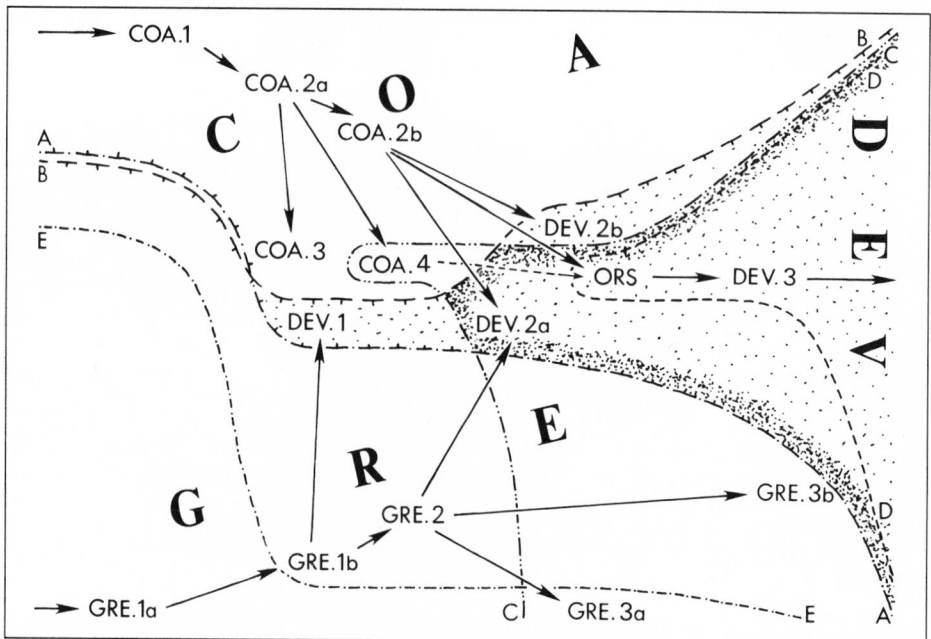

Fig. 15.4.   Schematic chart of the devlopment of the Devonian controversy, amplifying fig. 15.2. The successive or alternative interpretations of Devonshire geology are those defined on fig. 15.3; the arrows between them represent lines of interpretative development. The DEV domain (between boundaries A and B) is stippled, as in fig. 15.2, but here the line of denser stippling marks the edge of that domain on a stricter definition, enclosed within boundary C as well. The widths of the domains, and of their subordinate spaces, indicate impressionistically their changing consensual plausibility during the controversy.

grained" analysis represented visually in figure 15.4. It is clear that the finally successful interpretation DEV.3 was the outcome of processes of theorizing much more complex than those shown as a first approximation in figure 15.2. Even without bringing into consideration whatever lines of persuasive pressure may have linked the various interpretations, it is evident that DEV.3 was reached by no simple line of interpretative development. It can be regarded as having combined five distinct components, represented by its position in relation to the five interpretative boundaries. But with the partial exception of boundary D, DEV.3 was not unique in its possession of any one of those components. In other words, most of the other interpretative variants shared *some* of the components with the scheme that was finally successful. The interpretations cannot be sharply separated into right and wrong, correct and incorrect, even if those terms are defined merely in terms of the consensual outcome (i.e., DEV.3 itself) rather than in any suprahistorical sense.

More specifically, it is clear from figure 15.4 that the new DEV domain that opened up between the two most important boundaries (A and B) in the course of the controversy had a significant existence well *before* the formulation of DEV.3. Although it had low plausibility at first (and is therefore shown as a narrow space on fig. 15.4), it was occupied by DEV.2a and DEV.2b and, even earlier, by DEV.1. Indeed, once causal considerations are brought in, it might well be the case that the new domain at the rear of the initial battle lines A and B was first opened up *by* the apparently fruitless conjecture DEV.1, and then widened in plausibility by DEV.2a, which was the first scheme to be on the DEV side of boundary C as well. In other words, such theoretical schemes, although apparently taking no direct part in any supposed "triumphal march" toward the solution of the controversy, may in fact have had an indispensable role in its development. That causal interpretation can only be tested, however, by bringing the human actors onto the stage. For these schemes followed developmental pathways and crossed interpretative boundaries not in some disembodied Popperian world of ideas or theory, but in the lives and actions of *persons*. This phase in the analysis must therefore end by summarizing *who* it was who devised or adopted these various interpretations. This will also indicate in a "coarse-grained" way the lines of interpretative *pressure* that linked the various schemes but cannot be shown on figure 15.4 without losing the clarity of the diagram.

GRE.1a had a kind of "preexistence" in the widely held belief among geologists in the early 1830s that fossil plants similar (if not necessarily identical) to those of the Coal Measures were to be found in many regions in ancient Greywacke (pre-Old Red) strata. Late in 1834, De la Beche's explicit proposal of GRE.1a, specifically referring to Devonshire, precipitated the Devon controversy (§5.1, §5.2; fig. 4.8, D). Murchison and Lyell immediately proposed COA.1, in opposition to it, in order to salvage their long-standing conviction (which represents the "preexistence" of COA.1) that no such plants would be found in strata so much older than the true Coal Measures (§5.2). When Murchison and Sedgwick eventually went to Devon in 1836 to look for themselves, they found what Murchison took to be Lower Silurian fossils and confirmed their expectation that the Culm was altered Coal Measures at the top of the sequence (§7.2); this constituted COA.2a (fig. 7.1, A). But their failure to find convincing evidence for an unconformity below the Culm led

Murchison, while still in the field, to toy privately with the ephemeral conjecture COA.3 (§7.2; fig. 7.1, B). At the Bristol meeting immediately afterward, such doubts were concealed, and COA.2a was presented in public (§7.3; fig. 7.7, A). De la Beche at once conceded his critics' *structural* interpretation, which put the Culm in a trough at the top of the sequence, but he rejected their dating of the Culm and their alleged gap or unconformity; he thereby shifted to GRE.1b (§7.3; fig. 7.7, B). Phillips identified Mountain Limestone fossils from the older part of the Culm, thereby modifying COA.2a into COA.2b (fig. 7.7, C). Finally, and still at Bristol, the puzzling conflict of evidence between the GRE and COA proposals led Buckland to suggest DEV.1 (§7.3; fig. 7.7, D); this was the first scheme that publicly combined the COA claim that the Culm was Carboniferous with the GRE insistence that it lay conformably on the older strata.

After the Bristol meeting, De la Beche accepted Murchison's Lower Silurian dating for the youngest pre-Culm strata, and therefore assigned the Culm to the Upper Silurian; this constituted a shift to GRE.2 (§7.6; fig. 8.1, A). When Harding reported fossil plants in the putative Silurian strata, Murchison suggested COA.4 as a way of redating them as Carboniferous (fig. 8.1, C); but he dropped it again, or at least shelved it, when Sedgwick was unconvinced (§8.2). Meanwhile, Austen had become convinced that the fossils from the younger pre-Culm strata in south Devon were Mountain Limestone species; after displaying them in public late in 1837, he modified their dating to somewhere near the *base* of the Mountain Limestone, correlated some lower part of the sequence with the Old Red, and thus formulated DEV.2a (§9.6; fig. 9.2, A). De la Beche adopted Austen's correlation, but claimed to find structural evidence that the Culm strata were *older*, and not after all at the top of the sequence; he thus shifted into GRE.3a (§9.7, §10.5; fig. 9.2, B). Sedgwick, on the other hand, was determined to retain the older strata for his Cambrian; coupled with his growing unease about the elusive unconformity required on all COA schemes, he devised DEV.2b as an alternative solution (§9.8, §9.9; fig. 9.2, C).

Murchison's continuing unease about the plants in the older strata later caused him to shift privately toward a revival of his earlier conjecture, COA.4, while his doubts—shared with Sedgwick—about the elusive unconformity pushed him toward some kind of DEV scheme. He prepared the ground for a distinctively new interpretation by enlarging the Old Red into a system, thus shifting to the ORS position (§10.4). In the spring of 1839 he persuaded Sedgwick to join him as a nominal co-author of the article in which the "Devonian system" was proposed as the equivalent of the Old Red, and DEV.3 was thereby made public (§11.2; fig. 11.2). Austen with his DEV.2a, together with most of those who had previously adhered to various forms of GRE and COA, adopted DEV.3 either almost immediately (§11.4) or not long afterward, though without necessarily adopting the *term* "Devonian." Williams however only made the partial concession toward DEV.3 represented by GRE.3b (§11.5). In the summer of 1839 Murchison searched the Rhineland for better evidence that the "Devonian" fauna was indeed sandwiched between Carboniferous and Silurian, and that it constituted a major system (§12.1, §12.2). This first attempt to strengthen the case for DEV.3 ended in virtual failure; but Murchison's research in Russia in 1840 yielded more direct and persuasive evidence (§13.4, §13.6), and thereafter DEV.3 became the consensual solution of the Devonian controversy.

That brief review of who it was who devised the various interpretations of Devonshire and Devonian geology suggests one important conclusion that is surely valid far beyond this particular case study. The finally successful interpretation DEV.3 was undoubtedly powered by Murchison's dogged conviction, embodied in all forms of COA and DEV, that no land plants had existed as early as his Silurian period (boundary A). It also satisfied his intuitive conviction that the Old Red Sandstone was far too massive to be regarded merely as a basal subdivision of the Carboniferous, and that it must represent a major period in the history of the earth (boundary D). And it incorporated his and Sedgwick's joint perception that the Culm strata lay in a trough and were the youngest in the region (boundary E). But in addition, the construction of DEV.3 required Murchison to concede to De la Beche that the Culm lay with complete conformity on the older strata (boundary B). It also required him to exploit the conjecture that Austen had first brought forward, that some of the older rocks of Devon might be far younger than had previously been supposed (boundary C). Even leaving aside the less direct roles of these and other participants in exercising interpretative pressure on each other's constructions, it is clear that the finally successful outcome of the Devonian controversy was not simply the achievement of one brilliant and innovative *individual*, as the older heroic tradition in the history of science might have portrayed it. On the other hand, to characterize it instead as the achievement of a *group* of scientists would be equally unrealistic, unless it is made clear that that group was a collection of highly unequal individuals in far from harmonious interaction. As the detailed narrative of the controversy has shown, the myth of the egalitarian research collective is in this instance as inappropriate as the myth of the lonely genius.

15.3   A Pattern of Trajectories

The foregoing "coarse-grained" analysis has summarized the pathways by which all the various interpretations of the Devonian problem were developed, and it has briefly identified the participants who devised them. The next stage toward a "finer-grained" analysis, which at the same time will bring the actors fully onto the stage, is to plot the interpretative *trajectories* of the individuals who were most intensely involved, and thereby to chart in more human terms the development of the controversy from initial conflict to ultimate consensus. In the first instance, it is convenient to plot such trajectories as if each participant had tackled the Devonian problem in isolation, although in fact, as the narrative has shown, the chief participants were locked in an intensive network of mutual interaction from beginning to end. By a further simplification, it is convenient to combine evidence derived from the whole range of documentary sources—from field notebooks, letters, printed articles, and books—as if such differences between private and public statements were insignificant for the understanding of each individual trajectory. Figure 15.5 condenses into a single diagram the main features of the cognitive

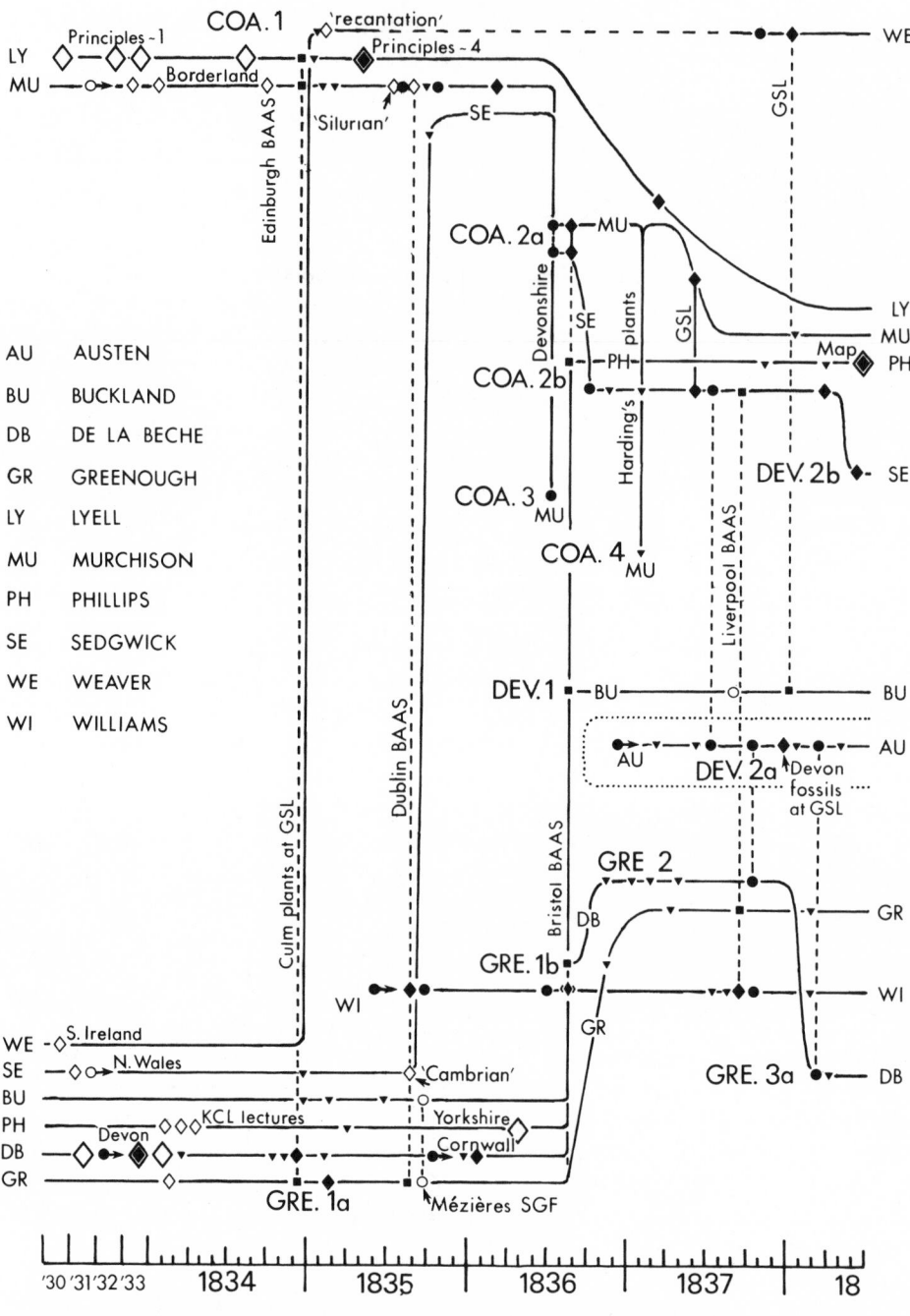

AU   AUSTEN

BU   BUCKLAND

DB   DE LA BECHE

GR   GREENOUGH

LY   LYELL

MU   MURCHISON

PH   PHILLIPS

SE   SEDGWICK

WE   WEAVER

WI   WILLIAMS

Fig. 15.5.   Schematic chart of the development of the Devonian controversy, modeled on fig. 15.4 but plotting the interpretative trajectories of the ten participants most continuously involved in the controversy. The historical timescale on the horizontal axis is now quantified (but note that the years before and after the controversy are greatly compressed, to save space). The vertical axis again represents an unquantifiable impression of theoretical distance (the minor differences of level required to keep the trajectories apart have no significance). To avoid confusing an already complex chart, only the boundary of the DEV domain is shown (by a fine dotted line), using the stricter definition represented by the line of denser stippling on fig. 15.4. The historical evidence that substantiates any point on these trajectories can be found by

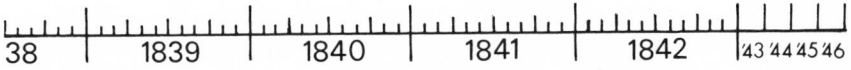

reading the relevant month on the timescale and referring to the corresponding section or sections of the narrative; only the more important documents are marked. The black symbols for books, articles, and fieldwork denote those directly relevant to the Devonian problem; the open symbols represent those with an indirect bearing on it. The category of "books" also includes major memoirs and maps; "articles" also includes papers published only in abstract, and directly relevant parts of larger, less directly relevant publications. Only a few selected points are shown for field and museum work. (The interpretation ORS on fig. 15.4 is not marked, but is represented by *Silurian System* just before the first formulation of DEV.3.)

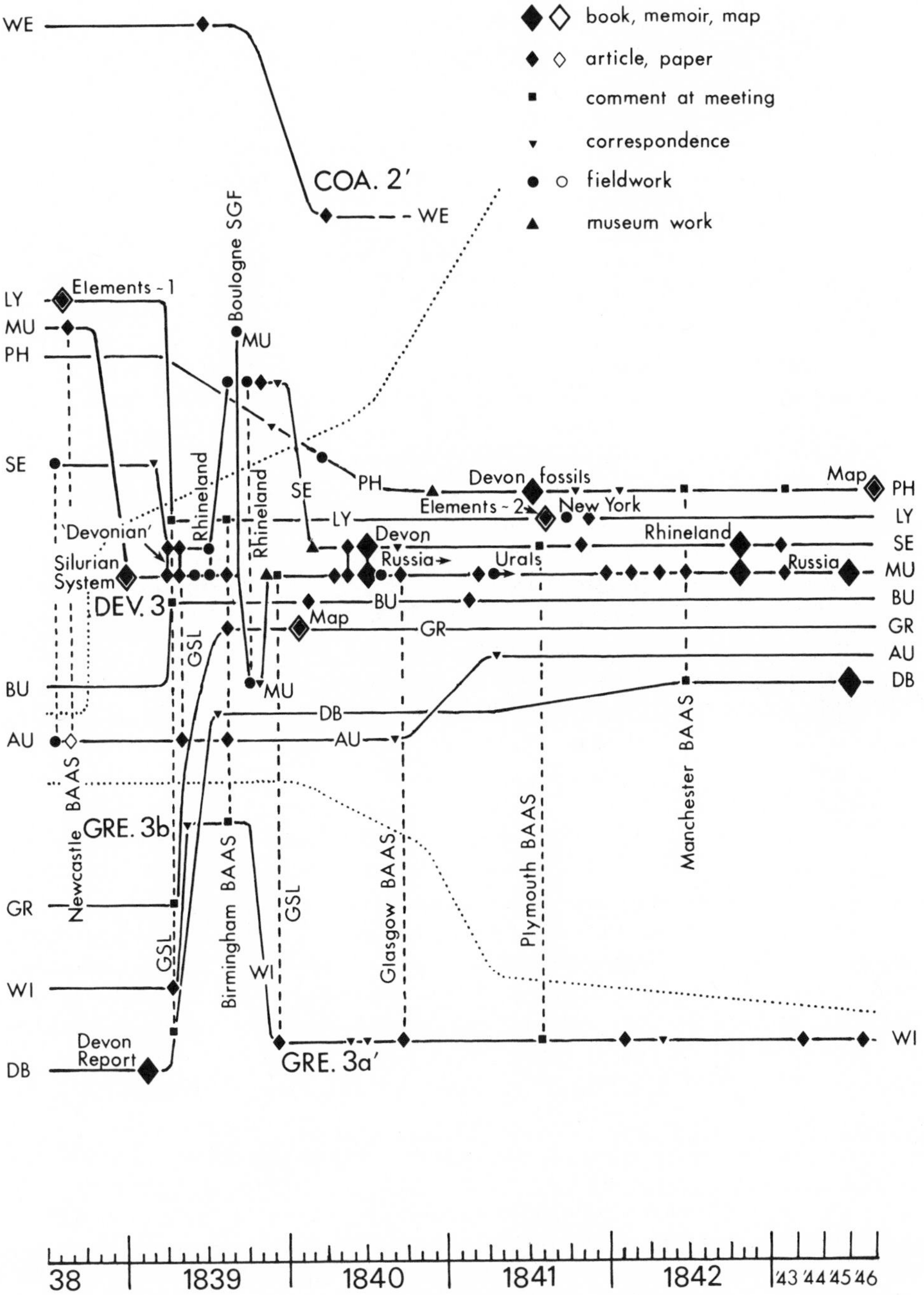

trajectories that have been described in the course of the detailed narrative, for the ten individuals whose involvement was most intense and most continuous. Not only does it amplify the schematic charts in Figures 15.2 and 15.4, it also uses the detailed narrative to plot the development of the controversy against an accurately *quantified* historical timescale. In terms of the detailed information it condenses into visual form, it is by far the most important diagram in this book.

Murchison's trajectory is certainly the most important, though hardly the simplest. His project to extend Smith's methods to the older strata (§4.1), his delineation of what he later termed the "Silurian" strata in the Welsh Borderland (§4.3), and his firm conviction that those strata contained no trace of fossil land plants (§4.4), all placed him squarely on the COA side of the argument, even before he explicitly proposed COA.1 late in 1834 in response to De la Beche's threateningly anomalous report of alleged Greywacke coal in Devon (§5.2). His arguments for COA.1 were strengthened by his fieldwork in 1835 on what he claimed to be the closely analogous case of Pembrokeshire (§6.2, §6.5). His fieldwork in Devonshire itself in 1836 (§7.2) confirmed that interpretation but modified it into COA.2a, which was presented in public at the Bristol meeting soon afterward (§7.3). But his private doubts about the reality of the unconformity required for COA.2a were expressed, while he was still in the field, in his fleeting conjecture COA.3 (§7.2). The following winter, Harding's report of fossil plants in strata that Murchison regarded as Lower Silurian threatened to destabilize his cognitive structure again, and he proposed in private another fleeting conjecture, COA.4 (§8.2). Publicly, however, he merely modified his interpretation into COA.2b, and later abandoned all claims to Silurian strata in Devon (§9.4).

Toward the end of 1838, as he completed *The Silurian System*, Murchison suddenly abandoned his previous emphasis on the sharp distinction between the Silurian and Carboniferous faunas (§10.4). He proposed enlarging the Old Red Sandstone into a system in its own right (ORS) and predicted that its equivalents elsewhere might contain an intermediate fossil fauna that would efface that sharp distinction. But it required the impact of the publication of De la Beche's *Report*, with its lack of acknowledgment of Murchison's research (§10.5), to spur him in the spring of 1839 into reviving his highly implausible conjecture COA.4, now with the crucial modification—the abandoning of any claim to an unconformity—that transformed it into DEV.3 (§11.1). Once launched in public within that context (§11.2), the consolidation of DEV.3 and its "Devonian system" became for Murchison a project in its own right, in parallel with the promotion of his Silurian system. In the summer of 1839 he spent a long fieldwork season on the Continent, searching for—and initially believing he had found—stronger evidence in favor of DEV.3, which had now expanded in significance far beyond Devonshire (§12.1, §12.2). But at the Boulogne meeting he reverted temporarily to a position similar to his earlier COA.2b (§12.4); subsequently, under the impact of Sedgwick's trenchant criticism in the field, he only recovered to the extent that he adopted what was in effect a Continental form of DEV.2a (§12.5). However, after his return to England a closer study of the relevant fossils revived his confidence in DEV.3 (§12.8), and in 1840 a long season in Russia provided him with what he regarded as conclusive evidence in its favor (§13.4). Thereafter he incorporated the Devonian into his wider ambition to bring order to the whole of the "Palaeozoic" era in the history of the earth, and to display its fossil record as one of continual change and increasing overall diversity.

Lyell's trajectory followed the main track of Murchison's. Although he was much less continuously involved, the resolution of the controversy was of the greatest importance to the theoretical synthesis he had set out in his *Principles of Geology*. As soon as the controversy began, his model of steady, piecemeal organic change made him a natural ally of Murchison on the COA side of the argument, since the contrary claims of GRE presented his theoretical synthesis with a serious anomaly (§5.2). He continued in his advocacy of COA throughout the next four years (§8.4, §9.8), although he had no occasion to follow the evolving argument in detail. As soon as DEV.3 was proposed, however, he adopted it enthusiastically, seeing in it a highly satisfactory resolution of the original anomaly. The consolidation of DEV.3 and the Devonian system provided by far the best exemplar for his model of gradual faunal change, which he himself had only been able to establish among the relatively insignificant strata of the Tertiary.

Little need be added to the summary of De la Beche's trajectory already given in the "coarse-grained" analysis. His report of fossil plants and coal in the ancient Greywacke of Devon (§5.2) was for him no anomaly at all, but simply another example (GRE.1a) of an already well-attested generalization; he therefore defended his report vigorously against Murchison's and Lyell's aprioristic COA.1 (§5.3). When confronted at Bristol with COA.2a, which unlike its earlier variant was at least based on fieldwork, De la Beche immediately made the concession represented by GRE.1b (§7.3), and on further reflection he shifted to GRE.2 (§7.6). After Austen's fossils had been pronounced similar to those of the Mountain Limestone (§9.3), De la Beche avoided that judgment's implicit support for COA by claiming field evidence that the true Culm and its plants were older, not younger as all other geologists maintained (§9.7); his consequent GRE.3a was obscurely incorporated in his official *Report* (§10.5). He was skeptical about DEV.3 when it was first proposed in public by his opponents (§11.4, §11.5), but within a few months he was considering it seriously, or at least its weaker variant DEV.2a (§12.3). By that time, however, his official duties had shifted to south Wales, and he had no inescapable occasion to give his opinion in public on the new interpretation; his adoption of DEV.3 only became unambiguous as the controversy finally subsided (§14.4, §14.5).

Greenough's trajectory followed De la Beche's in much the same way as Lyell's followed Murchison's. His previous work made him a natural recruit to the GRE side, and he defended GRE.1a when De la Beche's first report from Devonshire was discussed late in 1834 (§5.2). He continued to support the GRE side of the argument in the subsequent years, right up to the time of the first major discussion of DEV.3 in the spring of 1839 (§11.4). Yet within a few months he decided to adopt DEV.3 for the revised version of his great geological map of Britain (§12.3), although he did not adopt the term "Devonian" and was less than generous in his acknowledgment of those who had constructed the new interpretation (§13.1).

Buckland was less continuously involved in the controversy, but his distinctive trajectory through it was of great significance. Although firmly on the GRE side at the start of the controversy (§5.5), he was impressed at Bristol by the strength of the case for COA. In the discussion he proposed DEV.1 as a compromise (§7.3); this was the first interpretation to open up the DEV domain as a conceivable possibility. Although his suggestion was not pursued explicitly by himself or anyone else, it may have contributed to the ease and enthusiasm with which he accepted DEV.3 almost as soon as it was proposed in public (§11.4), and thereafter he promoted it as a highly satisfactory solution to the controversy (§13.2).

At the start of the controversy, Sedgwick, like Buckland, sided unhesitatingly with De la Beche and adopted GRE.1a, since he too saw nothing anomalous about the report of coal and plants in the Devon Greywacke (§5.4). But after Murchison explained his Pembrokeshire analogy in 1835, Sedgwick began to have serious doubts about GRE and tacitly adopted COA.1 (§6.4). That provisional conclusion was quickly confirmed during his fieldwork with Murchison in Devon in 1836 (§7.2); he was an enthusiastic proponent of COA.2a at Bristol (§7.3), and his conversion was strengthened by his fieldwork immediately after the meeting (§7.4). Accepting Phillips's opinion on the fossils from the lower part of the Culm, he later modified his interpretation to COA.2b. But he rejected the implications of Austen's fossils (§9.3) and continued to claim the older strata for his "Cambrian" system of ancient Greywacke rocks (§9.8). He developed DEV.2b, which he proposed publicly though obscurely in 1838 (§9.8), as a way of reconciling that position with his failure to find adequate field evidence for the unconformity that all variants of COA demanded. He was at first highly skeptical of Murchison's DEV.3 proposal for escaping from that dilemma (§11.1), since he considered that his collaborator's earlier conjecture to the same effect (COA.4) had been no more than an ad hoc device (§8.2). But he soon agreed to be a nominal co-author of the article in which DEV.3 was publicly launched in the spring of 1839 (§11.2). He joined Murchison in the Rhineland that summer (§12.2), but during their fieldwork he became increasingly skeptical about the Devonian, and by the end of the season he had rejected DEV.3, and reverted to a position similar to his earlier COA.2b (§12.5, §12.6). However, in the spring of 1840 the fossil evidence brought him back to DEV.3, and his confidence in it strengthened gradually, particularly in the light of Murchison's Russian research that summer.

Phillips's trajectory was somewhat similar in form to Sedgwick's. Although not closely involved at first, he supported De la Beche on the GRE side of the argument, until at Bristol he heard Sedgwick and Murchison present their case for COA.2a. He accepted that interpretation on the spot and added to its plausibility by citing extra fossil evidence that modified it into COA.2b (§7.3). As he became more deeply involved with the study of the fossils from Devon, he was privately puzzled by Austen's fauna in particular; but nonetheless he embodied COA.2b on his small-scale geological map of Britain, published in 1838 (§10.3). When DEV.3 was proposed in public a few months later, Phillips cautiously reserved judgment about it. But by the time he finished his official report on the Devon fossils in 1841, he had not only adopted DEV.3 but had also given it a Lyellian twist well beyond what its original proponents had conceived, although he remained reluctant to use the term "Devonian" itself (§14.1).

The trajectory of Weaver, who was the oldest of this set of ten participants, is also reminiscent of Sedgwick's, but with one important difference.
Weaver's 1830 paper on Ireland was in effect a foreshadowing of GRE, and
its mysterious suppression reflects the skepticism of influential critics of that
position, even before GRE was formulated with reference to Devonshire
(§4.4). Shortly after De la Beche proposed GRE.1a, Weaver "recanted" and
switched abruptly to an Irish analogue of COA.1, though this had little effect
on the plausibility of GRE in Devon (§5.6). After nearly three years off stage,
Weaver entered the Devon controversy more directly, strongly in support of
COA (§9.2, §9.4). But unlike Sedgwick and other proponents of COA, he was
never convinced by DEV.3. He only made a limited concession toward it,
suggesting that the contentious strata might somehow lie *between* the Silurian
and the Old Red (§13.5; COA.2′ on fig. 15.5). But in the eyes of other geologists
this was so implausible that he was ignored, and he soon faded from the scene.

In that final respect, Weaver's trajectory was a mirror image of that of
Williams, the other Somerset geologist in the controversy. Williams entered
the argument at the Dublin meeting in 1835 in strong support of GRE (§6.2),
and he continued to argue for that interpretation in the subsequent years (§7.3,
§9.1). Despite the scorn of the metropolitan gentlemen of geology, his detailed
mapping and the fine fossil evidence that he amassed made it impossible for
them to dismiss him as an incompetent amateur. It was Williams who in the
spring of 1839 provided the peg for the first expert discussion of DEV.3 (§11.4).
Although he himself rejected the new interpretation, he did make a partial
concession in its direction, shifting his position to GRE.3b. But later that year
he claimed to have found field evidence for a new structural interpretation
of the position of the Culm (GRE.3a′ on fig. 15.5), which isolated him much
further from the growing consensus around DEV.3. Undeterred by the increasing neglect of his opinions by other geologists, he continued to advocate
his own interpretation right up to the time that the controversy finally subsided.

Finally, the trajectory of Austen, the youngest of this set of ten participants, was one of the simplest and yet of great importance. Austen's papers
in the early 1830s established his reputation as a reliable local geologist, but
they contained no commitment to either side of the developing controversy.
In the course of 1837, however, he began to insist on the Mountain Limestone
character of the fossils from his local area (§9.2, §9.3). After showing that area
to both Sedgwick and De la Beche, and having his fossils discussed in London,
he began to formulate DEV.2a (§9.6), probably as a conscious synthesis between the older geologists' respective interpretations. DEV.2a was the first
interpretation that on a strict definition was truly in the DEV domain, in that
it combined the three crucial features represented by boundaries A, B, and
C (fig. 15.4); almost certainly it had a powerful though unacknowledged influence on Murchison's later formulation of DEV.3. Austen himself was at first
skeptical about DEV.3, but he eventually adopted it in much the same way
and for the same reasons as Phillips (§14.1), although his removal from Devonshire for family reasons had by then made him a marginal figure in the
controversy.

The plotting of these individual trajectories in fig. 15.5 has added human
flesh to the dry bones of the schematic fig. 15.4. It shows the complex period

of interpretative lability and pluralism that preceded the public formulation of DEV.3. It shows the shift of balance from GRE to COA by the defection or conversion of certain individuals, particularly the highly respected field geologist Sedgwick and the equally respected fossil specialist Phillips. It also shows in striking fashion how a stable consensus congealed around DEV.3 soon after its first proposal in public, leaving only marginal individuals outside the DEV domain. But above all, plotting the trajectories against the absolute historical timescale provided by the detailed documentation clearly reveals the contrast between certain *stable states* of interpretation, on the one hand, and the specific occasions that destabilized them and produced rapid changes in the opinions of particular individuals, on the other. It also shows the contrast between the longer-term interpretations and the occasional fleeting conjectures (COA.3, COA.4, DEV.1), the significance of which can only be seen in the perspective of the whole controversy. What the chart does not and cannot do, however, is to explain or even to depict the causal dynamics of either stability or rapid change in any single trajectory, or the causal dynamics of the formation of a consensus out of the interweaving bundle of trajectories. The intrinsic limitations of any analysis or visual representation in terms of separate trajectories must be obviated in other ways. The next stage of the analysis will set aside for the time being the emphasis on dynamic or diachronic development, and will consider in more static or synchronic terms the internal structuring of the group or set to which these ten participants and others belonged.

15.4   Mapping the Field of Competence

A first stage toward a fuller understanding of the dynamics of the Devonian controversy is to consider the diverse roles of the participants. Even a "coarse-grained" level of analysis has shown that the cast list of those who contributed in significant ways was not limited to star performers, such as De la Beche and Murchison. Minor actors, such as the provincial geologists Austen and Williams, could not be kept out of the story, even in brief summary; the full narrative has shown that the walk-on parts of local collectors, such as Harding and Hennah, were also far from negligible. In the foregoing summary of trajectories, the ten individuals who were most continuously involved in the controversy were treated almost as if their changing opinions all carried the same persuasive weight among their contemporaries. Yet the narrative has shown that no such equality reigned within the geological world of that period, any more than it does in any specialist scientific community of the present day. To analyze the roles of *all* the actors, including even the walk-on parts, a different approach is needed.

To talk of a geological "community" at the time of the Devonian controversy is misleading on many counts, not least because it suggests anachronistically a strong-boundaried professional group marked by standardized training and certification, with only the uninitiated lay public outside. It is questionable whether that image of a disciplinary "community" is valid even for a modern science. But possession of, say, a relevant Ph.D. may in practice be treated by others in a modern discipline as a mark of minimal competence, in the sense that a controversial knowledge claim from such a source will be given the benefit of the doubt, at least for a time. Although there was no such formal or standardized training in early nineteenth-century geology, membership in either of the two national geological societies functioned in practice in much the same way. As explained at the beginning of this book, the Geological Society of London and the Société géologique de France, with their overlap and substantial foreign membership, accounted for perhaps two-thirds of all those in any country who were regarded at the time in some minimal sense as *competent* geologists (§2.2, §2.3; fig. 2.3). Any unexpected or controversial claim that was made by a member of either society would, at least initially, be treated with respect.

As in a modern science, however, that merely implied that the formal qualification, represented by inclusion within the strong and unambiguous boundary of membership, coincided approximately with a much more important qualification that was informal and tacit. But that invisible and weak boundary of minimal competence (the outermost dashed circle in fig. 2.3) did not—and does not in modern sciences—enclose an undifferentiated interior of equal competence. It is merely the lowest or outermost of a series of invisible boundaries that separate various degrees of relative competence. Whatever the power—then or now—of an official myth of democratic equality within "the scientific community," practitioners knew then—as they know now—that some scientists are more equal than others, and that the formal hierarchies of position and influence are by no means coincident with what will here be termed the informal and tacit *gradient of attributed competence*. That phrase will now be expanded and explained with specific reference to the Devonian controversy, although there will continue to be much in the analysis that could be applied mutatis mutandis to modern scientific research.[1]

The relative scientific status of the geologists involved in the Devonian controversy can be recovered from the historical record, as it has been throughout the detailed narrative, by noting the ways in which they treated and referred to each other and to each other's work. Their correspondence is by far the richest source of evidence, because it was both informal and private; but published work is also revealing, at least if read between the lines. Scientific status was primarily expressed in terms of the *competence* of any individual geologist to deliver reliable information or ideas of specific kinds. Competence in this context is in no sense a suprahistorical absolute, still less a judgment awarded retrospectively by late twentieth-century standards; it was *attributed* to the individual at the time, by himself and by others. But differences of attributed competence did not divide geologists into sharply distinct categories of relative status; it sorted them out along a tacit *gradient*. No space need be wasted in arguing over the precise peck order of all the individuals ranged along the gradient; it is sufficient to locate them with rough approximation in a series of broad *zones*. These are like climatic or vegetational zones up a mountainside: perceptible through many clues, but not

1. This section is based on, but develops and modifies, a preliminary analysis of the Devonian controversy (Rudwick 1985) and a brief discussion of the early scientific work of a minor participant who was of major importance in other fields, namely, the young Charles Darwin (Rudwick 1982b).

sharply distinct. The weak boundaries between them are more like contour lines than stone walls: marked on the map but not directly visible to those who are moving around on the ground.

That last phrase indicates that any map of the topography of attributed competence must allow for the movement of individuals across it. In the case of the Devonian controversy, there was at any one time a high degree of tacit consensus about the position of most individuals on the gradient. Those to whom others attributed only a limited competence were generally content to accept that judgment and to work within their prescribed limits. Conversely, those who were looked up to as examples of outstanding competence were not unwilling to accept that attribution and to act accordingly. But this is too static a description of an evaluative exercise in which the relative status of any individual at any one time was the resultant of forces derived from himself and others, pushing him either upward toward greater esteem and credibility or downward toward dismissal and oblivion. For scientific status, like the stock market, could go down as well as up. The competence attributed to any individual could change markedly in the course of time, either upward through the production of work that others regarded as plausible and important, or downward by the production of work that was found implausible or trivial, or by simple failure to produce anything. Competence was not attributed once and for all, but earned and cultivated, worked for and maintained.

Three major zones can be distinguished along the gradient of attributed competence. They can be represented visually as a series of concentric rings, from the center or hub of the social world of geology to its periphery. As already emphasized, they were only separated by weak and tacit boundaries. In figure 2.3 the zones are shown superimposed on the Venn diagram of the strong-boundaried geological societies. In figure 15.6 a part of this circular map of the geological world is redrawn at a larger scale, to represent those who were more or less directly involved in the Devonian controversy, and to allow space for marking the places—or, in some cases, trajectories—of those individuals during the years of the controversy itself (see §15.5).

At the center or peak of attributed competence was the small number of those who, for want of a better word, will be termed *elite* geologists. They were men with a strong, indeed primary, commitment to geology rather than any other branch of science. They were highly active in the affairs of its institutions and in practical fieldwork, and usually highly productive in publication. They interacted intensively with each other, whether in cooperation or in rivalry and antagonism. Above all they regarded themselves, and were generally regarded by others, as competent arbiters of the most fundamental matters of both theory and method within the science. Among those who have featured in this account of the Devonian controversy, the zone of the elite clearly included Sedgwick, Murchison, De la Beche, and Lyell for a start. Greenough and Buckland also merit inclusion, although they were less active by the 1830s than the first four. Élie de Beaumont in Paris and von Buch in Berlin can stand as representatives of the elite geologists on the Continent. Polymathic men of science such as Whewell and Humboldt also deserve inclusion, because their achievements in other branches of science lent great weight to their opinions—and their far from negligible work—in geology itself. At the time of the Devonian controversy younger geologists such as Phillips

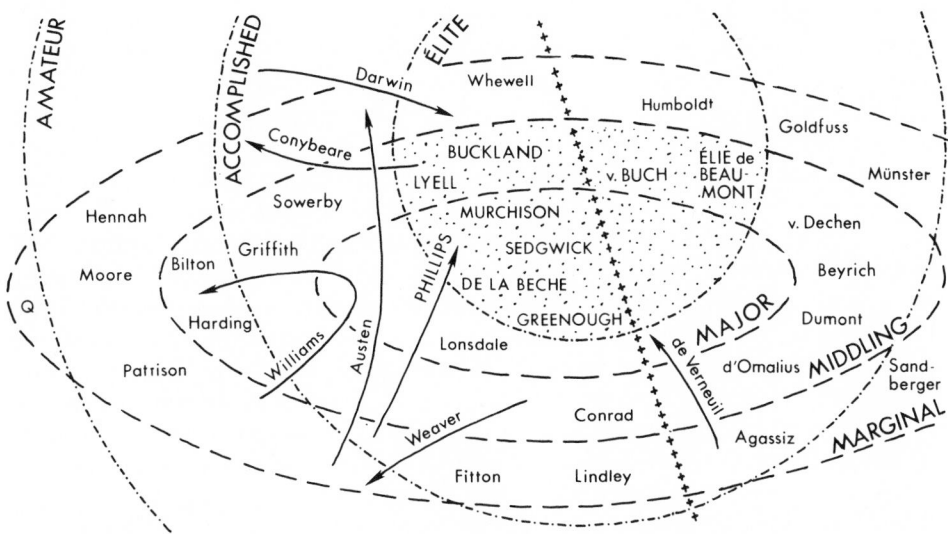

Fig. 15.6. Diagrammatic "map" of the tacit social and cognitive topography of the Devonian controversy. Superimposed on the three elliptical zones of relative involvement is part of the circular field for geology as a whole (see fig. 2.3). The names of participants are placed where they were during the years of the controversy itself; those whose position shifted significantly during that period are shown by arrowed trajectories. The domain of the core set is stippled, and the names of its members shown in capitals. The line of crosses separates the English-speaking geologists (left) from the Continental; "Q" denotes the anonymous quarrymen and other nongentlemanly fossil collectors. The exact positions of individuals *within* the spaces defined by any specific combination of boundaries is of no significance.

and Darwin were moving up into the elite; some older and now less active geologists such as Conybeare were in effect slipping from it. At this period no geologist outside Western Europe would have presumed to classify himself with any of these men. But within Western Europe the elite was highly international in both outlook and activity; frequent correspondence and travel made its members not unlike the intercontinental commuting elites of modern science, albeit on a smaller scale.

Grouping such men into one zone on the tacit topography of the geological world does not imply any cognitive consensus among them, except in the general sense of their shared acceptance of what made geology a science. Within that general framework, the zone of the elite was precisely the arena for all major conflicts on matters of high theory and fundamental method in the science. To cite the Devonian controversy itself, it was, for example, the clash between the elite geologists De la Beche and Murchison, seconded respectively by Greenough and Lyell, that made it seem a "great" controversy from the start (§5.2, §5.3). Sedgwick's change of sides began a more general swing from GRE to COA precisely because as an elite geologist his opinion carried such weight (§6.4, §7.3). Murchison was anxious to get his co-authorship in the proposal of DEV for the same reason (§11.1). Élie de Beaumont's authoritative French analogies strengthened GRE at the beginning (§5.5); toward the end, von Buch's acceptance of the validity of DEV in the German states and Russia hastened its adoption into international usage (§13.2). Even in cases of profound disagreement within the elite, as between De la Beche and Murchison, the antagonists showed by their actions that they tacitly acknowledged each other's place in that zone. Their rhetorical strategy was of course to deride their opponents' opinions and to impugn their competence. But they could not afford to ignore those opinions, and they were forced to meet them with every weapon they could muster.

The next zone outward from the small elite of the science was the zone of those who, again for want of a better word, can be termed *accomplished* geologists.[2] They included first those whose primary commitments and competence lay in other sciences, where they were regarded as belonging to the relevant elite, but whose expertise or scientific judgment impinged in some auxiliary manner upon geology. Such men might do little or no fieldwork and not publish much that was strictly geological, but many were quite active in the affairs of the science nonetheless. Among those who have featured in the Devonian controversy, it is worth mentioning here such men as Brongniart and Lindley the botanists, Agassiz the specialist on fish, and Sowerby the expert on mollusks. Just as no geologist, even in the elite, would have presumed to doubt the great Herschel on matters of astronomy, so likewise they accepted these specialists' identifications of fossils almost without argument. Thus at the beginning of the Devonian controversy neither De la Beche nor Murchison queried Lindley's opinion that the contentious plant fossils from Devonshire were of Coal Measures species (§5.1 to §5.3). Toward the end, no one queried Agassiz's identification of the fish that Murchison brought back from Russia as authentic species of the Old Red Sandstone (§13.6). De la Beche tried to minimize the implications of the Devonshire plants by emphasizing how poorly preserved they were and hence how small the number of securely identified species (§8.1); but this was a caveat that Lindley himself had already made (§5.1). Such comments were merely marginal reservations about conclusions that in the main were accepted on all sides as authoritative.

This category within the zone of accomplished geologists shades into one that comprised men whose primary commitments and interests were in geology rather than any other branch of science, but whose expertise was limited to a particular geographical region, a particular group of strata or fossils, or some other limited part of the science. The opinions of such men were highly valued by the elite, even in matters of theory, but with one crucial reservation. Their theoretical or interpretative opinions were only respected in so far as those ideas were related directly to areas of expertise that the elite could not match. The accomplished geologists were not regarded as competent to judge more fundamental or global questions of theory or method, and they were generally content to accept that assessment. Among those who have featured in the Devonian controversy, this category includes for a start such fossil specialists as Lonsdale and de Verneuil, who only differed from men like Lindley and Agassiz in that their expertise was limited to fossils and their main scientific commitments were to geology rather than to botany or zoology. Fitton belonged there too for his mastery of the Cretaceous strata: although this had no direct bearing on the Devonian issue, it did give his pronounce-

2. The term "accomplished" here replaces the term "accredited" (Rudwick 1982b, 1985), which carried unintended but potentially misleading overtones of formal accreditation.

ments on the controversy a weight of authority they would not otherwise have had. Likewise Phillips was highly respected for his work on the Carboniferous strata and fossils of Yorkshire, even at the time the controversy began. (His tacit place on the gradient then moved up into the elite, as he successfully enlarged his sphere of authority to include not only the evaluation of Carboniferous strata and fossils anywhere, but also the fundamental questions that that work entailed for the understanding of extinct organisms in past environments.)

Beyond the English Channel, Dumont was accepted tacitly into the ranks of accomplished geologists, at least in the francophone world, as soon as his memoir on the province of Liège was published. Even the skeptical English acknowledged his mastery of the solid geometry of those complex folded strata as soon as he had demonstrated it to them on the spot (§6.3). Likewise when Murchison and Sedgwick traveled later on the Continent, they were able to benefit from the regional expertise of a whole network of highly competent French- and German-speaking geologists. To name only a few, they included Goldfuss and Nöggerath in Bonn, Steininger in Trier, Hausmann in Göttingen, Münster in Bayreuth, and d'Omalius near Namur (§12.1 to §12.5). Returning to Britain, Austen can be seen to have moved steadily upward through the accomplished zone in the course of the Devonian controversy. When it began, he was being treated as little more than a local geologist of modest competence who, in Sedgwick's somewhat patronizing words, "wants nothing which a little practice will not give him" (§7.4). But as time went on, his opinions on the complex geology of his home area in southeast Devonshire were treated by others with increasing respect, even when his conclusions differed from those of the elite geologists (§9.2). At the same time he was enlarging his own conception of his work. He proposed a radical revision of the age of the limestones in his home area (§9.3); he extended his fieldwork to the rest of southwest England and to northern France, and started exchanging fossils with foreigners such as de Verneuil (§10.2); and he even projected still more theoretical work on the relations between fossil organisms and their environments (§10.1). But unlike Phillips's somewhat parallel trajectory, Austen's work was ultimately too slight for him to gain tacit admission to the ranks of the elite, at least during the period of the Devonian controversy.

Finally, the case of Williams illustrates more clearly than any other the notion that an individual's place on the gradient was the resultant of conflicting forces. Williams was the exception that proves the rule: a provincial geologist who repeatedly claimed the right to pronounce on the wider theoretical implications of his fieldwork and fossils from southwest England, but who was repeatedly pushed back down the invisible gradient of scientific status by the scorn of the elite geologists. At the time the Devonian controversy began, Williams was still new to geology and modest in his evaluation of his own competence (§6.2). But as time went on he enlarged the scale and scope of his research, and came to regard his conclusions—based as they were on detailed and prolonged fieldwork—as more trustworthy than those of briefly visiting celebrities such as Murchison and Sedgwick (§9.1, §11.4). Since their disagreements were over local details with far from local implications, those elite geologists tried to minimize the value of Williams's work by referring to him as a mere "zealous collector" of specimens. Williams was understand-ably incensed, and he continued to use the Geological Society and the British Association as arenas for gladiatorial combat with those he rightly saw were ranged against him. In the end, however, he overextended himself. Once he moved beyond his familiar ground of Somerset and north Devon, his touch became less sure. In south Devon and Cornwall his reinterpretation of the structure and sequence of the strata, reversing his own initial conclusions, was seen as highly implausible, and not only by his old antagonists (§12.7). His earlier credibility ebbed away, and his opinions were left on the sidelines, no longer combated but merely ignored (§13.6).

The third zone down the gradient of attributed competence was the one from which Austen and Williams, for example, had both pushed their way upward, and to which Williams was ultimately relegated. This was the zone of those who can best be termed *amateur* geologists. These were men—and here, at last, a few women too—whose competence was regarded as far from negligible or unimportant, but yet as more limited than even that of the accomplished geologists. They were generally *local* in their field of geological work, but in a more restricted sense than men such as Austen and Williams. They included country gentlemen and ladies, physicians, lawyers, and cler-gymen with an intimate knowledge of their home areas; and government officials, service officers, and professional and business men whose duties took them to unfrequented parts of the world. Those who traveled abroad were often valued as much for their reliable firsthand reports of distant events, such as earthquakes and volcanic eruptions, as for their collections of rocks, minerals, and fossils from distant countries. But it was above all as local *collectors* that those who remained at home could contribute to geology.

Such people might have impressively accurate local knowledge, par-ticularly of good localities for fossils, but the elite and accomplished geologists would only rely on that knowledge at a strictly "factual" level. Most of the amateurs accepted this limited role; if they presumed to offer opinions on theoretical matters, they were generally ignored (§8.2, §10.1). They rarely published anything scientific, except briefly and occasionally; from their own point of view, they were participating primarily in the widespread and self-sustaining vogue for local natural history.[3] But in relation to geologists from outside their home area, they saw their role explicitly as providers of speci-mens that *others* would interpret: for example, Hennah sent fossils to Sedg-wick "to enable *you* to decide the question hitherto in doubt" (§7.4; emphasis added). Among those who have featured in the Devonian controversy, the various local collectors of fossils from southwest England clearly belong in the amateur zone. They included men such as Harding, the ex-officer and country gentleman; the clergymen Hennah and Bilton; the physician Moore; and the country lawyer Pattison. A similar network of local amateurs existed on the Continent too: the schoolteacher Sandberger of Weilburg in Nassau can stand as an example.

Finally, beyond the pale of even the amateur zone was the general public. However important they might be in the marketing of popular geo-logical books and as a pool for the recruitment of amateurs to the science, any

3. Allen (1976).

knowledge claims that they presumed to make would be treated with caution and skepticism, at least initially, by those in *all* the higher zones of competence. Even on apparently factual matters such as the provenance of anomalous specimens or the effects of earthquakes, their reports would be regarded as "not proven" until they had been checked and corroborated. This was a reasonable precaution, since even a competent amateur geologist would know how easily the uninitiated could be deceived. A fossil picked up loose on the shore, for example, might have come from the nearby cliff; but alternatively, unless the finder knew it was essential to check the rock matrix on the spot, it might have come from the ballast of a passing ship, and originally from some quite distant region (§14.1). In such matters gentlemanly status gave no immunity from deception.

In the Devonian controversy, as in other geological research in the same period, one group of men below the gentlemanly social stratum was important in the provision of significant specimens. These were the quarrymen and miners, now anonymous to history, whose work gave them opportunities for collecting specimens for which their social superiors—whether gentlemanly local amateurs or visiting elite geologists—were prepared to pay handsomely (§2.6). The finding and the extraction of fine specimens of fossils were, as they still are, matters of skill and even flair, in which a practiced eye, manual dexterity, and ample time count for far more than advanced education. But unless the provenance of fossils was carefully checked, a purchaser could easily be duped; even the most skillful collectors among the quarrymen were therefore treated in practice as being outside the zone of the amateurs.[4]

In summary, this gradient of attributed competence is important not just because it describes the relative esteem in which various geologists were held by others. A map of the *social* field of the science is at the same time a map of its *cognitive* topography. In encapsulates in descriptive terms, and to some extent in explanatory terms, the ways in which the treatment of knowledge claims depended on their point of origin. The modest claim implied in finding and recording reliably the location of a significant fossil would be fully accepted, even if it originated with mere amateur collectors such as Harding and Hennah. The claim would then be incorporated into the work of those in higher zones of attributed competence. The more ambitious claims implied in the identification of the species comprising a fossil fauna or flora, or in the delineation of a local sequence of strata, would likewise be fully accepted, provided they originated with accomplished geologists with the relevant specialized or local expertise, such as Lonsdale, Lindley, Austen, and—for a time—Williams. Their claims would be incorporated without question into the still higher-level interpretative conclusions of the elite geologists. Those elite geologists considered that they alone were competent to put forward the most fundamental, theoretical, or global claims to geological knowledge; others generally concurred with, or at least acquiesced in, that definition of their respective roles in the science. In this way the gradient of attributed

---

4. One important collector from beneath the ranks of the gentry, whose finds were nonetheless accepted without question, was the celebrated Mary Anning (b. 1799), who made a living out of the sale of the fossils from the Secondary (Liassic) strata around Lyme Regis in Dorset; but she came from a respectable craftsman's family. The link between scientific credibility and social status is discussed by Westrum (1978), for the case of meteorite finds in the eighteenth century.

competence helps define the social topography within which knowledge claims of all kinds were handled, and helps account for the ways in which those claims arose, struggled for recognition, and were accepted or rejected, assimilated or forgotten.

15.5   Locating the Core Set

At the beginning of this book, stratigraphical geology was identified as the most flourishing of the broad collective enterprises or programs of geological research in the 1830s (chap. 2). Within that program at that period, the Devonian controversy was perhaps the most acute of all the *focal problems* or "hot spots" that disturbed the even tenor or "quiet background" of straightforward research.[5] The social and cognitive distribution of those who contributed to the debate on that focal problem can now be plotted in relation to the broader topography of the science as a whole. It is convenient to do so in rather the same way, but using this time a gradient divided into three weakboundaried *zones of relative involvement.* These were related to the zones of attributed competence in a complex manner that is most easily comprehended in visual terms (fig. 15.6).

The zone of *major* involvement comprised those for whom the Devonian controversy, during most of the time it lasted, was related to a major part of their total geological activity, as a matter of fairly continuous and intense concern. They are those who have been referred to as the "major actors" in the controversy. They included Murchison, Sedgwick, De la Beche, and Greenough among the elite geologists; Phillips, Lonsdale, de Verneuil, Austen, and Williams among the accomplished geologists. The zone of *middling* involvement comprised those for whom the controversy, however important its resolution was considered to be, formed only a minor and intermittent part of their geological activity. They are those who have been referred to as the "minor actors" in the controversy. They include Lyell, Buckland, Élie de Beaumont, and von Buch among the elite geologists; Weaver, Sowerby, Dumont, Griffith, d'Omalius, Beyrich, and von Dechen, for example, among the accomplished geologists; and perhaps such amateurs as Harding and Bilton. The zone of *marginal* involvement comprised those for whom the controversy rarely affected their scientific activity and was only of occasional concern; however important they may have considered it to be, they remained for most of the time in the audience rather than on the stage. These are the "walk-on parts" referred to earlier. They included the omnicompetent Humboldt and Whewell as, so to speak, honorary members of the geological elite; Conybeare, Fitton, Lindley, Agassiz, Goldfuss, and Münster, for example, among the accomplished geologists; most of the amateur collectors such as Hennah, Pat-

5. The term "focal problem" (Rudwick 1982b) formalizes Collins's (1981) informal references to the "transient hot spots" that disturb the "quiet background" research in a modern scientific specialty. Other important focal problems developing within geology at the same period as the Devonian controversy, but with only the loosest connection with it, included the debate on geologically "Recent" crustal mobility and the argument over the reality of "Recent" glaciation far away from existing glaciers.

tison, Moore, and Sandberger; and perhaps the anonymous quarrymen who collected fossils for geologists to purchase and interpret.

As before, no space need be wasted in debating whether every one of these individuals has been correctly assigned; it is sufficient that some such diverse degrees of involvement characterized participation in the Devonian controversy, and that they were far from coinciding with degrees of relative competence. The interpretative value of the two superimposed "topographies" of figure 15.6 is that they allow general geological competence to be distinguished from involvement in one particular focal problem; more specifically, they allow individual shifts in involvement to be distinguished from shifts in competence. Thus, for example, Austen can be shown shifting from a minor to a major degree of involvement, and simultaneously from amateur to accomplished status; but then, after his move from Devonshire, retaining that well-earned status while reverting to a minor degree of involvement in the Devonian controversy. Williams can be shown rising and then falling on both gradients. Darwin can be shown becoming involved in a minor way, during a period in which his major involvement in a quite separate focal problem (the issue of crustal mobility in the geologically "Recent" past) was shifting him upward from an accomplished to an elite geologist.[6]

Most important of all, however, these superimposed topographies allow the identification of the *core set* of the Devonian controversy. The term was coined for the analysis of controversies in modern scientific research.[7] But it is equally valuable for the analysis of one from the age of gentlemanly specialists. A core set is in effect the social correlate of the cognitive category "focal problem." It can be defined as the small set of individuals through whose changing opinions a focal problem is ultimately treated by the rest of the "scientific community" as having been settled. Once those few individuals have concluded that the problem has been solved satisfactorily, then it *has* been solved: not in any prescriptive sense, but in the sense that it is treated de facto as solved. Thus as soon as conflict and controversy within the core set for any focal problem are replaced by virtual consensus, the focal problem is at an end and the core set dissolves. Like the zones of competence and involvement, a core set is a weak-boundaried grouping, and argument over the inclusion or exclusion of specific individuals is hardly worthwhile. In the case of the Devonian controversy, neither the zone of elite geologists as a whole, nor the zone of major involvement, is adequate by itself to characterize the core set: it is better defined by a combination of categories from *both* topographies.

Those whose consensus did, as a matter of history, bring the Devonian controversy to an end were *the elite geologists with at least a middling degree of involvement*. That combination (shown by the stippled space on fig. 15.6) encompasses Murchison, Sedgwick, De la Beche, Greenough, Phillips, Lyell, Buckland, Élie de Beaumont, and von Buch. These were the geologists whose

6. The evidence for all these judgments of individual positions and trajectories on both gradients is scattered throughout the narrative and cannot be summarized briefly. No attempt is made here to place *all* the "middling" and "marginal" individuals mentioned in the narrative. For a fuller discussion of Darwin's trajectory, see Rudwick (1982b).

7. Collins (1981). Despite the exponential growth in the overall size of the "scientific community" in the intervening century and a half, the average size of core sets may not have changed at all.

adoption of DEV.3, and of the Devonian system as a category of international validity (even if they declined to adopt the name), did *in practice* bring the controversy to an end; it was their consensus that led to the general and routine use of that category among geologists everywhere. Conversely, that definition of the core set for the Devonian focal problem quite properly excludes many whose opinions on the matter did *not* have the same determining influence: elite geologists whose involvement was too slight for their opinions to sway the outcome (e.g., Whewell, Darwin); accomplished geologists whose adoption of the consensual solution was not regarded as adding to its plausibility, despite their major involvement in the controversy (e.g., Austen, de Verneuil, Lonsdale); conversely, those whose rejection of that solution did not detract from its perceived plausibility (e.g., Williams, Weaver); and any amateurs at all.

To put the matter more simply, only elite geologists were deemed competent to judge whether and when a problem with such profound theoretical implications had been satisfactorily solved or not; and within that category, only those who had been at least moderately involved in the controversy were regarded as qualified to judge it. Put that way, it may seem to be stating the obvious; but it is important to emphasize the point nonetheless. The Devonian controversy came to an end, and the Devonian focal problem was solved, not when some set of new theoretical concepts had been formulated, nor when some set of new observations had been made, though both were of course relevant. As a matter of history, the controversy ended as soon as a set of less than ten leading geologists in the early 1840s converged in a collective judgment that the problem *had* received a satisfactory solution and that a reliable new piece of natural knowledge had been shaped. Their judgment has been reinforced by the ever-increasing empirical content and theoretical coherence of the Devonian system and period in the work of later generations of geologists (see appendix B). But that still leaves as open questions the social and cognitive processes by which the original core set came to their collective agreement, and the epistemological status of their conclusion.

Chapter Sixteen

THE SHAPING
OF NATURAL
KNOWLEDGE

16.1   The Continuum of Relative Privacy

In the previous chapter the overall shape of the Devonian controversy was reviewed in two ways: first in diachronic terms of its development and resolution, by a complex series of interpretative moves embodied in the personal trajectories of individual participants; then in synchronic terms of the tacit "topographies" of relative competence and relative involvement, which defined the diverse roles of all the participants, and more specifically the small core set of those whose consensus brought the controversy de facto to an end. In the crude first approximation to the temporal shape of the controversy (§15.1), it was pointed out that the lines of interpretative development that linked successive theoretical schemes could only be understood in conjunction with the lines of interpretative *pressure* by which alternative schemes impinged on their rivals and forced them into modifications and concessions (fig. 15.2). When reviewing the array of alternative schemes more fully and identifying their proponents, it would have been confusing rather than illuminating to attempt to show how the complex developmental lines of interpretative schemes (fig. 15.4) and personal trajectories (fig. 15.5) were linked by equally complex cross-lines of interpretative pressure. Yet only some such mental image of a *fabric* composed of the interwoven threads of development and pressure, warp and weft, can begin to do justice to the extremely complex argumentation that has been traced in the narrative of the controversy. In this final chapter, the character of that argumentative weaving will be analyzed more closely, and the nature of the final product will be assessed.[1]

In converting a pattern of disembodied interpretative schemes into a pattern of personal trajectories, two simplifying assumptions were made. The trajectories were depicted not only as if they were free from any linking lines of interpretative pressure, but also as if all sources of evidence from the most private to the most public were equivalent (fig. 15.5). Although the limitations of both assumptions need to be made explicit, it is important at the same time to preserve some sense of the reality and integrity of the personal lifeways that those diagrammatic trajectories inadequately represent. Those who participated in the Devonian controversy must be treated as persons whose interpretations were the outcome of active and purposeful work. To trace the "set of evolving structures" in a person's "cognitive economy" is to acknowledge its internal organization in terms of "knowledge, purpose and affect."[2] Gruber's terminology was developed principally to make sense of the private theorizing of a young man of science, Charles Darwin, who at the same time was publicly a minor participant in the Devonian controversy; but it has a much wider relevance and applicability. However intense the argumentative pressures that were borne in on them, the participants in the Devonian controversy were manifestly no mere automata reacting to external stimuli. Each was pursuing his own long-term purposive "network of enterprise" in science, constrained of course by many external circumstances, but not reducible to them.

Yet those personal threads of purposive continuity cannot simply be "read off" from some privileged set of documents, although historians of science—and particularly biographers—are often tempted to believe they can be. The array of documentary evidence, from the most private to the most public, cannot be divided into traces of personal life and traces of social life. As their use in the narrative of the controversy has shown, all such documents are at one and the same time the record *both* of purposive cognitive projects at the individual level *and* of argumentative interactions at the social level. In place of the customary dichotomy between the "public" records of a scientist's finished work and the "private" records of his or her preliminary work or work in progress, what is needed is the concept of a *continuum of relative privacy*, covering an array of activities ranging from the most deeply private speculation to the fully public formulation of conclusions, and leaving as its historical traces a corresponding array of documents ranging from strictly private notebooks to published articles and books.[3] No single activity and no single class of documents gives privileged or even preferential access to the scientist's "real" thoughts or beliefs, since all alike are the embodiments of modes of thinking and doing that form essential components of the whole process of scientific work.

---

1. Any attempt to relate the Devonian case study to the full range of philosophical and sociological theorizing about science would have enlarged this chapter out of all proportion to the rest of an already lengthy book. It must be left to the proponents (and opponents) of specific theories to use the Devonian as they think fit—which they will doubtless do anyway, whatever their opinions about my own brief comments on the case.

2. Gruber (1980, 1981a, 1981b).

3. This section amplifies and extends an earlier brief treatment of this theme, taking Darwin's early work as an example (Rudwick 1982b).

The Devonian controversy, like all but the most exceptionally innovative scientific work, was carried on entirely within the structured field of debate and argument represented by the complementary diagrams of figures 15.5 and 15.6. Unlike Darwin's lengthy and exceptionally private theorizing about the origin of organic species and the nature of the human mind, which was recorded in his notebooks over approximately the same period of years, there is no documentary evidence that any speculations about the Devon problem ever remained deliberately private for long.[4] Even Murchison's ephemeral conjecture COA.3, which in retrospect seems a partial anticipation of his later and successful solution DEV.3 (see fig. 15.4), was probably shared on the spot with Sedgwick; indeed Murchison's brief record of it in his field notebook—the only record there is—may well be a trace of a conversation between them (§7.2). Certainly his later conjecture COA.4 was shared in that way, since it is only recorded in their correspondence (§8.2). Sedgwick's dismissal of it seems then to have pushed it down, as it were, into a more private layer in Murchison's mind. But it remained sufficiently a shared idea for Sedgwick to recall it at once, as soon as his collaborator began to propose a modification of it, namely DEV.3, under changed circumstances and with new and much improved evidential support (§11.1).

It should hardly be necessary to emphasize how great an interpretative gap separated the raw phenomena in the field from even the most immediate, solitary, and private record of how those "facts" were perceived. When examining an exposure of rock on the seashore or in a road cutting, a geologist did not have the same perceptions as a nearby fisherman beaching his boat or a villager passing in a cart. What the geologist perceived were the already interpreted features of strata with a measurable orientation, containing identifiable fossils, and capable of being integrated on the spot—however provisionally—into an imagined picture of vastly larger structures and sequences. A feat of interpretative construction lay behind even the bare and terse notes of a Sedgwick, and still more obviously behind the sketched field sections and more discursive reflections of a Murchison. Even the notes that geologists made most privately, and in closest contact with the raw materials of the natural world, were thus the products of *skilled* observation, already impregnated with the tacit knowledge of the established craft of geological fieldwork. Such notes can be regarded as a trace of a private and "intrapersonal" dialogue, in the sense that they record a geologist's internal debate about the significance of what he was seeing. But in no way can that process be divorced from the collective framework of practice that gave it its meaning.[5]

Furthermore, most of the recorded field observations that related to the Devonian controversy were not only more or less "theory laden," in the straightforward sense that most scientists as well as historians and philosophers of science now accept as a matter of course, but also "controversy laden." The particular observations made, and their immediate ordering in the field, were often manifestly directed toward finding empirical evidence that would

4. For a visual representation of the long "submergence" of Darwin's biological theorizing, and its contrast to his open participation in geological debate at the same period, see Rudwick (1982b, fig. 2). Darwin's private theorizing is analyzed in detail in, e.g., Gruber (1981a) and Kohn (1980).

5. Holmes (1981, 1984) makes a similar point with reference to the interpretation of laboratory notebooks in experimental sciences.

be not merely relevant to the controversy but also *persuasive*. Many of the most innocently "factual" observations can be seen from their context to have been sought, selected, and recorded in order to reinforce the observer's interpretation and to undermine the plausibility of that of his opponents. More explicitly interpretative notes, such as those that Murchison commonly made at the end of the day, were often designed even more clearly as rehearsals of arguments to be used against real or imagined critics. Likewise, Williams sometimes interspersed his field notes with drafts of speeches he hoped to make at future geological meetings, to justify the conclusions he was developing while in the field. Geological fieldwork, although an activity that belongs toward the private end of the continuum of relative privacy, already embodied the integration of solitary seeing, thinking, and doing into a larger fabric of social interaction. When striding alone across the countryside or scrambling along the shore, the geologist carried with him not only hammer, map, notebook, and specimen bag, but also an inward and invisible microcosm of the larger world of geological debate.

Significantly, however, some of the most decisive occasions of field work during the Devonian controversy were those in which the geologist was *not* alone, and in which the microcosm was outward and visible. Although Sedgwick began to have serious doubts about GRE as a result of Murchison's reinterpretation of the analogous evidence in Pembrokeshire, he was finally persuaded by seeing Devonshire in Murchison's persuasive company (§6.4, §7.2). Conversely, Murchison later had to cope in the field with Sedgwick's penetrating criticisms of the interpretation of the Rhineland that he had confidently constructed while he was still on his own (§12.4). Examples could be multiplied, but the point should be clear: joint fieldwork provided an occasion for debate that was all the more intense for being conducted in close confrontation with some of the most relevant evidence.

This was however only one example of a wider range of occasions for similar "interpersonal" argument. In concrete terms, such occasions might find a pair of geologists discussing their joint fieldwork at the *table d' hôte* in a village inn, but also arguing in the library of the Geological Society before the start of a meeting, or scribbling down their latest ideas before the mail coach left. Only the last kind of occasion has left any consistent documentary trace; fortunately there is every reason to suppose that the exchanges recorded in letters were similar in character to those that vanished with the ale and the pipe smoke. The narrative of the Devonian controversy has shown, however, that scientific letters need careful interpretation. The long historical tradition of "collected letters" has encouraged a literally one-sided conception of correspondence as the unproblematic expression of a writer's views, rather than as a record of exchanges in which one participant's letters were always to some extent adapted to the particular recipient.[6] "*He writes in one style to you and in another to me,*" Murchison complained to Sedgwick at one point, in anger at De la Beche's imagined duplicity (§8.3). But that is just what Murchison himself, like any prolific letter writer, was doing all the time, even if he was unaware of it. The ways he formulated his views to Sedgwick, as his closest collaborator, were of course different in both manner and content

6. Outram (1983) makes this point with reference to Cuvier's vast and varied correspondence.

from what he would tell his antagonist De la Beche. Differential presentation occasionally extended to the deliberate withholding of important information from actual or potential rivals; for example, De la Beche sent Greenough, but not Sedgwick, a drawing of his crucial evidence for the conformity beneath the Culm in north Devon (§8.6; fig. 8.2). Such information might then be reserved for use in public as a trump card on some occasion when it would have the greatest impact. But even if the contents were formally the same, the style and manner were invariably adapted to the particular recipient.

Local amateurs would not write to the "big guns" of the scientific world in the same style as those elite scientists would write to each other. The deference shown by the amateurs, the restriction of their letters to the supposedly factual, and their abstention from the theoretical are all signs of their tacit acceptance of their proper place within the scientific world (§15.4). But between geologists of greater attributed competence, letters with any substantial content were invariably persuasive in manner and rhetorical in style, if not openly then at least when read "between the lines," as they were certainly meant to be. In their letter writing, and doubtless in their unrecorded private conversations, geologists dropped the facade of dispassionate and theory-free fact collecting that most of them espoused in public; they set out their opinions with the more or less open intention of *persuading* those who read them. Letters were powerful weapons in the battle between opposing approaches and interpretations. As the narrative of the Devonian controversy has shown throughout, letters were used for consolidating alliances favorable to the writer and for attempting to break up alliances ranged against him; they were a means of enhancing the writer's *credibility* and of detracting from that of his critics; and on one important occasion they were given reinforcement by an effective caricature (§5.3; fig. 5.4). But the rhetoric through which the writer intended these results to be effected was no "mere" rhetoric: it entailed the skillful marshalling of relevant evidence in ways that would make the greatest persuasive impact on the recipient. It was not a stylistic "extra" tacked onto the scientific content of a letter: it *was* that content itself, in the only form in which it could be effectively communicated.

A more specific function of letter writing was as a preliminary stage in the construction of formal papers. This is most evident in cases of joint authorship, as between Murchison and Sedgwick, while they were ironing out—or, more commonly, papering over—their differences of opinion during the drafting of their joint papers. One of the few major lacunae in the exceptionally rich documentation of the Devonian controversy is the almost complete lack of early drafts or rough, corrected manuscripts that would give an insight into the process of drafting itself. Nonetheless it is important to consider what was involved in this process, to which analysts of science have given far too little attention. Modern practitioners of scientific research have often been deceived by the imposed conventions of their discipline into supposing that their papers are unproblematic reports of experiments performed, results obtained, and conclusions reached. Sociologists have discovered, to their apparent surprise, that the formal accounts of research given in published scientific papers differ markedly from those offered by the same scientists in informal interviews.[7]

7. See, e.g., Mulkay and Gilbert (1982).

But the contrast comes as no surprise to reflective practitioners of scientific research, among whom there is a healthy tradition of deriding and unmasking the pompous obliquities of formal scientific style. Nor should the contrast surprise any analysts of science who recognize the persuasive intentions and rhetorical character of *any* form of scientific discourse.

Once it is accepted that formal publication in science is a ritualized means of persuading others of the *plausibility* of the author's conclusions, in the light of his or her *credibility* as a competent practitioner, it follows that the drafting of a paper must always involve a profound transformation of the materials from which it is constructed, whether those materials are new observations or older "literature" or (usually) both. A scientific paper is designed, covertly if not openly, to present the best possible case in court: except fortuitously, its structure will therefore *not* reflect the actual pathway of research. It may be dressed up to resemble the policeman's deadpan testimony in the witness box; but the true analogy is with the barrister's advocacy, designed to sway the jury to a favorable verdict. Any scientific paper, if it is at all substantial in content, presents the favorable evidence with maximum rhetorical effect; weak links in the argument are glossed over or concealed; any antagonists' cases is attacked, and the force of their evidence subtly undermined or openly dismissed or even ridiculed.

The narrative of the Devonian controversy has given ample illustration of this general point. Although the format for papers had not been standardized to the extremes found in many modern sciences, the strong "Baconianism" that most geologists espoused in public imposed a tacit obligation to present their papers with a similar facade of dispassionate and atheoretical objectivity. On the surface, almost all the papers relevant to the Devonian controversy appeared to be straightforward reports of strata unambiguously observed, structures and sequences soberly pieced together, and correlations modestly proposed. Like modern scientific papers, they concealed systematically almost all that was speculative or controversial. Yet contemporaries were not deceived by this; they were well able to read "between the lines" of formal papers, as their recorded informal comments amply demonstrate. They were under no illusions about the persuasive and argumentative intentions of what they read or heard in formal settings.

In particular, the tradition of lively informal discussion that was such a celebrated feature of Geological Society meetings (§2.2) provided an exceptionally open arena for unmasking the covert implications of the papers that were read there. Consequently, a paper had to stand up at that stage to more bracing criticism than any it might already have received from chosen and friendly critics. A collective or "transpersonal" consensus about the conclusions of a particular paper often began to form provisionally at or soon after the meeting at which it was read, sufficiently to be noticed by relatively impartial observers (§11.5).[8] But a lasting consensus would only be consolidated after the more competent participants had studied the fully published memoir in detail. For this arena of expert debate was still less than fully public. At a Society meeting the case could be presented quite boldly, because

8. The term "transpersonal" is borrowed here from its original usage in group psychotherapy. Some such term, if not this one, is certainly needed to express the frequently reported perception of the "sense of the meeting."

no detailed citation of evidence was required immediately. In practice, it was possible for the paper to be modified afterward, taking account of criticism made in the discussion, in order to strengthen the case before it was published in full. That subsequent preparation of a full version—usually for the Society's *Transactions*—required a much more careful marshalling of the argument. For on publication the detailed evidence produced to support the case would be examined with a beady eye by its most knowledgeable critics. Nonetheless, it is a striking feature of the Devonian controversy that such completed papers were much less important in the process of persuasion and in the formation of a consensus than their less polished antecedents. This reinforces the point already made, which ought by now to be an axiom, that published papers—and a fortiori the citations they contain—are a most misleading guide to the dynamics of scientific research practice, unless they are examined with full attention to their rhetorical purposes.[9]

In summary, the narrative of the controversy can now be regarded as describing not only the trajectories of individuals, but also the trajectories of their specific projects through the continuum of relative privacy. At various times any such project might oscillate between the private depths of "intra-personal" dialogue, of which in this case the records of fieldwork are the least inadequate trace, and the middle layers of "interpersonal" exchange, of which scientific correspondence is an imperfect but richly revealing trace. Sooner or later, however, the project would emerge as a formal paper into the semi-public arenas of geological debate, where it might become the object of a "transpersonal" consensus among a small group of the more competent participants, even before it appeared in fully published form. An individual's gross shifts in interpretative position in the course of the controversy, as depicted in fig. 15.5, must therefore be regarded as the crude resultants of more finely textured shifts of formulation and presentation within the "space" of the continuum of relative privacy. But what is most important about that continuum is that it shows no sharp disjunction at any point between the individual and the social, still less between "real beliefs" and their rhetorical presentation. From its most private layers to the most public, the scientific activities that constituted the Devonian controversy were social as well as personal; and they were expressed throughout in terms of rhetorical argument.

16.2   Contests on the Agonistic Field

The pervasively rhetorical character of all the activities that constituted the Devonian controversy is no modern discovery by the historian. It was evident throughout to the participants themselves. The clearest evidence for this lies in the metaphors and images they used to describe what they were doing. As the narrative has shown, the most widespread and vivid imagery was that of military conflict. This was most obtrusive in the writings of the ex-soldier Murchison, but it was not confined to him. The controversy was repeatedly

---

9. On the systematically misleading character of sociometric methods, such as citation analysis, see e.g. Collins (1974) and Edge (1979).

likened to a field of battle, with attacks and counterattacks, frontal assaults and sudden movements sprung from the flank or the rear, tactical retreats, and open routs. There was heavy artillery to be deployed and ammunition to be brought forward. Sometimes the imagery was naval, and the anticipated effect of a bombardment was the sinking of the opponents' vessel; alternatively, it was feared that one's own vessel might be captured or sunk by one's opponents, who might even be characterized as pirates!

However, the controversy was not described only in terms of open antagonism. A vessel, "bark," or "bottom" represented an interpretative scheme that had been launched onto the rough seas of expert criticism, where it would fare better if manned by collaborators with complementary skills. Any such collaboration involved the investment of time and other resources, in anticipation of profitable returns in the future; at one point, Murchison used explicitly the economic metaphor of the joint stock company to express the risks and obligations entailed in his collaboration with Sedgwick (§8.1).[10]

At least one or two vivid metaphors were drawn from the world of sport. Murchison claimed at one point that De la Beche was a jockey riding with false weights, using his official position to gain an unfair advantage in the race between their competing interpretations (§10.6); and at a later phase of the controversy he referred to his own interpretation—by that time radically transformed—as a horse that was sure to win (§13.4). Austen debated the chances of success of his own interpretation as a fighting cock that would have to contest with others (§10.1); and Murchison later likened Sedgwick to a cockfighting enthusiast "basketted" because he was unable to pay his debts when his bird was defeated (§12.4).

Perhaps the most significant set of metaphors, however, was that related to the courtroom. An interpretative scheme put forward within the controversy was frequently termed a "case." Murchison commented on the fortunes of the Devonian interpretation (DEV.3) as being *"in re* Devoniana" (§13.3), and referred to its persuasive fossil evidence as his *"pièces justificatives"* (§12.8). Phillips referred to those same fossils as "silent witnesses," though in practice he clearly believed that under his own expert interrogation they could be eloquent indeed (§9.7). Cases were up for judgment, implicitly by the combined judge and jury of those around the table at the meetings of the Geological Society. De la Beche and his supporters complained when his case was tried instead before the unfair and inappropriately egalitarian "tribunal" of the British Association (§7.5, §8.3). Yet even at the Society it could seem to an impartial observer as though "De la Beche was on trial, Sedgwick and Murchison the prosecutors" (§11.3).

What belonged equally to the courtroom, although the legal imagery was not extended explicitly to it, was the use of analogues in the Devonian controversy. For these were treated exactly like *precedents* within the common law tradition.[11] When his GRE.1a interpretation was first attacked, De la Beche

10. On the risks of collaboration, regarded as a matter of investment, see Pinch's (1981) account of a delicate partnership between an experimentalist and a theoretician in modern physics.

11. Apart from their obsession with abstract formalisms, nothing has more seriously vitiated analyses of scientific work by philosophers of science than their tacit assumption that science is governed by procedures analogous to statute law rather than common law. Even Toulmin (1972 pp. 236–46) only invokes the analogy of common law for those episodes of scientific crisis in which "codified procedures" fail, *not* for the normal business of science.

cited the analogous examples of Greywacke coal on the Continent as relevant precedents that supported and justified his case; and his critics did all in their power to undermine the plausibility of those analogues in order to deprive that case of its precedents (§5.3 to §5.6). At a later stage Murchison presented his conclusions on Pembrokeshire as a thinly veiled analogue for the way in which he suspected De la Beche had been mistaken in Devonshire, and thereby staked out a new precedent in advance of his own fieldwork there (§6.5). Such precedents were used by both sides throughout the controversy to strengthen their cases and to establish the justice of an ultimate verdict in their own favor.

All these metaphors give clear historical support for Latour and Woolgar's notion of an *agonistic field* pervading the whole process of scientific work.[12] Open conflict or antagonism may only be sporadic, but there is an underlying structural conflict or field of competitive argument throughout. In the Devonian controversy feelings often ran high: "De la Beche is a dirty dog," Murchison exclaimed at one point with ungentlemanly bluntness (§8.3). But high affective intensity is no intrinsic characteristic of the agonistic field: indeed, emotional outbursts could well be considered to be as counterproductive in scientific debate as on the etymologically original agonistic field of athletic contest. "Pray *keep your temper*," Sedgwick had to instruct Murchison at one point, knowing how easily his collaborator's hasty anger could lessen the force of their case; "it is most important for you to be *calm*, but *resolute and decisive*" (§11.4). In fact, for much of the time the Devonian controversy, like the Geological Society which was its primary arena, followed the parliamentary model: the contestants would make up over a convivial supper for any hard words spoken in the meeting room (§8.3). But that gentlemanly civility did not weaken the underlying agonistic field of substantive argument, any more than it does in parliamentary politics. The rhetorical devices that pervaded both formal and informal exchanges throughout the Devonian controversy were merely the surface expression of the agonistic field, on which the relative *plausibility* of alternative interpretations, and the relative *credibility* of their proponents, were contested ceaselessly.

At the start of the controversy, De la Beche's report of Greywacke plants from Devonshire was manipulated on the agonistic field to appear either as an anticipated confirmation of a secure generalization or as an outrageously implausible anomaly (§5.2). Murchison and Lyell sought to brand his report with the objective-sounding label of "anomaly," but its anomalousness was constructed by those critics, not discovered in the fossil plants themselves. The report was only anomalous within the critics' own frame of interpretative reference (i.e., COA), not within that of De la Beche and his supporters (i.e., GRE). "Anomaly" was an *attribution* made for polemical reasons on the agonistic field, not something intrinsic in the evidence. Likewise Murchison and Sedgwick's later "discovery" of the Culm trough of Devonshire had that title bestowed on it by its proponents (§7.3). Its status was not intrinsic or self-evident, and its plausibility had to be argued vigorously.[13] One component of that alleged discovery, namely the correlation of the Culm strata with the Coal Measures elsewhere, remained highly controversial for at least three

12. Latour and Woolgar (1979, pp. 237–38 and passim).
13. On the attributed character of scientific discovery in general, see Brannigan (1981).

years. Until Murchison brought back from the Rhineland some persuasive analogical evidence in its favor (§12.3), his "discovery" was considered by many others to be an *error,* and it was widely expected that that latter attribution would be the one that would ultimately stick. Only in retrospect, and only by the consensual decision of those competent to judge the matter, did the Culm trough come to be regarded as having been a "discovery" at all.

In the case of a later "discovery," namely the DEV.3 or "Devonian" correlation of the older Devonshire strata with the Old Red Sandstone elsewhere, the "microgenesis" of the idea within Murchison's mind is of particular importance in its own right. The documentary record shows how—as in other examples of innovative scientific thinking—the crucial insight was reached by slow, sporadic, and hesitant steps (represented in part by COA.3 and COA.4), largely submerged below the public record, and putting the apparently sudden *Gestalt* switch (§11.1) into a quite different perspective.[14] Important though it is, however, the individual's cognitive pathway is only of relevance to the social achievement of "discovery" status in so far as it helps explain the persuasive advocacy of that insight on the wider agonistic field. Once again, the status of "discovery" was not intrinsic to Murchison's insight, but was attributed to it, first by him alone but ultimately by competent consensual agreement.

From the first eruption of the Devon controversy, in a report that was branded an anomaly, all the way to its resolution, in a consensus that in retrospect seemed "natural," the controversy was thus played out on the agonistic field of intensive social interaction. The contexts of discovery and justification are thus not only descriptively inseparable, but also analytically indistinguishable. The Devonian controversy can be seen in retrospect to have had a definite temporal structure; but the individual and collective activities and events that gave it that structure had the same agonistic character throughout.

16.3   Persuasive Powers and Social Interests

The metaphors of military conflict entail, or at least raise, the question of the role of power in scientific argument. If the Devonian controversy was contested and ultimately settled with rhetorical weapons deployed on the agonistic field, the question arises whether the consensual victory of one particular interpretation was due to its objective superiority in explanatory terms or to the superior rhetorical skill and firepower of those who advocated it. Of course those alternatives are not mutually exclusive, but the question cannot on that account be evaded. As the narrative of the controversy has shown, the congealing of a consensus around the Devonian interpretation was attended by some highly persistent rhetoric by Murchison himself; but the crucial question is whether or how far there was a *causal* connection between the two. The narrative also shows, however, that the question of power must be

14. There is an instructive parallel here with Darwin's partial "anticipations" of the famous moment of insight when he reread Malthus and conceived the explanatory potential of natural selection: see Gruber (1981a).

pursued much further. It must, for example, include the micropolitical power to influence the program of the Geological Society's meetings and the resolutions of its Council; the macropolitical power to influence the government on matters of scientific patronage; and, above all, the power to choose to devote time and material resources to the Devonian controversy or any other scientific problem.

In all those manifestations of power combined, if not in each separately, Murchison emerges from the narrative as unquestionably superior to any of his opponents or even his collaborators. Sedgwick was considered his superior in rhetorical skills, but Murchison himself could evidently be highly persuasive on occasion. His micropolitical power within the Geological Society was perhaps exceeded by Greenough's, but not by much. His influence with Tory politicians was probably less than Buckland's, and for handling Whigs he had to defer to Sedgwick; but his sedulous cultivation of powerful aristocratic connections was unrivaled. Above all, even before his wife's inheritance of a substantial fortune, and certainly afterward, his wealth and freedom from other commitments allowed him quite simply to spend *time* on an issue such as the Devonian controversy, and to pursue it in trips to Paris and on lengthy fieldwork expeditions as far away as the Urals, in a way that none of his rivals could match.

By contrast, only Greenough came anywhere near the same combination of potential power, but he was considerably older and no longer vigorously active in fieldwork. De la Beche was almost cruelly tied to remote areas by his need to scrape a living from his mapping work; even when his official position became secure, it bound him by bureaucratic rules that left him with little freedom of action. Phillips was likewise restricted by the need to earn a modest income, although his part-time duties for the British Association did give him important lines of influence throughout the world of British science. Buckland and Sedgwick were both tied by academic and ecclesiastical duties; and Sedgwick, although free from the demands of a large family, repeatedly found it difficult to give the Devonian problem the attention that his collaborator Murchison expected.

It seems an inescapable conclusion, therefore, that the campaign to establish a consensus around the Devonian interpretation was immeasurably facilitated, at the very least, by the superior power of its original and chief proponent. On the other hand, it does not follow that the victory of that interpretation was nothing more than a manifestation of a silver tongue, political clout, and an ample bank balance. This is where it is important to recall the cognitive structure of the controversy, particularly its temporal dimension (fig. 15.2). For the ultimate triumph of DEV.3 was no straightforward victory. In the earlier phases of the controversy Murchison was deploying much the *same* powerful resources in the service of *other* interpretations, yet he was *not* able to cajole other geologists into forming any stable consensus. On the contrary, his opponents' GRE interpretations, although put forward with inferior resources of time, oratory, money, and clout, forced Murchison into making successive modifications of his COA interpretation, and in the process he even lost Sedgwick's wholehearted agreement (fig. 15.5). It was only the persistent stalemate between the two main positions that forced him to consider the possibility of a new class of interpretation (DEV) that would embody a crucial concession to his opponents. It was *that* interpretation (in its variant DEV.3), with its unacknowledged debt to what Murchison had previously rejected with all the power at his disposal, that eventually became the object of a successful campaign of persuasion and consensus. Seen in the light of the cognitive structure of the controversy, therefore, Murchison's undoubted power can hardly be regarded as a—let alone *the*—determinant of its outcome, but only as a set of resources that facilitated and perhaps hastened the consensual adoption of the solution he eventually proposed.

That conclusion, however, does not lessen the importance of Murchison's example for the understanding of another aspect of the Devonian controversy. On occasion the participants themselves distinguished clearly between substantive geological questions and the assignment of credit for priority in discovery (§11.4); but in the ordinary course of argument—and frequently in the course of a single letter—they were mixed almost inextricably. The narrative of the controversy has shown, perhaps to the point of tediousness, how dominant and obtrusive the issues of credit and priority were in the work of some of the participants, and evidently, by inference from the documentary record, in their emotional lives too. Murchison's apparently insatiable appetite for public recognition of his achievements is the most obvious example, but it is unique only in its intensity. His determination not to be robbed of what he himself ultimately termed his "monument" (§14.3)—the series of major systems of strata that he had delineated and named—may evoke from the psychoanalytically inclined the suspicion of some connection with his childlessness, and of some craving for another form of vicarious immortality. But to those more sociologically inclined another (and not wholly incompatible) explanation will come readily to mind.

As outlined at the beginning of this book (chap. 2), the Devonian controversy took place mainly within the social milieu of *gentlemen* of science, most of whom were not primarily dependent on their science for their livelihood. But the pursuit of scientific research was far from being securely established as an appropriate or socially acceptable occupation for a gentleman. Many of those involved in the Devonian controversy—and in other scientific problems being debated at the same period—were able to make their scientific work acceptable by hitching it, in effect, to some existing social role. Austen's fossil collecting and local geological mapping fitted readily into the tradition of the country squire combining the management of his estate with an interest in the local antiquities and natural resources. Sedgwick and Buckland could both align themselves with the venerable "Gilbert White" tradition of the parson devoted to natural history, though both reinforced it with that of the conspicuously eccentric don. Greenough imported the more typically Continental model of the wealthy patron of all the arts and sciences, becoming, as it were, a cultured German princeling on the Regent's Park. Lyell could adapt to the service of geology the existing model of the widely read and cosmopolitan man of letters, and lose no social status by being substantially dependent on his royalties as an author.

For two other major participants in the Devonian controversy, the problems of social identity were not so easy to solve. Murchison had failed as a soldier and as a conventional country gentleman, and was attempting instead to establish a place for himself as a full-time gentlemanly *specialist* in science.

Most of the other participants just mentioned were almost equally specialists, in their degree of commitment to geology, and some followed their science with equally full-time dedication. But Murchison was almost unique in his intensive pursuit of one highly specific project *within* geology.[15] Others spread their risks more widely; for Murchison it was a matter of make or break. If such highly specialized scientific work were to seem acceptable as an occupation for a gentleman of independent means, the outstanding importance of his achievements had to be acknowledged unambiguously by other men of science, and that recognition had to be made widely known to his social equals and superiors outside the world of science. Murchison's pressing need of such recognition, for the establishment of a social role without exact precedents, is ample explanation of his almost obsessive concern with receiving due credit for his work on the Devonian problem and other aspects of his broader project in Palaeozoic stratigraphy.

More particularly, it explains his concern that his achievements should be recognized not only among the highly competent geologists at the Geological Society and abroad, but also and indeed primarily in the wider social circles of Westminster, Whitehall, and Buckingham Palace, and among the landed gentry of "Siluria" and his aristocratic neighbors in Belgravia. If the would-be Lord Caractacus was adopting any existing social role, it was that of the nobleman with vast estates scattered across the country and even abroad: if not by outright ownership, then at least by their coloring on a geological map, the tracts of country recognized as Silurian and Devonian became, in his words, his "own property" (§13.4).[16]

While Murchison was striving to develop that quasi-aristocratic style of full-time science, De la Beche had been forced by his sudden poverty to explore a radically different way of maintaining social respectability. His establishment of a new position as *the* government geological surveyor was slow and cautious. It had to be achieved by small and discreet steps, to avoid frightening the authorities with the prospect of an open-ended commitment to indefinite expansion. With an adroit handling of the bureaucratic machine, De la Beche spread his risks as soon as he could, by maintaining his mapping under one government department (the Board of Ordnance), but starting a public collection of specimens under another (the Department of Woods and Forests). But this process of establishing himself as what later came to be termed a *professional* scientist was extremely precarious, particularly in its earlier stages. The government were scared of being "jobbed" or taken for a ride by De la Beche, as they had been by MacCulloch. Any rumor of alleged incompetence that came to the ears of the politicians and civil servants could have ruined De la Beche's new and promising career.

As the narrative has shown, this precarious situation accounts for De la Beche's almost paranoid alarm at Murchison's repeated attempts to propagate just such damaging rumors. Conversely, De la Beche's growing success readily explains the virulence of Murchison's attacks. What was expressed emotionally as extreme mutual jealousy and antipathy was powered by a real

---

15. It would be misleading to give the impression that Murchison did *no* significant geological work outside his Silurian and (later) more broadly Palaeozoic project, but to his contemporaries his other research seemed by comparison conventional and even dull.

16. Secord (1982) rightly emphasizes this territorial aspect of Murchison's ambitions.

underlying conflict of interests. That the interests were causally prior to the emotions, and that the antipathy was not primarily a matter of personalities, is clear from the history of the relationship between the two men. They were friendly toward one another before De la Beche embarked on his course toward professionalism, and De la Beche claimed—probably in all sincerity— to be surprised and dismayed by Murchison's antagonism when the Devonian controversy began (§5.3). Eight years later, as soon as De la Beche acknowledged in public and in practice what Murchison had achieved in the course of the controversy, peace was at once restored (§14.4). Only between those limits, while their contrasting career interests were in conflict, was their relationship one of apparently implacable hostility.

The question then arises, however, whether this conflict of career interests between the two major combatants in the Devonian controversy was reflected in any way in the substantive issues involved. Any gross correlation between victory in the controversy and a successful career pattern might seem to be ruled out: for Murchison's was the successful solution to the controversy, but De la Beche's was the career that became the dominant pattern for the future. It could however be argued that at least in the short run the correlation held, since until the later decades of the century the aristocratic model did seem to be triumphant in British science, not least through Murchison's exemplification of it. But in fact this kind of correlation is beside the point. As in the matter of power, so here too the cognitive structure of the controversy implies that the ultimate success of the Devonian or DEV.3 solution *cannot* be attributed to any such "external" or totally nonintellectual factor. For the successful solution borrowed decisive elements from *both* the preceding series of provisional solutions, GRE and COA, which had been championed by De la Beche and Murchison, respectively. In substance, though not of course in public acknowledgment, the victory in the Devonian controversy belonged as much to the professionalized government surveyor as to the "King of Siluria."[17]

There remains however the question of the possible influence of other forms of wider social interests on the course and outcome of the Devonian controversy. This is not a matter to be decided a priori, either by philosophers determined to maintain the cognitive purity of science against social contamination, or conversely by self-styled radicals determined to unmask the roots of all scientific knowledge as ideological.[18] The degree of permeability or insulation between scientific knowledge and wider social interests is a matter for empirical investigation, and it would be surprising if it did not vary widely between one specific situation and another. This is no place to discuss the possible influences—which are likely in any case to have been two-way— between the general features of early nineteenth-century geology and equally general features of the contemporary social world. But three specific features of the Devonian controversy do deserve comment at this point.

17. The title "King of Siluria" was not explicitly bestowed on Murchison until 1849, on a spectacular public excursion to Dudley during the second BAAS meeting in Birmingham (Secord 1982), but the sentiment certainly predates that event.

18. Shapin (1982) rightly insists on the primacy of *empirical* work in any assessment of the explanatory potential of historical studies in the sociology of scientific knowledge, and he rightly rejects attempts by philosophers to dismiss it a priori.

One of the most important long-term effects of the successful solution of the controversy was to consolidate an essentially *gradualistic* model of the history of the earth and of life, over against what had previously been the dominant model of occasional, sudden "révolutions du globe." At first sight, there might appear to be a striking compatibility between the formulation of that gradualistic model, primarily by geologists in England, and that country's successful if narrow avoidance of major social unheaval; and conversely between the continuing attractions of the alternative model among Continental geologists, and their history, particularly in France, of repeated political revolution.[19] It would be surprising if social experience were not at times drawn upon by scientists as a metaphorical resource, both for the construction of new concepts and for the assessment of their plausibility. The compatibility between English gradualism in geology and political life might therefore seem to go some way to explain both the formulation of the highly gradualistic Devonian interpretation by the staunchly Tory Murchison, and its perceived plausibility and rapid acceptance among a wider range of British geologists, all of whom, whatever their party loyalties, had a clear stake in political stability and gradual social change. But once the temporal dimension is introduced, this transnational correlation becomes far less impressive. Murchison, for example, espoused a relatively paroxysmal or "catastrophist" model of faunal change until shortly before he formulated the Devonian interpretation (§10.4); conversely, although Élie de Beaumont retained a strongly paroxysmal view of mountain building, he was persuaded by the plausibility of the Devonian interpretation into adopting a much more gradualistic view of faunal change than he had previously held (§14.5). A broader survey would only reinforce such individual examples. At best, therefore, the suggested structural correlation between knowledge and society must be regarded as "not proven."

Second, it is clear that Murchison hoped his Silurian research would provide a scientifically validated criterion in the economically vital search for new deposits of coal (§4.4). It is also clear that this had a profound effect on his attitudes and actions in the course of the Devonian controversy. It powered his attack on De la Beche's report of coal in the allegedly pre-Silurian strata of Devonshire (§5.4, §5.5); his tentative "anticipation" of a somewhat DEV-like interpretation (COA.4) was an ad hoc move to save it from empirical defeat (§8.2). It even led him to a bold conclusion in the Boulonnais, which he later had to retract and explain away with some awkwardness (§12.4, §13.3). No other geologist (with the possible exception of Scrope) shared his dogmatic conviction that no land plants had existed in Silurian times, and therefore that no coal deposits would ever be found in Silurian or still older strata anywhere. He violated his own (and other geologists') canons of reasoning by confidently interpreting the negative evidence of the absence of plant fossils as a conclusive sign of the nonexistence of plants at that period; and he was even inconsistent in not interpreting the analogous absence of fossil fish in the same way (§6.1).

It seems, therefore, that this was a clear case of scientific judgment being swayed by strictly extrascientific considerations: for if Murchison's claim

---

19. The fruitful development of paroxysmal theories on the Continent later in the century is explored by Greene (1982), though not in terms of the sociology of knowledge.

proved well-founded, it would vindicate his gentlemanly career in geology, in the eyes of those concerned with the economic prosperity and industrial future of the country, as few other geological discoveries could do. Yet this cannot be claimed as an example of the social determination of scientific knowledge. Unquestionably it helped to power Murchison's trajectory through the controversy, giving him an intense personal reason for persisting in his attack on the interpretations (GRE) that would have invalidated it. But seen in retrospect—and indeed as he himself would have seen it—his claim about coal deposits had been a two-edged weapon; it had led him astray as well as down the road to victory. The Devonian (DEV.3) interpretation, which in its "precursor" form (COA.4) he had first devised in order to salvage his precious criterion for coal prospecting (§8.2), was later adopted consensually on *other* grounds and did not entail the acceptance of that criterion by other geologists. In the course of time, and as a corollary of *other* evidence, they did come to agree that the Devonian strata contained the earliest known land plants, and to infer that the Silurian period had probably predated the first land flora anywhere. But there is no evidence that their judgment on those matters was swayed, as Murchison's had been, by a concern for the public justification of geological science, for in their eyes that was already sufficiently secure.

Third and last, the resolution of the Devonian controversy required the mobilization of those following a career pattern distinct from either Murchison's or De la Beche's. The clinching evidence was not only the result of Murchison's competent but quite conventional fieldwork in Devon, the Rhineland, and Russia, but also the product of fossil specialists such as the impoverished Lonsdale and the less than gentlemanly Sowerby and Phillips, all of whom were attempting to utilize their expertise for their livelihood. The consensual resolution of the Devonian controversy was unquestionably facilitated by the specialized skills deployed by such more-or-less *professional* palaeontologists. As long as major systems or *terrains* were considered to be delimited by *séparations tranchées*, geologists could hope to develop a good enough "eye" for characteristic fossils to recognize major systems in the field. But once those sharp distinctions were broken down into insensibly gradual transitions, closer scrutiny by fossil specialists became indispensable. Phillips found it worthwhile to do his own intensive collecting in the field (§13.5); but no fossil specialist could give fully authoritative opinions without prolonged indoor work, comparing new collections with old, and specimens with published illustrations. Thus the resolution of the Devonian controversy was found to require the cooperation of the professional fossil specialists, and it certainly signaled their rise into a newly prominent position in the geological world.[20] Its successful outcome unquestionably furthered their career interests by demonstrating the value of their specialized knowledge. But those interests cannot be regarded as having been, in any but a trivial sense, a determinant of that outcome. For as the full narrative has shown, the relevant expertise was contributed not only by Lonsdale, Sowerby, and Phillips—the vanguard of what was to become an army of specialized professional geologists later in the century—but also, for example, by a moderately wealthy *bourgeois* such as de Verneuil. It might be suspected that the professionals would have had

20. On this development in the natural history sciences generally, see Allen (1985).

a greater vested interest in multiplying the number of species that they alone could authoritatively distinguish and identify; but this is not borne out by any evident correlation between professional status and a propensity toward taxonomic splitting. When the Devonian fauna was first discussed, Lonsdale was indeed perceived as a taxonomic splitter, but his fellow-professional Sowerby as a lumper (§9.3). A single contrasted pair such as this cannot, of course, establish the point conclusively; but there is no evidence that the career interests of the professionals had any systematic effect on the *substance* of their expert judgments of the fossils. There is nothing to suggest that, if *all* the relevant specialists had happened to be, like de Verneuil, gentlemen of independent means, the outcome of the Devonian controversy would have been any different.

16.4   The Transformation of a Paradigm

The Devonian controversy was not a major "scientific revolution" or episode of "extraordinary science"; nor was it merely a minor puzzle within a humdrum and static tradition of "normal science," the solution of which left that "paradigm" essentially unchanged. As historians of science have long recognized, the Kuhnian terms are difficult to apply to real examples of scientific change without somewhat procrustean adjustments. In retrospect, perhaps the most important of Kuhn's insights was his emphasis on the socially embodied and socially sustained character of the traditions within which virtually all scientific work is and has been carried out. It is the activities of *persons*, not disembodied ideas, concepts, theories, or "research programs," that constitute research traditions.

The narrative of the Devonian controversy has provided a vivid illustration of that point, which ought by now to be an axiom. The practice of geology was sustained and developed during that period by a quite specific set of social activities, ranging from fieldwork, through intensive correspondence and discussion, to the publication of conclusions in definite conventional forms (chap. 2). In an age of gentlemanly specialists, without formalized modes of training, the cognitive traditions of the science were not maintained through the exposure of neophytes to the congealed wisdom of textbooks. Instead, they were passed on through the practice of fieldwork in the company of more experienced practitioners, and through witnessing the spectacle of expert debate in arenas such as the Geological Society. It was not from textbooks that the correct meaning and application of technical terms and concepts was learned, but rather from witnessing their *use*—whether routine or still controversial—by authoritative and established practitioners. The open-air expanses of fieldwork and the indoor arenas of expert debate were at the same time schools for ostensive learning.[21]

Within geology, the dominant collective enterprise of stratigraphy was sustained and developed by a shared collection of practical principles, much of it in the form of the tacit knowledge of the craft of geological fieldwork.

21. On the role of ostensive learning in science, see Barnes (1982, chap. 2). On paradigms and normal science, see Kuhn (1970, first published 1962).

The relatively stable core of this body of practical knowledge was learned ostensively by newcomers to the science, while more seasoned practitioners were continually refining and modifying the application of even the most well-established categories. Thus structural terms such as "saddle" and "trough," and the practical skills needed to recognize instances of them, were already stable by the start of the Devonian controversy, and remained so throughout; yet in an exceptionally difficult case, such as the alleged "Culm trough" of Devon, the application of the term was highly controversial and the consensual recognition of that trough was treated retrospectively as an important discovery (§16.2). "Slaty cleavage" was new enough at the start to distinguish the accomplished from the amateur geologists, and it was difficult enough for even elite geologists to disagree in some cases about its correct application (§7.3); but within a few years its use and recognition had become routine and stable. Such examples show how even the shared meanings and assumptions that formed the backdrop to the Devonian drama were stable only in a relative sense and on a short timescale. Seen in a longer historical perspective, even the most routine terms were subject to continual shifts of meaning, owing to modifications in their application and use by established practitioners.

Perhaps the best example of this from the Devonian controversy is the crucially important term "system." At the time the controversy began, the word was used casually and loosely as a convenient term for all formations of strata having roughly the same character and orientation *within a limited area.* But during the period of the controversy the term developed two more specialized senses by a kind of semantic divergence. In both cases the word came to denote all the formations with a certain distinctive character, *wherever they were found.* But in the French usage promoted by Élie de Beaumont, a *système* had folds and other structural features with a distinctive orientation; whereas in the English usage promoted by Murchison, a "system" had a distinctive set of fossils. In both cases the term came to imply an attribution to a specific period in the earth's history; but a *système* referred to the period of elevation or folding of the strata, whereas a system referred to the period in which they had been deposited. But the important point is that this contrast was not established once and for all by explicit definition; it emerged gradually and perhaps unexpectedly in the course of practical use. When Murchison first named his Silurian system, he seems to have meant it primarily as a *système,* and Élie de Beaumont seems to have understood it that way (§6.1). It was only in the course of use that Murchison gradually shifted his emphasis from structure to fossils, until "system" became a virtual synonym not of *système* but of *terrain* (§14.3).

When Murchison proposed the "Devonian system," the new term added the ambiguity of "Devonian" to that of "system." When the controversy began, "Devonian" simply denoted *any* rocks or fossils from Devonshire; right to the end, phrases such as "the Devonian fossils" could still bear that original meaning. The term was first used in a more specific sense to denote a part of the sequence of pre-Culm formations in north Devon; these "Devonian" strata were at that time attributed to a system intermediate between the Cambrian and the Silurian, or else correlated with part of the Cambrian itself (§7.3). That usage met with no approval and was quickly dropped; but it was resurrected nearly three years later to denote almost *all* the pre-Culm strata of Devon, now drastically redated as the equivalents of the sharply contrasting Old Red Sandstone elsewhere (§11.2). Owing to its original and continuing use in a strictly geographical sense, this new meaning of "Devonian" was at first far from clear. Only gradually did other geologists realize that Murchison was using it to denote *all* the formations *anywhere* that contained a distinctive fauna intermediate between the Silurian and the Carboniferous, and that he was attributing this "Devonian system" to a specific period in the history of the earth and of life.

Examples such as these illustrate the indefinite *malleability* of all the technical terms that constituted the language in which the course and outcome of the Devonian controversy was expressed. As in any other scientific debate, the crucial technical terms, and the new concepts they were used to express, did not have their meanings defined for all time: in Barnes's apt phrase, they did not come "prepackaged" with explicit and unchangeable rules for their future application.[22] On the other hand, their meanings were not simply fluid or plastic, in the sense that they could be modified to any extent and applied at will. The metal-working term "malleability" is a more appropriate metaphor by which to express the deliberate, strenuous—and sometimes noisy!—intellectual work by which meanings were modified and concepts shaped or forged. Generally the meaning changed only a little at a time, and often so little it was scarcely noticed by those who used the term. But in the long run, terms and concepts were capable of withstanding major changes in shape without breaking, and of being shaped into forms adapted to quite new uses.

The malleable meanings and changing applications of the terms used in the Devonian controversy point to the pervasive element of *interpretation* that participation in the debate demanded. The crucial questions of correlation that were the central issues of the controversy depended on interpretative acts, by which sense and meaning were extracted from a confusing array of empirical materials, and understanding achieved. For the cognitive goal of the participants was *understanding*, and only subordinately prediction and control. They were concerned primarily with understanding the puzzling rocks of Devonshire, and ultimately others from the Urals to New York and beyond, in terms of the history of the earth, the structure of its crust, and the sequence of life at its surface. Stratigraphical geology, like any other branch of scientific research though more patently than most, was (and is) intrinsically a hermeneutic activity. It is for this reason that the term "interpretation" has been used so freely throughout this narrative and analysis of the Devonian controversy, and terms such as "theory" so sparingly. That stylistic preference not only expresses the character of what was going on in the controversy, seen in retrospect; it also reflects the participants' own evident awareness that interpretation is what they were engaged in.

Like any other metaphors, malleability and hermeneutics have their limitations. One that they share is that they both suggest a solitary activity: the craftsman forging an artifact on his anvil or the biblical scholar wrestling with the meaning of his text. In fact, as the narrative of the controversy has shown, it was not only the changing uses of technical terms and the modified

22. Barnes (1982, chap. 2).

conceptual meanings they expressed, but also the very reality of alleged "facts" that were shaped through argument and debate. The best example of this is the interpretation of the crucially important relation between the Culm and the older strata of Devonshire, and the hotly debated reality or otherwise of an unconformity at that point in the sequence. Sociologists have often used the term "negotiation" to express this kind of collective process, but that diplomatic metaphor has its own important limitation in that it suggests an element of conscious bargaining and deliberately exchanged concessions. Yet although the concessions were rarely overt, the Devonian controversy did unquestionably involve something akin to diplomatic negotiation. The proponents of GRE and COA dug their heels in about, respectively, the integrity of the sequence of strata in Devon and the Coal Measures age of the Culm plants. Those "battle lines" (§15.1) were in effect the entrenched or nonnegotiable positions between the two sides. Yet mutual concessions were in fact made, as shown by the successive variants of GRE and COA, even if their character *as* concessions was not acknowledged in public—as is often the case in diplomacy—or even perhaps not consciously recognized. The role of formal published papers in relation to informal argument during the controversy could aptly be compared with the role of occasional—and generally unrevealing—press releases during the real hard work of diplomatic negotiations behind closed doors. And the finally consensual Devonian interpretation (DEV.3) could be regarded as the analogue of a successfully negotiated treaty, precisely because it incorporated the nonnegotiable positions of both sides and found a reconciliation between them.

All these metaphors jointly express the way in which the course and outcome of the Devonian controversy embodied a subtle, gradual, and yet ultimately radical change in the collective research tradition within which it took place. The "paradigm" of stratigraphical geology was real enough, but it was not static in character. It was not defined by any specific past achievement or by any set of them. Indeed, the only past "exemplar" that was cited repeatedly, namely William Smith's ordering of the English Secondary strata by means of their fossils, was used with clearly rhetorical intentions as a weapon on the agonistic field; its invocation was generally tendentious and its applicability far from agreed. In fact the paradigm was not defined at all, and its operation can only be discerned in terms of contemporary practice. The importance of the Devonian controversy is that its successful resolution *transformed* that practice, and hence also the paradigm that it embodied, and that it did so without any break, rupture, or revolution.

If the establishment of the Devonian system had merely brought order and clarity to a previously chaotic and obscure part of the stratigraphical record, it could have been regarded as the solving of a routine puzzle within a static paradigm. But, in fact, far more was entailed in the resolution of the controversy. In the first place, although Murchison's Silurian system had already embodied a claim to have brought order to the "chaos" of an undifferentiated ancient Greywacke, that claim remained highly precarious until the new Devonian system could be seen to link the Silurian to the securely established orderliness of the Carboniferous and still younger strata. Only then did it become clear that the practice of stratigraphical geology could in principle extend over the whole range of the recorded history of the earth, and that, in Buckland's vivid metaphor, the process of *enclosing* "the common field of geology" had no intrinsic limits (§13.2).

Second, although Murchison's Silurian strata in their original area had already suggested that even ancient Greywacke rocks could locally put on the appearance of much younger Secondary strata, the tacit link between rock type and age was not severed decisively until the new Devonian interpretation gained plausibility. For the startling and initially implausible redating of the older rocks of Devon involved attributing Greywacke of very ancient appearance to the same period as the Secondary formation of the Old Red Sandstone. That inference was then confirmed by Murchison's conclusion that the flat-lying strata in Russia, which were so reminiscent of the younger Secondary strata that Smith had first mapped in England, were in reality of Palaeozoic age (§13.4, §13.6). Only as a result of that striking example did it become clear that the practice of stratigraphical geology would have to abandon what had seemed one of its most heuristic rules of thumb: rock type could no longer be regarded as a rough guide to relative age.

Third, to compensate for the loss of that criterion, the outcome of the Devonian controversy vindicated but also extended and transformed Murchison's confident use of fossil evidence. Once again, he had already claimed the superior reliability of fossils when outlining his Silurian system, but the claim only became convincing when the Devonian interpretation resolved the puzzlingly anomalous character of the Devon fossils, and brought them into line with similar fossils elsewhere. Significantly, however, that vindication of the fossil criterion entailed a major change in its application, which Murchison himself had not anticipated. Lyell had already attempted to order the scattered Tertiary deposits by tracing piecemeal changes within a broadly similar set of fossils; but the validity of that work was not unquestioned, and anyway the Tertiary deposits were insignificant in thickness—and inferentially in their period of formation—in comparison with the older major groups or systems (fig. 3.4). It was only with the establishment of the transitional character of the Devonian fauna, linking those of the equally vast systems of Silurian and Carboniferous strata, that this Lyellian principle of gradual and *piecemeal* change came to be incorporated into stratigraphical practice. That practice was thereby drastically modified. For polemical purposes the Devonian continued to be presented as a vindication of William Smith, but in practice Smith's criterion of characteristic fossils had been transcended and transformed almost beyond recognition.

Finally, in what appears deceptively like a contradiction of the fossil criterion, the Devonian interpretation implied for the first time that major formations with radically different rock types *and fossils* might need to be assigned to the same system and hence to the same inferred period in the earth's history. The striking contrast between the older rocks of Devonshire and the Old Red Sandstone elsewhere, and between their respective fossils, was initially a major stumbling block in the way of the acceptance of the Devonian interpretation. In order to make the case persuasive, Murchison had to minimize what in retrospect was one of the most fruitfully innovative aspects of his interpretation. He had to claim that the contrast was not without parallel; yet the only known analogous instances were far smaller in scale, as for example between the normal Oolitic limestones of southern England and the Midlands, and the coaly strata of the same age in Yorkshire (§4.1). Murchison's confidence in the equivalence of the Devon rocks and the Old Red Sandstone was eventually vindicated in Russia, where he found the contrasting rock types and fossils in close association (§13.4). Although the causal explanation of the contrast remained obscure until long afterward, the resolution of the Devonian controversy entailed a drastic modification of stratigraphical practice, for it involved a further transformation in the application of the criteria by which correlations were to be established from one region to another.

In all these ways, the resolution of the Devonian controversy left the practice of stratigraphical geology profoundly changed from what it had been before. But this major transformation was accomplished with none of the symptoms of a "scientific revolution." It was achieved *within* an established paradigm, or rather within a socially embodied research tradition. It entailed the *gradual* modification of the meanings and uses of technical terms, methods, criteria, concepts, and so forth. The paradigm, like the terms and concepts in which it was expressed, proved to be gradually but indefinitely malleable and adaptable.[23] The Devonian controversy shows how in scientific research, no less than in politics, art, and religion, human traditions continually provide from *within* themselves the resources for creative renewal.

16.5   Social Construction and Natural Knowledge

One crucially important question remains to be addressed in this final section. The "great Devonian controversy" was settled through a social process of argument among the proponents of alternative interpretations, and eventually by their virtual consensus around an interpretation that had not been anticipated by anyone at the outset. But did the resolution of the controversy entail anything *more* than some experts' agreement to agree; did the outcome *also* embody a genuine improvement in human understanding of the structure and prehuman history of the planet? The concept of a distinctive Devonian system and of its corresponding period was unquestionably a human and social construction; but was it—and is it—*also* a piece of reliable natural knowledge?

The question may seem as absurd to modern practitioners of geology as it will seem out of place to many historical and sociological analysts of science. But the question of external reference or epistemological validity should not be "bracketed" or evaded, and certainly it cannot be by anyone with insider's experience of both the practice of scientific research and its reflective analysis. The explanatory power of the successful solution to the Devonian controversy was such that in retrospect it can seem both inevitable and *natural*, particularly to those who now use it as a matter of everyday routine. But the narrative and the earlier sections of this analysis have illustrated the complexity, contingency, and *non*-inevitability of the scientific arguments by which that consensual interpretation was achieved. Only such a historical reconstruction can retrace the stages by which the Devonian concept

23. Ibid., chap. 4.

was shaped. Only in this way can the subsequently concealed or forgotten evidence be recovered, to show how the finished model ship was inserted into the bottle and made to look so "natural" there (§1.6). Conversely, however, the end product of scientific research can seem so obviously *artifactual*, particularly to those who analyze the process of construction from the outside, that the question of its relation to the natural world can too readily be dismissed as of marginal interest, or—less excusably—as beyond the analyst's terms of reference. The modes of construction of model ships and their modes of insertion into bottles then become objects of study in their own right, and the makers' claims to be building representations of real ships that exist in the outside world are dismissed as an illusion or ruled out of order.

The Devonian controversy is an instructive case study with which to try to transcend this sterile dichotomy. It can be used to explore the way in which a consensual product of scientific debate can be regarded as both artifactual *and* natural, as a thoroughly social construction that may nonetheless be a reliable representation of the natural world. It is an instructive case not least because the participants themselves would probably have conceded, and their modern intellectual descendants would certainly agree, that the end product of this historical controversy was and is both artifactual and natural. The controversy subsided when the core set of major participants agreed that certain strata, which some (but not all) of them called the Devonian system, had been formed during a certain period in the past history of the earth, characterized by a unique and distinctive assemblage of animals and plants. Since the term Devonian was said to denote a distinctive major period, marked for example by the first flourishing of vertebrate animals and terrestrial plants, its claimed reference to a real sequence of prehuman events was unequivocal. It was put forward as an assertion about the natural world; it referred to events that were claimed to be as real as the Norman Conquest or Columbus's discovery of America, events equally inaccessible to direct experience but equally recoverable from historical records correctly interpreted.

On the other hand, Murchison would have agreed that he might have been deprived of his "monument" of the Devonian system; another geologist might have been implicitly commemorated, and another region might have provided the name for that portion of the record of the history of the earth. There was an element of contingency in the choice of Devonshire as the eponymous region, and in the ultimately universal adoption of the term "Devonian." As Murchison was well aware, he had to campaign for his Devonian system with all possible speed in order to forestall claims that, for example, "Rhenish system" would be a more appropriate term for strata that were somewhat poorly exposed and confusingly complex in Devonshire itself. In that terminological sense at least, the Devonian was implicitly recognized by all the participants to be an artifactual entity, despite its claimed reference to the natural world.

Moreover, at least some of those original participants would have agreed with most if not all their scientific descendants that the Devonian was also artifactual in a deeper sense. Those geologists (e. g., Élie de Beaumont) who claimed that the history of the earth had been punctuated by occasional global *révolutions* could also logically expect to find the strata divided by *séparations tranchées* into unambiguously discrete chunks, among which the Devonian system—under that or any equivalent name—would have the status of a *natural* entity. Even Murchison himself never wholly assimilated the implications of his adoption of a more gradualistic conception of organic change, and he continued to treat the Devonian and his other systems as ultimately natural units: "the [fossil] types are clear and distinct," he assured Sedgwick as the controversy subsided, "& birds of *passage* are not to frighten us" (§14.1). But those who did work fully through the implications of the gradualistic view (e.g., Lyell, Phillips) realized that the contingency that had attended the construction of a "Devonian system" went further than the name. If *révolutions* were not sudden and globally synchronized, but gradual and localized, and if organic change were likewise gradual and piecemeal, *any* division of the naturally seamless fabric of the earth's history was bound to be a matter of convention and convenience. Phillips certainly realized that the definition of his "Middle Palaeozoic"—as, for just this reason, he preferred to call the Devonian—was ultimately arbitrary (§14.1). He hoped to retain a more "natural" status for the highest level in his classification of the earth's history (e.g., his "Palaeozoic"); but he clearly recognized that the subordinate units, even if as vast in inferred duration as the major systems or *terrains*, were no less conventional in character than Lyell's much smaller but explicitly arbitrary subdivisions of the Tertiary (e.g., "Miocene").

Once the ultimately conventional character of any such stratigraphical divisions was recognized, the establishment of the Devonian and similar categories was implicitly acknowledged to have been a matter of constructing a series of artifacts. Yet that did not—and does not—dissolve the question of their natural reference, any more than does a publisher's decision to launch a series on European history by commissioning separate volumes on conventional "periods" such as the Reformation and the Industrial Revolution. An acknowledgment that the Devonian system is an ultimately conventional division of the stratigraphical record, and that it owes both its name and its defined limits to a highly contingent series of events in early nineteenth-century Europe, certainly amounts to an acceptance of its artifactual status as a social construction. But it does not and cannot settle the issue of its reliability as a representation of a portion of a real past history of the earth.

Two extreme proposals for settling that issue are equally unconvincing, indeed hardly credible. One would be the naively realist view that all that the establishment of the Devonian required was for geologists to go and look at the relevant sequences of strata and collect the fossils from them. The complexities of the controversy would then have to be attributed to the distorting effects of theoretical preconceptions that were unwisely clung to by individuals who ought instead to have faced the facts. That view, however naive it may sound, was embodied in the caricature that De la Beche drew to express the opening phase of the controversy (fig. 5.4). But his caricature was clearly rhetorical in intent; at other times De la Beche, like all the other participants, showed that he was well aware that more was involved than merely looking self-evident facts in the face. As the narrative of the controversy illustrates, even the simplest observations were necessarily embedded in a matrix of tacit knowledge, and dependent for their meaning on the socially acquired practice of stratigraphical geology. Some modern geologists might be found to defend an extreme realist view of the kind that De la Beche used for polemical purposes; but they would certainly not be the most reflective practitioners, and probably not the most original or creative ones either.

The opposite extreme has better intellectual credentials but is no more credible in the light of this particular case study. This is the extreme social constructivism espoused by some sociological analysts of modern science. They claim to have found, by detailed participatory observation of scientific research, that the processes of argument in the laboratory are so pervasively agonistic in character, even down to the "microprocessing" of instrumental readings and so on, that external "reality" cannot be seen to have *any* clearly discernible effect on the outcome. The claimed knowledge that is the end product of research appears to have been "manufactured." A research laboratory is regarded as just what the etymology implies, namely, a workshop for the social production of deceptively solid "facts." For it is claimed that that apparent solidity is *only* a social construction, and that "reality" is attributed to previously debatable statements when they pass irreversibly through a socially defined point of "stabilization" or "inversion" and are adopted into routine use. The consensus that establishes that routine use is said to depend on the definition, by the *winning* side in the controversy, of what *counts* as a competent experimental result and as relevant clinching evidence. The external point of reference in this pervasively social process is thus not a world of natural reality, but at most the criterion of technical success: the "knowledge" produced cannot claim more than instrumentalist status.[24]

This account of scientific knowledge as virtually *nothing but* a social construction may be held to have some plausibility when it is based on forms of modern research in which the "raw" natural world is almost excluded, or its possible effect minimized, by the systematic use of purified chemicals, laboratory strains of organisms, elaborate instrumentation, and so on. It is much less plausible as an account of a branch of science such as that in which the Devonian controversy took place, where the apparent disorder and un-expected contingency of the natural world was never far from the arenas of interpretative debate, and where technical success was a marginal issue in comparison with the search for meaning and understanding. Furthermore, the consensus that congealed around the Devonian interpretation was manifestly *not* dependent on any decision by the "winning" side that only that inter-pretation could *count* as a satisfactory solution. On the contrary, the consensus only formed when the apparent "losers" (the proponents of GRE) agreed that *their* criteria for a satisfactory solution had at last been met (in the proposal of DEV) in ways that had *not* been met in their opponents' earlier proposals (i.e., COA). And the narrative has revealed no single or decisive "point of stabilization" in the formation of a consensus, nor any evidence that it might not have been reversed and dissolved had some sufficiently persuasive counter evidence been produced. Despite the rapid initial formation of a provisional consensus around DEV.3 soon after it was first proposed (§11.4), its nominal co-authors both lost confidence in it while trying to consolidate it (§12.4 to

24. On the minimal effects of "reality," see Collins (1981, 1983); the metaphor of "manufacture" is used by Knorr-Cetina (1979, 1981); for the concept of "stabilization" or "inversion," see es-pecially Latour and Woolgar (1979). No attempt is made in this section to give a general evaluation of the increasingly important literature on "laboratory studies," but the relevance of the Devonian controversy as a contributory case should be obvious.

§12.6); it is only in retrospect that a stable consensus appears—deceptively—to have been both inevitable and irreversible (see fig. 15.5). The Devonian controversy does indeed suggest that "stabilization" takes place as a conjecture hardens into a conclusion that is taken to be reliable. But it is more a process than a point; however unlikely the dissolution of the new knowledge may come to seem, the process of stabilization remains in principle reversible.

In place of the implausible extremes of either naive realism or social constructivism, what is needed is not a compromise but a way of transcending the dichotomy: at a "metalevel," the interpretation of the Devonian controversy needs to mirror the structure of the controversy itself.[25] One key to the metaequivalent of DEV is the term "representation." The establishment of the Devonian interpretation, and all that it entailed and made intelligible, was like the drawing of a satisfactory *map* of that portion of the stratigraphical record and of the history of the earth. It was satisfactory in that it proved to be a reliable representation in ever more extensive contexts of use; yet it remained a representation, and not a replica. Bookish people with no practical experience of mapping often assume that a map is an unproblematic replica of reality, or merely a miniaturized version of what one would see from the air. Those who make intensive use of cartography know on the contrary that any map is a pervasively conventional representation. They also know that an indefinite number of different maps of the same area can be made for different purposes, yet all may be equally valid representations of the same natural reality. Even where such maps prove to be mistaken, they are always corrigible; but it makes no sense to talk of ever achieving a uniquely "perfect" representation, or a complete "correspondence" with reality, since different kinds of map are designed for different uses, and there is no limit to the further representations that may be needed for new and unforeseen purposes.

It should be clear that the analogy of mapping yields a way of retaining the constructivists' insistence on the social processes that went into the making of a piece of scientific knowledge such as the Devonian, while also allowing the realists' insistence that the real natural world of rocks and fossils, and the real history of the earth that originally brought them into existence, had a more than marginal effect on that claimed knowledge. Treating the Devonian as a reliable representation of a portion of the earth's history allows it to be recognized as a social construction through and through, just as much as a map is an artifact of paper and ink and a product of a social tradition of cartography. Yet the Devonian can also be treated as a reliable representation of a reality that existed before it was known, just as a map represents reliably, albeit conventionally, a topography that was there before the country was explored and before the surveyors arrived.

To put the point another way, neither "discovery" nor "construction" is *by itself* an adequate metaphor for the production of scientific knowledge. The outcome of research is neither the unproblematic disclosure of the natural world nor a mere artifact of social negotiation. The metaphor of *shaping*—or, in the original sense of the term, *forging*—has been used allusively throughout this book, as a less inadequate image. For the Devonian controversy shows

25. Recent discussions of the underdetermination of theories by evidence have in effect opened up a range of tenable positions *between* the extremes just outlined: for two essays derived from contrasting traditions but showing unexpected convergence, see the introductions to Hesse (1980) and Knorr-Cetina and Mulkay (1983).

how new knowledge is shaped from the materials of a real natural world, malleable yet often refractory; but it becomes knowledge only as those materials are forged into new shapes with new meanings, on the anvil of heated argumentative debate.

While the Devonian was shaped by social processes of actively interpretative work, however, what was being interpreted comprised a *cumulative* array of empirical evidence. Experimental psychologists have devised a vivid illustration of the active character of perception, which can be borrowed as a further analogy at this point. A photograph can be reduced to a pattern of dots, somewhat as it is in a newspaper, and a random selection of those dots presented to experimental subjects. As progressively more dots are added, and the subjects wrestle with the possible meaning of what they see, they may begin to guess what the original photograph represented. But different subjects may hazard quite diverse conjectures, depending partly on their previous experiences and interests. As still more dots are added, and the image becomes less grainy and more sharply defined, some of the early conjectures lose their plausibility while others gain; and in the end a virtual consensus may be reached. Thus throughout the process the actively interpretative work of perception is increasingly *constrained* by the addition of further information, yet at no time is it *determined* by it.[26]

Like any analogy, this one has its limitations, not least that it makes interpretation far more an individual than a social achievement; but it does suggest how the interpretative work that went into the shaping of the Devonian can be related to the cumulation of relevant empirical evidence. For a historical narrative of the Devonian controversy shows up what no static and ahistorical study of modern laboratory research can do, namely, the cumulative impact of *new* evidence on the course of the debate when traced over time. Much of that new evidence was of course put forward by participants with an interpretation to defend, but in no way does that detract from its significance. Both sides generally agreed that the new "dots," as it were, were reliable enough to be added, even if there was vehement argument about the meaning of the consequent improvement to the still ambiguous image. Even over the contentious and crucial issue of the alleged unconformity beneath the Culm strata, De la Beche's improved field evidence did eventually convince Sedgwick and Murchison, however reluctantly, *against* their theoretical expectations, leaving only the marginal Weaver unpersuaded (§8.6, §9.4, §11.2). Furthermore, some of the most significant new "dots" were added by participants who had *no* interpretative axe to grind, and whose finds could not therefore be suspected of bias: the best examples are the fossils found by amateur collectors such as Harding and Hennah, which repeatedly forced elite geologists on *both* sides of the controversy to reassess their interpretations. Eventually, however, the cumulative addition of new "dots" made one particular "reading" of the image consensually persuasive; but significantly it was one that had not been guessed at by *either* party while the image was still grainy and obscure.

In this way, it is possible to see the cumulative empirical evidence in the Devonian debate, *neither* as having determined the result of the research in any unambiguous way, as naive realists might claim, *nor* as having been

26. See, e.g., the illustrations in Gregory (1973, pp. 72–82); for the analogy with mapping see, e.g., Ziman (1978, pp. 82–84).

virtually irrelevant to the result of the social contest on the agonistic field, as constructivists might maintain. It can be seen instead as having had a *differentiating* effect on the course and outcome of the debate, *constraining* the social construction into being a limited, but reliable and indefinitely improvable, representation of a natural reality.[27]

To return, in conclusion, to the earlier metaphor of a court of law, which the participants in the Devonian controversy found so appropriate to express their own activities, the Devonian "case" was argued out in courtrooms such as the Geological Society by the persuasive advocacy of major participants such as Murchison and De la Beche. But other competent geologists, who in effect constituted the jury in the case, were not swayed into a consensual verdict only by the rhetorical skills of those presenting one side of the case, still less by being bribed out of court by promises of advancement in their careers. On the other hand, they would not have reached their verdict just by being left in the courtroom to study the exhibits that each side produced to back its arguments. What swayed them was of course the combination of rhetoric and evidence, persuasive argument and *pièces justificatives*. Like a court of law, and unlike, for example, a debating society, the geologists were concerned not with evaluating rhetorical performances but with reaching a justifiable conclusion about concrete past events in a real world. Significantly, both sides were obliged to shift their position during the protracted hearing, in response to the other's arguments and evidence, and the final verdict was based on a case that *neither* side had anticipated when the hearing began. That implies once more that the outcome was indeed constrained in unexpected ways by the evidence produced during the hearing; it suggests that justice was ultimately done to the natural world, which was the subject of the hearing and the raison d'être of the court itself.

Throughout this account of the Devonian controversy, the reflective opinions of the participants themselves have been given prominence, on the grounds that the self-understanding of scientific natives should never be lightly set aside by historical, sociological, or philosophical analysts of their activities. It is therefore appropriate to end this book by referring back to a remarkably perceptive comment by one of the major participants, which was quoted as an epigraph at the beginning (p. vii). To borrow De la Beche's prescient words, written several years before the Devonian controversy erupted, "the collision of various theories" about Devonshire geology, within the social arena of geological debate, not only spurred the contestants to strengthen their positions by exploring new regions and reexamining familiar ones. It also led to the finding of evidence that did "not suit either party," which obliged the contestants to devise "new theories" to explain it. And so, "in the end, a greater insight into the real structure" and history of the earth was indeed achieved.[28] What had been *shaped* by the social processes of argument and debate among gentlemanly specialists in the course of the "great Devonian controversy" had been a new piece of reliable scientific *knowledge*.

27. This formulation is derived in part from Bowker's (1973) important analysis of the much "harder" case of *religious* knowledge. His work points to the impossibility of "bracketing" the question of external reference, in a way that is directly applicable, by way of analogy, to the study of *scientific* knowledge.

28. De la Beche (1830, p. iii); the comment referred to debates that were current *before* the Devonian controversy erupted.

Appendices

A. "And They Lived Happily Ever After"

This note summarizes—in briefest outline—the subsequent careers of those who have figured most prominently in this account of the Devonian controversy. In reality, of course, the story had no fairy-tale ending: for several of the older participants, the controversy was one of their last pieces of active geological work, while others went on to work of quite different kinds.

Greenough (d. 1855), the oldest of them, continued to follow his geological and geographical interests up to his death, which occurred, appropriately enough, at Naples near the start of an ambitious journey to the East. De la Beche (d. 1855) continued to direct the steady expansion and consolidation of the Geological Survey and the Museum of Economic Geology (the latter now the Geological Museum), and he led the movement to found a School of Mines in London (now part of the Imperial College of Science and Technology). With such administrative responsibilities, he left the practical business of geological research increasingly to his growing staff of governmental geologists. He was awarded the Wollaston medal, somewhat belatedly, in the year of his death. There were then fears among geologists that a nongeologist might be appointed as his successor, but in the end Murchison (d. 1871) accepted the position and ruled the Survey in characteristically military fashion. Ironically, therefore, the chief champion of gentlemanly geology followed his earlier opponent in becoming a professional, at least in official position, though he managed to preserve the gentlemanly style of his earlier career. He was knighted in 1846 and created a baronet—at long last!—in 1862.

In his bitter dispute with Sedgwick over their Silurian and Cambrian systems, Murchison used his official position to the full to impose his own views on geological practice in Britain. Sedgwick (d. 1873) was pushed more and more into a defensive position; he never completed the large-scale memoirs that would have summarized his life's work adequately. He took an active part in the mid-century reform of his university and continued to teach into old age; but he saw his science changing rapidly under the new pressures of professionalism, and in the end became a kind of living fossil among his junior colleagues. His Oxford counterpart, Buckland (d. 1856), dropped out of active geological research much earlier, after his appointment (in 1845) as Dean of Westminster. In that position he remained prominent in public life, doing much to foster a respect for the natural sciences within the established Church and, in particular, to urge the practical value of geology in the nation's life.

Phillips (d. 1874) became a shining example of the new professionalism, rising from his less than gentlemanly origins to become Buckland's successor at Oxford. He continued to be highly active and creative in geological research.

457

Among other lines of work, he developed his earlier interest in the age of the earth. Although his estimates (roughly 100 million years for the known fossil record) were fairly modest by modern standards, they were a salutary brake on the undisciplined use of virtually unlimited time by Lyell and Darwin. Lyell (d. 1875) continued to follow a successful career as a scientific author, assuming the role of supreme theoretical arbiter in geology to the point where a disgruntled junior could epitomize his attitude as "la Géologie, c'est moi"! His chief opponent in France, Élie de Beaumont (d. 1874), continued to champion a paroxysmal interpretation of the physical history of the earth. Although he elaborated his theory to the point where its plausibility ultimately collapsed, it had by then become the inspiration for a highly fruitful tradition that dominated tectonic or structural geology for the rest of the century.

Of the provincial geologists who were important in the Devonian controversy, Weaver (d. 1855) had faded from the scene through ill-health even before it finally subsided, and Williams (d. 1850) published nothing more after the mid-1840s. Austen (d. 1884) outlived them all, became prominent in the Geological Society, and—in an undeniably "thin" period in its history—was awarded the Wollaston medal in 1862. (On the death of his father-in-law in 1853 he adopted his wife's family name, appearing in all later records as *Godwin*–Austen.)

Among those who played minor roles in the Devonian controversy, only Darwin (d. 1883) need be mentioned here. Having moved out of London in 1842, adopting a somewhat reclusive life in Kent for the sake of his health, he soon became less prominent at the Geological Society and did no further geological fieldwork. His major geological books, based on the *Beagle* voyage, appeared during the 1840s. When he returned to public prominence, with the *Origin of Species* (1859), it was of course as a biologist.

B.   The Devonian Modernized

Readers who are unfamiliar with modern geology may be curious to know about the present status of the "successful" solution to the Devonian controversy. This note gives a very brief and nontechnical summary.

The Devonian system and period have been universally adopted by geologists as an important major segment of both the sequence of strata and the history of the earth that that sequence represents and records. In both forms (as rocks and as time) the Devonian continues to be used in modern geology with its meaning recognizably congruent with what it became in the 1840s, though of course with its contents immeasurably amplified and refined as a result of further research throughout the world. As in the nineteenth century, it is bounded above by the Carboniferous and below by the Silurian. (The Carboniferous is now known in North America as a combination of the Mississippian below and the Pennsylvanian above; the Silurian now has a more restricted scope than in Murchison's definition, since the disputed strata that formed his Lower Silurian and Sedgwick's Upper Cambrian became—at a rough approximation—the modern Ordovician.) All these systems or periods, with Murchison's Permian at the top or in youngest position, together constitute the Palaeozoic era: a definition virtually unchanged from Phillips's extension of Sedgwick's original usage.

Murchison claimed that the Devonian, in view of the thickness of its constituent formations and the distinctive character of its fossils, would prove to be a *major* period in the history of the earth, comparable to the Silurian and Carboniferous. His intuition has been amply confirmed, in the opinion of modern geologists, by the ever-increasing accuracy of radiometric methods of dating. Making a realistic allowance for error, recent estimates assign to the Devonian the 50 million years lying between about 410 and 360 million years before the present. Its crucial place in the overall history of life has likewise been confirmed: it is regarded as the first period in which terrestrial vegetation and vertebrate animal life became relatively abundant and diverse. Detailed scrutiny of its invertebrate fossils is also regarded as having confirmed the broadly gradualistic view of faunal change on which the Devonian was first founded; the later interpretation of that gradualism in terms of organic evolution has greatly illuminated but not fundamentally altered that view. (The recent "punctuated equilibrium" theory merely reinstates a somewhat Lyellian view of transspecific change; it is too early to say whether recent speculations about occasional "catastrophic" episodes will succeed in reinstating Élie de Beaumont's view as well!)

The subdivisions of the Devonian that are now universally accepted by geologists reflect a sequence of faunas that connects the youngest Silurian through to the oldest Carboniferous faunas. Beyrich's confidence in the outstanding value of the goniatites in the stratigraphy of these rocks has been fully vindicated, for these cephalopod mollusks are now regarded as some of the most consistent and reliable fossils for the international and intercontinental correlation of Devonian strata. It is no accident that the subdivisions of the Devonian that form the agreed reference series for such correlations are all based on formations in Belgium and the Rhineland, rather than Devonshire. In order, from oldest to youngest, they are named after Gedinne, Siegen, Ems, the Eifel, Givet, Frasnes, and the Famenne (see fig. 12.2; Gedinne and Frasnes, which are not marked, lie respectively southeast and southwest of Givet). For although the "Rhenish massif" is now considered to be even more strongly folded and disturbed in structure than Dumont ever suspected, it can be unraveled more confidently than southwest England (or, for that matter, northwest France). All these regions are now regarded as the deeply eroded roots of a mountain chain that arose across northwest Europe around the end of the Carboniferous, in the "Hercynian orogeny" (named after the Harz). But Devonshire and Cornwall are considered to have been crumpled, sheared, and altered during those crustal movements to a degree that renders them inappropriate as a "type" region for the Devonian.

Despite many unsolved local problems, however, the rocks of southwest England are now correlated confidently with less disturbed sequences elsewhere. Murchison's insistent claim that the main (upper) part of the Culm must be of Coal Measures age has been vindicated by detailed study of far better collections of its fossil plants than were available to Lindley. Phillips's suggestion that its basal black limestones represented the Mountain (now, Carboniferous) Limestone has been similarly confirmed by the analysis of its invertebrate fossils. The Culm as a whole is regarded as representing a fairly complete Carboniferous sequence, correlated with "type" sequences named (in order from oldest to youngest) after Tournai, Visé, Namur, and Westphalia (see fig. 12.1; Tournai and Visé, which are not marked, lie respectively southeast of Lille and northeast of Liège). De la Beche's dogged insistence that the Culm strata lay with perfect conformity on the older series has also been vindicated, in the opinion of modern geologists, and his favorite locality (Fremington Pill) is still regarded as one of the best places in Devon to study the passage from youngest Devonian into oldest Carboniferous strata. Most of the Devonian subdivisions have been recognized in Devon itself from a detailed study of fossil collections far larger than those available to Phillips; the main limestones are now regarded as falling mostly within the Middle Devonian (Eifelian and Givetian), thus vindicating Austen's perception of their faunal affinity to those on the Continent.

The greatest problem still outstanding when the Devonian controversy subsided was to account for the striking contrast between the "ordinary" Devonian and the Old Red Sandstone that was agreed to be its temporal equivalent. It is characteristic of geological theorizing, in contrast to some other branches of natural science, that this lack of a *causal* explanation did not stand in the way of the general acceptance of that initially implausible correlation. (A more recent parallel would be the wide acceptance, among twentieth-century geologists and palaeontologists, of the plausibility of continental displacement on a global scale, *before* geophysicists devised a causal theory for it and in the face of their dogmatic insistence that it was physically impossible.) It was not until long after the general adoption of the Devonian interpretation that one of the many tentative explanations for the contrast came to be regarded as far more plausible than the rest. This was the suggestion that the Old Red Sandstone comprised sediments deposited in *nonmarine* conditions, in more or less land-locked basins on a large continent, whereas the "ordinary" Devonian comprised sediments deposited simultaneously in various marine environments. In the opinion of modern geologists, that explanation has been overwhelmingly confirmed by a variety of detailed studies, not least by the Lyellian comparison of sedimentary structures in the Old Red with their modern counterparts in nonmarine environments.

Although this brief summary necessarily glosses over many important problems in the modern understanding of the Devonian, it should be clear that the consensual outcome of the Devonian controversy in the mid-1840s has been incorporated into modern geology, and has been retained there without strain, because it has proved itself in practical usage as a valuable conceptual tool for ordering and understanding the history of the earth and of life.

C.   Note on Prices and Incomes

It is notoriously difficult to convert the prices and incomes from any period in the past into their modern equivalents. Major changes in the kinds and amounts of the goods and services required to sustain any given standard of life, in the standard expected by any given social group, and, more recently, in the character and scale of taxation, make any comparisons extremely hazardous and potentially misleading. Nevertheless it seems worthwhile to give readers who are not historians some guide, however rough, to the modern equivalents of the prices and incomes mentioned in this book. As a very crude approximation, it would seem that a conversion factor of about 40 in the value of sterling is on average the least misleading way to compare the monetary values of the 1830s with those of the mid-1980s. (In the following examples a *modern* exchange rate of two U.S. dollars to one pound sterling is applied, as being more realistic than the official rate at the time of writing.)

Spending a shilling to accept delivery of a long-distance inland letter would thus cost a British geologist of the 1830s the equivalent of about £2 or $4. By the time the Devonian controversy was over, however, the penny post had drastically reduced the cost—now borne by the sender—to much the same in real terms as at the present day. An ordinary scientific book selling for half a guinea would represent an outlay of about £20 or $40, whereas to subscribe five guineas for a copy of a finely illustrated monograph would cost ten times as much. The three guineas due annually for membership of the Geological Society—indispensable for any London geologist in the 1830s—represented about £125 or $250 *per annum,* but it brought access to one of the finest geological libraries in the world, to good geological collections, and to many of the facilities of a social club.

With such expenses involved, a modest income of £200 *per annum,* representing in modern terms perhaps £8000 or $16,000, was scarcely enough to support a gentleman in a life of science, unless he were unmarried and also in receipt of free housing and other benefits in kind (the proviso was met, for example, by the conditions of Fellowship at Oxford and Cambridge colleges). On the other hand, an annual income of £1000, roughly equivalent to £40,000 or $80,000, was quite ample to support a gentlemanly style of life and a family too; while an income of £3000 was positively princely. Again, it should be emphasized that these equivalents are mentioned here only as a *very* rough guide.

Bibliography

Manuscript Sources

This note summarizes the character of the manuscript documentation on which this account of the Devonian controversy is based. Abbreviations in parentheses are those used in the footnotes and refer to the libraries and archives listed at the end of this note.

Correspondence is by far the richest category of manuscript material. Although the handwriting of some of the participants in the controversy was execrable, many of them habitually dated their letters in full, thereby obviating most problems of dating; postmarks are usually available as a further check, since envelopes were rarely used even at the end of the period. (The rare cases of undated or misdated letters are mentioned in the footnotes; where postmarks are also missing—in letters delivered by hand—watermarks and internal evidence can be used to date them approximately.) There are very large collections of Murchison's incoming letters, both scientific and personal, in London (GSL and BLL, respectively); a smaller collection, including the important series he wrote to his wife during some of his expeditions abroad, is in Edinburgh (ULE). A very large collection of Sedgwick's incoming letters is in Cambridge (ULC); the only important gaps are in the series of Murchison's letters to him, many of which only survive as defective transcripts made later in the nineteenth century (ULC: AS, IIID). An equally large collection of Greenough's incoming letters is also in Cambridge (ULC); a smaller collection, and also much miscellaneous material such as loose notes, diaries, Geological Society ephemera, etc., is in London (UCL). A very large collection of De la Beche's incoming letters is in Cardiff (NMW), where there are also the late Dr F. J. North's invaluable transcripts of official letters and memoranda about the foundation and early years of the Geological Survey, the originals of which were destroyed in Southampton during the Second World War (a copy of these transcripts is held in London at BGS). A large collection of Phillips's incoming letters is in Oxford (UMO). Most of Lyell's incoming scientific letters are in Edinburgh (ULE), with a small but important collection in Philadelphia (APS); I have not seen his family correspondence at Kinnordy House (KHK), which also contains scientific material. Buckland's incoming correspondence has been widely scattered; small collections in London (GSL) and Exeter (DRO) have been used for this book, but the large collection in Oxford (UMO) became accessible to me too late to be exploited fully. Unfortunately, little of Élie de Beaumont's incoming correspondence is preserved at the École des Mines (EMP), apart from some official letters; but very full notes for some of his lecture courses are preserved at the Académie des Sciences (ASP). The most serious gap in the correspondence relating to the

463

Devonian controversy is that the incoming letters of the provincial geologists have not been preserved, or at least not located (a few letters to Williams are in SAS), so that their role must be reconstructed from their letters to more prominent geologists. Other collections of letters, of less central importance for the controversy, are listed below.

The Victorian "life and letters" compilations that were published after the deaths of many of the major participants have only been cited in this book where the original letter has not been located or is not accessible, or for other special reasons. Where comparison is possible, the published version often proves to have been drastically tidied up, or even bowdlerized. A heavily formalized punctuation that destroys the spontaneity of the original, and abridgements (often undeclared) that eliminate much of the "technical" scientific content, severely limit the value of these printed sources.

Almost complete series of the relevant field notebooks of Murchison and Sedgwick are in London (GSL, and one in BGS) and Cambridge (SMC, with one of Murchison's), respectively. Many of Murchison's and Greenough's manuscript geological maps are also in London (BGS and GSL, respectively). The complete series of Lyell's notebooks at Kinnordy House (KHK) is exceptional in giving a continuous record of his indoor research as well as his fieldwork; I have only seen the notebooks from the period covered in the published first volume (1972) of Wilson's biography. Unfortunately, the few extant field notebooks of De la Beche (at BGS) do not cover the period of the Devonian controversy; I heard too late that some of his field maps survive in the Geological Survey's Exeter office. The almost complete series of Williams's notebooks in Taunton (SAS, and one in SRO) is of special importance as the best extant record of the fieldwork of one of the provincial geologists involved in the controversy. Sedgwick usually dated his notes daily; in other cases enough dates are scattered through each notebook to allow other entries to be dated approximately,

Early or discarded drafts of papers read at the Geological Society and other meetings have not been preserved, and the full manuscript versions only rarely (there is one by Williams in SAS). But the Society's official summaries are recorded in manuscript in the ordinary minutebooks (GSL: BP, OM) and printed—not always identically—in the *Proceedings*. (It should be noted that the printed versions were normally published within about a month of the last meeting recorded in each "number," and *not* necessarily in the year printed on the title page of the subsequently collected volume.) The ordinary and Council minutebooks (OM, CM) record, respectively, the names of those who introduced guests to each meeting and the names of those who attended the Council meeting on the same day, thereby providing a minimal roster of the members present; the attendance books for this period, which would have given a complete list, have been lost. The Society's books of incoming letters (LR) incorporate, besides much correspondence about subscriptions and other routine matters, a valuable record of Lonsdale's activities as an informal clearinghouse for geological information among the leading members. The Council minutebooks record the progress of papers through the refereeing system, and a volume of collected referees' reports (COM P4), dating from the period of the controversy, fortunately preserves the reports—usually signed—for many of the relevant papers.

Although they are not manuscripts, it is convenient to mention here the reports of the meetings of the British Association, which were printed promptly in two weekly newspapers. These are important because the official *Report* of each meeting was not published until several months later, and its summaries of papers cannot be relied on as an accurate record of what was actually read at the time. The *Athenaeum* printed detailed summaries of the papers *and* of the subsequent discussions, which are invaluable because they were generally checked before publication, if not actually written, by a competent geologist (usually Phillips). The reports in the *Literary Gazette* are generally less detailed and less reliable, but they give a better impression of how the events struck an ordinary "lay" participant. For the meetings of the Société géologique, I have had to rely on the summaries of papers, and of their discussions, printed in its *Bulletin*.

This note has cited only the more important manuscript sources used in this book; other archives, used only for an occasional document, are included in the list below. Notwithstanding the gaps that have been noted, the documentation is on the whole remarkably complete. Probably the only missing documents that, if discovered, might throw significant *new* light on the Devonian controversy are (a) the incoming letters of provincial geologists, such as Austen; (b) the field notebooks of De la Beche; and (c) the missing letters from Murchison to Sedgwick, some at least of which were still extant when their "lives and letters" were being compiled in the late nineteenth century.

List of Archives

APS   Library of American Philosophical Society, Philadelphia
           CL   Charles Lyell papers

ASP   Archives of Académie des Sciences, Institut de France, Paris
           LEB   Léonce Élie de Beaumont papers

BLL   British Library, London
           RIM   Roderick Impey Murchison papers (Add. 46125–8)

BGS   Library of British Geological Survey, London
           DLB   Henry Thomas De la Beche papers
           GSM   Geological Survey and Museum papers
           RIM   Roderick Impey Murchison papers

BML   Department of Prints and Drawings, British Museum, London.

DRO   East Devon Record Office, Exeter
           WB   William Buckland papers (138 M)

EMP   Library of École supérieure des Mines, Paris
           LEB   Léonce Élie de Beaumont papers

FMC   Library of Fitzwilliam Museum, Cambridge (Perceval Bequest)

GSL   Archives of Geological Society, London
           BP   Business papers of Geological Society (BP archives)
           RIM   Roderick Impey Murchison papers (M archives)
           WB   William Buckland papers

HHO   Archives at Harcourt House, Oxfordshire
     WVH   William Vernon Harcourt papers

HLH   Houghton Library, Harvard University
     LA   Louis Agassiz papers

KHK   Archives at Kinnordy House, Kirriemuir, Angus
     CL   Charles Lyell papers

NLS   National Library of Scotland, Edinburgh
     HM   Hugh Miller papers

NMW   Department of Geology, National Museum of Wales, Cardiff
     DLB   Henry Thomas De la Beche papers
     TS   Transcripts of Geological Survey papers destroyed in Second World War.

NPG   Archives of National Portrait Gallery, London

RSL   Library of Royal Society of London
     JCS   James de Carle Sowerby papers

SAS   Library of Somerset Archaeological and Natural History Society, Taunton
     DW   David Williams papers

SAU   St Andrews University Library
     JDF   James David Forbes papers

SMC   Archives at Sedgwick Museum, Department of Earth Sciences, Cambridge
     AS   Adam Sedgwick papers

SRO   Somerset County Record Office, Taunton

TCC   Library of Trinity College, Cambridge
     WW   William Whewell papers

UCL   D. M. S. Watson Library, University College London
     GBG   George Bellas Greenough papers

ULC   University Library, Cambridge
     AS   Adam Sedgwick papers (Add. 7652)
     CD   Charles Darwin papers (DAR)
     GBG   George Bellas Greenough papers (Add. 7918)

ULE   University Library, Edinburgh
     CL   Charles Lyell papers
     RIM   Roderick Impey Murchison papers

UMO   Archives at University Museum, Oxford
     JP   John Phillips papers
     WB   William Buckland papers

Printed Sources

Agassiz, Louis. 1833–43. *Recherches sur les poissons fossiles* . . . . 4 vols. Neuchâtel.
———. 1834a. Über das Alter der Glarner Schiefer-Formation, nach ihren Fische-Resten. *N. Jahrb. Miner. Geogn. Geol. Petr.-Kunde* 1834:301–6.
———. 1834b. On a new classification of fishes, and on the geological distribution of fossil fishes. *Proc. Geol. Soc. London* 2 (37): 99–102.
———. 1842. On the succession and development of organised beings at the surface of the terrestrial globe; being a discourse delivered at the inauguration of the Academy of Neuchâtel. *Edinburgh N. Philos. J.* 33:388–99.

———. 1843. Report on the fossil fishes of the Devonian system or Old Red Sandstone. *Rept Brit. Assoc. Adv. Sci.* 1842 (reports): 80–88.

———. 1844. *Monographie des poissons fossiles du Vieux Grès Rouge ou Système Dévonien (Old Red Sandstone) des Îles Britanniques et de Russie.* 2 vols. Neuchâtel and Soleure.

Ansted, David T. 1838. On a new genus of fossil multilocular shells, found in the slate-rocks of Cornwall. *Trans. Cambridge Philos. Soc.* 6:415–22, pl. 8.

———. 1840. On the Carboniferous and Transition rocks of Bohemia. *Proc. Geol. Soc. London* 3 (66): 167–70.

d'Archiac, Étienne, and Édouard de Verneuil. 1842a. Sur les fossiles des terrains anciens des bords du Rhin. *Bull. Soc. Géol. France* 13: 257–62.

———. 1842b. On the fossils of the older deposits in the Rhenish provinces, preceded by a general survey of the fauna of the Palaeozoic rocks, and followed by a tabular list of the organic remains of the Devonian system in Europe. *Trans. Geol. Soc. London,* ser. 2, 6 (2): 303–410, pls. 25–38.

Arrowsmith, Aaron. 1830. *A map showing the geological position and commercial distribution of the coal of England and Wales.* London.

Austen, Robert A. C. 1834. An account of the raised beach, near Hope's Nose, in Devonshire, and other recent disturbances in that neighbourhood. *Proc. Geol. Soc. London* 2 (37): 102–3.

———. 1836. On the part of Devonshire between the Ex and Berry Head and the coast and Dartmoor. *Proc. Geol. Soc. London* 2 (46): 414–15.

———. 1838a. On the geology of the south-east of Devonshire. *Proc. Geol. Soc. London* 2 (53): 584–89.

———. 1838b. On the origin of the limestones of Devonshire. *Proc. Geol. Soc. London* 2 (57): 669–70.

———. 1839a. Considerations on geological evidences and inferences. *Rept Brit. Assoc. Adv. Sci.* 1838 (trans.): 93.

———. 1839b. On the structure of south Devon. *Proc. Geol. Soc. London* 3 (63): 123–24.

———. 1840. Note on the organic remains of the limestones and slates of south Devon. *Rept. Brit. Assoc. Adv. Sci.* 1839 (trans.): 69.

———. 1842. On the geology of the south-east of Devonshire. *Trans. Geol. Soc. London,* ser. 2, 6 (2): 433–89, pls. 41, 42.

Babbage, Charles. 1830. *Reflections on the decline of science in England and on some of its causes.* London.

Beyrich, Heinrich Ernst. 1837. *Beiträge zur Kenntniss der Versteinerungen des Rheinischen Übergangsgebirge.* Vol. 1. Berlin.

———. 1838. Mémoire sur les Goniatites qui se trouvent dans les terrains de transition du Rhin. *Ann. Sci. Nat. (Zool.)* 10:65–91, pls. 6, 7A.

———. 1839a. Considérations sur les roches fossilifères du terrain de transition du Rhin. *Ann. Mines.* ser. 3, 15:51–78.

———. 1839b. On the Goniatites found in the Transition formations of the Rhine. *Ann. Nat. Hist.* 3:9–20, 155–65, pls. 1, 2.

Boase, Henry S. 1834. *A treatise on primary geology; being an examination, both practical and theoretical, of the older formations.* London.

Bowman, J. E. 1842. On the upper Silurian rocks of Denbighshire. *Rept Brit. Assoc. Adv. Sci.* 1841 (trans.): 59–61.

Brochant de Villiers, André. 1835. Notice sur la carte géologique générale de la France. *Comptes-Rendus Séances Acad. Sci.* 1835:423–29.

Brongniart, Adolphe. 1828a. Observations sur les végétaux fossiles des terrains d'anthracite des Alpes. *Ann. Sci. Nat.* 14:127–36.

———. 1828b. *Prodrome d'une histoire des végétaux fossiles.* Paris.

———. 1828–37. *Histoire des végétaux fossiles ou recherches botaniques et géologiques sur les végétaux renfermés dans les diverses couches du globe.* 2 vols. Paris.

Brongniart, Alexandre. 1823. *Mémoire sur les terrains de sédiment supérieurs calcairo-trappéens du Vicentin, et sur quelques terrains d'Italie, de France, d'Allemagne, etc., qui peuvent se rapporter à la même époque.* Paris.

Bronn, Heinrich G. 1848–49. *Index palaeontologicus, oder Übersicht der bis jetzt bekannten fossilen Organismen.* 3 vols. Stuttgart.

von Buch, Leopold. 1836. Extrait d'une lettre de M. Léopold de Buch à M. Élie de Beaumont. *Bull. Soc. Géol. France* 7:155–58, pl.

Buckland, William. 1820. *Vindiciae geologicae; or the connexion of geology with religion explained, in an inaugural lecture delivered before the University of Oxford, May 15, 1819, on the endowment of a Readership in Geology by His Royal Highness the Prince Regent.* Oxford.

———. 1821. Notice of a paper laid before the Geological Society on the structure of the Alps and adjoining parts of the Continent, and their relation to the Secondary and Transition rocks of England. *Ann. Philos.*, n.s. 1:450–68.

———. 1823. *Reliquiae diluvianae; or, observations on the organic remains contained in the caves, fissures, and diluvial gravel, and on other geological phaenomena, attesting the action of an universal deluge.* London.

———. 1836. *Geology and mineralogy considered with reference to natural theology.* 2 vols. London.

———. 1840. Address to the Geological Society, delivered at the anniversary, on the 21st of February, 1840. *Proc. Geol. Soc. London* 3 (68): 210–67.

———. 1841. Address delivered on the anniversary, February 19th. *Proc. Geol. Soc. London* 3 (81): 469–540.

Buckland, William, and William D. Conybeare. 1822. Observations on the south-western coal district of England. *Trans. Geol. Soc. London* ser. 2, 1 (1): 210–316, pls. 32–38.

Buckland, William, and Henry T. De la Beche. 1835. On the geology of the neighbourhood of Weymouth and the adjacent parts of the coast of Dorset. *Trans. Geol. Soc. London*, ser. 2, 4 (1): 1–46, 3 pls.

Burat, Amedée. 1834. *Traité de géognosie ou exposé des connaissances actuelles sur la constitution physique et minérale du globe terrestre.* Paris.

[Chambers, Robert.] 1844. *Vestiges of the natural history of creation.* London.

Conrad, Timothy A. 1838. Report of the Palaeontological Department of the [Geological] Survey. *State of New York*, Communication no. 200, 107–19.

———. 1840a. On the Silurian system, with a table of the strata and characteristic fossils. *Amer. J. Sci.* 38: 86–93.

———. 1840b. Third annual report, on the Palaeontological Department of the [Geological] Survey. *State of New York*, Communication no. 50, 199–207.

———. 1842. Observations on the Silurian and Devonian systems of the United States, [with] descriptions of new species of organic remains belonging to the Silurian, Devonian, and Carboniferous systems of the United States. *J. Acad. Nat. Sci. Philadelphia* 8: 228–80, pls. 12–17.

Conybeare, William D. 1833. Report on the progress, actual state, and ulterior prospects of geological science. *Rept Brit. Assos. Adv. Sci. 1831–32:365–414.*

———. 1839. *An analytical examination into the character, value, and just application of the writings of the Christian Fathers during the ante-Nicene period.* Oxford.

Conybeare, William D., and William Phillips. 1822. *Outline of the geology of England and Wales, with an introductory compendium of the general principles of that science, and comparative views of the structure of foreign countries.* Vol. 1. London.

Cuvier, Georges, and Alexandre Brongniart. 1811. Essai sur la géographie minéralogique des environs de Paris. *Mém. Classe Sci. Math. Phys., Inst. Imp. France,* 1810, pt. 1.

Darwin, Charles, ed. 1839–42. *The zoology of the voyage of H.M.S. Beagle under the command of Captain Fitzroy, R.N., during the years 1832 to 1836.* 4 vols. London.

von Dechen, Heinrich. 1823. Geognostische Bemerkungen über den nördlichen Abfall des Niederrheinisch-Westphalischen Gebirges. In *Das Gebirge in Rheinland-Westphalen*, edited by Jacob Nöggerath, 2:1–151, table 1. Bonn.

———. 1839. *Geognostische Übersichts-Karte von Deutschland, Frankreich, England und den angrenzenden Laendern. Nach den grösseren Arbeiten von L. v. Buch, É. de Beaumont und Dufrénoy, G. B. Greenough*. Berlin.

De la Beche, Henry T. 1826. On the geology of southern Pembrokeshire. *Trans. Geol. Soc. London*, ser. 2, 2 (1): 1–20, pls. 1, 2.

———. 1829. On the geology of Tor and Babbacombe Bays, Devon. *Trans. Geol. Soc. London*, ser. 2, 3 (1):161–70, pls. 18–20.

———. 1830. *Sections and views, illustrative of geological phaenomena*. London.

———. 1831. *A geological manual*. London.

———. 1833a. *A geological manual. Third edition, considerably enlarged*. London.

———. 1833b. *Manuel géologique*. Paris.

———. 1834a. *Researches in theoretical geology*. London.

———. 1834b. On the anthracite found near Biddeford in north Devon. *Proc. Geol. Soc. London* 2 (37): 106–7.

———. 1835. *How to observe. Geology*. London.

———. 1836. [On the rocks of north Cornwall.] *Proc. Geol. Soc. London* 2 (43): 225–26.

———. 1839. *Report on the geology of Cornwall, Devon and West Somerset*. London.

———. 1846. On the formation of the rocks of south Wales and south-western England. *Mem. Geol. Surv. Great Britain* 1:1–296.

Dufrénoy, Pierre, and Léonce Élie de Beaumont. 1841a. *Carte géologique de la France*. Paris.

———. 1841b. *Explication de la carte géologique de la France*. Vol. 1. Paris.

Dumont, André-Hubert. 1832. Mémoire sur la constitution géologique de la province de Liège. *Mém. Couronnées Acad. Roy. Sci. Belle-Lettres Bruxelles* 8, no. 3.

———. 1836. Rapport fait à l'Académie Royale des Sciences et Belle-Lettres de Bruxelles, sur l'état des travaux de la carte géologique de la Belgique. *Bull. Soc. Géol. France* 8:77–82, pl.

———. 1838. Rapport sur les travaux de la carte géologique, pendant l'année 1838. *Bull. Acad. Sci. Bruxelles* 5:634–43.

———. 1839. On the equivalents of the Cambrian and Silurian systems in Belgium. *Philos. Mag. J. Sci.*, ser. 3, 15:146–52.

———. 1848. Mémoire sur les terrains ardennais et rhénans de l'Ardenne, du Rhin, du Brabant et du Condros. *Mém. Acad. Roy. Sci. Bruxelles* 2:221–451.

Eaton, Amos. 1831. Observations on the Coal formation in the State of New York; in connexion with the great coal beds of Pennsylvania. *Amer. J. Sci.* 19:21–26.

Ehrenberg, Christian G. 1839. Über die Bildung der Kreidefelsen und des Kreidemergels durch unsichtbare Organismen. *Abh. Kön Akad. Wissenschaften Berlin* 1838:59–148, pls. 1–4.

Eichwald, Edouard, 1825. *Geognostico-zoologicae per ingriam marisque Baltici provincias nec non de trilobitis observationes*. Kazan.

———. 1840. Geognostische Uebersicht von Estland und der Nachbar-Gegenden. *N. Jahrb. Miner. Geogn. Geol. Petr.-Kunde* 1840:421–30.

Élie de Beaumont, Léonce. 1828. Notice sur un gisement de végétaux fossiles et de bélemnites, situé à Petit-Coeur près Moutiers, en Tarentaise. *Ann. Sci. Nat.* 14:113–27.

———. 1829–30. Recherches sur quelques-unes des révolutions de la surface du globe, présentant différents exemples de coïncidence entre le redressement des couches de certains systèmes de montagnes, et les changements soudains qui ont produit les lignes de démarcation qu'on observe entre certains étages consecutifs des terrains de sédiment. *Ann. Sci. Nat.* 18:5–25, 284–416; 19:5–99, 177–240.

Fitton, William H. 1827. Additional notes on part of the opposite coasts of France and England, including some account of the Lower Boulonnois. *Proc. Geol. Soc. London* 1 (1): 6–10.

———. 1828. Address delivered on the anniversary, February 1828. *Proc. Geol. Soc. London* 1 (6): 50–62.

[———.] 1839. [Review of] *Elements of geology* . . . by Charles Lyell . . . *Edinburgh Rev.* 69: 406–66.

[———.] 1841. [Review of] *The Silurian System* . . . by Roderick Impey Murchison, F.R.S. . . . *Edinburgh Rev.* 73:1–41.

Geological Society of London. 1808. *Geological inquiries.* London.

Goldfuss, August. 1826–44. *Petrefacta Germaniae* . . . . *Abbildungen und Beschreibungen der Petrefacten Deutschlands und der angränzenden Länder unter Mitwirkung des Herrn Grafen Georg zu Münster.* Düsseldorf.

Greenough, George B. 1819. *A critical examination of the first principles of geology; in a series of essays.* London.

———. 1820. *A geological map of England and Wales.* London.

———. 1834. Address delivered at the anniversary meeting of the Geological Society, on the 21st of February, 1834. *Proc. Geol. Soc. London* 2 (35): 42–70.

———. 1835. An address delivered at the anniversary meeting of the Geological Society of London, on the 20th of February, 1835. *Proc. Geol. Soc. London* 2 (39): 145–75.

———. 1840a. *A physical and geological map of England and Wales.* London.

———. 1840b. *Memoir of a geological map of England; to which is added, an alphabetical index to the hills, and a list of the hills arranged according to counties.* London.

———. 1840c. Memoir to accompany the second edition of the geological map of England and Wales. *Proc. Geol. Soc. London* 3 (67): 180–85.

Griffith, Richard. 1838. *Outline of the geology of Ireland.* Dublin.

———. 1839. On the geological structure of the south of Ireland. *Rept Brit. Assoc. Adv. Sci.* 1838 (trans.): 81–84.

———. 1843. Statement of the fossils which have been discovered in the several members of the Carboniferous or Mountain Limestone of Ireland, with a view to show the zoological identity of the whole series . . . . *Rept Brit. Assoc. Adv. Sci.* 1842 (trans.): 51–53.

———. 1844. On the Old Red Sandstone, or Devonian and Silurian districts of Ireland. *Rept Brit. Assoc. Adv. Sci.* 1843 (trans.): 46–49.

Hardy, Philip D. 1835. *Proceedings of the fifth meeting of the British Association for the Advancement of Science, held in Dublin* . . . . Dublin.

[Harless, C. F. and J. Nöggerath.] 1836. Versammlung der Naturforscher und Aerzte zu Bonn im September 1835. *Isis* 1836: columns 641–810.

Hennah, Richard. 1817. Observations respecting the limestone of Plymouth. *Trans. Geol. Soc. London* 4 (2):410–12.

———. 1828. Additional remarks on the nature and character of the limestone and slate, principally composing the rocks and hills round Plymouth. *Trans. Geol. Soc. London,* ser. 2, 2:405.

———. 1830. On the animal remains found in the Transition limestone of Plymouth. *Proc. Geol. Soc. London* 1 (14): 169–70.

———. n.d. *A succinct account of the lime rocks of Plymouth, being the substance of several communications read before the members of the Geological Society, in London, and partly printed in their Transactions.* Plymouth.

Herschel, John F. W. 1830. *Preliminary discourse on the study of natural philosophy.* London.

Heywood, James. 1843. *Illustrations of the Manchester meeting of the British Association for the Advancement of Science, June 1842.* Manchester.

Hoffmann, Fredrich. 1830. *Geognostische Atlas vom Nordwestlichen Deutschland.* Berlin.

von Humboldt, Alexander. 1823. *Essai géognostique sur le gisement des roches dans les deux hemispheres.* Paris.

Hutton, James. 1795. *Theory of the earth, with proofs and illustrations.* 2 vols. Edinburgh.

de Koninck, Léopold. 1842–44. *Description des animaux fossiles qui se trouvent dans le terrain carbonifère de Belgique.* Liège.

Lindley, John, and William Hutton. 1831–37. *The fossil flora of Great Britain; or, figures and descriptions of the vegetable remains found in a fossil state in this country.* 3 vols. London.

Lonsdale, William. 1840a. On the age of the limestones of south Devon. *Proc. Geol. Soc. London* 3 (69): 281–86.

———. 1840b. Notes on the age of the limestones of south Devonshire. *Trans. Geol. Soc. London*, ser. 2, 5 (3): 721–38.

Lyell, Charles. 1830–33. *Principles of geology, being an attempt to explain the former changes of the earth's surface, by reference to causes now in operation.* 3 vols. London.

———. 1834. *Principles of geology: Being an inquiry how far the former changes of the earth's surface are referable to causes now in operation.* 3d ed. 4 vols. London.

———. 1835. *Principles of geology. Being an inquiry how far the former changes of the earth's surface are referable to causes now in operation.* 4th ed. 4 vols. London.

———. 1836. Address to the Geological Society, delivered at the anniversary, on the 19th of February, 1836. *Proc. Geol. Soc. London* 2 (44): 357–90.

———. 1837. Address to the Geological Society, delivered at the anniversary, on the 17th of February, 1837. *Proc. Geol. Soc. London* 2 (49): 479–523.

———. 1838. *Elements of geology.* London.

———. 1841a. *Elements of geology.* 2d ed. 2 vols. London.

———. 1841b. A letter addressed to Dr Fitton, dated Boston the 15th of October, 1841. *Proc. Geol. Soc. London* 3 (82): 554–58.

———. 1845. *Travels in North America; with geological observations on the United States, Canada and Nova Scotia.* 2 vols. London.

Mantell, Gideon A. 1839. *The wonders of geology; or, a familiar exposition of geological phenomena.* 3d ed. 2 vols. London.

Miller, Hugh. 1841. *The Old Red Sandstone; or, new walks in an old field.* Edinburgh.

———. 1847. *Footprints of the Creator; or, the Asterolepis of Stromness.* Edinburgh.

zu Münster, Georg. 1832. *Über die Clymenien und Goniatiten im Übergangs-Kalk des Fichtelgebirges.* Bayreuth.

———. 1834. Mémoire sur les Clymènes et les Goniatites du calcaire de transition du Fichtelgebirge. *Ann. Sci. Nat. (Zool.)* 2:65–99, pls. 1–6.

———. 1840. Die Versteinerungen des Übergangskalkes mit Clymenien und Orthoceratiten. In *Beiträge zur Petrefacten-Kunde,* edited by Georg zu Münster, 3: 33–121. Bayreuth.

Murchison, Roderick I. 1827. On the coal-field of Brora in Sutherlandshire, and some other stratified deposits in the north of Scotland. *Trans. Geol. Soc. London*, ser. 2, 2 (2): 293–326, pls. 31, 32.

———. 1833a. Address to the Geological Society, delivered on the evening of the 15th of February, 1833. *Proc. Geol. Soc. London* 1 (30): 438–64.

———. 1833b. On the sedimentary deposits which occupy the western parts of Shropshire and Herefordshire, and are prolonged from N.E. to S.W., through Radnor, Brecknock, and Caermarthenshire, with descriptions of the accompanying rocks of intrusive or igneous characters. *Proc. Geol. Soc. London* 1 (31): 470–77.

————. 1834a. On the Old Red Sandstone in the counties of Hereford, Brecknock and Caermarthen, with collated observations on the dislocations which affect the northwest margin of the south Welsh coal-basin. *Proc. Geol. Soc. London* 2 (34): 11–13.

————. 1834b. On the structure and classification of the Transition rocks of Shropshire, Herefordshire and parts of Wales, and on the lines of disturbance which have affected that series of deposits, including the valley of elevation of Woolhope. *Proc. Geol. Soc. London* 2 (34): 13–18, table.

————. 1834c. On the Upper Greywacke series of England and Wales. *Edinburgh N. Philos. J.* 17:365–69.

————. 1835a. On the Old Red Sandstone and the formations beneath it. *Rept Brit. Assoc. Adv. Sci.* 1834 (trans.): 652.

————. 1835b. On the Silurian system of rocks. *Philos. Mag. J. Sci.*, ser. 3, 7:46–52.

————. 1836. On the geological structure of Pembrokeshire, more particularly on the extension of the Silurian system of rocks into the coast cliffs of that county. *Proc. Geol. Soc. London* 2 (43): 226–30.

————. 1839. *The Silurian system, founded on geological researches in the counties of Salop, Hereford, Radnor, Montgomery, Caermarthen, Brecon, Pembroke, Monmouth, Gloucester, Worcester and Stafford; with descriptions of the coal-fields and overlying formations.* 2 vols. London.

————. 1840a. On the Carboniferous and Devonian systems of Westphalia. *Rept Brit. Assoc. Adv. Sci.* 1839 (trans.):72–73.

————. 1840b. Sur les *roches dévoniennes*, type particulier de l'*old red sandstone* des géologues anglais, qui se trouvent dans le Boulonnais et les pays limitrophes. *Bull. Soc. Géol. France* 11:229–56

————. 1841a. On a section and a list of fossils from the State of New York by James Hall, Esq. *Proc. Geol. Soc. London* 3 (77):416–17.

————. 1841b. First sketch of some of the principal results of a second geological survey of Russia, in a letter to M. Fischer. *Philos. Mag. J. Sci.*, ser. 3, 19:417-22.

————. 1842. Anniversary address of the president. *Proc. Geol. Soc. London* 3 (86):637–87.

————. 1843a. Notice of a memoir on the geology of the western states of North America, by David Dale Owen, M.D., of Indiana. *Rept Brit. Assoc. Adv. Sci.* 1842 (trans.):44–45.

————. 1843b. On the geological structure of Russia (delivered at an evening lecture). *Rept Brit. Assoc. Adv. Sci.* 1842 (trans.): 45–46.

————. 1843c. *Geological map of England and Wales.* London.

Murchison, Roderick I., and Édouard de Verneuil, 1841a. On the stratified deposits which occupy the northern and central regions of Russia. *Rept Brit. Assoc. Adv. Sci.* 1840 (trans.): 105–10.

————. 1841b. On the geological structure of the northern and central regions of Russia in Europe. *Proc. Geol. Soc. London* 3 (76): 398–408.

————. 1842. A second geological survey of Russia in Europe. *Proc. Geol. Soc. London.* 3 (88): 717–30.

Murchison, Roderick I., Edouard de Verneuil, and Alexander von Keyserling. 1842. On the geological structure of the Ural Mountains. *Proc. Geol. Soc. London* 3 (89): 742–53.

————. 1845a. *The geology of Russia in Europe and the Ural Mountains.* Vol. 1. London.

————. 1845b. *Géologie de la Russie d'Europe et des Montagnes d'Oural.* Vol. 2. Paris.

Nöggerath, Jacob, ed. 1822–26. *Das Gebirge in Rheinland-Westphalen nach mineralogischem und chemischem Bezuge.* 4 vols. Bonn.

d'Omalius d'Halloy, Jean-Baptiste. 1828. *Mémoires pour servir à la description géologique des Pays-Bas, de la France et de quelques contrées voisines.* Namur.

————. 1831. *Éléments de géologie.* Paris.

Outhett, John. 1840. *Laurie's new map of London.* London.

Pander, Christian H. 1830. *Beiträge zur Geognosie des Russischen Reiches*. St Petersburg.

Peach, Charles W. 1844. On the fossils of Polperro in Cornwall. *Rept Brit. Assoc. Adv. Sci.* 1843 (trans.): 56–57.

Phillips, John. 1829. *Illustrations of the geology of Yorkshire; or, a description of the strata and organic remains of the Yorkshire coast. Part I. The Yorkshire coast*. York.

———. 1834. *A guide to geology*. London.

———. 1836. *Illustrations of the geology of Yorkshire; or, a description of the strata and organic remains of the Yorkshire coast. Part II. The Mountain Limestone district*. London.

———. 1837–39. *A treatise on geology*. 2 vols. London.

———. [1838]a. *An index geological map of the British Isles. . . .* London.

———. 1838b. Geology. *Penny Cyclopedia* 11:127–51.

———. 1839. Remarks on a note in Prof. Sedgwick and Mr Murchison's communication in the last number. *Philos. Mag. J. Sci.*, ser. 3, 14: 353–54.

———. [1840]a. *New index geological map of the British Isles, and adjacent coast of France. . . .* London.

———. 1840b. Palaeozoic series. *Penny Cyclopedia* 17:153–54.

———. 1841. *Figures and descriptions of the Palaeozoic fossils of Cornwall, Devon and west Somerset; observed in the course of the Ordnance Geological Survey of that district*. London.

———. 1846. *Geological map of the British Isles and adjacent coast of France*. London.

Portlock, Joseph. 1839. On a small tract of Silurian rocks in the county of Tyrone. *Rept Brit. Assoc. Adv. Sci.* 1838 (trans.): 84.

Rogers, Henry D. 1834. Some facts in the geology of the central and western portions of North America, collected principally from the statements and unpublished notices of recent travellers. *Proc. Geol. Soc. London* 2 (37): 103–6.

Römer, Ferdinand. 1844. *Das Rheinische Übergangsgebirge. Eine paläontologisch-geognostische Darstellung*. Hannover.

Rozet, Claude A. 1828. *Description géognostique du bassin du Bas-Boulonnais*. Paris.

———. 1830. Notice géognostique sur quelques parties du département des Ardennes et de la Belgique. *Ann. Sci. Nat.* 19:113–53.

Sandberger, Guido. 1842. Vorläufige Übersicht über die eigenthumlichen bei *Villmar an der Lahn* auftretenden jüngeren Kalk-Schichten der älteren (sog. Uebergangs-) Formation. *N. Jahrb. Miner. Geogn. Geol. Petr.-Kunde* 1842: 379–402.

[Scrope, George Poulett.] 1839. [Review of] Murchison's *Silurian System*, &c. *Quart. Rev.* 64:102–20.

Sedgwick, Adam. 1820. On the physical structure of those formations which are immediately associated with the primitive ridge of Devonshire and Cornwall. *Trans. Cambridge Philos. Soc.* 1:89–146.

———. 1831. Address to the Geological Society, delivered on the evening of the anniversary, Feb. 18, 1831. *Proc. Geol. Soc. London* 1 (20): 281–316.

———. 1832. *A syllabus of a course of lectures on geology*. 2d ed. Cambridge.

———. 1833. *A discourse on the studies of the university*. Cambridge.

———. 1835a. Remarks on the structure of large mineral masses, and especially on the chemical changes produced in the aggregation of stratified rocks during different periods after their deposition. *Trans. Geol. Soc. London*, ser. 2, 3 (3):461–86, pl. 47.

———. 1835b. Introduction to the general structure of the Cumbrian Mountains; with a description of the great dislocations by which they have been separated from the neighbouring Carboniferous chains. *Trans. Geol. Soc. London*, ser. 2, 4 (1): 47–68, pls. 4, 5.

———. 1837. *A syllabus of a course of lectures on geology*. 3d ed. Cambridge.

———. 1838. A synopsis of the English series of stratified rocks inferior to the Old Red Sandstone—with an attempt to determine the successive natural groups and formations. *Proc. Geol. Soc. London* 2 (56): 662; 2 (58): 675–85.

————. 1841. Supplement to "A synopsis of the English series of stratified rocks inferior the the Old Red Sandstone," with additional remarks on the relations of the Carboniferous series and Old Red Sandstone of the British Isles. *Proc. Geol. Soc. London* 3 (82): 541–54.

————. 1843. Outline of the geological structure of north Wales. *Proc. Geol. Soc. London* 4 (96): 212–24.

[————.] 1845. [Review of] *Vestiges of the Natural History of Creation. Edinburgh Rev.* 82:1–85.

Sedgwick, Adam, and Roderick I. Murchison. 1829. On the structure and relations of the deposits contained between the Primary rocks and the Oolitic series in the north of Scotland. *Trans. Geol. Soc. London*, ser. 2, 3 (1): 125–60, pls. 13–17.

————. 1836. On the *Silurian* and *Cambrian* systems, exhibiting the order in which the older sedimentary strata succeed each other in England and Wales. *Rept Brit. Assoc. Adv. Sci.* 1835 (trans.): 59–61.

————. 1837a. A classification of the old slate rocks of the north of Devonshire, and on the true position of the Culm deposits in the central portion of that county. *Rept Brit. Assoc. Adv. Sci.* 1836 (trans.): 95–96.

————. 1837b. On the physical structure of Devonshire, and on the subdivisions and geological relations of its old stratified deposits. *Proc. Geol. Soc. London* 2 (51): 556–63.

————. 1839a. Classification of the older stratified rocks of Devonshire and Cornwall. *Philos. Mag. J. Sci.*, ser. 3, 14:241–60.

————. 1839b. Supplementary remarks on the "Devonian" system of rocks. *Philos. Mag. J. Sci.*, ser. 3, 14:354–58.

————. 1839c. On the classification of the older rocks of Devonshire and Cornwall. *Proc. Geol. Soc. London* 3 (63):121–23.

————. 1840a. On the classification and distribution of the older or Palaeozoic rocks of the north of Germany and of Belgium, as compared with formations of the same age in the British Isles. *Proc. Geol. Soc. London* 3 (70):300–311.

————. 1840b. on the physical structure of Devonshire, and on the subdivisions and geological relations of its older stratified deposits, &c. *Trans. Geol. Soc. London*, ser. 2, 5 (3): 633–704, pls. 50–58.

————. 1842. On the distribution and classification of the older or Palaeozoic deposits of the north of Germany and Belgium, and their comparison with formations of the same age in the British Isles. *Trans. Geol. Soc. London*, ser. 2, 6 (2): 221–301, pls. 23–28.

Smith, William, 1815. *A delineation of the strata of England and Wales with part of Scotland. . . .* London.

Sowerby, James [and James de Carle]. 1812–46. *The mineral conchology of Great Britain.* 7 vols. London.

Steininger, Johann, 1822. *Gebirgskarte der Länder zwischen dem Rhein und der Maas.* Mainz.

Strangways, W. H. T. Fox-. 1822. An outline of the geology of Russia. *Trans. Geol. Soc. London*, ser. 2, 1 (1): 1–39, pls. 1, 2.

Swainson, William. 1834. *A preliminary discourse on the study of natural history.* London.

de Tchihatcheff, Pierre. 1845. *Voyage scientifique dans l'Altaï orientale et les parties adjacentes de la frontière de Chine, fait par l'ordre de l'Empereur de Russie.* 2 vols. Paris.

Ure, Andrew, 1829. *A new system of geology, in which the great revolutions of the earth and animated nature, are reconciled at once to modern science and sacred history.* London.

de Verneuil, Édouard. 1838. Sur les terrains anciens du bas Boulonnais. *Bull. Soc. Géol. France* 9:388–96.

———. 1840. Sur l'importance de la limite qui sépare le calcaire de montagne des formations qui lui sont inférieurs. *Bull. Soc. Géol. France* 11:166–79.

———. 1847. Note sur le parallélisme des roches des dépôts paléozoiques de l'Amérique septentrionale avec ceux de l'Europe. *Bull. Soc. Géol. France*, ser. 2, 4:646–709.

Weaver, Thomas, 1830. On the geological relations of the south of Ireland. *Proc. Geol. Soc. London* 1 (17): 231–34.

———. 1835. [On coal in Ireland.] *Proc. Geol. Soc. London* 2 (38): 118–19.

———. 1838a. On the geological relations of the south of Ireland. *Trans. Geol. Soc. London*, ser. 2, 5 (1): 1–68, pls. 1, 2.

———. 1838b. Geological relations of north Devon. *Proc. Geol. Soc. London* 2 (53): 589–90.

———. 1839. On the older stratified rocks of north Devon. *Philos. Mag. J. Sci.*, ser. 3, 15:109–29.

———. 1840. On the mineral structure of the south of Ireland, with correlative matter on Devon and Cornwall, Belgium, the Eifel, &c. *Philos. Mag. J. Sci.*, ser. 3, 16:276–97, 388–404, 471–76.

Whewell, William. 1837. *History of the inductive sciences from the earliest to the present time*. 3 vols. London.

———. 1838. Address to the Geological Society, delivered at the anniversary, on the 16th of February, 1838. *Proc. Geol. Soc. London* 2 (55): 624–49.

———. 1839. Address to the Geological Society, delivered at the anniversary, on the 15th of February, 1839. *Proc. Geol. Soc. London* 3 (61): 61–98.

Williams, David. 1834. On the several ravines, passes, and fractures in the Mendip Hills and other adjacent boundaries of the Bristol coal-field, and on the geological period when they were effected. *Proc. Geol. Soc. London* 2 (36): 79–80.

———. 1836. On certain fossil plants from the opposite shores of the Bristol Channel. *Rept Brit. Assoc. Adv. Sci.* 1835 (trans.): 63.

———. 1838. On some fossil wood and plants recently discovered . . . . low down in the Grauwacke of Devon . . . . *Rept Brit. Assoc. Adv. Sci.* 1837 (trans.): 94–95.

———. 1839a. On the classification of certain geological formations in Devonshire. *Philos. Mag. J. Sci.*, ser. 3, 14:358–59.

———. 1839b. On as much of the Transition or Grauwacke system as is exposed in the counties of Somerset, Devon and Cornwall. *Proc. Geol. Soc. London* 2 (63): 115–17.

———. 1839c. on the geological position of the culm and plant-bearing beds of Devon and Cornwall. *Philos. Mag. J. Sci.*, ser. 3, 15:293–96.

———. 1840a. On the geology of Devon and Cornwall, with reference to a paper read before the Geological Society on December 4th, 1839. *Philos. Mag. J. Sci.*, ser. 3. 16:59–64.

———. 1840b. On as much of the great Grauwacke system as is comprised in the group of west Somerset, Devon, and Cornwall. *Proc. Geol. Soc. London* 3 (66): 158–62.

———. 1840c. On the geological horizon of the rocks of S. Devon and Cornwall, as regards that section of the great Grauwacke group comprised in the counties of Somerset, Devon and Cornwall. *Rept Brit. Assoc. Adv. Sci.* 1839 (trans.): 68–69.

———. 1841. On the older strata of Devonshire. *Rept Brit. Assoc. Adv. Sci.* 1840 (trans.): 103–4.

———. 1842a. Plausible reasons and positive proofs showing that no portion of the "Devonian system" can be of the age of the Old Red Sandstone. *Philos. Mag. J. Sci.*, ser. 3, 20:117–35.

———. 1842b. Supplementary notes on the true position in the "Devonian system" of the Cornish Killas. *Philos. Mag. J. Sci.*, ser. 3, 21:25–29.

———. 1844. On the Killas group of Cornwall and south Devon; its relations to the subordinate formations in central and north Devon and west Somerset; its natural subdivisions; and its true position in the scale of British strata. *Philos. Mag. J. Sci.*, ser. 3, 24:332–46.

———. 1846. On an important slate term in the Killas series of Cornwall and south Devon, not sufficiently adverted to in my former classification of that group in 1843; with additional proofs in confirmation of the true geological position of the Ocrynian or Devonian system. *Trans. Geol. Soc. Cornwall* 6:334–37.

Young, George, and John Bird. 1822. *A geological survey of the Yorkshire coast: Describing the strata and fossils occurring between the Humber and the Tees, from the German Ocean to the plain of York.* Whitby.

Secondary References

Agassiz, Elisabeth Cary. 1885. *Louis Agassiz. His life and correspondence.* 2 vols. Boston (Houghton Mifflin).

Allen, David E. 1976. *The naturalist in Britain: A social history.* London (Allen Lane).

———. 1985. The professionals in natural history in the early nineteenth century. In *Natural history in the early nineteenth century,* edited by Alwyne Wheeler. London, in press.

Altick, R. D. 1957. *The English common reader: A social history of the mass reading public 1800–1900.* Chicago (University of Chicago Press).

Andrews, S. M. 1982. *The discovery of fossil fishes in Scotland up to 1845.* Edinburgh (Royal Scottish Museum).

Ayckbourn, Alan. 1975. *The Norman conquests.* London (Chatto and Windus).

Barlow, Nora, ed. 1858. *The autobiography of Charles Darwin, 1809–1882. With original omissions restored.* London (Collins).

Barnes, Barry. 1974. *Scientific knowledge and sociological theory.* London (Routledge and Kegan Paul).

———. 1977. *Interests and the growth of knowledge.* London (Routledge and Kegan Paul).

———. 1982. *T. S. Kuhn and social science.* London and Basingstoke (Macmillan).

Barnes, Barry, and Steven Shapin, eds. 1979. *Natural order. Historical studies of scientific culture.* Beverly Hills and London (Sage).

Barrett, Paul H. 1974. The Sedgwick-Darwin geologic tour of north Wales. *Proc. Amer. Philos. Soc.* 118:146–64.

Bartholomew, Michael. 1973. Lyell and evolution: An account of Lyell's response to the prospect of an evolutionary ancestry for man. *Brit. J. Hist. Sci.* 6:261–303.

———. 1979. The singularity of Lyell. *Hist. Sci.* 17:276–93.

Berman, Morris. 1978. *Social change and scientific organization. The Royal Insitution 1799–1844.* London (Heinemann) and Ithaca (Cornell University Press).

Birembaut, Arthur. 1964. L'enseignement de la minéralogie et des techniques minières. In *Enseignement et diffusion des sciences en France au XVIIIe siècle,* edited by R. Taton, 365–418. Paris.

Bowker, John. 1973. *The sense of God. Sociological, anthropological and psychological approaches to the origin of the sense of God.* Oxford (Clarendon).

Bowler, Peter J. 1976. *Fossils and progress. Paleontology and the idea of progressive evolution in the nineteenth century.* New York (Science History).

Brannigan, Augustine. 1981. *The social basis of scientific discoveries.* Cambridge (Cambridge University Press).

Brock, W. H. 1980. The development of commercial science journals in Victorian Britain. In *Development of science publishing in Europe,* edited by A. J. Meadows, 95–122. Amsterdam, New York, and Oxford (Elsevier).

Butterfield, Herbert. 1957. *George III and the historians.* London.

Campbell-Smith, W. 1969. A history of the first hundred years of the mineral collection in the British Museum, with particular reference to Charles König. *Bull. Brit. Mus. (Nat. Hist.), Hist. Ser.* 3: 235–59.

Cannon, Walter F. [Susan Faye Cannon]. 1964. History in depth: The early Victorian period. *Hist. Sci.* 3: 20–38.

———. 1978. *Science in culture. The early Victorian period.* New York (Science History).

Clark, John Willis, and Thomas McKenny Hughes. 1890. *The life and letters of Adam Sedgwick.* 2 vols. Cambridge (Cambridge University Press).

Cleevely, Ron J. 1974. The Sowerbys, the *Mineral Conchology,* and their fossil collection. *J. Soc. Bibliogr. Hist.* 6:418–81.

Coleman, William. 1964. *Georges Cuvier zoologist. A study in the history of evolution theory.* Cambridge, Mass. (Harvard University Press).

Collins, H. M. 1974. The TEA set: Tacit knowledge and scientific networks. *Sci. Stud.* 4:165–86.

———. 1975. The seven sexes: A study of the sociology of a phenomenon or the replication of experiments in physics. *Sociol.* 9: 205–24.

———. 1981. The place of the "core-set" in modern science: Social contingency with methodological propriety in science. *Hist. Sci.* 19:6–19.

———. 1983. An empirical relativist programme in the sociology of scientific knowledge. In *Science Observed,* edited by K. D. Knorr-Cetina and M. Mulkay, 85–113. London, Beverly Hills, and New Delhi (Sage).

Cowell, F. R. 1975. *The Athenaeum: Club and social life in London 1824–1974.* London (Heinemann).

Craig, G. Y. 1971. Letters concerning the Cambrian-Silurian controversy. *J. Geol. Soc.* 127:483–500.

Cumming, D. 1985. John MacCulloch, high priest, blackguard and thief, reassessed. In *Natural history in the early nineteenth century,* edited by Alwyne Wheeler. London, in press.

De Beer, G. R., ed. 1958. *Darwin and Wallace. Evolution by natural selection.* Cambridge (Cambridge University Press).

———. 1960–61. Darwin's notebooks on transmutation of species. *Bull. Brit. Mus. (Nat. Hist.), Hist. Ser.* 2:27–200.

Dott, R. H., Jr. 1969. James Hutton and the concept of a dynamic earth. In *Toward a history of geology,* edited by Cecil J. Schneer, 122–41. Cambridge, Mass.

Douglas, J. A., and J. M. Edmonds. 1950. John Phillips's geological maps of the British Isles. *Ann. Sci.* 6:361–75, pls. 5, 6.

Edge, David. 1979. Quantitative measures of communication in science: A critical review. *Hist. Sci.* 17:102–34.

Edmonds, J. M. 1975a. The geological lecture courses given in Yorkshire by William Smith and John Phillips, 1824–25. *Proc. Yorkshire Geol. Soc.* 40:373–412.

———. 1975b. The first geological lecture course at the University of London, 1831. *Ann. Sci.* 32:257–75.

———. 1979. The founding of the Oxford Readership in Geology, 1818. *Notes Rec. Roy. Soc. London* 34:33–51.

Edmonds, J. M. and J. A. Douglas. 1976. William Buckland, F.R.S. (1784–1856) and an Oxford geological lecture, 1823. *Notes Rec. Roy. Soc. London* 30:141–67.

Elkana, Yehuda. 1981. A programmatic attempt at an anthropology of knowledge. In *Sciences and cultures,* edited by Everett Mendelsohn and Yehuda Elkana, 1–76. Dordrecht and Boston (Reidel).

Eyles, Joan M. 1969. William Smith: Some aspects of his life and work. In *Toward a history of geology,* edited by Cecil J. Schneer, 142–58. Cambridge, Mass.

———. 1985. Sir Joseph Banks, William Smith and the French geologists. In *Natural history in the early nineteenth century,* edited by Alwyne Wheeler. London, in press.

Eyles, V. A. 1937. John MacCulloch, F.R.S., and his geological map: An account of the first geological survey of Scotland. *Ann. Sci.* 2:114–29.

Fleck, Ludwik. 1979. *Genesis and development of a scientific fact.* Trans. Fred Bradley and Thaddeus J. Trenn. Chicago and London (University of Chicago Press).[First published as *Entstehung und Entwicklung einer wissenschaftlicher Tatsache.* Basel, 1935.]

Florilège des sciences en Belgique pendant le XIXe siècle et le début de XXe. 1968. Bruxelles (Académie Royale de Belgique).

Fourmarier, P. 1968. Jean-Baptiste d'Omalius d'Halloy. 1783–1875. In *Florilège des sciences* 431–38, pl. Bruxelles.

Fox, Robert. 1980. The savant confronts his peers: Scientific societies in France, 1815–1914. In *The organization of science and technology in France,* edited by Robert Fox and George Weisz, 241–82. Cambridge (Cambridge University Press) and Paris (Maison de l'homme).

Frank, Robert G., Jr. 1980. *Harvey and the Oxford physiologists. A study of scientific ideas.* Berkeley (University of California Press).

Garland, Martha M. 1980. *Cambridge before Darwin. The ideal of a liberal education, 1800–1860.* Cambridge (Cambridge University Press).

Geertz, Clifford. 1973. Thick description: Toward an interpretive theory of culture. Chap. 1 in *The interpretation of cultures. Selected essays.* New York (Basic Books).

———. 1976. From the native's point of view: On the nature of anthropological understanding. In *Meaning in anthropology,* edited by Keith H. Basso and Henry A. Selby. Albuquerque (University of New Mexico Press).

Geikie, Archibald. 1875. *Life of Sir Roderick Murchison . . . based on his journals and letters with notices of his scientific contemporaries and a sketch of the rise and growth of Palaeozoic geology in Britain.* 2 vols. London (John Murray).

Geison, Gerald L. 1981. Scientific change, emerging specialties, and research schools. *Hist. Sci.* 19:20–40.

Gerstner, Patsy, A. 1971. The reaction to James Hutton's use of heat as a geological agent. *Brit. J. Hist. Sci.* 5:353–62.

Gillispie, Charles C. 1951. *Genesis and geology. A study in the relations of scientific thought, natural theology, and social opinion in Great Britain, 1790–1850.* Cambridge, Mass. (Harvard University Press).

Golden, Jacqueline. 1981. *A list of the papers and correspondence of George Bellas Greenough (1778–1855) held in the Manuscripts Room, University College London Library.* London (University College London).

Goodfield, June, 1981. *An imagined world. A story of scientific discovery.* New York (Harper and Row).

Greene, Mott T. 1982. *Geology in the nineteenth century. Changing views of a changing world.* Ithaca and London (Cornell University Press).

Gregory, Richard L. 1973. The confounded eye. In *Illusion in nature and art,* edited by R. L. Gregory and E. H. Gombrich. London (Duckworth).

Gruber, Howard E. 1980. The evolving systems approach to creative scientific work: Charles Darwin's early thought. In *Scientific discovery: Case studies,* edited by Thomas Nickles, 133–30. Dordrecht and Boston (Reidel).

———. 1981a. *Darwin on man. A psychological study of scientific creativity.* [2d ed.] Chicago (University of Chicago Press).

———. 1981b. On the relation between "Aha experiences" and the construction of ideas. *Hist. Sci.* 19:41–59.

Hall, Marie Boas. 1981. Public science in Britain: The role of the Royal Society. *Isis* 72:627–29.

Hays, J. N. 1983. The London lecturing empire, 1800–1850. In *Metropolis and Province,* edited by Ian Inkster and Jack Morrell, 91–119. London (Hutchison).

Hendrickson, W. B. 1961. Nineteenth century state geological surveys [in U.S.A.]: Early government support for science. *Isis* 52:357–71.

Herbert, Sandra. 1977. The place of man in the development of Darwin's theory of transmutation. Part II. *J. Hist. Biol.* 10:155–227.

[Herries-] Davies, Gordon, L. 1968. The tour of the British Isles made by Louis Agassiz in 1840. *Ann. Sci.* 24:131–46.

———. 1969. *The earth in decay. A history of British geomorphology.* London (Macdonald).

———. 1980. Richard Griffith—his life and character. In *Richard Griffith*, edited by Gordon L. Herries-Davies and R. Charles Mollan, 1–31. Dublin (Royal Dublin Society).

Hesse, Mary B. 1980. *Revolutions and reconstructions in the philosophy of science.* Hassocks, Sussex (Harvester).

Heyck, T. W. 1982. *The transformation of intellectual life in Victorian England.* London and Canberra (Croom Helm).

Hodge, M. J. S. 1972. The universal gestation of nature: Chambers's *Vestiges* and *Explanations. J. Hist. Biol.* 5:127–51.

Holmes, Frederic L. 1974. *Claude Bernard and animal chemistry: The emergence of a scientist.* Cambridge, Mass. (Harvard University Press).

———. 1981. The fine structure of scientific creativity. *Hist. Sci.* 19:60–70.

———. 1984. Lavoisier and Krebs: The individual scientist in the near and deeper past. *Isis* 75:131–42.

Hooykaas, R. 1959. *Natural law and divine miracle. An historical-critical study of the principle of uniformity in geology, biology and theology.* Leiden (Brill).

———. 1970. Catastrophism in geology, its scientific character in relation to actualism and uniformitarianism. *Meded. Kon. Nederl. Akad. Wetenschappen, Afd. Letter-kunde*, n. ser. 7 (7).

House, M. R., C. T. Scrutton, and M. G. Bassett, eds. 1979. The Devonian system. A Palaeontological Association International Symposium. *Spec. Pap. Palaeont.* 23.

Inkster, Ian, and Jack Morrell, eds. 1983. *Metropolis and province. Studies in British culture 1780–1850.* London (Hutchinson).

Jordanova, L. J., and Roy S. Porter, eds. 1979. *Images of the earth. Essays in the history of the environmental sciences.* Chalfont St Giles, Bucks (British Society for the History of Science).

Knorr [-Cetina], Karen D. 1979. Tinkering towards success: Prelude to a theory of scientific practice. *Theory and Society* 8:347–76.

———. 1981. *The manufacture of knowledge. Towards a constructivist and contextual theory of science.* Oxford (Pergamon).

Knorr [-Cetina], Karen D., Roger Krohn, and Richard Whitley, eds. 1981. *The social process of scientific investigation.* Dordrecht and Boston (Reidel).

Knorr-Cetina, Karen D., and Michael Mulkay, eds. 1983. *Science observed. Perspectives on the social study of science.* London, Beverly Hills, and New Delhi (Sage).

Kohn, David. 1980. Theories to work by: Rejected theories, reproduction, and Darwin's path to natural selection. *Stud. Hist. Biol.* 4:67–170.

Kuhn, Thomas S. 1970. *The structure of scientific revolutions.* [2d ed.] Chicago (University of Chicago Press).

Latour, Bruno, and Steve Woolgar. 1979. *Laboratory life: The social construction of scientific facts.* Beverly Hills and London (Sage).

Laudan, Rachel, 1976. William Smith. Stratigraphy without palaeontology. *Centaurus* 20:210–26.

———. 1977. Ideas and organizations in British geology: A case study in institutional history. *Isis* 68:527–38.

———. 1982. The role of methodology in Lyell's science. *Stud. Hist. Philos. Sci.* 13:215–49.

Launay, Louis de. 1940. *Une grande famille de savants: Les Brongniart*. Paris (Rapilly).

Lyell [Katherine], ed. 1881. *Life letters and journals of Sir Charles Lyell, Bart*. 2 vols. London (John Murray).

McCartney, Paul J. 1977. *Henry De la Beche: Observations on an observer*. Cardiff (National Museum of Wales).

MacLeod, Roy M. 1983. Whigs and savants: Reflections on the reform movement in the Royal Society, 1830–48. In *Metropolis and province*, edited by Ian Inkster and Jack Morrell, 55–90. London.

Marchand, Leslie A. 1941. *The* Athenaeum. *A mirror of Victorian culture*. Chapel Hill (University of North Carolina Press).

Medawar, Peter. 1967. *The art of the soluble: Creativity and originality in science*. London (Methuen).

Millhauser, Milton. 1954. The scriptural geologists. An episode in the history of opinion. *Osiris* 11:65–86.

Morrell, J. B. 1971. Individualism and the structure of British science in 1830. *Hist. Stud. Phys. Sci.* 3:183–204.

———. 1976. London institutions and Lyell's career: 1820–41. *Brit. J. Hist. Sci.* 9:132–46.

———. 1983. Economic and ornamental geology: The Geological and Polytechnical Society of the West Riding of Yorkshire, 1837–53. In *Metropolis and province*, edited by Ian Inkster and Jack Morrell, 231–56. London.

Morrell, J. B., and Arnold Thackray. 1981. *Gentlemen of science. Early years of the British Association for the Advancement of Science*. Oxford (Clarendon).

Mulkay, Michael, and G. Nigel Gilbert. 1982. What is the ultimate question? Some remarks in defence of the analysis of scientific discourse. *Soc. Stud. Sci.* 12:309–19.

Murray, John, IV. 1919. *John Murray III, 1808–1892. A brief memoir*. London (John Murray).

Needham, Raymond, and Alexander Webster. 1905. *Somerset House, past and present*. London (Fisher Unwin).

Neve, Michael. 1983. Science in a commercial city: Bristol 1820–60. In *Metropolis and province*, edited by Ian Inkster and Jack Morrell, 179–204. London.

Oldroyd, D. R. 1979. Historicism and the rise of historical geology. *Hist. Sci.* 17:191–213, 227–57.

Opie, Redvers, 1929. A neglected English economist: George Poulett Scrope. *Quart. J. Econ.* 44:101–37.

Orange, A. D. 1971. The British Association for the Advancement of Science: The provincial background. *Sci. Stud.* 1:315–29.

———. 1972. The origins of the British Association for the Advancement of Science. *Brit. J. Hist. Sci.* 6:152–76.

———. 1975. The idols of the theatre: The British Association and its early critics. *Ann. Sci.* 32:277–94.

———. 1983. Rational dissent and provincial science: William Turner and the Newcastle Literary and Philosophical Society. In *Metropolis and province*, edited by Ian Inkster and Jack Morrell, 205–30. London.

Ospovat, Dov. 1981. *The development of Darwin's theory: Natural history, natural theology and natural selection, 1838–1859*. Cambridge (Cambridge University Press).

Outram, Dorinda. 1983. Cosmopolitan correspondence: A calendar of the letters of Georges Cuvier (1769–1832). *Archives* 16:47–53.

Pfetsch, F. R., and R. von Gizycki. 1975. Die Gesellschaft deutscher Naturforscher und Ärzte: Bildung von Sektionen und Abspaltung von Gesellschaften. In *Innovationsforschung als multidisziplinäre Aufgabe*, edited by F. R. Pfetsch and R. von Gizycki, 101–53. Göttingen.

Pickering, Andrew. 1981. The hunting of the quark. *Isis* 72:216–35.

Pinch, Trevor. 1981. The sun-set: The presentation of certainty in scientific life. *Soc. Stud. Sci.* 11:131–58.

Polanyi, Michael. 1958. *Personal knowledge. Towards a post-critical philosophy.* London (Routledge and Kegan Paul).

——. 1966. *The tacit dimension.* Garden City, New York (Doubleday).

Porter, Roy. 1976. Charles Lyell and the principles of the history of geology. *Brit. J. Hist. Sci.* 9:91–103.

——. 1977. *The making of geology. Earth science in Britain, 1660–1815.* London, Cambridge, and New York (Cambridge University Press).

——. 1978. Gentlemen and geology: The emergence of a scientific career, 1660–1920. *Hist. J.* 21:809–36.

Ravetz, Jerome. 1971. *Scientific knowledge and its social problems.* Oxford (Clarendon).

Renier, Armand. 1949. À propos du début des études géologiques en Belgique. B. L'oeuvre cartographique d'André-Hubert Dumont (1808–1857). *Bull. Acad. Roy. Belg., Sci.* 35:143–56, 709–29.

Robinson, Howard. 1948. *The British Post Office: A history.* Princeton (Princeton University Press).

Rogers, Emma, ed. 1896. *Life and letters of William Barton Rogers.* 2 vols. Boston and New York (Houghton, Mifflin).

Rudwick, Martin J. S. 1962. Hutton and Werner compared: George Greenough's geological tour of Scotland in 1805. *Brit. J. Hist. Sci.* 1: 117–35.

——. 1963. The foundation of the Geological Society of London: Its scheme for cooperative research and its struggle for independence. *Brit. J. Hist. Sci.* 1:325–55.

——. 1967. A critique of uniformitarian geology: A letter from W. D. Conybeare to Charles Lyell, 1841. *Proc. Amer. Philos. Soc.* 111: 272–87.

——. 1969. Lyell on Etna, and the antiquity of the earth. In *Toward a history of geology,* edited by Cecil J. Schneer, 288–304. Cambridge, Mass.

——. 1970a. The strategy of Lyell's *Principles of Geology. Isis* 61: 4–33.

——. 1970b. The glacial theory. *Hist. Sci.* 8:136–57.

——. 1971. Uniformity and progression: Reflections on the structure of geological theory in the age of Lyell. In *Perspectives in the history of science and technology,* edited by Duane H. D. Roller, 209–27. Norman, Oklahoma (Oklahoma University Press).

——. 1972. *The meaning of fossils. Episodes in the history of palaeontology.* London (Macdonald) and New York (American Elsevier).

——. 1974. Poulett Scrope on the volcanoes of Auvergne: Lyellian time and political economy. *Brit. J. Hist. Sci.* 7:205–42.

——. 1975a. Charles Lyell, F.R.S. (1797–1875) and his London lectures on geology, 1832–33. *Notes Rec. Roy. Soc. London* 29:231–63.

——. 1975b. Caricature as a source for the history of science: De la Beche's anti-Lyellian sketches of 1831. *Isis* 66:534–60.

——. 1976. The emergence of a visual language for geological science, 1760–1840. *Hist. Sci.* 14:149–95.

——. 1978. Charles Lyell's dream of a statistical palaeontology. *Palaeont.* 21:225–44.

——. 1979. Transposed concepts from the human sciences in the early work of Charles Lyell. In *Images of the earth,* edited by L. J. Jordanova and Roy S. Porter, 67–83. Chalfont St Giles, Bucks.

——. 1982a. Cognitive styles in geology. In *Essays in the sociology of perception,* edited by Mary Douglas, 219–41. London (Routledge and Kegan Paul).

——. 1982b. Charles Darwin in London: The integration of public and private science. *Isis* 73:186–206.

——. 1985. The group construction of scientific knowledge: Gentlemen specialists and the Devonian controversy. In *The kaleidoscope of science,* edited by Edna Ullmann-Margalit. Atlantic Highlands, New Jersey (Humanities Press), in press.

Rupke, Nicolaas. 1983. *The great chain of history. William Buckland and the English school of geology (1814–1849).* Oxford (Clarendon).

Schneer, Cecil J. 1969. Ebenezer Emmons and the foundations of American geology. *Isis* 60:439–50.

Secord, James A. 1982. King of Siluria: Roderick Murchison and the imperial theme in nineteenth-century British geology. *Vict. Stud.* 25:413–42.

——. 1986. *Controversy in Victorian geology: The Cambrian-Silurian Dispute.* Princeton (Princeton University Press), in press.

Seymour, W. A., ed. 1980. *A history of the Ordnance Survey.* Folkestone, Kent (Dawson).

Shapin, Steven. 1982. History of science and its sociological reconstructions. *Hist. Sci.* 20:157–211.

Société Géologique de France. 1880. Célébration du cinquantenaire de la Société. *Bull. Soc. Géol. France,* ser. 3, 8:i–lxxxii.

Speakman, Colin. 1982. *Adam Sedgwick. Geologist and Dalesman, 1785–1873. A biography in twelve themes.* Heathfield, East Sussex (Broad Oak Press).

Stone, Lawrence. 1979. The revival of narrative: Reflections on a new old history. *Past and Present,* no. 85: 3–24.

Thackray, J. C. 1978a. R. I. Murchison's *Geology of Russia. J. Soc. Bibliogr. Nat. Hist.* 8:421–33.

——. 1978b. R. I. Murchison's *Silurian System. J. Soc. Bibliogr. Nat. Hist.* 9:61–73.

——. 1979. T. T. Lewis and Murchison's *Silurian System. Trans. Woolhope Nat. Field Club* 42:186–93.

Thoreau, Jacques, 1968. André Dumont, 1804–1857. In *Florilège des sciences,* 439–51. Bruxelles.

Torrens, Hugh. 1979. Geological communication in the Bath area in the last half of the eighteenth century. In *Images of the earth,* edited by L. J. Jordanova and Roy S. Porter, 215–47. Chalfont St Giles, Bucks.

Toulmin, Stephen. 1972. *Human understanding. I. The collective use and evolution of concepts.* Princeton (Princeton University Press) and Oxford (Clarendon).

Turner, Victor W. 1974. *Dramas, fields and metaphors. Symbolic action in human society.* Ithaca (Cornell University Press).

Weindling, Paul J. 1979. Geological controversy and its historiography: The prehistory of the Geological Society of London. In *Images of the earth,* edited by L. J. Jordanova and Roy S. Porter, 248–71. Chalfont St Giles, Bucks.

——. 1983. The British Mineralogical Society: A case study in science and social improvement. In *Metropolis and province,* edited by Ian Inkster and Jack Morrell, 120–50. London.

Wells, J. W. 1963. Early investigations of the Devonian system in New York, 1656–1836. *Geol. Soc. America,* special paper no. 74.

Westfall, Richard. 1980. *Never at rest. A biography of Isaac Newton.* Cambridge (Cambridge University Press).

Westrum, Ron. 1978. Science and social intelligence about anomalies: The case of meteorites. *Soc. Stud. Sci.* 8:461–93.

Wheeler, Alwyne, ed. 1985. *Natural history in the early nineteenth century.* London (Society for the History of Natural History).

White, Hayden. 1984. The question of narrative in contemporary historical theory. *Hist. and Theory* 23:1–33.

Wiegel, E. 1973. Die Entwicklung der staatlichen geologischen Kartierung in Nordrhein-Westfalen vor 1873. *Fortschr. Geol. Rheinland-Westfalen* 23: 11–54.

Wilson, Leonard G. 1972. *Charles Lyell. The years to 1841. The revolution in geology.* New Haven (Yale University Press).

Woodward, Horace B. 1907. *The history of the Geological Society of London*. London (Geological Society of London).

Ziegler, Willi. 1979. Historical subdivisions of the Devonian. In *The Devonian system, Spec. Pap. Palaeont.* no. 23, edited by R. R. House, C. T. Scrutton, and M. G. Bassett, 23–47.

Ziman, John. 1978. *Reliable knowledge: An exploration of the grounds for belief in science*. Cambridge (Cambridge University Press).

Index

Académie des Sciences, 28, 136, 143

Accomplished geologists, 422–28

Agassiz, Louis: as accomplished geologist, 422; faunal change, concept of, 390–91, 394; fossil identifications by, 101–2, 130, 260, 352, 364, 390; his involvement in Devonian controversy, 426; Wollaston medal to, 101–2; other references, 133, 135, 254, 277, 366

Agonistic field, 15, 435–38

Alps: Agassiz fossils from, 101–2; Élie de Beaumont's fossils from, 106, 121, 122, 167, 197; Lyell's work in, 132–33, 200; other reference, 366

Amateurs in geology, 17–18, 40, 424, 433

Annales des Mines, 70, 273–74

Ansted, David Thomas, 245

Anthracite in Greywacke, question of, 60, 78–82

d'Archiac, Étienne, 370, 387

Asmus, Hermann, 377

Athenaeum (weekly), 26, 32, 160, 218, 335

Austen, Robert Alfred Cloyne: as accomplished geologist, 423–25, 428; analysis of actions by, 410–13, 417; background of, 149–50; at Birmingham meeting, 316, 317; and De la Beche, 272, 288–89; and Devonian system, 288, 298–300, 317–18, 369–70, 388–89; Devonshire and Cornwall fieldwork of, 171–73; Devonshire interpretation of, 222–31, 236–38, 245, 258–59, 264, 270; fossils, his concept of, 252, 372–73; fossils found by, 222–30, 236,

248, 258–59, 266, 278, 283, 345, 348, 353, 370, 410, 416; Geological Society papers of, 224–30, 298, 388; his involvement in Devonian controversy, 426–27; later career of, 458; and Lonsdale, his controversy with, 370; at Newcastle meeting, 251–53; at Plymouth meeting, 374; and social identity, 440; de Verneuil, his meeting with, 256–57; other references, 156, 247–48, 260, 277–79, 314, 315, 319, 332, 335, 436, 460

Ayckbourn, Alan, 13

Babbage, Charles, 25, 27

Baconianism, 24–25, 66, 434

Banks, Sir Joseph, 20–21

Baring, Sir Francis Thornhill, 125, 203

Basin, 49

Beach, Thomas, 69

Belemnites, 106, 121–22, 132

Belgium: and Cambrian system, 138–40; coal in, 57; and Devonian system, 319–21, 329, 337, 341, 352, 357, 459; Dumont's mapping of, 136–38; Sedgwick's fieldwork in, 312, 319–20; and Silurian system, 138–40, 148, 253–56, 319–21; *terrains* of, 136

Beyrich, Ernst, 274, 303–4, 307, 310, 311, 426, 459

Bilton, Revd William: fossils found by, 186–87, 211–12; other references, 206, 260, 291–94, 298–99, 331, 424

Birmingham, British Association meeting at, 316–19, 343, 348

Blumenbach, Johann Friedrich, 65–66, 309

Board of Ordnance. *See* Ordnance Trigonometric Survey

Boase, Dr. Henry, 88–89, 112, 133, 172

Boulogne, Société géologique meeting at, 321, 343, 348–49

Boulonnais, 255–56, 274, 321–23, 349, 354–55, 443

Bristol, British Association meeting at, 159–70, 173, 186, 191, 193, 201, 212–14, 222, 290, 410

British Association for the Advancement of Science: Birmingham meeting of, 316–19, 343, 348; Bristol meeting of, 159–70, 173, 186, 191, 193, 201, 212–14, 222, 290, 410; Cork meeting of, 392–93; De la Beche's criticism of, 176–77; Dublin meeting of, 130, 133–35, 140, 151, 169, 218, 254; Edinburgh meeting of, 86, 88, 133; Glasgow meeting of, 361, 364–68, 371; history of, 30–34; Liverpool meeting of, 216–18; Manchester meeting of, 389–91, 392, 396; Newcastle meeting of, 250–54, 262, 273; Plymouth meeting of, 373–76, 378, 380; other references, 37, 68, 73, 85, 312, 424, 436, 439

Brochant de Villiers, André Jean François Marie, 90–91, 106, 143

Brogniart, Adolphe, 106, 140

Brogniart, Alexandre, 47, 49, 53–54, 140

Brora coal field, 68

Brougham, Lord, 107

von Buch, Christian Leopold: at Bonn meeting, 140; and Cambrian and Silurian systems, 148; and

DATE DUE

DEMCO 38-296